THE NATURAL HISTORY

OF

APES, MONKEYS

AND

LEMURS

**WITH ARTICLES ON EVOLUTION, THE KINGDOM OF
MAMMALS AND MUCH MORE**

———

CHIEF CONTRIBUTORS
RICHARD LYDEKKER, F.R.S.
SIR HARRY JOHNSTON
PROFESSOR J. R. SINSWORTH-DAVIS, M.A.

———

British Library Cataloguing-in-Publication Data
A catalogue record for this book is available from
the British Library

ABOUT THIS BOOK

THE Harmsworth Natural History belongs to the group of educational publications which includes the Harmsworth Encyclopædia, the Harmsworth Self-Educator, the Harmsworth History of the World, the World's Great Books, and the Children's Encyclopædia. The present addition to this series covers a wide field which, of course, could only be briefly touched upon in any other of these kindred works.

THE Natural History is a popular survey of all living creation as it exists in the world to-day. It may be doubted if so comprehensive a book of Animal Life has ever been offered to the public at a popular price, and it may be confidently asserted that no popular work of this kind has ever before been furnished with such a remarkable gallery of animal paintings as this book contains. The colour plates, about 150 in number, are unique among animal pictures, and the hundreds of black and white drawings by the same famous artist, William Kuhnert, in addition to the photographs from real life—some of them reproduced by direct colour photography, so that we thus have Nature in her own colours— enrich the book with an unrivalled series of illustrations.

IT has been one of the chief aims of the editors to avoid the technicalities which make the conventional Natural History a dull and difficult book to read; and the aim constantly borne in mind has been to make this pageant of animal life in the world to-day as picturesque as pen and brush can make it.

AND yet, of course, this book is not merely a magazine of natural interest, it is an encyclopædia of living things ; it has the value of scientific authority and the interest of the story-book as well. If an appeal is made to names, the claim made for this book is that it has behind it the vast experience and learning of Mr. Richard Lydekker, whose name is among the greatest in Natural History. It has behind it the great authority of Prof. J. R. Ainsworth-Davis, whose work has carried his name into every well-equipped library of Natural History in the world. It has behind it not only these two unquestioned authorities, each responsible for a great part of the actual contributions, but the name of Sir Harry Johnston, a man eminently distinguished in many spheres, who writes in this book of wild life in its native haunts, of animals as he knows them.

IT is right that in a work bearing such high scientific authority there should be something more than mere descriptions of animals, beginning as the conventional Natural History begins, with the first class of Mammals ; and our book opens with a series of chapters important and interesting enough to make a notable volume in themselves. These chapters bring into the work many other names of great distinction in the various departments of Natural History, and we find among them Sir Herbert Maxwell, F.R.S. ; Dr. Alfred Russel Wallace, F.R.S. ; Prof. J. W. Gregory, F.R.S., of the University of Glasgow ; Prof. C. Lloyd Morgan, F.R.S., of the University of Bristol ; Prof. E. B. Poulton, F.R.S., of the University of Oxford ; Prof. J. A. Thomson, of the University of Aberdeen ; Prof. Turner, of the Royal Agricultural College, Cirencester ; Mr. R. I. Pocock, the Superintendent of the Zoological Gardens ; Mr. W. P. Pycraft, of the Natural History Department of the British Museum ; Mr. Aubyn Trevor-Battye ; Mr. W. H. Hudson ; Major F. T. Alexander, and others.

IT is not necessary that any more should be said of a book with such names as these, with such pictures as these, and with such authority as every page in these volumes has stamped upon it.

A. M.

THE MAN-LIKE APES · CHIMPANZEE AND YOUNG

TABLE OF CONTENTS

THE PAGEANT OF LIFE

A SURVEY OF MAN'S NEIGHBOURS IN THE WORLD

By Sir HERBERT MAXWELL, F.R.S.

AMONG the fairy stories with which Grimm enthralled the imagination of mid-Victorian childhood, there is none which lingers so agreeably as that which describes the experience of a youth who, I forget by what happy accident or device, acquired the gift of hearing and understanding the conversation of what are commonly, but incorrectly, described as "dumb animals." It would indeed be rash to commit oneself to the statement that any living creature above the rank of a jellyfish is incapable of oral communication with its fellows, or to apply to it the epithet "dumb" in a more literal sense than one addresses an unknown correspondent as "dear sir," or refers to a briefless barrister in the House of Commons as "my honourable and learned friend." It was Alphonse Karr—was it not?—who declared that his dog could do everything except speak, and thanked Heaven that the dear creature could not do that, else he would assuredly have bored his master by repeating the same observation over and over again.

Well, thus far the fairy gift of interpreting the language of beast and bird has been withheld from us; but recent research has given us a clearer understanding of our fellow-inhabitants of the globe by removing some of the mists which still hung round natural history, even after the far-seeing Linnæus had opened our understanding to the leading affinities of animated Nature.

In an early year of the twentieth century the enterprising editor of a daily newspaper addressed inquiry to a number of persons, inviting their opinion as to which English writer had most powerfully influenced the general trend of thought in the century that had just come to a close. Not a question to be answered off-hand, one will see; but, after giving it the consideration it deserved, I wrote without hesitation the name of Charles Darwin. Some there be who will demur to such pre-eminence being given to a mere naturalist over the great theologians, historians, philosophers, moralists, poets, essayists, novelists, and men of science of the later Georgian and Victorian eras, many of whom made enduring impression upon their respective provinces of intelligence; but none of these had to overcome such determined opposition from students in the other departments of human knowledge; none could claim a victory so decisive, or one that was to affect so profoundly the methods and conclusions of workers in widely different fields of intellectual energy.

This is no place for discussing whether Darwin placed correctly every link in the chain of evolution, or whether he laid upon the doctrine of natural selection more stress that it has been since found able to bear. These matters continue, and are likely for some time to continue, to afford material for discussion and even for dispute. What all men

acknowledge is that Darwin, surveying the products of scientific research accumulated by preceding generations, checked some, rejected others, and verified many of them by laborious personal observation, and, having carefully collated with them the result of man's deliberate interference with the reproduction of domestic animals, lifted the whole scheme of biology into a higher plane, presenting animated Nature in a novel and startling aspect.

It cannot be denied that such a revelation administered a violent shock to the ordinary citizen's habit of thought. Those who are old enough to remember events in the early 'sixties will not have forgotten that the new doctrine was authoritatively denounced as incompatible with—nay, subversive of—the Christian faith. It was repugnant even to robust intellects to be asked to acknowledge that man owned descent from a common ancestor with the vermin that preyed upon him—to admit that when "God formed man of the dust of the ground, and breathed into his nostrils the breath of life, and man became a living soul," it was identically the same breath of life that animated all other living creatures. All men felt that if this new teaching were to be accepted, the material world must be viewed from a standpoint very different from that whence their forefathers had surveyed it, and this could not be done without some modification in the established understanding of spiritual things.

Darwin, therefore, had arrayed against him, not only the preconceived theories of renowned biologists like Sir Richard Owen, but also the principles and prejudices of religious men and women, the instinct and sentiment of poets, and the passive conservatism of the mass of his fellow-men. To overcome all these called for more than a mere exposition of conclusions, however logical, founded upon facts, however demonstrable; it required the prophet's fire to burn them into unwilling minds. An article in the "Quarterly Review" for July, 1860, dealing with the "Origin of Species" published in the previous year, condemned it, not without ridicule, as a work "most dishonourable and injurious to science"; yet the writer could not be insensible to the magic of genius, whereby a treatise, of necessity highly technical, possessed all the fascination of a romance.

"We feel," he says, "as we walk abroad with Mr. Darwin, very much as the favoured object of the attention of the dervish must have felt when he had rubbed the ointment around his eye, and had it opened to see all the jewels and diamonds and emeralds and topazes and rubies, which were sparkling unregarded beneath the earth, hidden as yet from all eyes save those which the dervish had enlightened. But here we are bound to say our pleasure terminates."

D

I

A

And now, after the lapse of half a century, none who has given any serious and sustained thought to the problem of existence can regard human life as differing in essence from the vital principle in the humblest animals, or detect a break in the physical continuity between the oyster and the epicure.

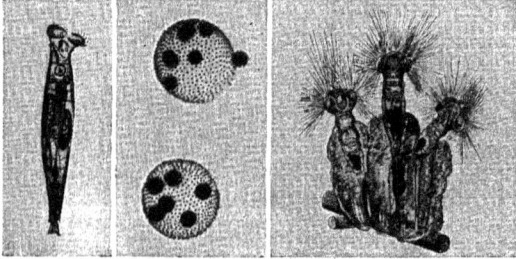

SOME EXAMPLES OF TYPICAL ANIMALCULES
On the left a common rotifer, volvox in centre, and floscularia on the right

Pious men who have been at pains to examine the doctrine of evolution find therein nothing inconsistent with the direction and control of a Supreme Power, and much that seems to postulate the existence of one.

The result, as affecting the ordinary, unscientific lover of Nature, has been to impart a deeper interest to the pageant of life, to give a new significance to the deadly struggle for survival which has been maintained for ages beyond our power of numeration, and to explain the purpose of the merciless suppression of unfit types and individuals. The pageant of life, said I? Nay, but a pageant is only a make-believe, whereas what we witness is actual, never-ending conflict of race against race, clan against clan, with no quarter given or taken. Nor are we, proudly as we claim the lordship of creation, mere spectators of the strife. Never may we lay down our arms; warfare is our lot—defensive war against creatures individually more powerful than ourselves or collectively more injurious, against the agents of disease and against our own kind—offensive war against other creatures, innocuous in themselves, but whose skins and feathers we covet for clothing or their flesh for food. Hordes of captives, too, we keep in bondage, some as slaves for forced labour, some for butchery as occasion serves .

THE SO-CALLED WATER BEAR
This animalcule, here greatly magnified, is less than a twenty-fifth of an inch in size, and yet has a wonderfully developed organism

With respect to individuals, we evade the penalty of unfitness by sedulous provision of hospitals and sanatoria, making successful effort to keep thousands of persons alive who Nature intended should succumb to disease, and prolonging the lives of others far beyond the natural term by medicine and invalid diet, not to mention such artful devices as spectacles, artificial teeth, and aseptic surgery. But woe betide those races of men which fail to hold their own—American Indians, Eskimo, Veddahs of Ceylon, Australasian aborigines! Unable to adapt themselves to changing environment, they disappear from an unkindly world as surely as the iguanodon of the Cretaceous weald or the hairy mammoth of Pleistocene times.

Ruthless stringency in eliminating the unfit results in the uniform vigour and energy of what we term the lower animals. Epizootics—diseases that from time to time decimate, and more than decimate, the more fecund races of animals—leave the survivors fitter than ever to maintain the struggle, and epidemics among human tribes in a state of barbarism have a like effect. One is apt to recognise in this a teleological purpose, the periodical elimination of a large percentage of consumers in the interest of the survivors and, through them, of the race. It is only in the human community that diseased or enfeebled individuals are permitted to exist and transmit their infirmities to their off-

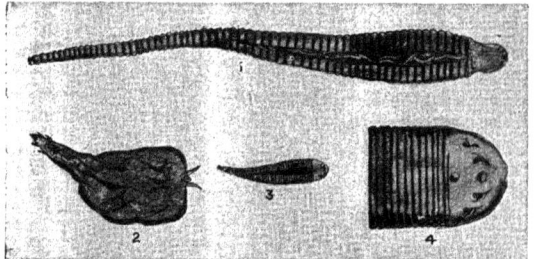

PENTASTOMIDS, STRANGE AND DANGEROUS PARASITES
1, female; 2, young larva; 3, male; 4, enlarged head of male

spring. When a wild creature loses the power of sustaining itself, the end is not far off. Animals that will freely risk their lives against any odds in defence of their young show the utmost indifference to the evil plight of their fellows, in some cases attacking a wounded comrade and hastening its demise.

As to life itself, what it is and whence it comes, the most advanced biologists can afford no explanation. Only one principle is universally recognised by men of science—*nullum vivum sine vivo*—" every living thing proceeds from something else that lives." There is no such process as spontaneous generation. Let us accept as the present limit of our understanding the frank confession of Professor Marcus Hartog: " Of the ultimate origin of organic life from inorganic life we have not the faintest inkling. If it took place in the remote past, it has not been accomplished to the knowledge of man in the history of scientific experience, and does not seem likely to be fulfilled in the immediate, or even the proximate, future."

But this much has been ascertained—a fact which has a humbling, though not a humiliating, bearing upon our own self-consciousness—that all the higher

2

animals, not excepting man himself, are built up of innumerable cells, which are separate organisms, exhibiting all the characteristics of living beings, namely, irritability, motility, digestion, nutrition, growth, and reproduction. We may even go a step further back, and recognise that vegetable and animal life are of common origin and nature, the cells building up the higher plants being mostly isolated in separate cavities connected by slender threads, the higher animals consisting of cells in close and universal contact with each other. These cells, known as Protista, are the most primitive form of life known to science ; they are the common source whence proceed the two great rivers of life—the animal and the vegetable.

It is only within recent times that modern appliances and research have brought to our knowledge how life in one form or another pervades every spot in this globe that is capable of sustaining it. It exists even under conditions which might excusably have been pronounced incompatible with

THE VENUS FLY-TRAP, AN INSECT-EATING PLANT
The teeth of the leaves close with a snap and imprison the insect in the trap

THE SUNDEW, A FLESH-EATING PLANT
On the left the sundew is bending towards a piece of meat ; on the right it has reached the food and is about to fasten on it

it. Mr. James Murray, biologist with Sir Ernest Shackleton's Antarctic expedition, sank a shaft through fifteen feet of solid ice to the bottom of a coastal lake, and brought up swarms of rotifers and water bears (Macrobiotus). Now, this lake never thawed during the two summers which the explorers spent there ; it may have been frozen solid for years or even for centuries, during which these creatures had been embedded in a temperature of forty degrees (Fahrenheit) below zero, incapable, of course, of discharging any of the functions of life. Yet no sooner were they brought to light and placed in water than the rotifers began scooping in floating atoms of food, and they, as well as the minute rhizopoda and water bears, set about at once the process of reproduction, either viviparously or through egg-laying.

Now, these microscopic creatures are far removed above the lowest forms of life. The so-called water bear, for instance, has a simple, but segmented and perfectly symmetrical structure, the sexes of individuals being distinct and separate. It has eight legs, armed with effective claws, teeth hardened at the tips by a calcareous deposit, a blood fluid containing numerous corpuscles, a stomach,

a brain and nervous system, an alimentary canal, an excretory orifice between the hindmost pair of legs, sometimes a pair of eyes, but never a heart. Although it never exceeds one millimetre in length ($25\frac{1}{2}$ millimetres = 1 inch), it is the host of many bacteria which infest its blood.

Of all segmented animals, macrobiotus, or the water bear, is the most widely distributed, some species inhabiting water, but the greater number moss, every tuft of which swarms with them. Moisture is essential to their active existence, but if that fails, the animal dries up into what appears to be a grain of sand. In this condition it may remain for years ; but the vital spark is not extinct. Moisten the atom, and it becomes plump and pliant once more, resuming its normal functions and interrupted pursuits as if nothing had happened. Wherever a scrap of moss manages to maintain existence on the desolate Antarctic rocks, Mr. James Murray found it to be the harbour of this lowly race. Often the moss and its inhabitants were only thawed into life for two or three hours on a few days at midsummer,

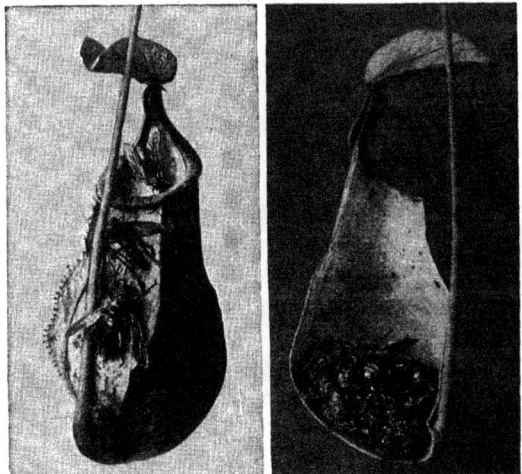

THE PITCHER PLANT, OR NEPENTHES, AN INSECT-EATER
On the left some flies are shown about to enter the pitcher ; on the right is seen a section of the pitcher, containing the dead flies

being frozen hard again each night, and remaining hard-frozen night and day for 350 out of 365 days.

No man need blush to own macrobiotus among his " poor relations," for there does not exist a more harmless race than these minute creatures, which prey — for nearly all forms of life prey upon other forms—only upon mosses and algæ, nibbling through the cell-walls of these humble forms of vegetation and imbibing the sap. But one cannot without a shudder recognise affinity with another class of animal, not distant of kin from the gentle water bear, suggesting the fancy that there have been two creative agencies at work—one, a benignant Deity, designing channels through which the vital force shall be moulded into more perfect beauty and nobler form ; the other, a malignant demon, intent upon devising subtle and corrupt instruments for defacing, maiming, and destroying the handiwork of the rival Power. The life-history of the pentastomids (see page 2) shows how the perpetuation of every living animal involves the affliction, and often the sacrifice, of some other form of life. These pentastomids are loathsome creatures, inhabiting the nostrils or lungs of dogs and other carnivorous animals. Their eggs or larvæ, being expelled by coughing or sneezing, fall on grass or other herbage, and depend for development upon being swallowed by some browsing creature.

When this happens, the larva, which is furnished with clawed limbs, bores through the walls of the stomach, and penetrates to the liver or some other principal organ, where it ensconces itself for several months. As the time for assuming its adult form approaches, the creature begins wandering through the tissues of its host, sometimes causing dangerous, and even fatal

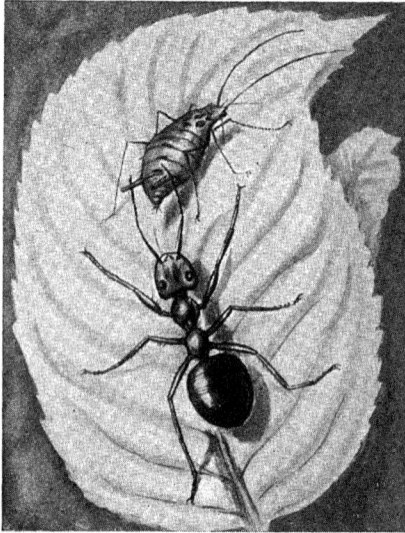

AN ANT MILKING AN APHID

disturbance. If the said host—be it sheep, deer, rabbit, or what not—is devoured by a dog or other beast of prey, the larva, when swallowed, travels into the lungs or nasal cavity of its second host, where it awaits its development into an adult, and the unlovely cycle is renewed. Note, by the by, that pentastomids are quite distinct from the cestodes, or tapeworms, which they greatly resemble in haunt and habit. The cestodes are far lower in the scale of animated Nature than the pentastomids, being hermaphrodite, destitute of a digestive tract, and absorbing nutriment by soaking in the dissolved food in the intestines of their host ; whereas the pentastomids are bi-sexual, the males moving about freely in quest of their mates, and live by sucking blood from the internal organs wherein they have acquired lodgment.

Such are some of the sordid and revolting passages in the pageant, inseparable from the unity of its design ; whereof, if one would have clear understanding, he must not shrink from accepting the stern truth that it is based upon universal rapine, and that no animal, from man down to macrobiotus, and lower, can exist save at the expense of other creatures.

It is vain to appeal to Holy Writ against this doctrine. " Behold now behemoth . . . he eateth grass as an ox." True it is that the hugest terrestrial mammals—elephant, rhinoceros, hippopotamus, and so on—are vegetarian in diet ; but the herbage and foliage whereon they subsist would be useless to them were it not alive or had it never lived. Animal and vegetable life being fundamentally identical in their origin, every modern textbook of zoology must take account of certain primitive forms of vegetation—Protophyta—because it has not been

THE COMMUNAL LIFE OF THE ANTS

An ant-hill showing insects emerging from and returning to their quarters. In the section of the nest are seen some ants waiting on the larvæ, which are diligently nursed and tended

possible to define a frontier between them and Protozoa, the most primitive forms of animal life. Plants must be regarded as beings, animated by a motor force indistinguishable from that which permeates the frame of the mightiest monarch and the meanest mollusc ; but, in obedience to their prescribed nature, they build themselves into cell-walls, forfeiting the power of locomotion. Having neither mouths nor stomachs, they receive nutrition by absorption instead of prehension; but they have a power denied to animals —namely, that of converting mineral and chemical substances into food ; whereas animals can only derive nutrition from proteids—that is, organic compounds of carbon, nitrogen, hydrogen, and oxygen prepared for them by the mediation of vegetable growth. An ox grazing in a meadow receives its proteids direct from the laboratory. Man, as his dentition indicates, was designed to do likewise by consuming fruit and grain ; but the possession of a restless brain and supple fingers soon made him impatient of spending more than half his waking hours in eating ; he got into the habit of employing humbler animals to do the browsing, gnawing, and crunching for him, and. by devouring their flesh, received a supply of proteids in concentrated form, thereby economising much valuable time. Nevertheless, when he consumes flesh as a staple instead of a supplementary diet, he does so at the cost of liability to certain forms of disease which are often induced by too nitrogenous food.

The chain connecting the human consumer with the vegetable laboratory may consist of many links, each one being a separate act of rapine. Suppose one making a supper of stewed eel. The eel is not known to eat any vegetable ; very likely it has been living on spawn deposited by a salmon. The salmon has spent the summer devouring herrings, which, in their turn, have accounted for billions of crustaceans, which may have been the ultimate agents in extracting proteids from vegetable matter, though perhaps they made their prey of still humbler organisms which were vegetarian feeders.

The vegetable kingdom is capable of reprisal upon animals, not merely in the vindictive sense of poisoning them, but by making them their prey.

SLAVE-DRIVING ANTS MAKING A RAID UPON ANTS OF PEACEFUL SPECIES

THE ANT-LION SNARING ITS PREY

Bacteria, which are a humble form of vegetable life, are the direct agents in many, probably in most, forms of acute disease. Also, several of the more highly organised plants retaliate upon animals by catching them and absorbing their juices for their own nourishment (see page 3). One of the most highly specialised of these plants is the North American Dionæa —Venus's fly-trap —of which the leaves are longitudinally hinged with three sensitive hairs on each segment. Let an insect but touch one of these hair triggers, the two segments close with a snap, the marginal teeth interlock, and when that leaf unfolds again, nothing remains of the unlucky trespasser but a pair of wings and an empty skin. Several native British plants are bloodsuckers—the common sundew of moorland bogs, the innocent-looking butterwort, and the crafty little Utricularia with its floating bladders, each with a trapdoor standing invitingly open until some tiny creature is tempted to enter, when the prison door closes smartly, and that insect or crustacean goes forth no more. Pitcher plants, several species of arum, and many others, extract and assimilate the juices of insects, which they are specially constructed to attract and detain for their sustenance.

It is not for man the omnivorous to condemn the furtive devices of plants applied to so practical an end, but it is difficult to discern the motive of certain plants which, by their scent or colour, attract flying insects and slay them, without deriving any advantage from their death. There stands in Lord Ducie's garden at Tortworth Court, Gloucestershire, a specimen of Tilia petiolaris, a species of lime-tree, the flowers of which attract thousands of bees and dipterous flies, and kill them in such quantity that there is a distinct ring of rank grass corresponding to the circumference of the branches where the ground has been fertilised by the dead bodies. Londoners may see in Kew Gardens a large specimen of this tree, which, it need hardly be said, acts in the same deadly manner.

Beekeepers must avoid planting the African torch lilies—commonly known as red-hot pokers—near their hives, for the honey of these brilliant flowers intoxicates the bees so quickly that they perish.

5

THE LIFE HISTORY OF A DRAGON-FLY FROM LARVA TO FLY

Photo. A. Forrester

1, larva in the water ; 2, larva climbing stem of plant ; 3-5, gradual emergence of the imago, or perfect fly ; 6, the fly spreading its wings for its first flight

Howbeit, life is not all ferocity. Killing is the universal means of subsistence—the system prescribed for the maintenance of the myriad actors in the pageant ; but in tracing upwards the development of living creatures from primitive and relatively simple forms one arrives at a point where more amiable traits begin to manifest themselves, ultimately affecting the habits and behaviour of the higher animals as powerfully as the craving for food.

It was said by Linnæus that "Nature is most marvellous in the smallest of her creatures." In no class of animals does one realise the truth of this so forcibly as among the insects. More numerous in their species than all other land animals put together, insects are usually diminutive in size, the largest of them not exceeding the smallest of birds in bulk ; yet it is in this group of invertebrate creatures, individually so insignificant, that one first encounters that passionate solicitude for offspring which ranks so high as a virtue in every human code of ethics. Also, leaving man out of account, it is almost exclusively among insects that cooperative communities are to be found, with complete systems of governance and subdivision of labour. Many vertebrate animals are gregarious, but these are merely sociable, not social. Some of the hunting creatures may combine for attack, but instances of deliberate co-operation, such as the construction and maintenance by beavers of a dam for the good of the colony, and, by the weaver birds, of a common roof for their crowded nests, are difficult to find.

Everybody knows something about the honey bee, and understands something about the perfect manner in which the inhabitants of a hive are divided into classes, each having its structure specialised for definite duties. It is nearly 3,000

THE ICHNEUMON FLY

On the left is a caterpillar from which ichneumon larvæ emerge ; in the centre is the ophion ichneumon, and on the right the long-tailed ichneumon. Photos by B. Hanley, D. English. and W. P. Dando

years since King Solomon urged his readers to consider the ways of the ant, but our knowledge of those ways is still very fragmentary, though the various races of ants have been ingeniously classified, and a full description of their structure, physiology, and nutrition may be found in any handbook. What has been ascertained about their social organisation cannot but stimulate our desire to learn more, for the industry of ants is not only incessant, it is of amazing range and variety, comparable only to the organised activity of a human community (see page 4).

The ants that keep herds of aphids, tending them with the utmost care, and milking them (see page 4) of that sugary fluid beloved of ants, are more than mere herdsmen or cowboys ; they are skilful stock-breeders. Every autumn they carefully collect the eggs of aphids, storing them in the ant-hill through the winter, and when the larvæ are hatched in March, they take the little things out and place them on the appropriate food-plant, to be milked in due course when they reach the adult stage. Well may Lord Avebury write of this as "a case of prudence unexampled in the animal kingdom."

Even more remarkable as an example of foresight, involving a mental process indistinguishable from reason, is the industry of the leaf-cutter ant of Central America. A colony of these, having tunnelled a subterranean passage for many hundreds of feet from the nest to the tree which they have marked out for stripping, send out troops of cutters and carriers, which bring the whole of the leaves into the nest, and hand them over to another set of workers. These chew the leaves into pulp and store them ; the mycelium of a certain fungus begins to spread throughout the mass, which they prune carefully to induce it to throw out white nodules,

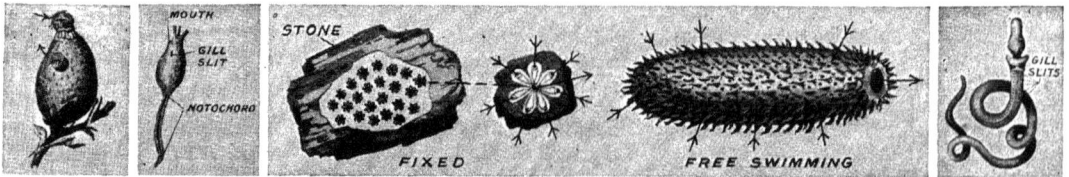

SIMPLE SEA SQUIRT AND LARVA   COLONIAL SEA SQUIRTS (ARROWS INDICATE THE WATER CURRENTS)   BALANOGLOSSUS

instead of the conidia and spores which would appear if the fungus were left alone. These white nodules form the food of the colony.

Some species, like the European rufescent ant, have given up honest work, if indeed they were ever capable of it, and live by slave-driving. It is exactly one hundred years since François Huber described a raid by an army of rufescents, which stormed a nest of the ash-coloured species and carried off all the larvæ and pupæ to rear them as slaves (see page 5). The rufescents are splendid fighters, formidably armed, but they have become so dependent on slave labour that they cannot even clean and feed themselves, much less construct their own dwellings or rear their young. When deprived of the services of their slaves, these doughty warriors quickly perish of starvation.

It was the practice of the old school of biologists to attribute to instinct such actions among the lower animals as appeared to anticipate or reflect some of the organisation and occupations of human society. This was satisfactory enough, so long as people did not press too curiously for a definition of instinct as an agency apart from reason. The absence of any such definition makes it difficult to regard these disciplined hordes as faultless automata, devoid of volition; nor may one lightly dismiss the conclusion arrived at by Lord Avebury after a prolonged series of characteristically patient experiments and close observations—namely, that the mental powers of ants and other co-operative insects " differ from those of men not so much in kind as in degree."

The social ants, bees, and wasps among the Hymenoptera, as well as the termites (so-called white ants) among the Neuroptera, are passive in the care of nurses during their larval and pupal stages; it is strange, therefore, to find that one of the deadliest insect enemies of ants is the larva of a neuropterous fly, the fly itself being perfectly harmless to them. This formidable larva, which is popularly known as the ant lion, marks out with its body a circle in dry sand, within which it digs a pitfall, throwing out the sand by jerking its broad head. It then buries itself at the bottom of the trap and waits until an ant, hurrying along on business intent, trespasses on the shifting slope of the pit. No nightmare can exceed in horror the fate incurred by that hapless insect. The monster, lurking below with wide-set eyes, flings showers of sand upon its victim, till it brings it within reach of its cruel jaws (see page 5).

It is pleasant to turn from contemplating the murderous proceedings of insects like these to enjoy the beauty of the Lepidoptera—the butterflies and moths—which, if as larvæ they exact a heavy tribute of living vegetable tissue, amply repay the debt in their perfect state by fertilising visits to flowers. No summer day is at its best without the fluttering of their jewelled wings. One might think that such exquisite beauty should be their warrant against persecution. But Nature cannot allow caterpillars to multiply unchecked, wherefore she has provided a special constabulary in the shape of about 6,000 species of ichneumon flies, of which some 1,200 species are native to Britain. Most kinds of ichneumon flies set about business in a singularly repulsive way. The female thrusts an egg into the body of a caterpillar, cocoon, or chrysalis, whence is hatched a maggot that gradually eats up the tissues of its host, turns into a pupa, and finally issues as a slender hymenopterous fly (see page 6).

Insects, in the opinion of Mr. David Sharp, must be regarded as the most successful of all the forms of terrestrial life, but very few of them find water a congenial home. It is true that very many flying insects pass their larval stages in shallow, fresh water; and the metamorphosis from a water-breathing larva to an air-breathing insect is one of the most remarkable processes in Nature, being instantaneous (see page 6); nor did the transformation escape the trained vigilance of Tennyson, who wrote:

> To-day I saw the dragon-fly
> Come from the wells where he did lie;
> An inner impulse rent the veil
> Of his old husk, from head to tail
> Came out clear plates of sapphire mail.
> He dried his wings: like gauze they grew;
> Through crofts and pastures wet with dew
> A living flash of light he flew.

Of all the myriad forms of life with which the ocean swarms, only very few kinds of insects succeeded in establishing themselves therein; but their absence is amply balanced by innumerable species of crustaceans, some of which people the waters of Polar seas as densely as swarms of locusts do the air. It is in the sea, moreover, that Ascidians, or sea squirts (see page 6), and their allies abound, presenting the first hint of a backbone in the shape of what is termed a notochord. They appear to be an intermediate form, possibly degenerate, but perhaps only arrested, between the worms, or other type of invertebrate animal, and the vertebrates. One of the

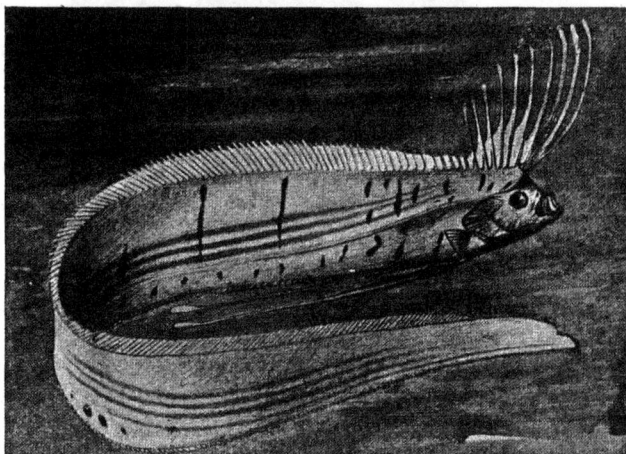

THE MYSTERIOUS OAR-FISH, WHICH ANGLES FOR ITS PREY

most remarkable forms of marine life is the balano-glossus, which studs the sand with worm-like castings. Nothing could be less suggestive of a backboned animal than these slimy creatures which most people would assume to be a kind of marine earthworm. Yet in these Hemichordata, so called because a notochord extends for half their length, must be recognised a direct descendant of extinct ancestors from which all the vertebrates trace their lineage.

Passing them by, we come to the fishes, earliest of all animals to develop a vertebral column, with its indispensable supplement—a head. No class of living creature is so generally carnivorous as the fishes are. Very few species content themselves with vegetable fare. From the lustrous salmon, ideally modelled for swift energy, to the hideous and sluggish angler or fishing frog, nearly all spend their lives in hunting or being hunted to the death. The blue shark, with its terrible armature of cutting teeth, and the still larger and more formidable Rondeleti's shark differ from the trout and minnows of the babbling brooks only in the relative size of their victims.

Low as fishes rank in the scale of vertebrates, some exhibit in their habits striking analogues to the inventions of man. Most species take their prey by hunting ; others are adept anglers, notably the king of the herrings, or oar-fish (see page 7), that mysterious denizen of abysmal ocean which there is reason to identify with the sea-serpent. The ventral fins of this great fish, situated close under the throat, are each reduced to a single, long, and flexible ray tipped with

THE ARCHER FISH SHOOTING WATER AT A FLY

an attractive crimson tag, and with these rods and baits Regalecus—to give the creature its scientific name—lures small fry within reach of a pair of serviceable jaws. The angler or fishing frog, the "wide-gab" of Scottish fishermen, buries its huge carcase in the ooze and lies in wait for fishes attracted by the baits displayed on its long dorsal rays. Then there is the archer fish, Toxotes jaculator, of Oriental rivers, which aims jets of water at insects on the bank, or on overhanging foliage, with such true aim as seldom to miss the mark. The invention by man of the electric battery has been anticipated in three genera of fishes—the electric

rays in both the Atlantic and Pacific Oceans, the electric catfish of African rivers, and the still more formidable gymnotus of the Orinoco, all of which employ this marvellous force at will, whether in self-defence or to disable or kill their prey.

Although the majority of fishes bring the cares of reproduction to a close with the act of spawning, many species have developed traits of higher intelligence by sedulous attention to nest-building, incubation, and attention to the fry when hatched. The pretty little gunnel, or butterfish, of the British coasts and the common stickleback of brooks and ponds are well-known exponents of domestic virtue, and among exotic fishes may be mentioned the huge Australian barra-munda, some of the cat-fishes, the South African lepidosiren, the North American bowfin, and the freshwater bass.

From the fishes, or at least from fish-like, water-breathing ancestors with fins and internal gills, were derived in a remote age the amphibians, provided with four limbs and external gills, either temporary or permanent, as well as internal lungs, and these in turn were the progenitors of the reptiles. Both these classes have fallen upon evil times ; our globe is no longer the congenial home it was for animals with their special requirements, and their modern representatives have dwindled from their pristine importance. Many races of amphibians have disappeared ; those that remain—frogs, toads, newts, and the like—deserve more than passing regard, for among their ancestry of the Carboniferous Age we recognise the earliest appearance on earth of four-footed creatures with five-fingered limbs, a structural design appropriated later for the higher mammals, including man himself. Indeed, a fossil labyrinthodon from the Upper Miocene, in Switzerland, was hailed by Scheuchzer in 1726 as the skeleton of antediluvian man, and remained labelled to that effect in the Haarlem Museum until Cuvier proved it to be a gigantic newt.

By imperceptible stages the amphibians pass into the class of reptiles—the only cold-blooded animals breathing by lungs alone. These likewise have been forced by changing terrestrial conditions to surrender the supremacy they enjoyed when the world was still young ; for although crocodiles and pythons yet make a respectable appearance in the Eastern hemisphere, and alligators and anacondas retain their places in America, they are all mere dwarfs compared with the ponderous dinosaurs which dominated the Mezozoic land. Indeed, if we leave aside the question between exoteric or inspired intelligence and esoteric mental evolution, there are some grounds for suspicion that, had the climate of this globe remained as it was when the lower Cretaceous beds were laid down, man would never have been in a position to claim the lordship of creation, and that the sovereignty would have been usurped by a reptile. Note, by the way, that it is recorded in Holy Writ that the serpent "was more subtle than any beast of the field." Dr. Robert Munro recognises the erect posture of man as one of the chief factors in his "subtlety," that is, in his mental ascendancy over other creatures. No other creature is exempt from the necessity of using its anterior limbs primarily for support and locomotion ; only in man have they been set free for manipulative industry. From what was once only a forefoot has been evolved the most perfect and most complex prehensile organ hitherto produced by Nature ; so perfect a servant of the will stimulates the brain, the brain responds by setting the hand to fresh tasks, and so hand and head perpetually react one upon the other, their energy being endless.

Now, the iguanodon of the Wealden Clay was an enormous lizard some thirty or forty feet long, walking on its hind legs in a posture as nearly erect as was consistent with carrying a long and heavy tail clear of the ooze. Fortunately for the prospects of man, iguanodons were starved or frozen to extinction before the freedom of their five-fingered hands could stimulate their brains to a sense of the possible destiny of their race ; and nowadays the only existing reptile that is known to resort to biped locomotion is the Australian frilled lizard, a strange little being that tries to frighten away assailants by making hideous grimaces at them.

Anybody who has closely observed the behaviour of animals among themselves can scarcely doubt that they are able to communicate with each other,

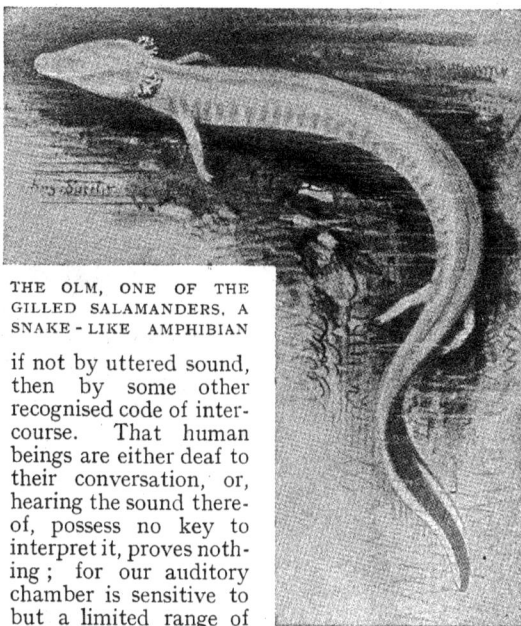

THE OLM, ONE OF THE GILLED SALAMANDERS. A SNAKE - LIKE AMPHIBIAN

if not by uttered sound, then by some other recognised code of intercourse. That human beings are either deaf to their conversation, or, hearing the sound thereof, possess no key to interpret it, proves nothing ; for our auditory chamber is sensitive to but a limited range of sound. There are many persons who have never heard so high a note as the squeak of a bat. Depend upon it, the sounds we hear wild animals utter, and many that they utter which we do not hear, are not mere noise ; they are charged with significance.

The pageant has not been altogether silent up to this point ; crocodiles grunt or bark, frogs croak, many fishes possess and exercise sound-producing organs ; but not hitherto has it been enlivened by a single note of music. With the advent of the birds, however, the air soon becomes vocal ; not very musical at first, for birds are but modified reptiles, and long ages had to elapse before they could bring themselves to discard teeth. Of all forms of wild animals, birds assuredly have found most friends and admirers among human bipeds. Their attractive forms and conspicuous behaviour, their ingenuity as architects, their exemplary care for their young, and, perhaps most of all, their loquacity, which, in many species, develops into a unique faculty of melody, have endeared them beyond all other creatures to all who have eyes to see and ears to hear withal.

It is unfortunate that affection for birds should manifest itself so generally by imprisoning them, thereby depriving them of the exercise of that faculty which is their chief glory, the power of flight—a faculty which man from earliest time has envied and applied all his ingenuity to emulate, hitherto with remarkably indifferent success. If birds be gifted with a sense of humour, what amusement must be provided to such masters of wingmanship as the swift and the falcon by the ponderous and noisy contrivances with which our most daring aeronauts manage, or fail to manage, a precarious transit of a few miles through mid-air!

FROGS AT PLAY ON THE BANK OF A POND

A FRILLED LIZARD RUNNING ON ITS HIND LEGS

The trade in cage-birds is a sorrowful subject of contemplation for humane persons, so high is the rate of mortality among the captives, and so inevitably must the survivors suffer from confinement and want of exercise. To realise the necessity of physical exercise to the health of all birds, one need but examine the daily habits of two extreme types. The albatross, with its solid weight of twenty pounds or more, sustained by a wing-spread of twelve or fifteen feet from tip to tip, will sail in wide circles round a ship, through rough weather and fine, for many days and nights without once alighting to rest. Does the bird remain awake all the time, or was Tom Moore right in referring to its " cloud-rocked slumberings " ? If it sleeps, how does it steer a course ? Professor Hutton has, rightly no doubt, explained that this bird's soaring flight is automatic, " combining, according to the laws of mechanics, the pressure of the air against his wings, with the force of gravity " ; but he seems to be wrong when he adds that " by using his head and tail as bow and stern rudders, the albatross is enabled to sail in any direction he pleases, so long as the momentum lasts." This suggests an initial momentum from wing-strokes, whereas the propelling force is continuous, produced by air-pressure on the wing-spread caused by the weight of the creature's body. Moreover, it is neither by head nor tail that the albatross steers, but by altering the plane of its wings.

In contrast to the majestic sailing of the albatross, consider the incessant diurnal activity of the smallest British bird—the gold-crested wren—a mite weighing about one ounce and measuring three inches and a half in length. That excellent ornithologist, George Montagu, moved into his study a nest containing eight young gold-crests, without interrupting the attentions of the parent birds to their brood. Closely observant, he ascertained that the little mother visited the nest on an average of more than once in every two minutes during a summer day of sixteen hours, thus making about 550 trips in the day. Allowing, on a moderate estimate, that she did not travel more than 100 yards in each trip, her day's journey amounted to 31 miles,

a distance exceeding the traveller's stature 622,614 times ; and this was performed on ten days continuously. The equivalent for a man of six feet high would be a daily march of 707 miles, in the course of which he must collect and distribute food among his children, besides spending time and energy in keeping the nursery scrupulously clean.

The elucidation of such a fact in the life-history of a wild creature is of infinitely greater importance to natural history than any contribution likely to be made by a mere bird-fancier, who, whatever care and affection he may bestow upon his favourites, must act primarily as their gaoler. One undertaking to study human nature will not succeed best by confining his observation to prisons and penitentiaries ; he must frequent scenes of free and busy social life.

Systematists hold that the modification of reptiles into birds bridged a narrower gap than that dividing any other two classes of vertebrate animals ; nevertheless, the change from cold-blooded creatures to those which, like birds, have a blood-temperature several degrees higher than that of man is a remarkable one. Certain it is that no class of creatures presents a more varied or more attractive subject for observation than do the birds. No animals exceed some of them in devotion to their young, or exhibit more dauntless courage in defending them. Intrepid human hunters go far afield to match themselves against big game, but where is the sportsman who will venture unarmed to encounter a creature 112 times his own bulk ? Yet on a day in last summer that was the exploit of a cock partridge, which flew at a friend of mine, and continued to peck violently at his boot until the mother bird had collected her brood and led them away to safety.

The colouring of birds commands admiration, not only for its range between the sable of the raven and the spotless candour of the swan, with every conceivable combination of iridescent or pigmentary hues, but also for the beauty and structural complexity of the plumage on which it is displayed. One may see yon fellow cleaning out his pipe with the tail-feather of a pigeon or the pen-feather of a gull, without bestowing a thought on the delicate adjustment of interlocking hooks that unite the fibres into a firm, elastic web.

Except among the falcons and hawks, the male usually has the advantage in size and brilliancy of plumage. The office of incubation being generally discharged by the female, it is important for her safety that her colouring should assimilate to her immediate environment. So closely do these correspond that it is commonly the bright eye of a sitting grouse, partridge, or ptarmigan that betrays the nest to a passer-by. In some families, however, the male bird takes exclusive charge of the nursery, and dresses for the part. Thus in the Eclectus genus of parrots, inhabiting the Moluccas and the Solomon

Archipelago, the females are usually attired in gorgeous scarlet. They lay their eggs, and, leaving their mates to hatch them, fly away to disport themselves in their fine raiment. The dutiful husbands, on the other hand, wear protective green plumage, nicely matching the foliage round the nest. It may further be mentioned that in this genus the disparity of the sexes is so extraordinary that they were at first classed as belonging to separate species.

We have loitered so long on the upward course of our pageant as to leave little time for noticing the highest group of vertebrate animals—namely, mammals, or creatures which suckle their young through teats, or exceptionally, as in that strange egg-laying mammal the duck-billed platypus, which nourish their young in a mammary pouch. The link connecting mammals with the reptilian vertebrates still awaits elucidation, although Professor Seeley has shown how closely the dentition of the fossile Theriodonts, an extinct group of large reptiles, approaches the characteristic specialisation of mammalian teeth.

In proportion as living organisms rise in the scale of being do they part with their plasticity, yet no class of animal has shown greater adaptability to environment than has the mammalian. Every acre of land capable of yielding vegetable food, and thousands of square miles, especially in Polar regions, which yield none, have been peopled by warm-blooded, more or less hairy, quadrupeds ; although man, the highest mammal, has exterminated some species and greatly reduced the multitude of others. So rapidly did mammals increase, when they obtained a footing on the earth, that the land proved incapable of sustaining them all. A branch of the Insectivora—that great order of which the hedge-hog, the mole, and the shrew-mouse are British representatives, being in danger of being crowded

A MOTHER PARTRIDGE PROTECTING HER BROOD

out, took to themselves wings and founded a new order—Chiroptera, or bats—which subsist by aerial hunting. The ocean would seem a most unsuitable home for creatures unable to inhale the free oxygen in water, yet it once swarmed with enormous mammals—whales, walrus, dolphins, porpoises, seals ; and even now, after men have been working destruction for centuries upon these marine herds, there still roams in the sea the most gigantic of living animals—Sibbald's rorqual, or blue whale, attaining a length of upwards of 85 feet. Longer creatures than this have trod the earth in the past—diplodocus Carnegii, for instance, which, from nose to tail-tip measured 120 feet—but none has ever equalled the rorqual in bulk.

By a remarkable coincidence, the two creatures which were the last to own the supremacy of man are respectively the largest and well-nigh the smallest that have come under his dominion—namely, Sibbald's rorqual and the pestiferous mosquito. The rorqual was the only animal in the world which no man, not even the most intrepid whaler from Dundee or Bergen, dared to attack, until the invention a few years ago of bomb-harpoons fired from a gun. But now this giant of the deep is hunted as diligently as any other whale, and its extinction may be not far off, for whales reproduce themselves very slowly. The mighty rorqual never did injury to the human race ; the slenderest antelope could not be more innocent of offence ; but the other creature which set man longest at defiance—the malarial mosquito—is accountable for more human mortality and suffering than all the beasts of prey that ever roamed the world.

By these crowning examples of the supremacy of mind over matter, the ascendancy of man over his fellow-denizens of the globe has been raised to the highest pitch hitherto attained, and with them the pageant of life, as now revealed to us, may be brought to a fitting close.

HERBERT MAXWELL.

A FRILLED LIZARD GRIMACING
Photographs by W. P. Dando

The artist has grouped in this striking picture some of the extinct winged reptiles whose formation may be imagined from their existing remain 1, pteranodon occidentalis; 2, pterodactylus crassirostris; 3, rhamphorhynchus; 4, dimorphodon macronyx; 5, 6, small pterodactyles

THE FIRST INHABITANTS OF THE WORLD

MASTERS OF THE EARTH BEFORE THE RISE OF MAN

By Professor J. W. GREGORY, F.R.S.

OUR knowledge of the former inhabitants of the earth is derived from those remains of animals and plants which are found in the rocks of the crust and known as fossils. We can only expect to find fossils in such materials as sand, sandstone, clay, and limestone that have been deposited in layers on land or in water. Such rocks as granite, lava, and statuary marble, that are composed of crystalline constituents, were either molten and solidified under conditions far too hot for the existence of life, or else, as in the case of marble, the rock has been so altered that any traces of life that may once have existed have been destroyed. Fossils, therefore, come from the rocks formed by the deposit of grains of sand and particles of clay, which, as they fell, buried the remains of animals or plants; and also from rocks, such as limestone, which, as we shall see presently, are composed of fossil shells or skeletons.

Accordingly, it is natural to turn to the oldest known rocks that could contain fossils in order to learn from them what were the first living creatures of which we can expect certain knowledge, and the nature of the climate and the geographical conditions under which they lived. Many of these old rocks have undergone such slight alterations that if animals with thick shells had been laid down in them, the fossil remains would certainly have been discovered. The oldest sandstones of the British Isles, the Torridon Sandstones of North-Western Scotland, are so little altered that they were for long regarded as part of the Old Red Sandstone, which is famous for its well-preserved fossil fishes, shells, and ferns. The Torridon Sandstone, however, has hitherto yielded no recognisable fossils; but life was present at the time of its formation, as it contains grains of phosphate of lime which retain traces of organic structures. In North America, around Lake Superior, in Scandinavia, and in South Africa, there

SIR RICHARD OWEN AND THE SKELETON OF A MOA
An extinct flightless bird of New Zealand

are immense beds of sandstone, of approximately the same age as the Torridon Sandstone, but they also have yielded no fossils. In Montana, however, some beds of this age have furnished a few ill-preserved fossils, including a crustacean known as Beltina, and these are the oldest known animals.

The rocks formed before the Torridon Sandstone and its contemporaries extend as a thick series that serves as the foundation on which all the later fossiliferous rocks have been laid down. As these underlying or " Archæan " rocks are composed of crystalline materials, it is not surprising that no fossils have been found in them; but they offer some evidence of the possible existence of life on the earth at the time they were deposited.

Most limestones are formed by the action of plants or animals which extract lime from water, and use it to make their skeletons or shells. It is true that some limestones have been formed chemically, but those due to living agencies are by far the more abundant; hence many geologists have believed that the thick limestones associated with the old crystalline rocks indicate the probable existence of living beings on the earth when they were deposited.

Again, graphite, a mineral composed of carbon, is abundant in these old crystalline rocks; and carbon, where found in the sedimentary rocks, is usually due to the action of plants. It is therefore thought likely that these ancient seams of graphite indicate the existence of plants, probably seaweeds.

Some of the old rocks may therefore be due to the action of living beings, of which no certain trace remains. For many years it was thought that a structure found in one of the ancient limestones of Canada was a fossil. It was named Eozoön, in the belief that it was " the dawn of life." But even if this limestone had been formed by some organic agency, no trace of the organism can have been preserved; and the Eozoön is really only a banded structure due to the alteration of the limestone by an invasion of molten rock. The geologist turns from these crystalline rocks feeling that some life was probably in existence at the time of their formation, but that the secret of its origin and nature, once confided to these rocks, has been lost for ever. The evidence of the Torridon Sandstone series is still more disappointing, because some of its rocks should have preserved any fossils deposited in them. In the next rocks laid down after the Torridon Sandstone there are abundant fossils — remains of highly organised creatures, including members of so many of the chief groups of animals that they must have had long chains of ancestors. The earlier rocks, therefore, do not represent a period without life, as is implied by the name of Azoic, often given to them.

The animal kingdom is divided into two sub-

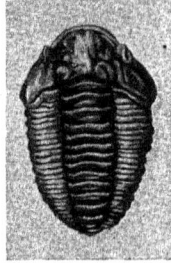

THE TRILOBITE
One of the most ancient of the crustaceans; the trilobite was indeed the king of the primordial seas

shrimps, centipedes, spiders, and insects; and the shellfish and their allies. The rest of the many-celled animals have backbones, and are therefore called the Vertebrates, including fishes, amphibians, reptiles, birds, and mammals. With the exception of the backboned animals, representatives of all these groups are known in the oldest Palæozoic rocks, and most of the groups are represented by their chief subdivisions. Fishes, however, certainly appeared soon afterwards, and thus members of all the chief divisions of the animal kingdom were living in the earlier half of the oldest group of fossil-bearing rocks; and if we follow the development of the animal kingdom, the classification based upon living animals serves for those of any previous period of the earth's history.

The climate of the earth as a whole seems to have been remarkably uniform, considering the extreme complexity of the causes that determine it. There is no indication from geological evidence

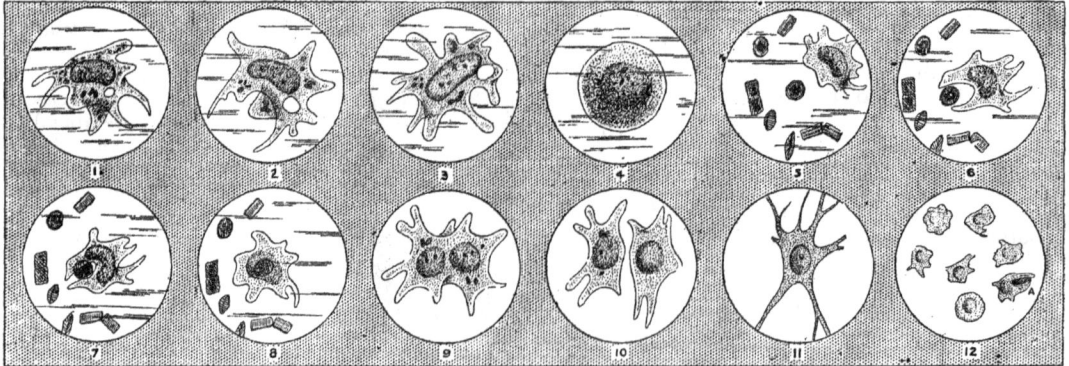

THE MINUTEST FORMS OF LIFE AS SEEN UNDER THE MICROSCOPE

1-3, an amœba floating in a tiny drop of water and the rapid changes it undergoes; 4, its sudden change into a globular mass, when touched; 5, an amœba approaching some diatoms, or minute single-celled plants; 6, about to swallow a diatom by, 7, folding part of itself over the food, and, 8, engulfing the plant in an improvised stomach; 9, about to divide, the nucleus separating first; 10, the division completed; 11, a nerve-cell with fibres, from the human spinal cord; 12, amœba-like cells in human blood which devour, destroy, or otherwise remove microbes: note the various shapes these cells assume and, at A, a cell eating a microbe

kingdoms, of which the first includes animals which consist either only of one cell or of several similar cells, united together into a colony. They are mostly minute in size, and from their simple structure are called the Protozoa. All remaining animals have a body composed of many cells, which are modified into different forms, so as to serve different purposes. These many-celled animals, or Metazoa, include six groups, which have no backbone, and are therefore Invertebrate. These groups are the sponges; the jelly-fish and corals; the starfish, sea urchins, sea cucumbers, and sea lilies; the worms and their allies; the animals with jointed legs, including crabs, lobsters,

THE ETERNAL RECORD OF THE ROCKS
Footprints of a labyrinthodon, a huge extinct amphibian of the early world, preserved in sandstone through ages of time

that the earth has been steadily cooling down from a more heated condition, for the climate indicated by the oldest sedimentary and fossiliferous rocks was not appreciably warmer than it is in the present times. There have been numerous great local variations of climate at different periods of the earth's history. One region has been colder at one time and warmer at another; but it is doubtful whether the mean climate of the whole world has been materially different from the present. At one time it was a common belief that the forests which produced the coal - seams indicated a much moister and hotter climate throughout the world whole world. The view, for example, has been expressed that the sun must then have been much larger and given out much more heat, in order to maintain the vegetation that grew in the British area when the coal-seams were being formed. But at that very period vast glaciers and snowfields existed in parts of India, Australia, and South Africa, so that, at least in parts of the great continent of which those lands were then parts, the conditions were much colder than they are now.

Again, at a time geologically quite recent, the climate of the British Isles was much colder than it is to-day. All the mountains were capped by perpetual snow, and glaciers flowed down their slopes and spread over the adjacent lowlands. Greenland, on the contrary, which is now covered by a vast ice-cap, at one time enjoyed a milder climate, as is shown by the vegetation and tree-ferns that once grew there. But, as Lord Kelvin has shown, these changes in the Arctic climate can be explained by a different geographical distribution of land and water.

A considerable change in the chemical composition of the atmosphere might

A FOREST IN THE CARBONIFEROUS AGE
The luxurious vegetation which the ages have transformed into coal

have had a great effect upon the animals and plants of the world, but fossils give no evidence of any change in the composition of the air. The minute structure of the plants that formed the Coal Measures shows that they lived under the same essential physical conditions as at present. The air was no doubt often moister in some localities, and drier in others, than it is in them to-day. But there does not appear to have been any sufficient change in the gases of the atmosphere seriously to have affected the structure either of animals or plants. The openings on leaves, by which plants breathed, were of the same size and proportions as are those of modern plants. The constituent of the air, other than moisture, that is most likely to have varied in amount is the gas carbonic acid; and it has been suggested that the former occurrence of glaciers in countries now too warm for them was due to a diminution in the amount of carbonic acid in the air, but no such

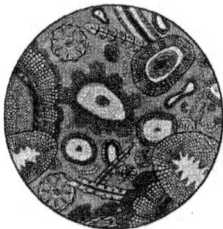

A SECTION OF A PIECE OF COAL

change has yet been demonstrated, and it appears physically improbable. It is one of the beneficent functions of the sea to preserve an atmosphere of almost constant composition. A great increase in the amount of carbonic acid in the air might have a very decided effect upon the life of the earth, and this gas is discharged into the air in vast quantities during volcanic eruptions. The sea, however, at once absorbs the major part of the gas thus added to the air; and similarly, if the amount of carbonic acid in the atmosphere were to fall below the normal, the sea would give off some of its supply, until the deficiency had been nearly made up.

The evolution of life on the earth has also been controlled throughout geological time by constant physical limitations imposed by the strength of materials.

Many animals now extinct were much larger than their nearest living relations, and the discovery of great bones has often given rise to exaggerated estimates of the size of the monsters to which they belonged. These mistakes were due partly to misleading comparisons with existing animals. The thigh bone of a kangaroo is much larger in proportion to the length of the body than that of a crocodile, which, as the largest of living four-legged reptiles, was naturally used by early naturalists as the best standard of comparison for large fossil reptilian remains. Hence, when the thigh bones of large reptiles were found, it is not surprising that the size of the animal to which they belonged was over-estimated. In other cases, the size was exaggerated in consequence of mistakes in restoring broken bones.

Similarly with birds. The largest known egg is that of the æpyornis of Madagascar, which is estimated as 150 times larger than the egg of the common fowl. If there had been the same difference in size between the two birds as between their eggs, the æpyornis would have been colossal. The early Arab sailors knew of the æpyornis egg, and, as the bird was already extinct, their lively imagination restored it as the famous roc in the tale of Sindbad the Sailor. The egg, however, was laid by a bird that could not fly, like all the biggest living birds, such as the ostrich and cassowary. Indeed, it is doubtful whether any heavy or large-bodied flying animal has ever existed. Most of the known extinct flying birds were small, and as birds increased greatly in size,

SKELETON OF DIPLODOCUS, AN EXTINCT REPTILE
This monster was 80 ft. long and stood 14 ft. high

like the moa of New Zealand (see page 13), or the ostrich, or the æpyornis, flight became mechanically impossible; so their wings decayed, and they were confined to running on the land.

The largest flying creatures that have ever existed were reptiles, such as the American pteranodon. It sometimes had an expanse of wing estimated at from 20 to 25 ft.; but its body was comparatively small. Dr. C. W. Andrews tells me that the length of the body was certainly not more than between 2 and 3 ft. The wings must have had a slow, heavy beat, and their main use may have been to act as a great gliding plane, with which the animal could skim through the air like a flying fox; they were less suitable for active flight than the wings of a bat. The modern albatross has an expanse of wing of 15 ft., and no larger extinct flying birds have been well established.

The limitation of the size of flying birds follows inevitably from the mechanical principle, to which Professor Barr has called attention, that the weight of an engine increases more rapidly than its power. More power can be obtained from a given weight of material by using it in many small engines than from one large engine. Hence, flying machines, which require the greatest possible power from the smallest possible weight, often use several engines instead of one large one. Nature has never produced a bird with many sets of wings, driven by independent muscles and as many distinct hearts; so the limit in the possible size of flying animals is small. There have been many land animals in the past

RESTORATION OF DIPLODOCUS TO SHOW ITS APPEARANCE
From the model at South Kensington Museum

larger than their existing descendants, but it is not the rule for modern animals to be smaller than their ancestors. Modern horses, dogs, cattle, camels, and giraffes are all larger than their extinct forerunners were. It is doubtful whether any animal that lived on dry land was much heavier than the

largest existing elephant; for animals, like bridges, are limited in size by the principle that in similar structures the weight increases more rapidly than the loads those structures can bear. Professor Barr, in calling attention to this law, has pointed out that the elephant's body is almost as large as the nature of the materials allows, since an animal having a body two or three times as long, as high, and as wide as an elephant's would require legs filling up nearly the whole space under its body. Animals with a heavy body like the elephant's must have very thick legs, necessarily straight, as they are thus best adapted for supporting a heavy load. Hence, those extinct reptiles that had bodies about as large as the elephant had also thick, straight legs. Various animals have been much longer than the elephant, attaining the length of 100 ft., or more; but they lived either in water or marshes, where much of the weight of the body was supported by the soft material of the swamp or by the water.

Aquatic animals may be much larger than land animals, but no extinct animal is known as large or as heavy as the biggest modern whales, which are 85 ft. long, and weigh as much as 70 tons. Former

SKELETON OF BRONTOSAURUS
This giant reptile was over 66 ft. long, and its body was 8½ ft. thick

amphibious animals, however, were much larger than the hippopotamus, the biggest existing swamp-dweller. There is, indeed, no living land animal as large as the giant reptiles that lived in Britain and North America at the time of the formation of the Oolitic limestones which now form the Gloucestershire hills and Portland Bill. The first known of these giants was the cetiosaurus, of which remains were found near Oxford and were referred by Sir Richard Owen to some colossal aquatic reptile.

The complete skeletons have since been found of two allied American reptiles. One of these, the diplodocus, is well known from the cast of its skeleton in the Natural History Museum at South Kensington. It is 80 ft. in length, and stood 14 ft. high, while some allied reptiles, of which only fragments are known, were even larger. The diplodocus had a very massive body, a long, tapering tail, with flattened sides, and a long neck, with a very small head. That the animal lived in water, rather than in swamps, is shown by its flattened tail, which was no doubt used for swimming. The aquatic habit of these monster reptiles is also indicated by the arrangement of their eyes. Unfortunately, the skulls

GUEREZA MONKEYS WITH THEIR SPLENDID MANTLES OF SNOWY WHITE HAIR

B

"NATURE RED IN TOOTH AND CLAW"   PYTHON KILLING A PARROT

BATTLES OF THE WILD: GIANT HERON AND SEA EAGLE IN DEADLY COMBAT

are usually crushed and less well preserved than the rest of the skeleton; but a skull of the morosaurus shows that it had a third eye on the top of its head. An eye gazing heavenwards would be of service to an animal creeping on the ground or floating on water, but of no use to a monster living on land, that towered above other contemporary animals. The diplodocus would have been too sluggish to dodge a falling thunderbolt, and its brain was too small to appreciate the beauty of the stars.

That these reptiles occasionally visited the land is obvious from the size of their limbs. In some, the legs may have been strongly bent, so that the animal could rest the full length of its body on the ground. On shore, the reptiles could only have moved slowly, and could not have defended themselves by the swift charge of the elephant or

the top of the back to the belly, and was raised on limbs of $8\frac{1}{2}$ to $10\frac{1}{2}$ ft. in length. The animal had a more massive and shorter neck, and a comparatively larger head, than the diplodocus had; its tail was shorter, and does not appear to have had the fin-like flattening. Its habits were probably also amphibious.

Land animals that belong to very different groups have often acquired similar forms owing to the influence of similar external factors. It is a great advantage to animals to stand erect. As Professor W. A. Osborne has pointed out, man is pursued by digestive troubles, owing to his alimentary canal, which was intended to be horizontal, being now placed vertically. But indigestion is, after all, a matter of detail in comparison with the advantages man has gained by his erect attitude.

HOW THE FACE OF THE EARTH WAS CHANGED AND THE EXTINCT ANIMALS WERE BURIED

This series of diagrams shows in a graphic way how the rocks have been formed by layer upon layer of fresh deposits washed down by the rivers, and how the remains of animals that had lived in the ages when the various deposits were formed became embedded therein, thus building up for us "the record of the rocks."

captured their prey with the nimble agility of lizards. They probably spent part of their time on land or in swamps, browsing, like the hippopotamus, on vegetation. If, as is probable, they laid eggs, they must have made nests on land, and stayed there for a while to protect their eggs and young.

Closely related to the diplodocus was the brontosaurus, of which the best impression can be formed from the skeleton in the Museum of Natural History in New York (see page 16). This specimen is of particular interest, as it consists of the actual fossil bones, and not of casts. The brontosaurus was at one time overestimated in size, and it could afford to be reduced, as its actual length was over 66 ft. Its height is less certain, as this would depend on the bending of its limbs. The body was $8\frac{1}{2}$ ft. thick from

Birds, which early adopted an erect habit, have been unable to acquire the fullest advantage from the change by the development of the fore limbs into hands, as these have been converted into wings. Several kinds of reptiles and mammals, however, have been more successful, and secured increased efficiency from their limbs by a subdivision of labour, using the two hind ones for walking and the two front ones as hands.

The fore end of land animals is often weighted by a heavy head, and the adoption of a more or less erect attitude requires a new balance that is often secured by the tail. The kangaroo is the best-known animal with this shape. Some living lizards occasionally run on their hind legs, with their front legs raised off the ground. A similar

SKELETON OF A GIANT SLOTH

specimen in the Natural History Museum at South Kensington illustrates how the iguanodon walked upon its hind legs, balancing itself by its tail and using its front limbs as paws, with which it probably collected and served into its mouth the foliage upon which it fed. That it was a vegetable-feeding animal may be seen by examination of its teeth. The head has a remarkable resemblance to that of the horse, so that no doubt the animal lived on similar food.

The iguanodon was one of many large reptiles that fed upon vegetation; and these herbivorous animals were preyed on by carnivorous reptiles, such as the megalosaurus in England and the more completely known ceratosaurus in Colorado. Both were probably kangaroo-like in gait; they had light, hollow bones, and were doubtless much swifter in their movements than their herbivorous relations on which they fed. They captured their prey by a rush or bound, and killed it by cutting its throat with their sharp, saw-edged teeth, as a tiger tears the throat of a cow, or their long, conical fangs punctured the skull and pierced the brain of their smaller victims.

In spite of the apparently endless variety of geographical conditions due to variations in temperature, moisture, and vegetation, the environments that are essentially different were very few. The chief were the differences between life on land, in air, or in water. Life in each of these habitats has remained under fundamentally the same conditions throughout geological time. Animals that

form and habit were developed in various extinct mammals and reptiles. Thus, the remains of the giant sloth of South America, the megatherium, at first presented a puzzling problem, as they unquestionably belonged to an animal allied to the sloths which haunt trees to feed upon their foliage; but the giant sloth was so colossal that no tree could have borne its weight. This animal, however, which was from 16 ft. to 18 ft. long, presumably lived upon the ground; when it desired leaves above its reach, it stood upon its hind legs, clasped the trunk of the tree with its front limbs, and, with a mighty heave, either broke the tree across or tore it up by the roots, when it could then browse upon the leaves at its ease.

The kangaroo-like habit was also adopted by a gigantic reptile that lived in Kent and Belgium at a period somewhat earlier than the deposition of the English Chalk. This animal was first known from scattered bones and teeth found in the Weald of South-Eastern England. It was called the iguanodon, from the resemblance of its teeth to those of the existing lizard, the iguana. The true iguana is a comparatively small lizard, from 4 ft. to 5 ft. long. If the difference in length between the iguanodon and iguana had been as great as that between their thigh bones, then the iguanodon would have been 45 ft. long. The animal, however, was only about 24 ft. long, as its hind legs were much larger in proportion to the body than those of the iguana. The cast of a Belgian

THE GIANT SLOTH AS IT LIVED ON THE EARTH

lived on land always had the same air to breathe; their dimensions have been controlled by the ever-acting force of gravity, and their shapes by the same strength of the wind and radiation of heat; and they had the same alternative of food—vegetation or other animals. Those that took to life in the air were limited by the mechanical difficulties in the flight of large bodies which have so long delayed the invention of flying machines heavier than air. Animals that live in as dense a medium as water are moulded to the shape that reduces to a minimum its resistance to their passage. The same physical properties that have shaped the modern torpedo gave a similar form to various aquatic reptiles and mammals. And as oars have been replaced in navigation by the screw propeller, so animals living in water have found the feet less efficient as organs of locomotion than the creation of currents of water by undulations of the body or by the lashing of the tail. Hence,

them ashore, where some of them finally made their homes. The growth of animals living upon vegetation was soon followed by the development of others that preyed on the vegetarians. Owing to the superiority of land plants to water plants

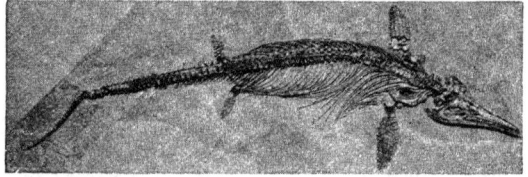

SKELETON OF THE ICHTHYOSAURUS

as food, most backboned animals are terrestrial, though the Amphibia, the lowest group above the fishes, still, as their name implies, live largely in water. Various members of the higher groups, the reptiles and mammals, have from time

THE ICHTHYOSAURUS AND PLESIOSAURUS IN THEIR NATIVE HAUNTS, RESTORED FROM EXISTING SKELETONS

the legs soon became reduced in size. Thus, in the crocodile, they are small, and only bear the weight of the body when actually moving on land; when stationary, the body rests on the ground.

SKELETON OF THE PLESIOSAURUS

In the struggle for existence, backboned animals have been often forced by competition to change their mode of life. Their original ancestor no doubt lived in water; but, as vegetation developed upon land, the new food of the fruitful fields tempted

to time returned to an aquatic life, while even some birds, such as the penguins, have lost the power of flight, and live more on sea than land.

The land animals that have returned to aquatic conditions were probably tempted by the desire for fish. The otter, with its flexible body and short legs, illustrates an early stage of the change in form produced by this change of life. Seals and sea lions have gone a stage further; they live still less on land, and so their limbs have been reduced to mere flippers, and the propeller-like tail drives the animal forwards, like an oar worked from side to side behind a boat. Finally, in the whales and dolphins, which cannot live ashore, the hind limbs have been either entirely lost or reduced to rudimentary bones, buried in the tissues of the body. They have a fish-like form, and are popularly regarded as fishes. Their teeth, if they have any, resemble those of fishes or reptiles more than those of mammals, since all the teeth in the jaw are of the same simple conical form, such as those of a crocodile, and they have lost the difference between the front and back teeth of their terrestrial ancestors. These conical teeth are more serviceable, as they at once pierce and hold the slippery fishes upon which they live; and as the dolphins bolt their food, without gnawing the flesh off the bones or grinding and chewing it up into pulp, they have neither the front cutting teeth nor the back grinding teeth necessary to most land animals.

There was, among the extinct reptiles, a similar series of forms due to adaptation to the same conditions of life. No living reptile is similar in form to the sea lion, but many such existed in the period called the Age of Reptiles. The best known of these sea-lion-like reptiles was the plesiosaurus, which differed in external shape from the sea lion mainly in the great length of its long, slender neck. This animal has been compared to the head of a lizard mounted on the end of a snake that has been strung through the body of a duck which had flappers like a seal. This animal lived most of the time in the sea, and its long neck gave its head such wide and swift movement that it could catch any fish within a long reach, and its sharp, conical teeth would hold its prey as firmly as a tooth-edged vice. The form of its legs indicates that the plesiosaurus sometimes visited the shore; and the permanent adoption of life in water by other reptiles led to their acquiring bodies similar in general form, the presence of fins and the absence of legs, to the whales and dolphins. The best known of these dolphin-like reptiles is the ichthyosaurus. It was entirely aquatic. That it dived to greater depths than the plesiosaurus is indicated by its larger eyes, which were expanded to admit as much as possible of the feeble light in the darker, deeper waters; and the eye was protected from the water-pressure by a ring of bony plates, which, when pressed inwards, formed a dome guarding the delicate structures beneath. The ichthyosaurus did not, like many land reptiles, lay eggs, for a reptile cannot lay eggs in the myriads produced by fishes, and the ichthyosaurus was ill adapted

THE EXTINCT GLYPTODON, A GIGANTIC ARMADILLO

for hatching eggs in a nest; so its young were born fully developed and active.

Though evolution has gone steadily forwards along the same main lines, the pace has not been uniform. There have been long, quiet periods, during which the rate of change was slow, and there have been intervening periods of widespread earth-movements and great geographical changes. These disturbances caused the extermination of many species, which had acquired fixed habits and structures, and the appearance of fresh species, better adapted to the new conditions.

The main progress of evolution, however, has not been due to change in environment so much as to the constant struggle for existence, owing to the influential fact that animals and plants tend to increase more rapidly than does the food supply. Codfish produce eggs and flowers produce pollen grains by the million, and plants produce vast quantities of seeds. It follows that only a small proportion of eggs and seeds have any chance of reproducing their kind. The unceasing stimulus of evolution has been the struggle for existence between closely allied forms of life. The progress of the animal kingdom has necessarily followed the development of plants, as all animals are dependent directly or indirectly upon vegetable materials for their food.

The struggle between animals has been strenuous, and even fierce, throughout geological history. Even in the earliest period of which we have any appreciable information as to its life, we find beds of slate in which nearly every slab contains the remains

THE PANGOLIN, SHOWING NATURE'S PROTECTIVE ARMOUR

THE ARMADILLO, A RELATION OF THE GLYPTODON

THE OTTER, AT HOME ON LAND OR IN THE WATER

of a trilobite (see page 14), showing that these animals swarmed in the waters in which the rocks were laid down. Some of the layers of rocks in the Old Red Sandstone are crowded with fossil fish remains; they are often so closely packed that it is difficult to separate the different individuals. They are crowded more tightly than sardines in a tin, for they are jammed together like sardines crushed into one confused mass.

These fishes had probably been caught in pools of water which gradually dried up, leaving all the dead fishes on the mud at the bottom; but they must have swarmed in the waters from which these pools were formed, and have waged a constant struggle for the available supplies of food. Ever since animals appeared upon the earth, the price of life has been the death of others. The fit have survived, and the less fit have either perished or been pushed back into positions where the conditions of life are less easy and less pleasant. Thus, sponges with a silicious skeleton were common on the shores of ancient seas, but they have been beaten in the struggle for existence, and now survive either in such sanctuaries as the deep seas or the cold shores of the Antarctic, which sponges of modern types have not yet invaded in force. Sea lilies once lived in shallow water around the British Isles, but they, too, have been driven from the coasts; they now live in the oceanic abysses with other animals too old-fashioned to adapt themselves to the fierce struggle for place and food in lighter and warmer waters along tropical and temperate shores Similarly with land animals, the

modern representatives of many of the more ancient forms of life live in the sheltered security of remote islands. Thus a three-eyed lizard (hatteria) survives in New Zealand, and the marsupials which once lived in Europe are now extinct in the Western hemisphere except in the continent of Australia.

In the struggle between the different animals for life and food, Nature has always shown a merciless indifference to pain. The jellyfish, impressions of which are found in the earliest fossiliferous rocks, no doubt had stinging cells like their modern descendants. The primitive sharks had jagged teeth that would have given most painful wounds. The jaw of a shark is very effective from the point of view of the shark, but it was not designed to save pain to its victim. Other sharks had barbed spines that would have inflicted severe tortures on the animals they lacerated. Scorpions, found among the Silurian rocks of Scotland, probably had poison-fangs, and perhaps gave as burning a sting as those of any modern scorpion.

The first defence adopted against these instruments of torture, and against the closely set, solid crushing teeth, by which the jaws of some early sharks could crush fragile shells and bones like a vice, was a thick external armour. Many ancient animals were protected by a powerful calcareous or bony covering. Thus the fishes of the Old Red Sandstone lakes and rivers wore an elaborate array of closely packed bony plates or thick overlapping scales. The soft parts of ancient sea lilies were enclosed in a thick-walled chamber, which

THE SEA OTTER, A RARE MARINE MAMMAL

was often protected, in addition, by strong, sharp spines. Many of the older shellfish had much thicker shells than have any of their modern representatives.

SEA LIONS, NOW OF SO AQUATIC HABIT THEY SELDOM VISIT LAND

Land animals were also protected by a bony armour. Thus, even the primitive amphibians, the group that includes frogs and newts, were protected by plates of bone, although all their modern descendants have especially soft skins, and are usually deficient in hard defensive structures.

Many land animals trusted to offensive rather than defensive methods of protection. Thus, they were provided with horns, of which the most specialised forms are those of cattle and deer, whose horns are now almost as large and elaborate as at any previous period in the earth's history. There is no living deer so large as the great extinct Irish elk, of which the antlers were sometimes 11 ft. across, and weighed nearly 100 lb.; but this deer has only become extinct recently, and was possibly exterminated by man. The type of defence by horn that has been more frequently adopted is that by upright horns, borne on the top of the head, or along the middle line of the back. This type is represented by the rhinoceros, which has one or two horns upon the front of the head, but many earlier land animals, such as the tinoceras, of Wyoming, in North America, had several pairs, though they were shorter, and only one pair was probably long enough to have been of much service in attack.

Extinct reptiles show the most elaborate development of defence by a middle line of unpaired horns and spines. Thus, the stegosaurus, which once lived in Colorado, and was 25 ft. long, had a row of massive spines along its back, and some single spines were 4 ft. broad and 3 ft. high. The triceratops had rhinoceros-like horns, one on its snout and a longer and sharper pair on its forehead, and it must have used them in fighting by an upward thrust, like the rhinoceros; and, as the triceratops had the largest known head of any land animal, the momentum of the blow would have given the horns great penetrating power.

Complete external armour is still maintained by the tortoises and turtles; but their shells are less elaborate than those of the extinct horned tortoises which once lived in Australia and South America, and had, in addition, long spikes. Armour-plating is, however, now out of date, and even among turtles many of the largest have lost their rigid armour, and have a leathery flexible skin. Among the mammals external armour is still retained by the small armadillos of South America and the scaly ant-eaters of India and Eastern Asia. But the existing armadillos are tiny compared with their predecessors; of which the glyptodon, for instance, had a solid rigid shell, 7 ft. in length. Armour in the strife between animals, as in human warfare, has been discarded, as it hampered rapid movement. Its weight handicapped animals that had to compete with agile foes, and it was less to be trusted than increased brain-power, which gives its possessors keener senses, quicker movement, more effective energy, and greater adaptability and resource. The bulky stegosaurus needed its weighty row of spines, but as it had a smaller brain in proportion to the size of the animal than any other known land animal, even its huge spines could not save it; and the triceratops, which holds the record for the lowest proportion of brain to skull among known reptiles, also found that safety does not depend on spikes. Animals with larger brains beat those whose trust was in armour, and those that could think swiftly inevitably replaced their more sluggish contemporaries.

J. W. GREGORY

THE WHALE, A MAMMAL THAT HAS BECOME MONARCH OF THE SEAS

THE RISE OF MAN ABOVE THE BEAST

"THE HIGHEST YET," AND THE LONG STRUGGLE UPWARD

By Sir HARRY· JOHNSTON

As we survey the mammalia of the past, where do we discern the first "pointings" of man? I wish to write of this and other subjects in Natural History reverently, because to me—and no doubt to many others—the study of Life, past and present, is rapidly becoming identical with a religion—an intense, passionate desire to get a little nearer to a conception of the truth regarding the universe.

In that universe which at night we see ranged about us in the sky, and of which we know our little globe to be an inconspicuous portion, the most accessible, and perhaps to us terrestrials the most fascinating, problems are, first of all, the origin of Life, of conscious matter, and, secondly, the differentiation of mere protoplasm into more and more complex forms, and the ascent and increase of this complexity until it reaches its highest development in man—man, who, without exaggeration, seems to us to be hovering on the verge of God-like development. What is the purpose in man? What is his pedigree? Why has he been selected to be the precious isthmus, the one link between an animal past and a spiritual future?

During the last period of the Primary epoch, the Permian, in South Africa, Eastern Europe, or elsewhere in the world, there arose from the earliest reptilian forms the Theromorpha, or "beast-like" reptiles. They retained about them some features of their amphibian ancestry, but in their development beastwards they had risen from the earth on legs that were longer and sturdier; the belly no more rested on the ground or dragged itself over mud and rock, the braincase became larger, and the bones of the skeleton more elaborate and better suited to rapid movements. And the teeth, instead of remaining a uniformity of simple spikes, differentiated into three types; there were the eight—or more, or fewer—incisors in the front of the mouth for seizing; the long, sharp-pointed canine tusks—most reptilian of all existing teeth, and used for stabbing, killing, tearing; and the three- or four-coned, broadened molars for champing and grinding.

From out of these theriodont reptiles arose the protomammal, an egg-laying creature with teeth very like those of its reptilian ancestor, except that the molars began to develop double roots, but with a larger brain, a higher temperature of the blood, and an altogether peculiar form of skin-covering—hair.

Terrible was the struggle for life against the reptile monsters that dominated the land and water at that period; and the early mammals, possibly smallish animals, could only escape their enemies by activity, and cunning above all. The dinosaurs and other reptiles of the Secondary epoch grew to enormous dimensions, in this respect being really the lords of the earth on which they dwelt; but their brains had not developed co-ordinately with their bodies. Thus they were unable to devour all their protomammalian rivals, or to avoid the bacteria, the flagellates, or the intestinal worms which caused diseases and devastating plagues, or to adapt themselves quickly enough to the changes in climate or food supply. When Nature's radiating, evolutionary plan had developed two new classes out of the reptile stock—the birds and the mammals—brain-matter, albeit in rudimentary form, had begun to count; for brain-matter would seem to be the germ of the soul and the link with the Deity.

But birds, in order to fly, lost the use of their hands; in order to bolt their fish or shellfish and the insects they began to pursue on the wing, they had neglected teeth, face, and nose, and were saddled for all time with a callous beak. The beak, it is true—like the elephant's trunk—became a very tolerable instrument, and in their later developments several groups of birds trained their hinder feet to act as hands.

Indeed, it is not altogether easy to understand why the birds missed the prize that ultimately fell to the mammals. Class for class, they have come in a very good second to the goal set up to record progress at the end of the Tertiaries. Our instincts may not be wholly at fault in associating birds with the Divine Principle, with, in fact, what we call Religion. Man from his earliest years of self-consciousness had a kindly feeling for the bird and a loathing for the reptile. If the bird has not been the chief agent in carrying out God's purpose on earth, it has at any rate received the tiny consolation prize of man's love and gratitude—gratitude for the display of the bird's ineffable beauty, exquisite song, delicious flesh, and average harmlessness.

Now, casting our eyes back to the protomammal, we realise that in its hunted existence, fleeing hither and thither before gobbling reptiles, it did not do as the bird had done—develop hands into wings and make its nest on the tree-tops, or lay its eggs on some inaccessible cliff or mountain. On the contrary, in one of those unconsciously brilliant ideas that characterise the evolutionary scheme, some protomammal and its descendants retained its eggs a longer and longer time within the body-cavity until it laid them half or wholly hatched. The living young was then gathered up, and carried or sheltered on the under side of the body between the thighs, much as the penguins of the Antarctic shelter their new-born chicks. Such an idea as the growth of the egg inside the mother's body was not new in Nature's repository, and has often occurred and re-occurred in the developmental history of invertebrates and vertebrates, especially in fishes and reptiles, living and extinct. The huge sea-dwelling ichthyosaurs produced their young in a fully developed condition, able to earn their own living, instead of laying eggs, which would need a long external development.

THE EXTINCT SABRE-TOOTHED TIGER, OR MACHÆRODUS

with very long, flat, and saw-edged tusks. And this with time and practice grew into the perfect machairodont, a creature in its South American exaggeration larger than the living lion, with immensely powerful fore limbs, huge retractile claws, and sabre - shaped teeth twelve inches in length outside the gum, a creature which may have got so extreme in its development that it no longer cared about chewing flesh, but fed mainly on gulping down the blood of bigger mammals, whose arteries it pierced with its dagger tusks.

But by this time Nature had grown sick of bloodshed, and removed her interest from the machairodont, which forthwith dwindled and vanished. She pondered over the mastodon for a time. This huge creature had not always been quite so big. Its first definite known ancestor, the moeritherium, was perhaps the size of a large bear. The moeritherium lived in Egypt, and there, or somewhere on the eastern shores of the Mediterranean, had grown and specialised from some primitive ungulate not unlike in appearance to the hyrax. The hyrax group is well worthy of study, as although, of course, like everything else, it has its modern peculiarities and specialisation, still, of all living mammals, it represents most nearly the group which may have arisen as early as the close of the Secondary epoch, when first the great orders of mammals seem to have become distinct from the primal stock. The remotest ancestors of the hoof-bearing group, of which the hyrax is the nearest living representative, branched off from the generalised marsupial-like Eutherians, not far from the first distinct ancestors of man—the lemurs—and also not far removed from the rodents and creodont carnivora. An early peculiarity of this ungulate-lemur group was a flattening of the

Gradually in the early mammals the actual laying of eggs—such as still takes place in the monotremes of Australia, the echidna and duckbill—changed into the growth of the egg inside the ovary and the birth of living young, which, still imperfect, were at first carried in the ventral pouch or fold of the skin. By degrees with the marsupials it became more beneficial to the welfare of the species to give birth to young sufficiently perfect to require no carrying about in an external pouch. We have marsupials at the present day whose young at birth are so much more grown up than those of the kangaroo—in which there has been a retrograde development—that the shelter in the inadequate pouch is little more than a formality.

In the middle or latter part of the Secondary epoch the time seemed to be at hand for the development of a mammal which could destroy other mammals, to say nothing of birds and reptiles, and be king of them all. And so out of some semi-marsupial like the thylacine of Tasmania, or an insectivorous type slightly resembling the solenodon of the West Indies, or the tenrecs of Madagascar, there appeared a beast with sharp-cusped molar teeth, long canines, and powerful fore limbs, in general aspect not unlike a hyæna or a cur dog. This was the creodont carnivore, from one of whose many differentiations sprang the porpoises and whales, from another the seals and the true carnivores of to-day. Casting aside the creodonts, Nature seized on one of their sections, improved its brain and limbs, modified the teeth, and gave the Earth a creature like a large dog, which in various aspects suggested also the civet, the raccoon, the bear, and the glutton.

Leaving the doggiest children of this dog type on one side—afterwards to produce dogs, raccoons, bears, and perhaps weasels—Nature rushed upwards along the line of rapidly improving civets, in the course of which she had thrown off the hyæna scavenger and was contemplating the beautiful but blood-thirsty cat. Then almost without a pause she started away again with an early type of cat, and fashioned it into a creature the size of a large leopard,

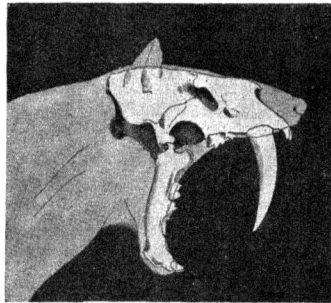

claws into nails or hoofs. It may be that man's thumb-nails and the nails of his great toes have descended without a break from the latter part of the Secondary epoch, when this flattening of the claws began in the creatures that afterwards developed into

HEAD OF SABRE-TOOTHED TIGER
This diagram illustrates the abnormal development of the jaws

ungulates, on the one hand, and lemurs, on the other. Anyone who examines the hyrax at the present time, or who sees it alive in the forests of Africa running up and down the branches, more like a squirrel than a monkey, will notice that, although its serviceable toes and fingers are reduced to three on the hind feet, with the vestige of a fourth, and four on the fore feet, with the rudiment of a thumb, they are much more finger-like than those of the other groups of ungulates, and in some way represent a compromise between the hands and feet of a lemur and those of an elephant or rhinoceros. Other features about the hyrax order suggest a connection in the early Tertiaries with ancestral forms of rodents and lemurs.

From out of the early types of the hyracine ungulates—which then possessed the full normal number of teeth in Eutherian mammals; that is to say, three pairs of molars and four pairs of premolars, one pair of canines, and three pairs of incisor teeth in both jaws—developed the elephant order, an early member of which was the moeritherium, which has just been referred to.

This clumsy-looking animal, the size of a tapir, had a rather long face, possibly even a snout developed like that of a tapir, six vertical incisor teeth in the upper jaw, the middle ones tusk-like in shape, four incisors only in the lower jaw, procumbent in position, a useless little canine tooth in the upper jaw only, three pairs of premolars, and three pairs of molars on each side of both jaws.

Evidently it was a creature that frequented banks of rivers, where it grubbed for roots. As it prospered, multiplied, increased, and varied, the head grew larger and longer, the incisors developed into

THE MOERITHERIUM, ANCESTOR OF THE ELEPHANT

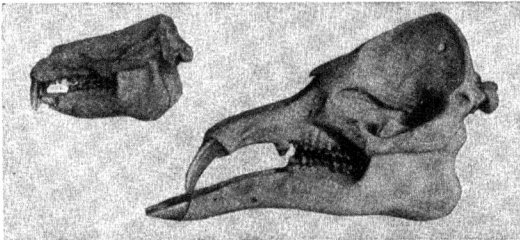
Skull of Mœritherium   Skull of Palæomastodon

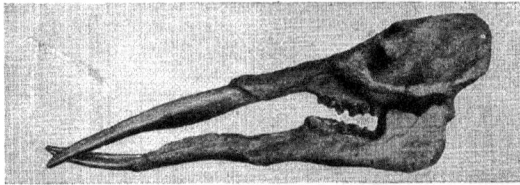
Skull of Tetrabelodon

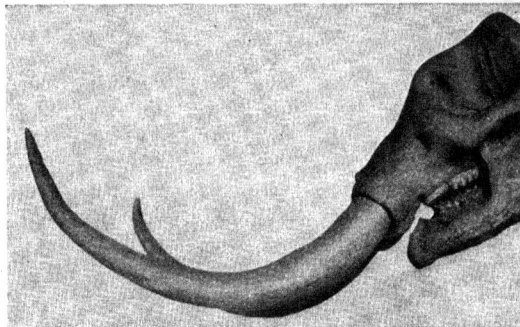
Skull and tusks of the Mammoth
THE EVOLUTION OF THE ELEPHANT'S TUSKS

extraordinary tusks, sometimes quite procumbent, at others produced downwards, and the nostrils at the end of the long snout were extended into a short proboscis. The bulk of its body increased in size, the limbs became stouter, longer, and straighter, the toes shorter, till at last they were encased in a sort of pocket of skin, tendon, muscle, and fat, like the fingerless feet of an elephant. Then, as the long incisor teeth became more and more useful as implements for digging and grubbing, or weapons of offence, the bones of the face grew shorter and shorter. Yet in outward appearance the face was as long as, or longer, than ever, because the under jaw was prolonged in a spoon-shape and the muscles of the face remained to control the proboscis, though there was no bone to support them. This gave the ancestors of the mastodon and elephant the appearance of having an immensely long nose. But the actual nose or nostril part is not really the whole of what we call the elephant's trunk. A good deal of it is the vestige of the long face and the upper jaw without the bones.

In this manner the mastodon became for a time a candidate for the position of supremacy on earth, and, in its desire to secure the post by priority of intellect, fashioned from itself under Nature's guidance the stegodont elephants, and they, in their turn, evolved into the modern Indian and African forms. But the race is not always to the bulky. The mastodon had the defects of its qualities to answer for, and there were anatomical considerations and peculiarities which caused it to fall out of favour. It was allowed a temporary place of splendour as a curiosity with prodigious tusks—the biggest teeth that any creature has yet developed—and then withdrew into the background, though, antediluvian veteran that it is, it still, in the guise of the elephant, lags superfluous on the stage.

So we have eliminated all but one of Nature's candidates for Earth-mastery—the lemur. It is a moot question whether the lemurs originated in North America or in the temperate regions of the Old World. The North American lemurs—of which there is absolutely no trace among the fossils of the West Indies or South America—seem to have been of a very generalised type, so generalised in some of the examples as to have been first of all classed among primitive rodents, ungulates, and even creodonts. Even in the existing lemuroids there are traces of affinities of a very remote kind with certain groups of the insectivora and the bats, and other points which show that the first beginnings of the lemuroid development cannot have been very distant from a marsupial type. Consequently, the first "pointings" of man, undoubtedly to be found in the suborder of the lemurs, showed themselves at a time when all the Eutherian mammals might possibly have been ranged within one great order, and that not far removed from the opossums.

In North America the generalised or aberrant lemurs appear to have become extinct at the close of the Eocene, and scarcely to have lingered into the Miocene. In Europe the lemuroid development was considerable, but during the early Miocene apparently made way for the real monkeys, the surviving lemurs passing on into Tropical Africa and also into Tropical Asia, though it is just possible that the existing East Asiatic tarsier-lemur may have been received still more anciently from North America across Bering Strait.

During the Eocene Period, and perhaps the early Miocene, the large island of Madagascar was united to the continent of Africa by a broad, curved isthmus, which probably included the Comoro Islands. Madagascar may also have had an independent isthmian connection with South-East Africa and a partial or sometimes complete connection with Ceylon or Southern India by large islands across the Indian Ocean. But the main land connection of Madagascar, till it ceased to have any land connection at all, was with the continent of Africa, and when the lemurs passed down into Africa from Europe, pressed southwards by the competition of superior forms, they entered the large peninsula of Madagascar and made themselves thoroughly at home in its forests. In continental Africa—where they are represented at the present day by the wide-spread genus Galago, and by the West Central African Pottos—they were no doubt for a time without a rival amongst the mammals

MAMMOTH APPROACHING CAVE BEARS
From the painting by F. Cormon

frequenting the trees, and Africa at this period of its geological history—the Miocene—may have been a particularly well-forested region.

In Africa, if not in the Mediterranean region, the lemur group prospered so considerably that it was able to develop higher types with larger brains, shorter faces, and eventually a slightly reduced number of teeth; in short, monkeys of a generalised type, the fossil remains of which in Madagascar, for this transitional form evidently passed from Africa to Madagascar before the connection between the two broke up, can hardly be styled lemurs, American monkeys, or Old World monkeys. The actual fossil types discovered in Madagascar by Dr. Forsyth Major, Dr. Grandidier, and others, are still lemurs in the majority of their characteristics, though on the borders of monkeydom. But in our speculations we may well imagine that, through transitional forms of this description, the lemurs created a generalised type of monkey somewhere in Africa, north, south, east, or west, and that this generalised type was the parent form of the New World and the Old World groups — the platyrrhines, or monkeys with upward-directed nostrils, and the catarrhines, or monkeys with downward-directed nostrils. To put it briefly, Africa, we may surmise, was once inhabited by what we should call nowadays American monkeys, and this as early as the very beginning of the Miocene Period; otherwise this group could not have colonised South America.

In our survey of the ascent of man it may seem waste of time and space to go off on this side issue of the American monkeys, but it is a problem so interesting and so closely allied to the enigma of human origin that it merits some consideration.

At the present day they exist most numerously and variously in the great "peninsula of Brazil," which was once almost separated from the rest of South America by an inland sea in the basin of the Amazons, and another inland sea in the Argentine. The platyrrhine monkeys are almost as abundant, however, in the Guianas and Colombia, and they penetrate past the Panama isthmus—in an ever-decreasing number of genera and species—to Nicaragua and the tropical seaboard regions of Central Mexico. Their farthest range northwards towards America, therefore, is in about 20° N.; southwards of Brazil they only extend to Uruguay, and westwards to the Andes. They are found in Ecuador, but not in Western Peru, Chile, or Patagonia. Possibly they may never have reached the Pacific coast west of the Andes, but although now absent from all the southernmost part of South America, they are found fossil in Patagonia. Still, the fossil remains in Patagonia and the rest of Argentina do not represent more than one or two genera, nor do they indicate such an abundance of development as in Brazil at the present day.

So far, not a single trace has ever been found in America, or even in Mexico, of any earlier forms of American monkeys. The fossil lemurs of North America were of a more generalised, less "ape-like" type than the later lemurs of Europe or of Madagascar, and there is no evidence to show that the American monkeys of to-day could have evolved in North America, in which continent, needless to say, there is not the slightest trace among its fossils of any real monkey such as inhabited Europe from the Miocene almost to the beginning of the human period. This is the more curious inasmuch as North America was in constant communication with North-Eastern Asia by way of the Aleutian and Kurile Islands—once a continuous isthmus—and received from Asia all kinds of mammalian types which must have co-existed in continental Asia and Japan with one or more genera of monkeys. A macaque monkey inhabits Japan and Corea even at the present day, and supports a climate that has far colder winters than has Great Britain.

If, therefore, the existing South American monkeys cannot be derived from North America, whence did they come but from Africa? The only other continental region of the Old World with which South America was connected in remote times was Australia; but, apart from the fact that Australia has never had any lemurs or monkeys, it has scarcely been connected through Antarctica with South America since the middle or late

MAMMOTHS, ONCE CANDIDATES FOR THE SUPREMACY OF THE EARTH
From a drawing by Sir Harry Johnston

Eocene. The conclusion, therefore, is irresistible, and is supported by much other evidence, that Africa has been connected across the Atlantic with South America by possibly two great isthmuses, one between Cape Verde and Brazil, and the other *via* Southern Angola and St. Helena to Patagonia and the Argentine.

It is most probable, therefore, that after the sub-order of monkeys was developed from the lemurs in Africa, it sent forth large numbers of these animals from West Africa to Brazil, where they developed into their existing types. And in South America, as among the extinct lemurs of Madagascar, are remains of a man-like creature. In the Tertiary formations of Patagonia fossil platyrrhines—namely, Homunculus and Anthropops—have been discovered whose skulls, by their exaggerated brain development, offer a strong but *superficial* resemblance to that of man.

In Africa, however, as the centuries succeeded one another, and hundreds of thousands of years were piled up, the generalised type—now specialised as American monkeys—had changed into the catarrhine or Old World monkeys as we know them to-day. And as Madagascar in this long lapse of time had become quite detached from the continent by a narrow strait of sea—not too narrow but that it might be crossed by hippopotamuses or pigs that were able to swim—it received no type of catarrhine monkey, and had to be content with the earlier transitional forms already referred to. From this generalised type of " American " monkey, therefore, there developed in Tropical and Northern Africa many of the genera with which we are acquainted.

at the end of the tail—suggest the beginning of the prehensility of the tail of the American monkeys. In the more specialised type of colobus, however, the coloration has become black with a little whitish grey, or black and white; the guerezas, the finest of all the species, have splendid mantles of snowy white hair. along the flanks, and an immense white plume at the end of the tail; or the whole tail may become one huge elongated plume of white silky hair. In some points about the skulls and interior anatomy of the colobus tribe are found more resemblances to the American monkeys than in the other Old World types They are also distinctly

THE FIGHTING DRYPTOSAURS, WEIRD CREATURES OF GREAT SIZE THAT FLOURISHED IN THE SECONDARY AGES

Perhaps the most " American-like " of these are the Colobus monkeys, now only found in Africa, if, indeed, they penetrated anywhere else. The colobus monkeys are so named because they are, to all appearance, " mutilated," in that they have seemingly lost the thumb on both hands. This loss is very recent. Apart from the fact that many of these creatures, even in the adult, show a tiny thumb-nail in the place where the thumb ought to be, the young colobus of one or two species will develop thumbs which gradually shrink and become absorbed by the rest of the hand as they grow older. Originally, these colobus monkeys appear to have been red, black, and grey, with very long tails, which in their sinuous writhings round the branches— and the more generalised species have no plumes

connected with the Semnopithecus monkeys—of which the best known is the hanuman—of ancient Europe, Southern, Central, and Eastern Asia.

The most numerous types of Old World monkey, however, are included in the sub-family of the Cercopithecines, the bulk of which are confined to Africa. It is from this group that most of the " beautiful " and pet monkeys are obtained, the mona and Diana being familiar examples.

The Baboons are now entirely African and West Arabian in their range, unless we may include among them the remarkable Black ape of Celebes. Here, indeed, is one of the most puzzling problems in the distribution and development-history of the Mammalia.

In the ancient and modern history of Asia there are no traces of any colobus, cercopithecus, or

SOME MONSTERS OF THE REPTILE AGE

TYRANNOSAURUS REX, ONE OF THE GIANTS OF THE REPTILE AGE

THE TRICERATOPS, A MONSTER REPTILE WITH THREE FORMIDABLE FIGHTING HORNS

cercocebus genus of monkey, nor has any of these types been found fossil in Europe. The semnopithecus monkeys of modern Asia, though they seem to be related to the colobus monkeys, apparently developed outside Africa, perhaps from an early colobine type, lived anciently in Europe, and nowadays are entirely restricted to the more southern, eastern, and even north-eastern parts of Asia. But they do not penetrate to the island of Celebes.

The macaques, whose range is from Gibraltar, Morocco, and Algeria, and thence, with an immense gap, through India to Malaysia as far south-eastwards as Java and Borneo, are perhaps distantly allied, on the one hand, to the mangabeys (Cercocebus) of Tropical Africa and to this isolated Black ape of Celebes, on the other. The baboons, which evidently sprang from a form like the mangabey, once extended their range from Saharan Africa through Arabia—where they still exist—to India. But there is apparently no fossil or living link between the baboons of India, which may have died out as late as the human period, and the baboon-like monkey, the Black ape of Celebes.

Now, here is the problem : to trace the line of man's ascent from the cercopithecine monkeys to some generalised form of anthropoid ape, not unlike the existing gibbons.

Beyond all question we are right, on the evidence we possess, in tracing man's ascent till we get as far as the cercopithecid family. Man came from the lemur, not from any other type of early Eutherian mammal. The lemur develops into a generalised monkey, and the generalised monkeys, by the early separation of Africa from South America, specialise into two big groups, the broad-nosed and the narrow - nosed. Man could not possibly have arisen from the American monkeys for a hundred and one reasons, which need not be given here. Nevertheless, Nature—which is not particular to a premolar or two, and can very soon get rid of a tail if it stands in the way— may have hesitated for some few thousands of years from which branch she would draw the man. In Patagonia, as we have already seen, the American monkeys developed a singularly human-looking type, the little homunculus, a creature with such a large development of skull and reduced face that certain Argentine naturalists were misled into thinking that it might really be one of the direct ancestors of man. Other genera or species of American monkeys have attained considerable cranial development, and, marching along parallel lines—for the student must by no

THE HYRAX OR " CONEY "
Looking like a rodent, this small animal may be related to the elephant

THE INDRI LEMUR OF MADAGASCAR
It is possible that the lemuroids, a very ancient group, may have furnished a remote ancestor of man

means overlook the fact that Nature is constantly repeating herself, copying and recopying her ideas— have also assumed a deceptively human aspect.

But somewhere, some time in the early or middle Miocene, Nature dropped them in the rather callous way in which she loses interest in her failures. The germ of man, therefore, lay in the great group of cercopithecid monkeys of Africa ; what route did it take to reach its final accomplishment ? To me this seems one of the most intricate puzzles in the study of the mammalia, but a puzzle not without precedent. What biologist can lay down without hesitation the similar track followed by the horse ? Sometimes he is impelled to ask himself if the type of the one-toed horse, with all the other equine characteristics, did not arise in duplicate, independently, after a certain stage —one true horse appearing in North America, and penetrating to South America, and another real horse arising in Europe or Asia. If so, are both these forms to be entitled to the same generic designation of Equus, or are they parallel types extraordinarily like one another, but only really connected if you go back to their grandparents ? Again, are they so separate that they would not breed together, and the cross prove fertile ?

In like manner, was man derived through creatures more resembling the gorilla, chimpanzee, and the extinct dryopithecus; then a gibbon and two or three unknowns, then a form like the Black ape of Celebes, and this again from the macaque, and the macaque from a mangabey, and so back to the cercopithecus genus ? Or, behind the gibbon, did he arise from those long-tailed monkeys of the semnopithecus group, with which the gibbon has in some respects such a strong external resemblance ? Some of the semnopithecines develop a nose strangely like the human nose, though less supported by bone. In such a case, though we have to get rid of a good deal of tail in the semnopithecines, we could trace man's ascent back to the colobus, and derive the colobus from the generalised group of monkeys which likewise gave birth to the American forms.

No doubt there has been a great deal of parallelism in these developments, and many of these types may have been very much in the running for the human ancestry down to the end of the Miocene Period. The baboons, for example, though lowly in some points of their anatomy, yet mentally, socially, and even anatomically, seem more allied to mankind than any other creature, except the highest forms of anthropoid ape. Of course, like everything else, at the present day they are specialised in their own way, have re-developed a very long muzzle, have exaggerated the brightly-coloured callosities of the hinder parts and of the face. But in intelligence over and over again they suggest the beginning of man. Perhaps of all the non-human creatures they come nearest to possessing an articulate language, with their mobile lips and tongue, yet they may be only an overplus of the human idea, developed independently from the macaque-like basis. But although the connection between the semnopithecus and the gibbons, which are the lowest of the anthropoid apes, may be only superficial, it seems very difficult to derive the gibbons from anything like a macaque or a primitive baboon; though it is easy to imagine such types as these two last passing,

THE BALD UAKARI
A broad-nosed monkey

through the Black ape of Celebes and certain extinct forms of ape, into the ancestor of man.

Were two types of humanity developed independently in later periods from two closely allied stocks of anthropoid ape? Yet were they not too far gone in difference to be able to mingle their blood, to unite after long separation, and perhaps widely different journeys through the world, and by uniting produce the genus Homo? Such a conjecture is not wildly impossible. Great as may be the differences, external as well as internal, between the Asiatic and African monkeys and baboons, it is remarkable to note—and this fact, of which I have often been an eye-witness in earlier days at the Zoological Gardens in Regent's Park, has never been sufficiently commented on by biologists — that a mandrill baboon from West Africa will hybridise with a rhesus macaque from India, and mangabeys and other cercopithecines will likewise unite in captivity and produce

GUEREZA MONKEY
This species has lost the thumb of both hands

offspring. It is scarcely stating the subject too broadly to say that but for differences in size—and perhaps leaving out the colobus monkeys—it is possible to obtain hybrids from all the very different genera and species of the cercopithecid family.

Unfortunately, the great anthropoid apes are so affected by captivity, so lose their vigour, and are so long in attaining maturity, that attempts to prove their fertility *inter se* have either never been made, or have produced no results. If such experi-

ments could be conducted by scientists—safeguarded, of course, by the various precautions to which scientific men are accustomed—and the results were that we we could gaze on a hybrid between a gorilla and an orang-utan, or a chimpanzee and an orang-utan, we should be better able to conceive the possibility of two independently developed human stems—equivalent in position to the one-toed horse of America and the one-toed horse of Europe, severally derived from three-toed ancestors—having united to form the genus Homo, or, further back still, Pithecanthropus.

The mention of this last creature, which lived in Java in the Pliocene Period, brings us finally to the consideration of the original birthplace of man. Was it in Africa, Europe, or Southern Asia? The Tertiary and Quaternary fossils of Tropical Africa

THE MAGOT, OR BARBARY "APE"
A narrow-nosed monkey

are so little known or so non-existent, owing to the geological history of that continent, that we have no means of determining at present whether Africa has ever been inhabited by anything lower in the scale of humanity than the Bushman or the Congo Pigmy. In North Africa we seem to find many traces, ancient and modern, of Homo primigenius, the heavy-browed man of Neanderthal, Corrèze, and Gibraltar. (The Bushman, Pigmy, and Negro belong to Homo sapiens). In Europe we find not only fossils of a great ape, Dryopithecus, but also—next to Pithecanthropus of Java — meet with the lowest known examples of humanity; but these emphatically belong to the genus Homo, and, though provisionally grouped as a separate species, Homo primigenius, are very near to Homo sapiens.

The aspect of the skull of this type of man superficially recalls the gorilla, with its huge brow-ridges, poor frontal development, prognathous face, very broad nose-opening, and feeble chin. The teeth, however, are not more gorilla-like than are those of the black Australians at the present day; in other words, there is only a very slight approximation to the teeth of the ape, as compared with the rest of known human teeth. The black Australians, indeed, seem to be almost the direct descendants of Homo primigenius, and in considering what was the outward aspect of this early human race in Europe, we need not go much further than the modern Australoid. The Negro, who is practically the only inhabitant of Africa between the Sahara Desert and Cape Colony—for the Bushman is only a variation of the Negro—preserves many anthropoid features, but was very early specialised as a human race by itself, derived, no doubt, directly from Homo primigenius, and probably evolved in India. Yet the negro is little more than a subspecies of Homo sapiens, and had as his remote parent the same ancestor that the white man and the Mongolian had.

The evidence in our possession so far supports to some extent the conclusion that humanity evolved from an anthropoid ape in or near India. In Northern India are found fragments of the skull of an ape which has been named Palæopithecus by Mr. Lydekker. The teeth of this creature are perhaps the most human in appearance of all the teeth of anthropoid apes. Palæopithecus was connected, on the one hand, with the chimpanzee and, on the other, with the orang-utan, and may have migrated eastwards to Java and there have given rise to Pithecanthropus. The skull of the young orang is extraordinarily human, but that of the mature animal evidences much recent divergence and degradation of the genus. The same may be said of

SKULLS OF MESOPROPITHECUS PITHECOIDES AND OF A YOUNG ORANG

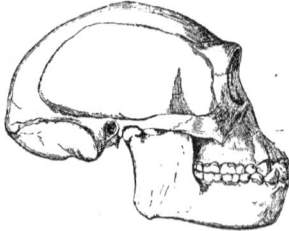

SKULL OF THE APE-MAN OF JAVA

the African gorilla. From some form like Pithecanthropus in the Malay Peninsula, in Further India, or in Java, may have originated the genus Homo.

This at first existed in a generalised type, named by some authorities Homo primigenius. India, which seems to be the matrix of fully half the Old World mammals, and which was quite possibly the scene of man's first emergence from the simian family, appears also to have witnessed the variation of Homo primigenius into Homo sapiens, and Homo sapiens into its modern subspecies; to have been, namely, the region in which the negro, the Proto-Caucasian—almost the equivalent of the black Australian—and the Mongol races arose divergently from the first truly human type, and migrated westwards, eastwards, north-westwards, and north-eastwards to conquer the world from the imperfect, because untutored, grasp of Homo primigenius.

Both Homo primigenius and Homo sapiens were of the family of man, because they had a brain-space in the skull of over 900 cubic centimetres, walked erect with the big toe more or less level with the other toes, had small canine teeth, no space or interval between any of the mature

CONJECTURAL RESTORATION OF THE CAVE MAN WHOSE SKULL WAS FOUND IN CORRÈZE, FRANCE

HOMO PRIMIGENIUS, THE MAN OF NEANDERTHAL AND OF PLEISTOCENE EUROPE
Drawn by Sir Harry Johnston to illustrate the probable appearance of the earliest type of man in Europe

teeth, a nose—not merely fleshy nostrils—a large development of the gluteal muscles, producing fleshy buttocks, a spinal column with a pronounced inward curvature, and specially human features in the liver, brain, the leg muscles, and sexual organs. They were also in all probability but scantily covered with hair on the body, though far more hairy than their modern descendants. Both sexes in early man had whiskers and a thin moustache and chin-beard, and the head-hair was long and abundant, to atone for the disappearance of the hair on the back aspect of the body from neck to heels.

Even Pithecanthropus, we may

A TIERRA DEL FUEGO FAMILY OF THE PRESENT DAY, SHOWING HOW PALÆOLITHIC MAN LIVED

suppose, used the implements provided by Nature—stones to throw, shells to cut, sticks to fling, strike, prod, or thrash. But Homo sapiens soon improved on these natural implements, and turned the unaltered materials around him into serviceable articles; deliberately shaped the weapons and tools with which he emerged from his struggle for feeding-grounds and shelters first with his primigenic brother, and next with the beasts of the forest and field.

Thus the clumsy primigenic man was originally driven from his Asiatic home by the swarming and successful Homo sapiens, and blundered into

37

Australasia, into Europe and North Africa, perhaps also North-Eastern Asia, repelled, however, from too northern a trend in his wanderings by the waxing cold of the Pleistocene Period.

But Homo sapiens was stopped by nothing, not even by those glacial periods which sealed and fortified his manhood by compelling him to think, to master many elementary arts and industries, to make fire, to strip off the skins of fur-bearing animals for his own covering, and to build shelters against the inclement weather or the too aggressively attentive wild beast.

What were the special foes of early man among the lower creation ? As soon as he took to a ground life, and ceased altogether to be an ape, he had good cause to dread the attacks of intestinal worms like the hook-worm, which deliberately enters through the pores of the skin and fastens on to the viscera ; or the filarial worms which may be absorbed in drinking-water ; or leeches and ticks coming from the grass ; or the still more terrible bacteria and protozoa introduced through the punctures of fleas, mosquitoes, flies and other insects. Many of these worms and insects were attracted by the exuviæ which accumulated round the primitive human settlements, for early man was very dirty and insanitary in his habits, if we may judge him by the lowest peoples of to-day. No doubt the bare skin of humanity was partly produced by an innate desire of the terrestrial ape that there should be as little harbourage as possible for these invertebrate pests which threatened him with diseases and anæmia and even extermination.

His vital energy being great, he accustomed

A NEGRO
The typical native of Africa from the Sahara to the Cape

himself to many of these intrusive micro-organisms, and his restless migratory habit further saved him. As to his visible enemies—putting aside lightning, floods, landslips, volcanoes, and blizzards—they were not many so long as he kept himself to himself. In his original Asiatic home he had the dangerous leopard to contend with, and perhaps the incoming tiger and lion from the north-east and north-west. Snakes occasionally bit or strangled him if he got in their way, crocodiles lay in wait for him by the riverside, and a solitary wild boar, wild bull, or buffalo, elephant, rhinoceros, or stag, may have pursued him out of wanton rage.

When man emerged into temperate Europe, Asia, and Africa, he found the grizzly and the cave bears, the cave lion, cave hyæna, and troops of wolves ready to attack and devour him unprovoked ; but against these enemies his social life of the clan or tribe, his use of weapons and of fire finally prevailed. When he himself became a fierce carnivorous hunter of other mammals he had to suffer occasionally in a losing battle with an infuriated bison or auroch, mammoth or megaceros, but the contest was one which he had deliberately sought, and which he could almost always avoid. These were among the dangers and drawbacks of his lot and he was prepared to face them. He had achieved his indisputable superiority over the rest of creation from the day on which he conceived the use of weapons outside his own person, and the principle of acting in concert with his fellow-men.   H. H. JOHNSTON.

A TITANIC CONFLICT BETWEEN PRIMITIVE MEN AND A TRAPPED MAMMOTH
From the painting by V. M. Vasnetzoff in the Historical Museum at Moscow

THE TAMING OF THE WILD

HOW MAN GAINED DOMINION OVER THE ANIMAL WORLD

By Sir HARRY JOHNSTON

UNTIL the human genus was fully established, with an average cranial capacity of 1,000 cubic centimetres, almost, indeed, until the human type had, in Southern Asia or Central Europe, developed into the existing species, Homo sapiens, there was little room in the mind of man for altruistic conceptions in regard to other animals. For the most part he regarded his fellow mammals as things to be hunted and eaten, or creatures which might hunt and eat him. From the anthropoid ape stage in his career to that of full humanity he had passed through a severe struggle to survive the attacks of leopards, tigers, lions, jaguars, enormous pythons, the great crocodile of the marshes, the vipers and cobras of the brushwood, and the petulance of bad-tempered bull-elephants or buffaloes. Here and there, no doubt, he had fierce combats with his long-armed, long-toothed gorilla cousins before they yielded and fled before him into the recesses of the African forest.

But when man's position was secured, his self-consciousness established, and his soul born, he looked on the world outside him in a different aspect. Tacitly he acquiesced in the existence of wild dogs, hyænas, jackals, vultures, and such-like creatures, hanging round the outskirts of his caverns in the hills or his forest patch in the middle of a savannah or a marsh. They fed on the offal round his home, the uneatable fragments of his repasts—skins, bones, legs, and viscera—and they in their turn, as they do even in the present day, warned him of the approach of their common foes—python and leopard, rogue elephant and child-slaying eagle.

Out of fear, and from some unconscious adherence to an unspoken pact, they left his children pretty much unharmed, unless a child wandered too far from the centre of the clan. There is a sense of tabu, or "things forbidden," even amongst beasts, birds, and reptiles. The cobra that is daily fed with a saucer of milk would have to be much and wantonly provoked before it flashed its fangs into the baby of its host; crocodiles—as I can testify from the evidence of my own eyes—leave unharmed the plovers that pick the leeches from their open jaws, and also the great water birds that frequent the sandbanks; these on their part warn by cry or flapping of wings the sleeping crocodile when its only enemy, man, approaches.

When the human herd shifted its dwellings and went to seek fresh hunting-grounds, it would be followed at a more or less respectful distance by its attendant troops of wild dogs, jackals, hyænas, and vultures. Thus I have seen myself, in the early days of East Africa, the Masai warriors followed from encampment to encampment by the same creatures, with the exception of the wild dog.

At last a more definite community of feeling would spring up between some of these animals and man. Vast herds of buffaloes, which then covered so much of man's first home in the tropics of the Old World, would soon grow accustomed to, and even benevolent towards, these troops of perfected apes. Man in those days would have hesitated to attack a buffalo or to excite the anger of a bull or cow by the abstraction of a calf. On the other hand, both had cause to dread the great predatory cats. The shouting, shrieking, and banging of tree-trunks, with which the quasi-humans announced the presence of a leopard or a sabre-toothed tiger, would give the buffaloes the necessary warning in time to be on their guard and gather their calves within the mass of the herd.

But unquestionably the dog was the first to be man's friend, ally, companion, and subject. The affinity between this generalised carnivore—which has been able to remain generalised in all but the development of its brain—with its conception of association and tribal combination, began, no doubt, far back in apehood. There is a singular mental affinity and affection latent in almost all monkeys for dogs, and in almost all dogs for monkeys. The fiercest captured baboon, longing to bite and tear any human being within reach, will suddenly soften at the sight of a terrier; and the most truculent bull-terrier will develop unexpected friendliness towards a ridiculous gambolling monkey. Of all the pets that can be given to anthropoid apes to console them in captivity, if consolation they require, there is none to which they take more readily than to a dog. This attraction of the one type to the other may even be traced as low down as the lemurs.

But this affinity of feeling between primate and canine scarcely goes beyond the dog. It may be that the dog subgenus alone has reached a degree of mental development which, as in the case of the parrot, the anthropoid ape, and the cat, raises attraction above the instinct of kind associating with kind to that of soul associating with kindred soul.

Among savages of the present day in those regions where dogs and even wolves abound, it has been not an uncommon incident for a mother who is grieving over the loss of her child to bring up a wolf cub or a puppy instead; still more strange, it is averred not only by legend, but by modern observations, that wolves—and therefore possibly wild dogs—will suckle motherless children instead of their lost puppies. Anyone who has studied the wild or the savage people in Africa, Asia, and America has noted the affection evidenced by the pariah and tribal dogs for the children of the clan. In some parts of Africa a child and a puppy are brought up together, and become lifelong friends. Few things in Nature, in fact, from our present outlook on the world, appear to us more beautiful than the devotion of the dog to man. It is so complete that the dog entirely surrenders himself to the will of a creature who is removed from him in relationship and

DOGS HELPING EARLY MAN TO HUNT THE BOAR

child, in the dread that their cries would attract a large concourse of men with bludgeons and missiles. For the early dog dreaded the throwing powers of humanity far more than the impact of a club or a sharp flint derived from a hand-to-hand struggle. It was the capacity of this creature to *throw* which made such an impression on the mind of the dog. You might avoid contact with the powerful hands or fists of early man, and his teeth had already become a negligible quantity; but if you merely stood at a respectful distance and bayed, you could still be struck and perhaps killed by the heavy stones or crooked sticks flung at you by his long, strong arm with ever-increasing accuracy of aim.

So the man and the dog made an unwritten and unspoken bargain. When the dogs saw that the humans were sallying out to hunt, they followed, and before long they ran alongside. Man wounded the game, and the dogs ran it down. Man kicked the dogs aside and took the best of the slain beast, but he always left behind enough to satisfy the hungry pack; and even if he did not, then and there, he raised no objection to the removal of the offal from his camp by the dogs which frequented it. Consequently, the more man prospered, the happier became dog life, until the camaraderie not only resulted in the development of two or three hundred different breeds from the original, a creature very like a dingo, but also established what is undoubtedly a great and precious principle—that man in his mastery of the earth need not necessarily exterminate every other living creature which he cannot use for food or clothing, as a beast of burden or traction, or as a creature to provide ornaments or manure.

development by three or four million years, though the development along some lines has been parallel. At the behest of the man, woman, or child, to whom the dog has attached itself, everything is to be attacked—the rest of the dog-world and all alien humanity not excepted. Yet the dog will not hesitate to recognise as his companions and sedulously respect all such other humans, beasts, birds, and reptiles as are placed under his owner's protection.

It was, however, mainly attachment to the principle of co-operation on both sides that so specially drew the dog and man together as allies in the dawn of humanity. In those days man was often pitiless towards wife and child. But for his dawning consciousness of the value of tribal organisation he would have slain the comrade and snatched the comrade's widow. Without any remorse he would let his mother die of hunger when she was no longer a useful adjunct of the community; and when the father became a nuisance in his tyranny, brothers might band together to slay their parent and peaceably feast on his remains, sleeping afterwards without remorse until dreams, provoked by indigestion, initiated the terrors of religious belief. If the dog had gone solitary, like the fox, man would have overlooked him, or have killed him for his coat, or even have devoured him, when the frightful pressure of the struggle for existence turned him into a bloodthirsty carnivore. But if he attacked a troop of dogs, they all stood by one another. On the other hand, the dogs feared to attack a lonely woman or

THE SHEPHERD'S DOG AT HIS MASTER'S GRAVE
From the painting by Sir Edwin Landseer, R.A

40

THE DOG AS MAN'S HELPER IN THE CHASE

THE BEAR HUNTED BY DOGS

THE WILD BOAR HUNTED BY DOGS

From the paintings by Frans Snyders in the Dresden Art Gallery

CANINE HELPERS IN POLAR EXPLORATION: ESKIMO DOGS DRAWING SLEDGES IN THE ARCTIC REGION

One likes to think that the dog—not always necessary now that we have developed electricity as a house guard—is, nevertheless, still being kept and maintained because of its community of mind with man ; because of its beauty, its half a million years of faithfulness ; its participation in man's struggles, defeats, joys, and triumphs. Has it not had the honour at the North Pole and the South Pole—associated in the last region with another of man's ancient allies, the horse—to assist in enabling mankind to view the last undiscovered portions of his present home ? Indeed, about these recent Polar discoveries there are two very significant facts which should tend to the softening of racial and of animal prejudices. A black man as well as a white man, according to the white man's story, stood alone on the patch of ground in the Farthest North, where the earth is neither north, east, south, nor west, but where the central point of its axis faces alternately perpetual day or perpetual night. And neither of these men could have got there with the present means at man's command but for the vigour, endurance, and obedience of Eskimo dogs. Likewise, when Sir Ernest Shackleton got within a hundred miles or so of the Antarctic Pole, he did so not only with the help of equally faithful dogs drawn from the Far North, but by using the muscles and feeding on the flesh of ponies—ponies, perhaps, not widely different in appearance from the first horse tamed by man.

An immense interval in time, no doubt, separates man and the dog from man and the domesticated ox, sheep, goat, horse, ass, pig, buffalo, goose, llama, cat, camel, silkworm, reindeer, duck, peacock, hawk, goldfish, turkey, guinea-pig, rabbit, ferret, guinea-fowl, pheasant, swan, quail, and canary—a list which represents the principal beasts and birds that have been more or less permanently domesticated by the patience of man, and gives, as nearly as possible, the imagined order of their domestication.

Force of circumstances made man, as soon as he *was* man, a great hunter and fisherman. Troops of humans wandered along the sea-coasts of Tropical Asia, gathering, opening, and eating shellfish. With their hands, and later with implements, they fished in rivers and shallow sea pools. They built dams of sticks and trunks across the bed of a rivulet, and caught the fishes thus arrested on their way down stream. They first of all killed large animals, which fell into natural pitfalls or stuck in marsh mud. Another leap of the intelligence invented the trap and the springe. The elastic rind of palm fronds, and later the fibres of innumerable plants, gave the strings which were used in these traps, and were afterwards applied to a pliant stick, which became the bow. In fact, the pitfall led to the trap, and the trap led to the bow and arrow.

But at last increasing numbers, competition, hunger, and other causes drove man northwards from that part of Tropical Asia in which he had been born —Malaysia or India. Southwards and eastwards were great islands. Some of these he and his successors reached by floating logs or some temporary uprising of the sea-bed, so that a straggling of a low type of humanity penetrated to Australia and Tasmania. But the main progress of man from his Asiatic centre was northwards and westwards. With his powers of intellect and of artifice he soon overran the whole habitable world, but always and everywhere as a hunter, as a carnivore, with one domestic animal, the dog.

Century followed century, millennium millennium, and still found man hunting, feasting on flesh, fish, and shellfish, with such roots, mushrooms, fruits, nuts, leaves, and wild grain as he could obtain directly from Nature's hand. His life as a hunter of big beasts led him to follow the wild horses for their succulent flesh, and, above all, the bison and cattle. Every now and again he would kill some cow with a

AN ELK SLEIGH IN ARCHANGEL

TEAM OF REINDEER IN NEWFOUNDLAND

defenceless calf, and would drink not only the blood of the slain cow, but would press the milk from the full udder, and drink that likewise, and the calf would be driven home to the kraal with some vague intention of preserving it for future consumption when it was older. As there *is* pity in an ape or a monkey, so still more there should have been pity in the mind of primitive woman ; and one such, two such, twenty such no doubt took compassion on the half-grown calf, and signified that it should not be killed.

With the uprise of Neolithic Man came, no doubt, the great development of taming and domesticating beasts and birds, and agriculture—which, in point of fact, is the domesticating of vegetable forms.

Around the cave entrances or the rude huts of a Palæolithic settlement the seeds of certain edible plants grew more luxuriantly in the manure and offal of the human settlement than outside among weeds and stones. This attracted the attention of the more intelligent type of man which was rising up in Asia or Europe ; and seeds of such plants were deliberately planted in favourable spots where they might grow luxuriantly. In this way agriculture began, but before man tilled the ground he was a herdsman, just as he was a hunter first of all. And

North, were regarded with a new-found awe, due not only to admiration of palpable strength and bulk or sheer beauty, but to a consciousness of motives, symbolism, intangible forces, embodied or typified by some of these creatures. This bird heralded the coming of spring, with vivifying rains and warmer sunshine. This beast, if let alone, multiplied at the time of harvest, and became accidentally associated with productive crops ; that reptile dwelt in the places where their dead were buried, and so might be a warden of the spirit world.

So that a good deal of the domestication of beasts and birds, and licence allowed to reptiles, arose not only from a matter-of-fact belief in their value as food or means of transport, but in feelings of superstition and religion. It is doubtful, for example, whether the cat, first domesticated in Egypt from a wild African species, was tamed so that it might catch mice and rats, or because it seemed the dwelling-place of a goddess, and for that reason ought to be protected and induced to dwell in the habitations of the people by whom it was worshipped.

But the ox (Bos taurus), though afterwards worshipped as perhaps no other beast has been in Europe, Asia, and Africa, was certainly first drawn

CARAVAN OF CAMELS CROSSING A DESERT IN THE NORTH-WESTERN PROVINCES OF INDIA

the first herdsmen arose not in America—which is rather a side issue and an overflow of humanity than a chapter on the main line of man's ascent—and perhaps not in Europe, but in Western Asia and nearly simultaneously in Northern Africa.

These Neolithic people had, with increased security, food supply, and leisure to think, greatly advanced in their ideas of religion. In a sense, a wider humanity opened to them. More distinctly and more firmly than did Palæolithic Man, they attributed souls, and even dominant souls, to the beasts and birds, the reptiles and the insects around them, and to the great forces of Nature in river and sky, harvest and fire, disease and its remedies, the begetting of life, and the process of death. Not only did they fear the lion and the bear, but they deified these brother beasts. The hippopotamus typified a goddess, the elephant a god, the cat was protected, and made tabu in virtue of its " slimness," its beauty, its large saucer eyes. Even that persevering scarabæus beetle rolling up the ball of dung and dust which contains its eggs was looked upon as the embodiment of a great principle, and no longer was crushed under the heel of the sandal. The ibis, vulture, eagle, hawk, fish, and wolf, the leopard, and no doubt the mammoth and hairy rhinoceros, lingering in the

into the fold of man for a utilitarian purpose Cattle were domesticated to create a reserve supply of meat against hard times ; then came the idea of using the superfluous milk that was more than sufficient for the rearing of the calf; and in some directions the milk and dung of the cattle assumed such importance in the eyes of the people that the flesh was tabued as sacred. The bison, on the other hand, for some reason we do not understand, never lent itself to domestication, though it abounded in enormous herds in Spain, France, Germany—on the verge of which it still exists—and Russia, in the time of man of the New Stone Age.

After a first attempt at taming ibex and audad, we find that domesticated sheep and goats of ancient Egypt were derived from the same Eurasian species as those which have supplied Europe, Asia, Negro Africa, and the whole world with their present stocks of sheep and goats. Sheep of the earlier Neolithic type in prehistoric Europe and Asia seem to be little else than a tamed mouflon, in many respects like the wild mouflon of Sardinia and Corsica, which once had a far greater range in various subspecies over Central and North-Western Europe, even reaching to Britain, in perhaps three different subspecies—possibly early domesticated forms brought by the first Neolithic tribes. Mr. J. G. Millais thinks that one of these still survives in the remarkable little Soa, or Soay, sheep of an islet in the St. Kilda group, west of the Hebrides.

Curiously enough, the Hottentots who crossed the southern part of Africa from east to west two thousand or fifteen hundred years ago, coming apparently from what is now German East Africa, brought with them the Egyptian type of ox and the North-east African fat-tailed sheep, but possessed no goat ; though there are many reasons for supposing that the goat permeated Negro Africa before the sheep, which last is still rather a new animal in parts of the Congo basin.

The horse loomed very considerably in the scope of vision of early man in the Old World. The equine species indigenous to North America may have become extinct soon after the arrival of the first human beings in that continent—more likely from the ravages of a tsetse fly, found fossil in recent American deposits, than from the attacks of man ; and man certainly co-existed with a clumsy, red - haired, white-spotted equine in South America, the Hippidium by name.

We have no reason to suppose that any lingering horse species or genus was of particular importance to man in America before the coming of the European. But in Europe, and possibly Central Asia, Palæolithic Man thought much of the horse as an article of food, no doubt rounded up the herds by a concentrated tribal attack, and drove them into blocked-up gullies, or other arranged traps, much as the wild negroes of Africa do, or used to do, with zebras and other game. Then the usual procedure would follow ; foals would be caught, which were not immediately required for eating. Boys of the clan would make pets of them, perhaps climb on to their backs, and gradually the idea of the horse for transport or riding might occur to the mind perhaps of the first Neolithic men in Europe. But the real domestication of the horse seems, like that of most other beasts and birds, to have taken place in the as yet unlocated original home of Neolithic civilisation, somewhere in south-easternmost Europe or Western Asia.

The true horses of, let us say, 7000 B.C.—as apart from the asses and zebras—were natives of all Eurasia south of the ice cap, and of much of sub-Tropical Asia. There may have been a wild horse in North Africa as well, but this is not certain. For the most part, the horses of Europe and of temperate Asia were of a type resembling either an underbred cart-horse of the present day, or the very similar existing wild horse of Tibet. But in Asia Minor, Syria, Persia, North-Western India, and Northern Arabia—it may be also right across North Africa to Mauretania—there seems to have been a subspecies differing from its northern congener in the possession of a more elegant shape, a smaller head, a smoother, sleeker coat, and, most important feature of all, a tail set on with a fine arch and heavily plumed from its very base ; whereas the tail of the northern horse is what one sees in the Przevalski wild horse, set closer to the body and not beginning its expanding plume of hair until some inches away from the base.

This southern horse was no doubt the origin of the Barb and Arabian breeds. It was the horse ridden by the Asia Minor Hittites, and either inhabited Cyprus from the time of that island's connection with the mainland, or was early imported thither as a domestic animal. The Cyprus horses represented in the ancient art of that region are obviously of the Arabian type. In Northern India, the southern and the northern breeds may have mingled in domesticity, and the result of that cross-breed has found its way in various pony shapes to different countries, such as Burma, China, Japan, and right down south as far as the island of Timor.

BREAKING BISON INTO HARNESS

The horse was relatively late in its arrival as a domesticated animal in Egypt, but was soon employed extensively by the Egyptian kings for their chariots in war. In the domesticated state the horse has had a career of quite seven thousand years in Europe, if not longer. In the cave drawings of France, which may be fifteen thousand years old, the horse is occasionally shown as though covered with a dressed skin as a saddle, and with indications of reins or head-stall. The horse of the Greeks was mainly of northern origin, with its hog mane—the long, drooping mane is a feature of the southern horse—its large head, and stout legs. But in Asia Minor the northern and southern breeds mingled, as may also have been the case with Northern Africa and Southern Spain. In Britain, a generalised type of domesticated horse was probably created from

MAN'S BURDEN BEARERS IN EASTERN LANDS

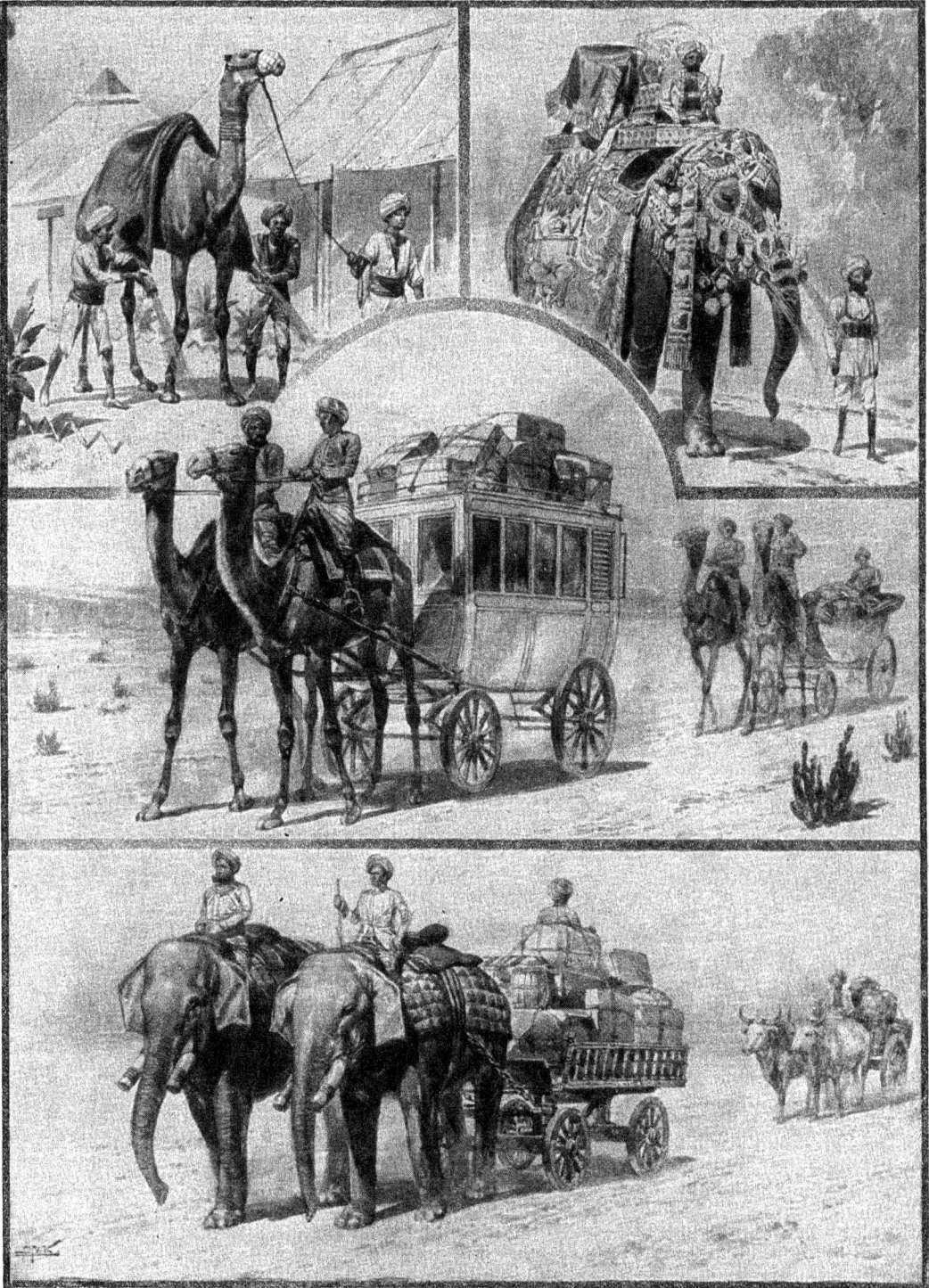

CAMELS, ELEPHANTS, AND OXEN EMPLOYED AS BEASTS OF BURDEN IN INDIA

The top picture on the left shows the camel as a substitute for a watering cart in Bikanir, and that on its right the Rajah of Nabha's elephant employed in watering. The centre picture shows a camel omnibus conveying luggage, and in the lower one elephants and oxen are drawing waggons

the wild stock by Iberians and Celts long before the arrival of the Romans, and the main progenitor of British domesticated breeds down to the opening of the sixteenth century was evidently the indigenous British wild horse, scarcely differing from the northern type of Europe and Asia, though quite conceivably it may have been smaller in size.

The Egyptians first domesticated the African ass; and almost simultaneously the Asiatic ass, or onager, was tamed by the people of Mesopotamia, but does not appear to have been so successful a domesticated animal as was the African. Still, there has evidently been an interbreeding between the Egyptian and the Persian or Syrian wild asses in captivity, and this mingling has produced the handsome, large-sized riding donkeys of Eastern Arabia and Southern Persia, which have played a considerable part in the civilisation of Moslem East Africa and of Egypt,

into those regions, whence it passed on to Malaysia. Here, mixed with a domesticated local species, it became the most important homestead animal of the Malay Archipelago and the many Pacific islands, being carried nearly everywhere by the restless half-Caucasian Polynesian race, as far away from the Malay Peninsula and Sumatra as Hawaii and Samoa, but never to lonely, sulky, savage, Palæolithic Australia, wherein—and in Tasmania likewise—man never knew any other domesticated animal but the sheep-farmer's plague, the dingo; though Melanesia and New Guinea possessed the pig and perhaps the fowl.

The wild boar of Europe is the original parent, together with a later admixture of the Asiatic race from China, of all modern domesticated breeds. It was, no doubt, first tamed by that remarkable nucleus of Neolithic Man somewhere in Western

A TEAM OF WHITE BULLOCKS DRAWING AN INDIAN RAJAH'S GUN-CARRIAGE

and even, perhaps, of Europe of the Middle Ages; for they attracted the attention of the Crusaders, and were imported by them into Europe.

Mules seem to have first come into existence about 1500 B.C. in Asia Minor. They proved at once of great use for transport, especially in the mountains, but did not attain great popularity as riding animals until the beginning of the Mohammedan era. The Crusades made them much more widely known and used in Europe, though the Arabs had firmly implanted mule-breeding among the industries of Spain. In France, mule-breeding did not begin till the beginning of the eighteenth century.

The pig was probably domesticated in several parts of the world independently, perhaps first of all by the early Aryan invaders in Northern India, in which case the wild parent would be the Asiatic wild boar. Similarly, the prehistoric Caucasian invaders of Indo-China carried the tame pig

Asia or Eastern Europe from which all civilisation seems to have radiated. But the process was probably repeated over and over again from Ireland to Japan, and North Africa to Lithuania, wherever a pig of the *Sus* genus could be caught young and brought up as a plaything and village scavenger. Even in Palæolithic, savage Africa, the superior negro types in the Western and Central Sudan, dwelling on the northern limit of the great forests, tamed the indigenous red river-hog. It is interesting to note that almost the only creature the negro has ever attempted to domesticate from out of the wild animals of his own land is the red river-hog of Equatorial West Africa. On the verge of the Egyptian Sudan, in the forests of Liberia, and here and there in the Congo basin, the natives capture the young of this pig and bring them up by hand, so that they run about the village tame, or are even kept in styes for food. There is no reason

TAMED ELEPHANTS DRAGGING HOME WILD ELEPHANTS CAPTURED IN THE INDIAN JUNGLE

whatever why this very handsome animal should not be added to our list of domesticated beasts, except the desire of the European sportsman to slay, and slay, and slay without stopping to think.

In the British Islands, the wild boar is thought to have existed in a smaller, longer-legged form than on the Continent, and has been provisionally named, in scientific language, Sus scrofa britannicus. There may have been another local subspecies in Ireland, the ancestor of the now nearly-extinct greyhound pig of domestication. The domesticated swine of Britain of the Roman period and Middle Ages were very like miniature wild boars, and only became shorter-headed in the seventeenth century, when from the Continent there began to be imported the "Polish" breeds. These had reached Eastern and Central Europe from China, in the train of the Tatar conquerors before they became Mohammedans. The Chinese were the first to pay attention to pig-breeding and to develop a plump, short-headed breed.

By whom was the Indian elephant first tamed? The impulse possibly arose among the Aryans, who first crossed the Himalaya and invaded India and Further India from ten thousand years ago onwards. And "Aryan," as we grope our way back into the past, only seems to be another and interchangeable name for the first Neolithic people who started the New Civilisation, without which Palæolithic Man might have remained a useless, destructive brute. No doubt the first white-faced, brown or yellow-haired, grey-eyed Aryans who crossed the Hindu Kush and entered a Negroid, Mongoloid, Australoid India, found elephants roaming all over that vast peninsula without let or hindrance; whereas now they are mainly confined to a few forest districts in the east and south, to Ceylon, and to the Malay Peninsula. Perhaps the invading Aryan, who brought the

horse with him, and, it may be, the northern ox, set to work to domesticate not only the wild cattle of Northern India, but captured young elephants and reared them "out of fun," as is occasionally done by negro peoples at the present day. and, having reared them, endeavoured to turn their strength to some practical account as an engine of war, a living turret for archers, a sentient battering-ram. So, as a great adjunct in warfare, and a royal present from one Eastern king to another, the Indian elephant passed to Persia, Asia Minor, Syria, Egypt, and Carthage. The Egyptians once possessed the African elephant in their own land at the dawn of their history, and later imported examples from the Sudan, but seem to have made no attempt to domesticate this animal. African elephants were imported from Algeria and Morocco, where they still lingered, to Rome for the spectacles of the circus. But the Romans and the Egyptian Greeks found, as others have done, that after arriving at maturity the African elephant is unmanageable.

Of course, in one respect, the elephant, like the parrot, has never become a really domesticated animal, for with very rare exceptions it will not breed in captivity; therefore the supply has to be constantly recruited from the herds of wild animals. This has given man very little opportunity for modifying the race as might otherwise have been possible. If, even yet, we were able to induce the elephant to breed regularly in captivity, what stupendous results might be achieved! If we can make in two or three thousand years anything so splendid as a cart-horse out of the rough, shaggy, middle-sized horse of Northern Europe, what might we not do with an African or an Indian elephant? Inversely, we could perhaps breed a pony or dwarf elephant, as Nature did during the later Pliocene in Malta.

The camel family originated in North America. Two or

ELEPHANTS HELPING TO BUILD A RAILWAY BRIDGE IN CEYLON

TROTTING MATCH BETWEEN AN OSTRICH AND A HORSE IN AMERICA
Drawn by H. W. Koekkoek, from a photograph by Brown Brothers, New York

three species penetrated in course of time to South America, where they still remain in the form of llamas, huanacos, and vicunias. In North America the whole race died out absolutely before the coming of man, but not before it had exported, by way of Bering Strait, at least one genus, Camelus, to North-Eastern Asia. The camel type travelled westwards, and made a new home in Central Asia, where it developed a local species with double humps, the Bactrian camel of to-day. The original type, still passing westwards, inhabited the plains of Southern Russia, the mountains and plains of Asia Minor, Syria, and Mesopotamia, Egypt and Mauretania. It is possible that in the far-back, prehistoric times it even penetrated up the eastern valley of the Nile to the semi-deserts of Gallaland and Somaliland, which it may have inhabited as a wild species till quite recently. The wild camels of Algeria and Tunis died out almost before the advent of man. It is difficult to understand, however, why wild camels should not have lingered on in the Sahara, a country so thoroughly suited to their habits. Apparently, however, after the extinction of the North African camels, with the doubtful lingering of a wild camel in Gallaland, the camel became unknown to Africa.

In Mesopotamia or Syria it was domesticated first about four or five thousand years ago, so far as one can guess from the evidence of monuments. The early Arabs derived it from these regions, and carried the camel with them all over Arabia, and imported it as a domesticated animal into the Abyssinian coast and Somaliland. The ancient Egyptians knew the camel, but not, apparently, at a very early date in their history. It was very slightly employed by them until perhaps the Roman era, and really only assumed its present importance after the Arab invasion. The Hebrews or kindred tribes who dwelt to the west of Egypt, in Syria and Palestine, possessed domesticated camels, as is evident from the frequent references to them in the oldest of the books of Scripture. But the Berbers of North Africa apparently knew nothing about them—whether or

not there were wild camels still in the Sahara—and only employed oxen, horses, or asses as a means of transport. It was not, though the theory may yet be upset by new discoveries, until the great Arab invasions of North Africa, between about A.D. 670 and 1040, that the Berbers of the desert so vigorously adopted the camel in their conquest of the Sahara. The possession of this animal enabled them to cross the Sahara in all directions, and no doubt had a far reaching effect on the opening up of Negro Africa.

The two-humped camel of Central Asia seems to have been first domesticated, not by the Tibetans or other Mongols, but by the Aryans, or at least Caucasians speaking Aryan languages—ancestors of the present Persians, Afghans, and Armenians. Eventually, however, the two-humped camel came into use as a domestic beast between North-East China and the Crimea.

The Arabs introduced the one-humped camel into Spain—in a portion of which country it now runs wild—and Arab influence also brought it to Italy in the Middle Ages. The nineteenth century witnessed its range extended to Australia, the United States, and Mexico.

The American camel forms, the huanaco and vicunia, were domesticated by the Incas of Peru, or their still more mysterious predecessors in the same Andean tablelands. This achievement may have taken place four or five thousand years ago. The llama, or tame huanaco, was used chiefly as a beast of burden for transport purposes, and also, to some extent, for food, though the flesh was very unpalatable. But considerable importance was attached to this beast, and a white llama was regarded as a specially fit sacrifice to the gods. The vicunia was tamed and bred for its woolly coat, and in its domesticated form was known as alpaca, or paco. Both these creatures were really domesticated in that they were bred under man's control, and, in fact, so changed and varied under many centuries of separation from the wild stock that until recently they were considered specifically separate, and are, as it is, quite different in colouring.

The association of man with the deer tribe has always been rather a close one from Palæolithic times. Venison was good to eat, and deer were more easily run down and killed with the aid of dogs than were horses, and they were not such terrible beasts to tackle at close quarters as the urus or the mammoth. The reindeer of Europe was a favourite quarry of primeval man, whose imagination was much impressed by its form, so that he made it the subject of his interesting drawings and carvings. It may be, indeed, that the domestication of the reindeer began, then, on the verge of the New Birth, the First Renaissance, the Neolithic Period; and that, as the hunter races withdrew northwards before the competition of the new agricultural people, they drove the reindeer before them, and continued to keep them in domestication as beasts of burden and traction, providers of milk and cheese, until at last the boreal regions of Scandinavia, Russia, and Siberia were reached in which the reindeer is, with the dog, the only domesticated animal.

The cat has disdained, one might almost say, to make herself generally useful. The dog sought out man, and thrust itself upon him as a companion; but the cat most reluctantly consented to leave the jungle to become a goddess by the banks of the Nile. No doubt the ancient Egyptians did with their indigenous cat as did the early Hindus with other small felines—picked up the kittens for their attractiveness, and, as they grew up, something in the cat's eyes or nature suggested affinity with one of those spiritual beings by which they personified some natural principle or force; in this case the goddess Bast (of Bubastis) one of the several manifestations of the great sun-goddess, Isis. Perhaps the large, round, golden eyes of pussy, whose very name is said to come to us from Bast—or from Pasht, another sister and rival goddess—suggested a resemblance to the sun deities. It is also obvious that the cat was a smaller, cheaper, and more manageable substitute for the lioness. Certain Egyptian paintings show that—probably under severe compulsion—the cat so far demeaned herself in some families as to become useful as a huntress and a retriever of birds. It was, however, somewhere about the time of the Greek invasion of Egypt that the cat attracted special attention as a vermin-killer. About that period Northern Egypt began to experience the beginnings of the rat invasion, which, from the little we know, came out of Central Asia to vex all the world. The house mouse has been in Europe, Asia, and Africa for many thousand years, but probably did not become a nuisance till men and women had grown particular about their homes, and thrifty as to their granaries. It caused great annoyance, however, in Egypt, Greece, and Rome for some hundreds of years before the Christian era. The Greeks and Romans tamed—perhaps domesticated—the marten, and employed it to rid their houses of mice, calling it respectively by names which they subsequently applied to the domestic cat. The Egyptians, however, found their sacred African cat to be a splendid mouser. Wild Egyptian cats were exported to Syria, Persia,

GROUP OF OSTRICHES ON AN OSTRICH FARM IN SOUTH AFRICA
From a photograph by J. Russell & Sons

Asia Minor, Greece, and Rome, and founded the world's stock of the fireside friend.

It is impossible in the space at my command to enumerate in detail the lesser triumphs of man in the domestication of animals. I can only allude to his creating breeds of domesticated carp, such as goldfish, veiled-tail fish, telescope fish; insects, like silk-producing moths, dye-yielding coccids; birds for exhibition, plumage, food, or song; the domestic fowl, peacock, turkey, guinea-hen; the ostrich, goose, duck, swan; the pigeon, turtle-dove, and canary-finch; rabbits for food and fur; guinea-pigs, rats and mice, for toys; and polecats, in the form of ferrets, for destroying vermin.

So far we have confined ourselves almost entirely to the animals that have been domesticated by man, that is, brought so much under his control that they have bred in confinement under his supervision, and in most cases have permitted him so to regulate their multiplication that he has created new types by deliberate selection in mating, and by altering the conditions of life. In these achievements, as in his agriculture and horticulture, he has had to fight recalcitrant Nature. No sooner has man attempted to invade a new district with his animal serfs than for them, as for him, there have sprung up scores of foes, chiefly invisible to the naked eye: germs, fungi, protozoa, bacilli, intestinal worms, and insect or spider parasites. Or the food has not suited, or the climate, or carnivorous birds and beasts have disputed possession and increased his labours.

Especially has it been when he has achieved his initial great success in some experiment that the blow falls. The silk trade brought about an immense aggregation of silk-spinning caterpillars in France in the early part of the nineteenth century. Consequently there broke out the terrible silkworm disease, which threatened to annihilate the whole lot. The history of Pasteur's patient struggle

and his final triumph is one of the romances of science. The various cattle plagues, originating in Africa, threatened to denude the Old World of oxen, and annihilated something like a million of these indispensable creatures at different times in Africa alone. But a way round was found, a cure, a prophylactic; and here we have triumphed over and over again with the resources of science. In the same way we are conquering the diseases spread by the tsetse fly, by the mosquito, by the tick, and by intestinal worms.

But besides the domesticated beasts, birds, fishes, and insects, there are many creatures which have been made man's servants in the individual, though they have refused to beget offspring in his keeping and so permanently enslave their race. The elephant to a great extent should really be included in this class, because when man wishes to recruit fresh conscripts he has to capture them in the forest. Still, the Indian elephant does sometimes breed in captivity, and freely in semi-captivity; and it is quite possible that the African elephant, humoured a little as to surroundings, might do the same. But the parrots usually refuse. No parrot has become truly domesticated except the Melopsittacus parakeet. Yet there are signs that this bird-equivalent to the monkey—is yielding, as witness the nesting which is going on in the outdoor aviaries at the Zoo. If I were a millionaire I would attempt to breed macaws in a vast aviary, and carry the eyes of mankind to extreme developments of colour as yet undreamt of.

The great apes, so far—though they are often tamed—have never brought forth young in captivity. Monkeys breed freely in menageries, but something has hitherto prevented mankind from domesticating his nearest relations on the animal side. The ancient Egyptians apparently tamed the baboon and trained it to ascend palm trees and

TEAM OF SIX OXEN WHICH ARE DRIVEN AS HORSES BY AN AMERICAN

LIONS IN CAPTIVITY ARE BRED AS FREELY AS IN THE OPEN
From a photograph of a group of lions in the Dublin Zoological Gardens

throw down the fruit. But there is silence in research as to the final results of this experiment. At any rate it came to nothing, for the baboon is still a wild gipsy of the desert and the bush.

In Western and Central Asia men first conceived the idea of utilising birds of prey as food-getters. Young hawks, falcons, and even eagles were taken from the nest, tamed, dressed, and "flown" at birds and beasts, from quails to bustards, hares and marmots to antelopes and deer. But they were never induced to procreate young in captivity, and always had to be recruited anew from the ranks of the wild. So they gave birth to no modifications under man's control. Likewise, the great chita, an aberrant hunting cat, was, and is, often tamed as a cub by natives of India and Somaliland—though no use is made of it there—and employed to hunt down antelopes. But, curiously enough, it very seldom breeds in captivity.

On the other hand, lions breed as freely in a cage as in the open—tigers less so, leopards and jaguars and pumas very seldom—and there is no want of inclination on the part of the lion to join the tide of human society. In Egypt, in Senegal, in Mesopotamia, perhaps once in India, he has been tamed and allowed his liberty, apparently without disastrous consequences, if he or she were regularly fed. The same thing has occurred less markedly with the tiger. But the chita, puma, and the ounce of Central Asia are superior to these large felines in their placable temperaments and the complete absence of danger to human beings in their domestication, which might be more actively pursued.

The otter has been tamed in India and taught to catch fishes for its master, but does not breed in captivity; and this is true also of the cormorant. In Japan and China, as of old in Britain, the cormorants fish for the people who train and keep them, but do not breed up young to similar service.

The nilgai is tamed by the Hindus, and seemingly breeds without reluctance in a captive state; yet somehow this large, handsome beast has not become a domesticated animal like the ox and buffalo. The webs of certain spiders, the cocoons of some moths, the flesh of many species of crabs, crayfish, and lobsters are utilised by man for his manufactures or his choicest food, yet the producers have never been domesticated like the coccids, the silkworm moths, the oyster. We have domesticated the trout and the carp, and the climbing perch, but not as yet the unspeakably lovely parrot fishes, the sea bream, or even the salmon or the sturgeon. Yet from what I have seen in that most wonderful of all existing aquariums, at the Battery, in New York, it is conceivable that in a millionaire's house, with a sufficiently large atrium, sharks may yet become household pets and sturgeons be kept on the premises to produce fresh caviare.

Indeed, having done so much since the Improved Stone Age, we demand to know why now, with our enormous advantages over Neolithic Man, we do not do more. Why do we not domesticate and breed in large herds the bushbuck? If there is one meat in the world more delicious than another, something which is like the perfection of Welsh mutton, it is the meat of the bushbuck, of various species, all over Africa. The Negro, we know, has practically domesticated nothing; he has simply accepted the domestic animals of the European and the Asiatic. Still, why should not the many civilised Negro peoples of Africa to-day, and the still more enterprising white races, try to preserve and domesticate the bushbuck? And the eland, still more so? And the kudu? Why has it been so easy to domesticate the ox and so difficult to domesticate the African buffalo? Why should the Indian elephant have been turned to man's purposes in Asia 5,000 years ago, and the

LLAMAS TRAINED FOR RIDING

A PIG TRAINED TO DRAW A CART

African elephant be still an unreclaimed wild beast, except in one tiny corner of the Belgian Congo? Why can we breed so easily the wild Madeira canary in cages and aviaries, and not the much more beautiful English goldfinch? These are questions which mankind might set themselves to solve, in order to increase the varied food products and utilitarian resources of the world, and, above all, to add to our store of beauty and wonderment.

As we review the past of man's taming of the wild, we seem to see one principal agency initiating these achievements—that mysterious Neolithic, or possibly Aryan, renaissance of man: the wonderful White man, almost, in his effect on the history of the world, like some special inbreathing of a Divine spirit. He seems to have arisen in some region north of the Caucasus, and to have spread across Central Asia, after becoming the dominant race in Europe and North Africa, till he reached Korea and Japan, where his vestiges remain, in the shape of the Ainu and the semi-Caucasian Japanese. He permeated the north of China and something of Tibet, and penetrated through the jungles of Burma till he reached Southern Siam, the Malay Peninsula, Sumatra, and Polynesia. In a much diluted form, by way of the Aleutian Islands, he reached North America; perchance also he may have got to South America by way of the Pacific archipelagoes and have started such original civilisation and taming of the wild as made their separate manifestations there. He turned the sub-continent of India from a vast forest of mainly Negro savagery into, at one time, the most civilised region of the world. From Egypt and from Mauretania he penetrated Palæolithic, utterly barbarian Africa; and gave the Negroes most of their languages and arts, domestic animals and religious ideas. Perhaps, therefore, he is the primal inspirer, or God's great channel of inspiration. At any rate, the White man, since the beginning of the Neolithic Age, has been the chief, if not the only, agency in this Taming of the Wild.

H. H. JOHNSTON.

LION AND TIGER IN CAGE WITH THEIR TRAINER IN CARL HAGENBECK'S MENAGERIE

THE BLUE-AND-YELLOW MACAW

"British Guiana has splendid macaws, the blue-and-yellow species presenting one of the finest examples of brilliantly contrasted colouring in Tropical South America."—*Sir Harry Johnston*

E

THE DUCKBILL OF AUSTRALIA

"The British Empire owns exclusively the most primitive of the egg-laying mammals—the duckbill, in Australia and Tasmania.
—*Sir Harry Johnston*

54

A GROUP OF FLAMINGOES

"The rosy Flamingo is one of the sights of the Bahama Islands in the British West Indies.'—*Sir Harry Johnston*

55

POLAR BEARS IN THEIR REALM OF ICE AND SNOW

"Within striking distance of the unconscious seal, the Polar bear kills its quarry with a blow of its ponderous paw."—*Aubyn Trevor-Battye*

GEOGRAPHY OF THE ANIMAL WORLD
DISTRIBUTION OF WILD LIFE IN TIME AND PLACE
By ALFRED RUSSEL WALLACE, F.R.S.

THERE is probably no branch of natural history about which so little information is given in popular works as that which discusses, and very largely explains, the causes of the varied and apparently erratic distribution of animals over the earth's surface. Yet there is none which to the general reader is so puzzling, and to many so interesting. This is partly due to the fact that the complete explanation is not only fundamentally based upon the laws of evolution, but that it requires also a frequent reference to the facts of geology and palæontology, and to many comparatively recent discoveries in physical geography.

Darwin himself laid the foundations of the study by his observations and deductions as to the total absence of both mammalia and amphibia from oceanic islands, and also by his extensive experiments and observations on the modes of dispersal of plants and animals; and it is only by combining all these sources of information with those of recent inquirers as to the nature of the deposits in the deep oceans that any intelligible view of the subject has been arrived at.

It may be as well first to state briefly some of the curious facts and apparent anomalies that have to be explained. Why, for instance, are deer entirely absent from Africa — except on the southern shores of the Mediterranean — though found over the whole of the other great continents from Patagonia to Greenland and from Britain to Japan? Why are camels found only in the great desert-belt of North Africa and Central Asia, and their nearest allies, the llamas and alpacas, only in the Andes south of the Equator and in Patagonia?

The old idea that animals are found where the conditions are fitted for them is disproved by the fact that the entire horse tribe is now limited, in a wild state, to Africa and Asia, but when introduced into both North and South America by the Spanish conquerors they ran wild and increased enormously. Most extraordinary is the case of the curious tapirs, animals something like small rudimentary elephants, of which two or three species are found in Tropical America, and the only other species—the Malayan tapir—in Borneo and the Malay Peninsula. Among birds we have the humming-birds swarming all over America from Patagonia to Labrador and Alaska, but in no other part of the world. The equally wonderful birds of paradise, of which about fifty different species are now known, are found only in New Guinea and a few adjacent islands. Lastly, we have the great mammalian order of Marsupials, limited to Australia and the adjacent islands as far as Celebes and the Moluccas on the one hand, while one family of the same order, the opossums, abounds in South America and as far north as California and Virginia.

We may also look at these differences from another point of view—that of the similarity of the animals in countries far apart, and the diversity that is often found between those which are comparatively near together.

If a person travels by any of the ordinary routes from England to Japan or the Amur, he will have gone very nearly half round the globe; yet when he reaches those remote countries he finds in the forms of life a wonderful similarity to those of his native land. The fields and woods are tenanted by titmice, hedge-warblers, wrens, wagtails, larks, redbreasts, thrushes, and buntings, some absolutely identical with his feathered friends, others so nearly resembling them that it needs a close examination to detect the difference between them. Even among the insects he will notice many butterflies, beetles, and so on, that will be familiar to him, though they may prove to be distinct species. In the Amur valley he will find even the mammals to be very similar. The fox, the badger, the weasel, the otter, the roedeer, the hedgehog, the mole, the squirrel, and the Irish hare, are all the very same as the British species; while the wolf and the bear are identical with those that inhabited the British Isles within the historical period, and are still common in Europe.

But if a settler in Australia goes to New Zealand, a distance of about 1,300 miles, he finds himself in a country zoologically and botanically entirely unlike the country he has left. Kangaroos, wombats, and the other varied forms of marsupials are absolutely wanting. There are no wild mammalia except two

TWO WONDERFUL BIRDS OF PARADISE

These birds number about fifty species and are confined to New Guinea and the adjacent islands. The top picture represents the Gorget, and the lower the Twelve-wired species

kinds of bats; the birds are almost all quite new to him; while insects are very scarce and altogether unlike the many handsome or curious species found in his own country.

Even more remarkable, perhaps, is the case of two small islands, Bali and Lombok, in the Malay Archipelago, only fifteen miles apart, yet strangely different in their productions. Their birds differ more than do those of Britain and Japan. Bali has red and green woodpeckers, barbets, weaver-birds, black-and-white birds like large wagtails, and many others quite unknown in Lombok, where we find screaming cockatoos and friar-birds with bald heads, and the strange mound-building megapodes, all allied to Australian species and entirely absent from both Bali and Java. Again, the great island of Borneo is much more like Sumatra and the Malay Peninsula, both in its mammals and birds, than is the much nearer island of Celebes, whose productions are more nearly allied to those of India or Africa.

Now, these and most of the other peculiarities in the distribution of animals are either directly explained or rendered intelligible by the application of the facts and general principles already enumerated; and I therefore propose now to explain as clearly and simply as possible the nature of those principles, which, when once thoroughly understood, will afford the means of solving almost all the problems of distribution both of plants and animals.

PERMANENCE OF OCEANIC AND CONTINENTAL AREAS. I begin with what is perhaps the most fundamental and far-reaching of the phenomena on which the interpretation of the existing distribution of plants and animals depends, the true importance of which was first perceived by Darwin.

Before the "Origin of Species" appeared, the common belief was that almost all the islands scattered over the great oceans were the remnants of former continents which had sunk beneath the waters, while our existing continents rose up to take their place. This view was accepted almost as a matter of course, and seemed, at first sight, to be quite in accordance with the fact that most of the older rocks in all parts of the world were of marine origin. But, during the voyage of the Beagle, Darwin was able to visit many of these islands, and he was struck by the poverty of their forms of life.

He also noticed the fact that all of them were either volcanic or of coralline formation, and they never contained any of the stratified rocks characteristic of continents and of islands that have once formed parts of them. He therefore concluded that they had all been formed in the ocean itself, though some of them appeared to be very ancient; and the total absence of native land mammalia could thus be easily explained, as these animals, though often good swimmers, could never cross wide seas or oceans. He further showed by experiment that the eggs of batrachians (frogs, newts, etc.), which were equally wanting, were quickly killed by salt-water.

Some years later, when electric cables began to be laid across the oceans in various directions, and when the voyage of the Challenger showed us the contour of the ocean floor and the nature of the deposits forming upon it, a striking confirmation of Darwin's views was obtained. For it was then for the first time shown that the floor of the great oceans did not form hills and valleys and mountain ranges, as had often been assumed; but was really an enormous, slightly undulating plain at a depth of from ten to thirty thousand feet, out of which plain the continents and islands usually rose abruptly; so that the 100-fathom line closely approached their shores, and even the 1,000-fathom line only left a narrow belt around them, in no way altering their general outline.

Most important consequences result from these facts. Sir John Murray, of the Challenger expedition, has calculated that the mean height of the land of the globe is 2,250 feet above sea-level, while the mean depth of the ocean is 13,860 feet below it. But the area of the land is little more than one-third that of the oceans, whence it follows that the bulk of the oceans in cubic miles is more than thirteen times that of the land. In my "Darwinism" (p. 345) I have given a diagram, showing at a glance these proportions; and I pointed out, I believe for the first time, that these great differences

THE VIRGINIAN OPOSSUM
This is the only marsupial that exists outside Australia and the adjacent islands

in the bulk of land and water rendered it almost impossible for any interchange of continents and oceans to have occurred during geological time. There must have been some approach to equality in the simultaneous rising and sinking of the solid crust of the globe. But if one of our smallest continents were to subside, and an adjacent portion of the ocean floor on either side of it to rise, the former would almost wholly disappear before the latter had approached within many thousand feet of the surface. And if a similar change went on successively, one continent after the other would sink, while the counter-balancing rise would hardly have rendered the vast extent of ocean any shallower than it was before. The contour of the ocean bed, and the comparative bulk of land and water, alone serve to indicate that the great and often-repeated vertical movements of the land over all the continents through-out geological time must have occurred within or on the margins of the areas they now occupy. If they had not been thus limited, if the gigantic forces of ele-vation and depression which have been always at work in the continents had been also at work beneath the ocean floors, the latter must have exhibited some of the abrupt inequalities of the former, instead of being entirely different from them in their contours.

These conclusions are quite in accordance with the broad features of stratigraphical geology in every part of the world. Everywhere we find the successive strata of which the earth's crust is formed occurring in long, irregular, but more or less parallel belts overlapping each other ; and in numerous cases exhibiting intercalated land surfaces or fresh-water deposits. This may be well seen by looking at any large geological map of England, or of any part of the Continent. This feature is perfectly explained by the minute examination that has now been made of the ocean floor near the land, as well as across its greatest depths. It has been found that the larger part of the débris of continents brought down by rivers is deposited mainly within a distance of from 50 to 100 miles from the coasts, while the very finest mud, which, of course, sinks more slowly, is deposited within 150 miles, and only in rare cases extends so far as 200 miles. All these are termed " shore deposits," and they form, when consolidated and uplifted, the successive strata of the earth's crust.

But beyond these limits, and extending for thousands and tens of thousands of miles over the whole floor of the deep oceans, is found a very different set of deposits termed " oozes." The most extensive is that termed globigerina ooze, because it consists very largely of the shells of a minute organism of that name, and it is found at almost all

THE TREE KANGAROOS
This species of tree-inhabiting marsupials is only found in the dense tropical forests of New Guinea and Queensland

depths from 250 to 2,000 fathoms. The remains of many other surface organisms occur among it, as well as of those that live at the bottom ; but no true land débris. There are, however, numbers of vol-canic particles, either derived from pumice, which floats for enormous distances, or from volcanic dust carried by the wind, which in the case of the Krakatoa eruption fell on the decks of ships nearly

1,000 miles from the volcano ; while, what is still more extraordinary, a small proportion of meteoric dust can always be found. This gives an idea of the extreme slowness with which the oozes are formed, since such dust can hardly ever be detected on land. Another indication of the same thing is the large number of sharks' teeth, and the ear bones of whales, the only parts of these animals which escape complete disintegration.

In these vast areas of deep-sea oozes covering the whole ocean floor we have another contributory proof that oceans and continents are not interchangeable. For if they had been so, as many writers still suppose, then each upheaved continent must rest on a basis formed of such consolidated oozes, with their characteristic fossils and inorganic particles. Yet in not a single one of our continents has any such base-rock been found, while our greatest geologists are almost unanimous in declaring that such consolidated oceanic oozes occur nowhere in the whole series of strata in any part of the world.

It is a very suggestive fact, as indicating that the law of subsidence balancing elevation is a real one, that all the greatest depths yet found in the ocean are in the vicinity of lofty volcanoes or ranges of mountains. Thus a depth of 23,250 feet has been found a little north of Porto Rico, West Indies ; 21,930 feet a little south of the Aleutian Islands ; 27,930 feet a little north-east of Japan ; 18,140 feet a little north of the Sandwich Islands ; 19,000 feet a little north-west of Bermuda. Very few of these very deep soundings have, however, been taken, as they are exceedingly laborious and costly.

The conclusion, therefore, of many of our greatest geologists, that the whole series of the stratified rocks present evidences of having been formed near to land, is supported by so many weighty arguments, derived from various sources, that it must be considered to be a well-established deduction of science.

Yet we still find many palæontologists insisting on land connections across the deepest oceans in order to explain similarities in their extinct animals. But there are other ways of explaining such facts much more in accordance with probability, and with the great generalisation I have here indicated.

EXISTING CONTINENTS AND THEIR PROBABLE EXTENSIONS. If we look at a globe or a map of the world in hemispheres, and consider the position of the continents and larger land masses, we find that they, even at the present time, are wonderfully connected ; that it would be actually possible for a man to traverse the whole of the continents, starting from Cape Horn, without ever going out of sight of land or requiring any other vessel than a small open sailing-boat. If we now look at a map showing the 1,000-fathom line around all the continents (such as that given in my " Darwinism "), and taking this as roughly indicating the amount of extension of the land during the whole Tertiary period of geology, we shall find all the continents united towards the north by wide stretches of land, offering ample facilities for the migrations of land animals at successive epochs.

WARMER CLIMATES IN PAST GEOLOGICAL PERIODS. But there is another consideration which is continually left out of account by those who still claim direct land connections to account for real or supposed affinities of the extinct animals of South America and South Africa, and that is the much warmer climate that prevailed in northern and Arctic regions throughout Tertiary times, and even very much earlier. This is proved by abundant remains of fossil plants found in Miocene strata all round and within the Arctic Circle, and of such a character as now grow a long way further south. On the west coast of Greenland, in 70° N., are beautifully preserved specimens of such trees as chestnuts,

THE AMERICAN TAPIR

THE MALAYAN TAPIR

The tapirs were distributed over North America, Europe, and Asia in the ages when these lands were warmer than at present. Four species are now found in Central and South America and one in the Malay Peninsula, Sumatra, and Borneo. None is found in any other part of the world

THE LLAMA OF SOUTH AMERICA   THE BACTRIAN CAMEL OF CENTRAL ASIA

These two animals are allied in all essentials of structure, even to the peculiar stomach adapted to water-storage. They are to-day as remote from each other as are the tapirs, and have no kin between South America and the desert regions of North Africa and Western Asia

sassafras, oaks, planes, beeches, plums, vines, and even a magnolia, all closely resembling trees and shrubs which now grow 20° or 30° further south in North America. Even so far north as Spitsbergen, one of the most barren and most inhospitable regions on the globe, a rich fossil flora has been found, indicating a climate fully as mild as that of the warmer parts of Canada at the present time, comprising such familiar plants as hazel, ash, and walnut, with water-lilies and an iris. As there was probably continuous land between Europe, America, and Asia, at least as low as the latitude of Stockholm, while the land a little further south had a warm temperate climate, it became possible for every kind of mammal of the temperate zones and many of the Tropics to migrate from one continent to the other.

There is also evidence that still warmer climates prevailed in the Cretaceous period, and, therefore, almost certainly in the early Tertiary, when all the chief types of mammalian life were well developed. In strata of this period in Greenland, not far from those of Miocene age above referred to, not only many of the trees of the later period occur, but, intermingled with them, such indications of warmer conditions as figs, myrtles, cycads, and tree-ferns. The only justification for postulating direct communication is that a shorter route is more effective, and therefore more probable, than a longer and more circuitous one. But this argument supposes two things which no geologist will admit—first, that the geological record itself is at all near to a full and continuous one ; and, secondly, that our knowledge of the actual record approaches to completeness. It is, on the contrary, held to be certain that the record of the animal life of all former periods contained in the rocks is extremely imperfect and fragmentary, and that our knowledge of

what exists, much of it buried deep in the crust of the earth, is wholly inadequate.

It is interesting to note that the very changes in the height of land and elevations of sea-bottoms in the northern parts of the Atlantic and Pacific oceans, which would facilitate the migration of land animals between the eastern and western continents, would themselves greatly assist in the production of the warm climates we find proofs of. It is universally admitted that warm ocean currents, aided by the warm winds that carry moisture from them, are the great sources of the mild and equable climates that even now prevail in many parts of the north temperate zone. The British Isles, as well as Norway, owe their mild winters to the Gulf Stream and the warm and moist westerly winds that are so prevalent ; while Labrador, in the same latitude, has a thoroughly Arctic climate, due to its being washed by a Polar current, and being exposed to cold winds from the Arctic regions of America.

But when the North Atlantic was elevated about 1,000 fathoms, the heating power of the Gulf Stream would be greatly augmented by two causes. It would be more concentrated by being kept out of the Polar seas, and would circulate more rapidly around the northern shores of the Atlantic. It would also be completely free from the Arctic icebergs which now pour into it as soon as the ice breaks up in the Polar basin, causing that wintry weather which is regularly experienced in Great Britain during the first half of May. And when such a mild climate prevailed as is shown by the abundant tree and shrub vegetation within ten degrees of the Pole, there would probably have been very small accumulations of ice, which would have no more serious effect on the climate than have the glaciers of Switzerland or of New Zealand in their respective cases.

EVIDENCE OF STILL EARLIER WARM ARCTIC CLIMATES. Before quitting this part of our subject, it will be well to note the abounding evidence of equally warm climates in the northern hemisphere throughout much, if not all, the Secondary period of geology. I will first notice the suggestive fact that, even more recently than the Cretaceous beds of Greenland above mentioned, in the London Clay (Eocene) of Sheppey, Bournemouth, and a few other localities, abundance of fossil plants has been found closely allied to living tropical or sub-tropical forms. Fruits of palms allied to those of the Nipa of Indian and Bornean river swamps are frequent, as well as the fruits of a proteaceous plant whose allies are now found in Australia. There are also species of acacia and anona, both warm-country forms; and, associated with these are found teeth and bones of crocodiles, a large sea-snake, and several mammals of the hog and tapir tribes. There are also numbers of marine shells now characteristic of tropical seas, including the nautilus.

Throughout the Secondary period similar conditions appear to have prevailed, the sea-shells being, both in size and structure, almost wholly of the type of those now confined to the Tropics. The abundance of large and often gigantic reptiles—terrestrial, marine, and aerial—which are found in these strata throughout Europe and America are equally indicative of tropical conditions. And these conditions must have prevailed also within the Arctic Circle, since the very early Triassic strata of the Parry Islands, in 77° N., contain many ammonites as well as the vertebræ of the large marine reptile ichthyosaurus; while, in the Triassic beds of Spitsbergen, ammonites and nautilus are found, and still earlier in the Coal Measures of that country are lepidodendrons and large fronded ferns.

We thus see that during the whole range of geological time between the Middle Tertiary and the Carboniferous ages there is almost continuous evidence of such mild and uniform climates far into what are now Arctic lands as would suffice for the temporary residence and continuous migration of most tropical animals between the Eastern and Western continents. No adequate reason, therefore, can be given for assuming an enormously difficult if not impossible bridging over of the great oceans, in order to account for the existing distribution of animal life; and still less in order to explain its apparent distribution in remote geological ages, as to the whole conditions of which we have such very scanty and imperfect knowledge.

Even now many wholly tropical animals live very well and breed freely in the whole temperate zone, such as domestic fowls and peafowl; while monkeys are found as high as 11,000 feet in parts of the Himalaya. As late as the Pliocene period they lived in France and Britain. Even the hippopotamus, an almost tropical animal now, inhabited South Europe, and ranged as far north as Yorkshire during the summers of the Glacial period!

THE POWERS OF DISPERSAL OF ANIMALS. The distribution of animals over the earth's surface of course depends in part upon their individual powers of locomotion—on land, through the waters, or through the air—but this is by no means the only factor of importance in determining their actual habitations, and is often entirely neutralised by other causes. In most continents, and in many of the larger countries or islands, birds exhibit a very similar restricted range to mammals, notwithstanding their very superior powers of locomotion. Reptiles also exhibit little difference in this respect, except that they are more dependent on temperature and become scarce in temperate and almost absent from very cold countries. The amphibia—frogs, toads, newts, salamanders, etc.—are in some respects more restricted than reptiles in their ranges, and more extended in others. This is due to the fact that they can withstand a lower temperature, their eggs being often frozen without injury; but, on the other hand, they are killed by salt water, and this explains the interesting fact that they, as well as mammals, are wholly absent from all true oceanic islands. This class of islands may be defined as being situated in the great oceans far from any extensive land masses, and being surrounded by depths of more than a thousand fathoms. All such islands, as was first pointed out by Darwin, are entirely without any of the older stratified rocks, consisting either of volcanic rocks or of coral formations which have been deposited upon them. Such are the Azores, Madeira, St. Helena, Mauritius, and all the remoter islands of the Pacific Ocean, none of which possesses either mammalia or frogs, and the most remote from land no reptiles or freshwater fishes. These are entirely absent from the Azores and St. Helena, while the Sandwich Islands possess two lizards—one very widespread over the whole western Pacific, and therefore presumably conveyed accidentally in canoes; the other a peculiar gecko, whose ancestors were probably introduced at some remote epoch by exceptionally favourable circumstances, such as large floating trees from more westerly islands. Within the Tropics especially, masses of trees and floating vegetation are often carried out to sea, and under favourable conditions may be driven by winds and currents for many hundreds or even thousands of miles, and carry with them small lizards or their eggs, insects, and land shells. These latter often hide in crevices or under bark, while some snails, when dormant, will stand immersion in salt water for twenty days without injury. We thus see an explanation of the curious fact that in the remotest islands, even when every kind of vertebrate except birds is absent, a few land shells and terrestrial insects, especially beetles, are always to be found.

But the more remote the island, and therefore the rarer and more exceptional the chances that bring these latter creatures to it, the more peculiar we find the existing species to be. Thus, in the very remote Sandwich Islands, with their rich vegetation and favourable climate, there are about 500 different kinds of land shells, almost the whole of which are peculiar species; and the moderately rich insect fauna is equally peculiar. In the Azores, however,

A COLONY OF BROWN PELICANS NESTING ON THE BANKS OF INDIAN RIVER. FLORIDA

BRANDT'S CORMORANTS IN THEIR HAUNTS ON THE ROCKS OFF MONTEREY, CALIFORNIA
From reconstructed groups in the American Museum of Natural History, New York

which are less remote from the nearest continent, and much more subject to violent storms—bringing numerous birds which are both seed and shell carriers—shells and insects are less numerous, and a much larger proportion belong to European forms.

All the facts connected with these oceanic islands, some of which form groups of great extent and of considerable geological age, and are scattered over all the great oceans, strongly support the view that the oceans themselves are permanent features of the earth's surface ; while the complementary fact, that the great continents in their main features and position are equally permanent, is demonstrated, first, by the occurrence in all of them of an almost identical series of stratified rocks of all the geological periods, and, secondly, because these rocks everywhere show plain indications of having been deposited very near the margins of the lands from the denudation of which they have been built up.

With the amount of preliminary knowledge here sketched out, we are in a position to explain most of those curious examples of discontinuous distribution which often appear so puzzling.

THE DISTRIBUTION OF TAPIRS. One of the most prominent of the puzzles of distribution to the earlier naturalists was that of the tapirs, the two commonest species of which inhabit Tropical South America from Brazil to Paraguay, and the Malay Peninsula, Sumatra and Borneo. Three other species also inhabit Tropical America, one in the high Andes and two in Central America from Panama to Mexico ; but none is found either in Africa or continental India.

But the course of geological discovery during the nineteenth century has completely explained this apparent anomaly. First, in 1825, Cuvier described the skeleton of the Palæotherium from the early Tertiary beds of Paris as being allied to the living tapirs ; and later, both in France and Germany, the remains of true tapirs were found in the middle Tertiary (Miocene) and late Tertiary (Pliocene) strata, both in France and England. Fossil tapirs have also been found in China and in North America from Carolina to California.

Now, always keeping in mind the extreme imperfection of the geological record, these discoveries clearly indicate that many species of tapirs inhabited warm and temperate Asia, Europe, and North America in middle and late Tertiary times. But during these epochs warm and even sub-tropical conditions prevailed in the Northern hemisphere as far as the Arctic Circle, accompanied by an abundant and luxuriant vegetation ; and we have also seen that these conditions were probably brought about, in part, by a greater extension of land in the North Atlantic, admitting of more or less easy communication between the Eastern and Western hemispheres. The problem of the tapirs is, therefore, completely solved.

DISTRIBUTION OF THE CAMEL TRIBE. The case of the camel is perhaps even more interesting than that of the tapirs, which it somewhat resembles. Camels are now inhabitants of the desert regions of Western Asia and North Africa, and of no other parts of the world. But in the Andes and temperate plains of South America is a group of small animals—the llamas, alpacas, and guanacos—which in all essentials of structure, even to the peculiar complex stomach adapted for water-storage, belong to the camel tribe. These two allied groups are, therefore, now almost as remote from each other as are the American and Malayan tapirs, with no allies whatever in intervening regions. But here, too, geology has furnished the solution. First, in the late Tertiary deposits of the Siwalik hills of North-west India remains of true camels have been found ; but, during the last thirty or forty years, a wonderful series of fossils has been discovered in North America, in late, middle, and early Tertiary strata, by which the whole gradual development of camels from a smaller and more primitive type has been exhibited.

The same strata have shown us early forms combining the characters of all the hoofed animals, from the swine and the hippopotamus to camels, cattle, deer, sheep, and antelopes, gradually becoming specialised into those varied forms. It is, therefore, clear that in all probability the camel and llama tribes originated in the Central United States, where, towards the end of the Tertiary period, they became extinct. Previous to this catastrophe, however, some of the true camels migrated to the Eastern hemisphere, probably by the way of continuous land in the North Pacific, and have left as their only survivors the camel and the dromedary. About the same time, and probably driven to migrate by the same adverse conditions which led to the extinction of so many of their allies, the llama group passed southwards along the central mountain ranges into South America, where they have found suitable conditions for their survival south of the Equator, in the high Andes, and on the arid plains of Patagonia. Here, again, the problem is satisfactorily solved.

THE DISTRIBUTION OF THE HORSE TRIBE. The horses, asses, and zebras constitute a distinct and very remarkable tribe of mammals, being the only members of that great class which possess a single functional toe on all four feet, forming the well-known horse's hoofs. These animals are now strictly limited in a wild state to Africa and Western Asia, though in quite recent times they ranged over Europe and the British Islands ; and many allied groups are found in still earlier times, back to the early Tertiary, showing the gradual transformation of a four-toed animal of small size, step by step, into the true one-toed horse.

But, what is more curious still, a similar and even more complete series of ancestral forms existed in North America, where the true horse was also developed, its fossilised remains being found over both North and South America nearly up to the Glacial period. Then suddenly all became extinct. But after the great Ice Age had come and gone, with the many changes of animal life that accompanied it, the two continents again became well adapted to horse life, as proved by the fact that, as we have said, when introduced by the early Spanish conquerors they ran wild, and increased enormously both in the western plains of the north, and in the llanos and pampas of the south.

It is curious that the series of these ancestral horses runs almost parallel in America and Europe, though the species are not exactly the same ; and this seems to imply that the North Atlantic land connection existed for a long period, and that inter-communication from America to Europe and from Europe to America frequently occurred. It is, in

mild climates as are indicated by other evidence really existed throughout a considerable part of the Tertiary period.

This strange history has been made use of by a talented American writer, Mrs. Charlotte Perkins Gilman, to point a moral against those who oppose the possibility of the social advancement of humanity. It is one of three such illustrations in a poem under the heading of " Similar Cases " in a volume entitled " In This Our World." The name " Eohippus " was given by Professor O. Marsh to one of the less-known remoter ancestors of the horse.

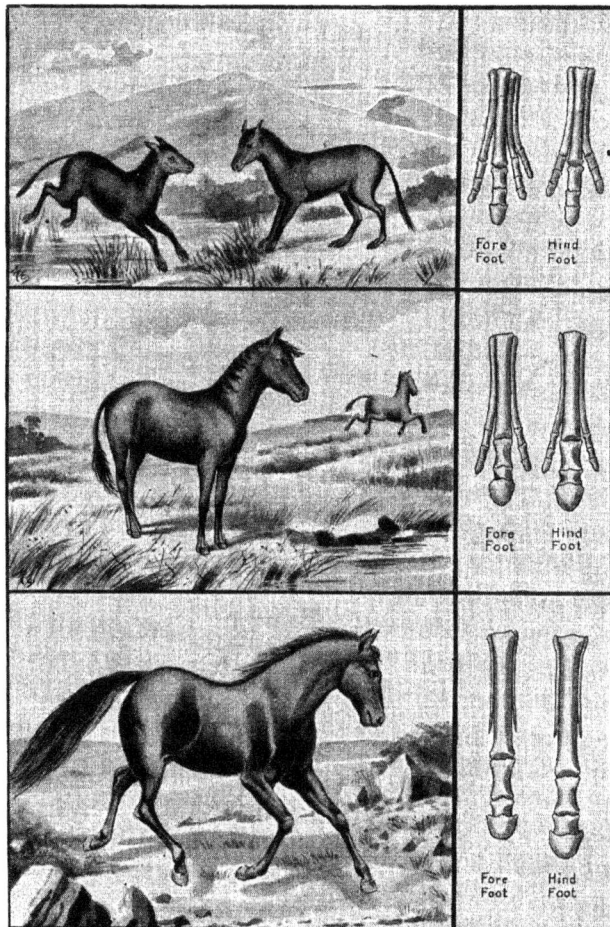

THE EVOLUTION OF THE HORSE

The top picture shows Eohippus, which was about the size of a fox and ran on four and three divided toes. The centre picture is Miohippus, which ran on three toes, and the bottom picture shows how all the toes but one have disappeared in the present horse

" SIMILAR CASES."

There was once a little animal
No bigger than a fox,
And on five toes he scampered
Over Tertiary rocks.
They called him Eohippus,
And they called him very small,
And they thought him of no value—
When they thought of him at all ;
For the lumpish Dinoceras
And Coryphodont so slow,
Were the heavy aristocracy
In days of long ago.

Said the little Eohippus :
" I am going to be a horse,
And on my middle finger-nails
To run my earthly course ;
I'm going to have a flowing tail,
I'm going to have a mane,
I'm going to stand fourteen hands high
On the psychozoic plain."
The Coryphodont was horrified,
The Dinoceras shocked ;
And they chased young Eohippus,
But he skipped away and mocked.

Then they laughed enormous laughter,
And they groaned enormous groans,
And they bade young Eohippus
Go and view his father's bones.
Said they : " You always were as small
And mean as now we see,
And therefore it is evident
That you're always going to be."
" What ! Be a great, tall, handsome beast
With hoofs to gallop on ?
Why, you'd have to change your nature ! "
Said the Loxolophcdon.

They considered him disposed of,
And retired with gait serene—
That was the way they argued
In the early Eocene.

fact, difficult to tell in which continent the true horses now living actually originated. In both there was an animal about the size of a fox which seems to be the earliest direct ancestor of the modern horses and asses. This intermingling of the ancestral forms of a group so well adapted for migration both in structure and habits renders it certain that such connecting lands and such

ELEPHANTS AND THEIR ANCESTORS. The elephants are of special interest for two reasons. Their distribution in middle and late Tertiary times was very similar to that of the horses, and they disappeared from America, Europe, and Northern Asia at as late or even later a period and with equal suddenness. The huge mammoth with its enormous curved tusks lived in Northern Europe and Asia as well as in North America down to the human period. Many other species are of late Tertiary date and died out a little earlier ; and among these were some curious small elephants from three to five feet high, whose remains are found in the caves of Malta and Cyprus. The allied mastodons have a similar distribution at a somewhat earlier period, and these ranged over both North and South America and survived there till the appearance of man.

None of these numerous forms gives any clue to the early stages of the elephant type so strikingly different from all other mammalia. But within the last ten years remarkable discoveries have been made, in the early Tertiary beds of the Fayum valley in Middle Egypt, of an interesting group of fossils of mammals which are believed to be the ancestral forms of the entire elephant tribe. The earliest of these is an animal not very different from the ancestral swine and tapirs, and of very small size. The whole series is not yet complete, but the following short account, from the "Guide to Fossil Mammals" in the British Museum, very clearly explains its bearing upon the problem.

"The fossils, so far as known, show that the earliest forerunners of the elephants were small marsh-dwellers which lived on a succulent food in the African region. They gradually increased in size without essentially altering their limbs or bodies; but as their legs lengthened and their neck shortened, their face and chin gradually became elongated to reach the ground for browsing. When this strange adaptation had reached its maximum degree, the chin shrivelled, leaving the flexible toothless face without any support. Thus arose the unique proboscis of the elephants, which has become prehensile by stages which cannot be traced, because soft parts are not preserved in ordinary geological formations."

In the "Guide to Recent and Fossil Elephants," 1908, there is a diagram showing the proportionate sizes of the three chief types of fossil elephants previously known, together with the two recent discoveries in Egypt, which is very instructive. In these we see the progressive shortening of the chin, the straightening of the tusks, then their bending downwards instead of upwards, and lastly their reduction to ordinary-looking teeth in both jaws. It is also very suggestive that this change occurred at a period when Africa was cut off from the rest of the world by a sea extending from the Atlantic across Northern India. These early forms were thus subject to special conditions from which they could not escape by migration, and thus gave full play and ample time for the action of variation and natural selection to initiate the main elephantine characteristics.

When, in the middle Tertiary, free land connections were formed to Asia, Europe, and North America, elephants had attained such a size and such completeness of organisation that they were enabled to spread over the two hemispheres, becoming modified in each into special forms adapted to the new conditions.

Then, after a long period of success, a number of competing tribes arose, such as the extensive and varied hoofed ungulates culminating in the cattle, deer, and antelopes, as well as a host of large and powerful carnivores culminating in the great sabre-toothed tiger, when the elephants everywhere became extinct, except in the two most favourable localities for their maintenance—that is to say, in the equatorial forests of Africa and Asia.

WHY ARE DEER AND BEARS ABSENT FROM TROPICAL AND SOUTH AFRICA? The Tropical African region affords us one of the most interesting problems in zoological geography. In many of its larger mammals and birds it presents a considerable resemblance to those of Tropical Asia, yet there are some most remarkable deficiencies. The most singular is the absence of the whole of the deer and the bear tribes, which are found in all the other continents, together with the camels, goats, and sheep which abound in the adjacent parts of Asia.

Now, the explanation of these peculiarities is to be found, first, in the existing fauna of the great island of Madagascar; and secondly, in that isolation of Tropical and South Africa from the rest of the world during the early portion of the Tertiary period, which is proved by a vast extent of early Tertiary marine strata to the north of it; and which has been already mentioned as having led to the early development of the elephants.

The interest of Madagascar lies in the fact that the evidence all points to its having been formerly united with Africa, but only at a time when that continent possessed a much more restricted fauna than now. For although Madagascar contains about seventy different species of mammals, these are mostly of low type, while none of the larger and more highly developed African groups is found there. Though there are nearly forty of the lower quadrumana (the lemurs), there is not a single monkey or ape. The great carnivora of Africa—the lions and leopards, the hyænas and lynxes—are absent and are represented only by eight civets, all of peculiar genera; the very numerous rodents of Africa, including squirrels, porcupines, and hares, are absent, the order being represented by a few peculiar mice; while the ever-present antelopes of Africa—nearly 100 species—as well as its zebras, rhinoceroses, and elephants, are quite unknown. In place of all these we find only a river-hog, while a small hippopotamus seems to have recently died out, both of which, being semi-aquatic in habits, might, in very favourable circumstances, have crossed the Mozambique Channel. The reptiles show similar singularities, most of the genera being peculiar; while two families of snakes, both abundant in Africa, are wholly wanting. Even the birds exhibit the same phenomena. The genera and some of the families are quite peculiar, but they seem to be related to Indian as much as to African types. A number of specially African birds, such as hornbills, plantain-eaters, and barbets, are absent.

We thus find that all the facts go to prove that the fauna of Africa must have been very poor as regards variety of its higher animals at the time when Madagascar was separated from it; and that epoch is demonstrated to be a rather remote one by the distinctness of the two faunas now, and also by the number of peculiar species and genera of the higher mammalia and birds characterising that continent. We conclude, therefore, that the separation from Africa dates from the early Tertiary period.

For the later history of animal life in Africa we are indebted to the geological record in other continents. We find that all the larger and more

THE LARGEST AND THE SMALLEST MAMMALS IN THE WORLD: AFRICAN ELEPHANT AND PIGMY SHREWMOUSE
Between the elephant's fore feet is a model, in exact proportion, of the pigmy shrewmouse, which is enlarged in the inset
From a photograph by J. L. Clarke of the models in the Natural History Museum, South Kensington

characteristic of the great mammalia of Africa inhabited Southern Europe and Northern India in the middle and late Tertiary periods. Such are the great carnivora—lions, tigers, and hyænas, the extensive and varied hoofed animals, from the zebras, rhinoceroses and swine, to the antelopes, the giraffes, and even the okapi—all of which have left their bones or skeletons in middle or late Tertiary strata.

It becomes clear, therefore, that not long, geologically speaking, after Madagascar was separated from Africa, the Saharan sea was partially elevated so as to open one or more free passages into the African continent. Then the highly organised mammals just enumerated would migrate into the new and inadequately peopled land, would there increase rapidly, and develop into a variety of distinct species fully adapted to the new conditions.

Then came the third and last great change, the simultaneous subsidence of the Mediterranean area, and the elevation of that of the Sahara and of Arabia, bringing about the desert conditions of these countries, and thus isolating Tropical and South

Africa. Thenceforth the typical African fauna was left to develop uninfluenced by immigrations from the extensive Euro-Asiatic continent, leading to that endless variety and abundance of "great game" in which this region is pre-eminent.

And now, at the end of this strange series of geographical changes and zoological migrations, we find the solution of the problem of those deficiencies which are so remarkable a feature of this continent. For the varied and widespread deer tribe only reached their full development in the later Tertiary period ; while the true bears seem to have originated about the same epoch, and to be in fact at their fullest development in our own time. These two groups, therefore, either did not exist in the adjacent regions, or were so comparatively weak in structure as not to be fitted to take part in the great rush of migration to the newly opened land. The absence of camels is probably due to the fact that, as already shown, they originated in North America and entered Asia from the north-east. They had reached North-west India towards the latter part of the Tertiary period, and were too remote from Africa to take part in the great incursion. The goats and sheep seem to be among the very latest developments of the large tribe of Bovidæ, or hollow-horned ruminants, and as they are pre-eminently mountain animals, they would not have been in a position to take part in a low-land migration, even had they been in existence at the time, which they probably were not.

As the whole subject is both vast and complex, I will here re-state the fundamental facts and conclusions on which a true explanation of all cases of anomalous distribution must be based.

(1) The most fundamental of all is the proof, by the agreement of several distinct lines of argument, that the great continental and oceanic areas are not interchangeable, but are permanent features of the earth's surface, subject only to marginal extensions and reductions within fairly well-defined limits.

(2) The slow but important changes of climate that appear to have been always going on. The more important of these for our special inquiry is the proof of an almost tropical climate in much of the temperate and sub-Arctic zone in the early Tertiary, changing gradually to warm-temperate towards its close, and then merging into the oncoming Glacial period. This must never be overlooked, as it often is, because, with moderate land extensions in the northern parts of both the Atlantic and Pacific Oceans, it offers ample means of migration, even of warm-country animals, between the several continents, and entirely obviates the supposed necessity for repeated bridging of the tropical and sub-tropical oceans, which has been shown to be unsupported by any evidence, opposed to a large body of facts, and almost, if not quite, physically impossible.

Other factors of secondary importance are : the various powers and facilities of dispersal by the different groups of animals ; the evidence of geology as to the preceding forms of life in the same areas and elsewhere ; the frequent changes of climate and the nature of the vegetation ; lastly, the continuous action of the competition of better-adapted forms of life, rendering it impossible for newcomers to maintain their existence, even if occasionally introduced into a new country.

We thus see that the present distribution of the whole vast complex of living organisms is the final outcome of the entire course of evolution from the earliest dawn of life. The locality and range of every animal and plant may therefore be looked upon as one of its vital characters. It is what it is and where it is because it has been developed in close adaptation to the physical and organic environment in which it lives, and where alone, except in very rare cases, it is able to maintain its existence.

The higher and more completely adapted forms—the dominant and aggressive species, as Darwin termed them—will always be able to maintain themselves, sometimes at the expense of native forms, when they gain access to a country previously occupied by lower types, as was the case when the highly developed animals of the Euro-Asiatic continent gained access to Tropical Africa. We see the same thing occur in the case of the European sparrow introduced into North America and the common rabbit into Australia, both of which have become pests in their new homes. But they have in both cases been assisted by man's interference with the native fauna and flora, and it is very doubtful if the same results would have followed any chance introduction of these species into countries where man had not already interfered with the balance of competing life.

The Division of the Globe into Zoological Regions. In order to describe and compare the distribution of the various species, genera, and families of animals, it has been found necessary to mark out a certain number of extensive areas characterised by distinctive forms of animal life. The older naturalists usually adopted the great geographical and racial divisions—Europe, Asia, Africa, America (North and South), and Australia, or the still broader Arctic, Temperate, and Tropical Regions. They also used the general term "India or the Indies" for all Eastern tropical lands, sometimes including all the Tropics, as in our still common but misleading term "West Indies."

The first thoroughly scientific attempt to establish a series of regions that should accurately summarise the main facts for any extensive class of animals was made by Dr. P. L. Sclater in 1857 for the class of birds, at that time, as it is now, probably the best-known of all the more extensive groups of animals. He was the first to point out that Europe and Asia do not correspond to primary divisions of animal life, as shown by the striking similarity and considerable identity of both the birds and mammals across the whole of Europe and temperate Asia. This formed his first great region, which he termed the "Palæarctic," as including all Old World northern lands. Then came the tropical portion of Asia, which possessed hosts of altogether peculiar species, genera, and whole families of birds not known in the temperate zone. This he termed the "Indian Region," because India in a wide sense

MAP OF THE WORLD ILLUSTRATING THE GEOGRAPHICAL DISTRIBUTION OF ANIMALS

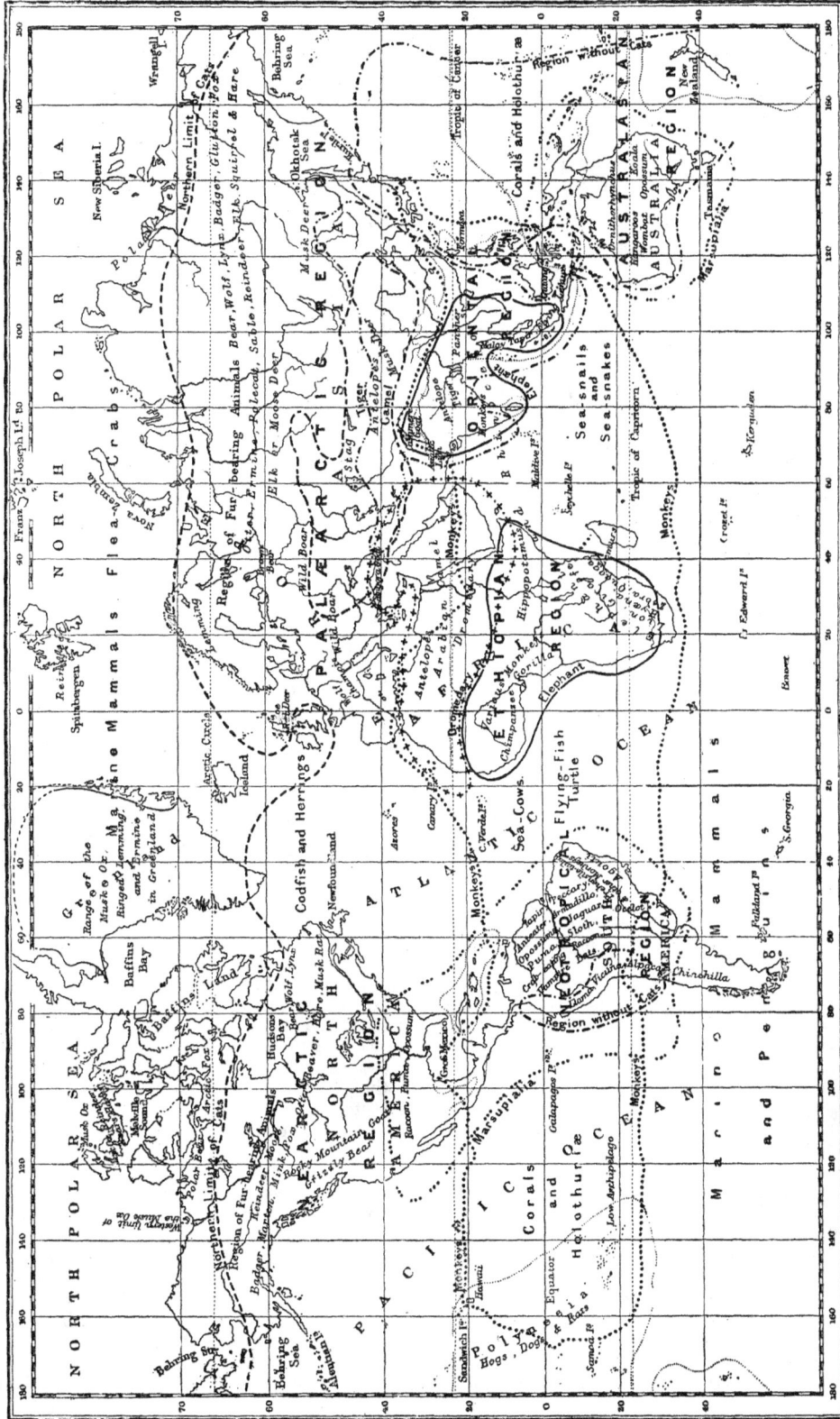

This map is based on the distribution scheme proposed by Dr. Philip Lutley Sclater in 1857—originally for the class of birds, but afterwards adapted to the animal kingdom—and includes the modifications and extensions introduced by Dr. Alfred Russel Wallace. It shows the six great regions into which, for this purpose, the earth has been divided—namely (1) the Palæarctic, embracing the Northern Old World and Africa to the Sahara; (2) the Oriental, or India, Burma, Siam, and the Malay Islands; (3) the Australasian, comprising the island continent, New Zealand, and most of the Pacific Islands; (4) the Ethiopian, that is, Central and South Africa; (5) the Neotropical, composed of South America, the West Indies and Central America as far as Mexico; and (6) the Nearctic, including Temperate and Arctic North America

formed the bulk of it. This term in my " Geographical Distribution of Animals " I altered to " Oriental Region "—perhaps unnecessarily—because it included Burma, Siam, and all the great Malay islands, and this name has been widely adopted. It may be mentioned that the Palæarctic Region also included North Africa as far as the Sahara, all its chief productions, both animals and plants, being closely allied to those of Europe or Western Asia. The remainder of Africa, possessing a large proportion of peculiar types, and being thoroughly isolated from the rest of the world, constituted the " Ethiopian Region." The fourth was the " Australian Region," perfectly characterised by its very peculiar marsupial mammals, as well as an immense number of peculiar genera, and of several remarkable families of birds. The eastern half of the Malay Archipelago belongs to it, as do also New Zealand and most of the Pacific islands. Then came South America, including the West Indies and Central America as far as Mexico, forming the " Neotropical Region," because it included the whole tropics of the New World. Temperate and Arctic North America constitute the " Nearctic Region " of Dr. Sclater. It has relations with both the Neotropical and Palæarctic regions, but has sufficient special features to be kept distinct.

It will be seen that the past geographical history of the several regions, as we have been able to trace it both by stratigraphical and palæontological evidence, enables us to understand why each of these regions possesses broad distinctive features while presenting certain resemblances to several others. By laying undue stress on these resemblances the primary zoological regions have been reduced in number and altered in constitution by some writers, so as to render them of little use or interest to the general reader, or even to the general student of zoology. Professor Huxley, for example, proposed a threefold division—Arctogæa, including the whole Northern hemisphere, with Africa and Malaysia ; Notogæa, for South America and Australia ; while New Zealand formed a region by itself. This division was founded on the prevalence of a low type of birds and of mammals in Notogæa, and on the lowest existing reptile—the curious lizard-like hatteria, found only in New Zealand, and there almost extinct. But these all depend on features of anatomical structure only appreciated by specialists, while a single region—including Europe, Asia, Africa, and North America—is absolutely useless for any practical purpose. Mr. Andrew Murray united India and Africa as one region, and North and South America as another ; while several modern writers insist that North America is too much like Europe and North Asia to be separated from them, and thus form what they term the " Holarctic Region."

As this last is rather widely adopted owing to its having been advocated and taught at Cambridge for many years by Professor Alfred Newton, especially as regards the class of birds, I will give a few of the main facts which show how great and fundamental are the differences of the two regions.

No fewer than twelve out of the thirty-six families of land birds inhabiting the Palæarctic region are not found in the Nearctic ; which latter has eight families not in the Palæarctic. Of genera, there are 118 which are Palæarctic but not Nearctic, and 113 Nearctic but not Palæarctic. Stating some of these facts in popular language, we find in North America three very common families of birds— the wood warblers (Mniotiltidæ), the hang-nests (Icteridæ), and the tyrant birds (Tyrannidæ)— containing together nearly 100 species spread over the whole country, all three of which families are totally unknown out of the American continent. In Europe and North Asia we have the hedge warblers (Accentoridæ), the fly-catchers (Muscicapidæ), and the starlings (Sturnidæ), quite unknown in America ; besides more than twenty genera with over 100 species of warblers and their allies, as well as twelve genera with fifty species of pheasants and their allies, all equally unknown in America. Summarising the whole, we see that, in either region, the genera of birds which it has in common with the other do not amount to one-third of the whole number it possesses ; while the species common to both, mostly Arctic, only amount to twenty. So far as the mammalia are concerned, there is a rather larger amount of resemblance, but even in their case the diversity is still ample.

But, further, when we consider the very important part North America has taken in the development of the higher animals, especially of the great tribe of ruminant mammalia, including such specialised types as the camels and the horses, which seem to have been peculiarly its own, any division of the earth into zoological regions which obscures or ignores this past history by uniting this vast country either with Eurasia or with South America to form a single region must be equally inadmissible. The same objection applies to uniting Africa with India. Each has its peculiar families, and a very large number of peculiar genera of all the higher animals ; but, because these are often small and not generally known, the groups of larger and more conspicuous mammals and birds which are common to both are allowed to obscure the more fundamental diversities. Even if we do, against the weight of evidence, make any such compound regions, we must immediately separate them again as " sub-regions," or we lose all the advantage of easy and intelligible classification of the many peculiarities of the Ethiopian fauna in all its groups from mammals to insects.

It is for these reasons, which I have given in full detail in my works on the subject, that I still thoroughly hold the scheme of six primary zoological regions proposed by Dr. Sclater to be in all respects more accordant with the great facts of the existing distribution of animals as well as with their geological history, and the past geographical changes of the earth, than any which has since been proposed. As the most accurate and most useful classification for the student, as well as the most intelligible and most illuminating for the general reader, it holds the field against all attempts hitherto made to improve it.   ALFRED RUSSEL WALLACE

ANIMAL LIFE IN THE BRITISH EMPIRE

A UNIQUE ASSEMBLAGE OF THE WORLD'S FAUNA

By Sir HARRY JOHNSTON

THERE are two reasons why it is important to try to realise the more interesting features of the fauna characteristic of the many countries in every clime which Fate has placed under the British flag. The first reason is the scientific, commercial, and æsthetic value of this unrivalled assemblage, such as no other nationality controls, of beasts, birds, reptiles, fishes, insects, spiders, crustaceans, molluscs, worms, echinoderms, corals, and sponges found within the limits of the British Empire. And the second reason is the necessity, for the prevention of disease, that old and young, poor and rich, black man and white man, should learn the main facts about harmful insects, life-draining intestinal worms, and, most of all, in regard to those infinitely minute yet infinitely powerful protozoa which lie at the base of the Tree of Life, but which in some parasitic forms can be far more dangerous to the bodies of man and other mammals than tiger, cobra, viper, or shark.

Since, however, it would be impossible in some cases, and difficult in others, to marshal for review forms that are either microscopic or externally small and insignificant, the protozoa, worms, echinoderms, molluscs, insects, and crustaceans are better left to be considered in detail in the section of this work devoted to the nonvertebrate forms of life. When we talk of the "fauna of the British Empire," it is usually only the animals which "catch the eye" that people think of, and these will be the theme of the following "abstract and brief chronicle."

The most ancient of the backboned creatures are the fishes, and among these the sharks count for much in popular imagination. Perhaps the most remarkable form of this subclass of cartilaginous-skeletoned fishes is the Port Jackson shark, found on the south-east coast of Australia, not dangerous to man, but of extraordinary appearance, with a weird resemblance to a human being in its eyes and nose. The sharks which follow ships and haunt harbours and surf-bound coasts in the Tropics for the chance of devouring human beings falling into the water belong to the genus Carcharias. The "hammer-head" sharks are occasionally captured off the east and south-west coasts of Great Britain. Basking sharks, which reach to a length sometimes of 40 ft., have been caught on the coasts of the United Kingdom. The strange saw-fishes, distant relations of the sharks, are found in the estuaries of East and West African rivers belonging to Great Britain. An awful-looking creature, justly dreaded by the negro fishermen of the West Indies and Central America, is the eagle ray. This enormous skate has developed its pectoral fins into immense wing-like surfaces measuring 20 ft. to 30 ft. across the body. The eagle ray is said to pursue the pearl-divers of the Cayman Islands and the Bahamas, to envelop them under water with its bat-like fins, and then devour them; but it is a little difficult to understand how this last act can take place, as the ray's teeth are flat, and adapted only to crushing shellfish.

The common sturgeon frequents the British coasts and is occasionally found in the Severn and the Trent, and used to be obtained in the Thames. Ever since an Act of Edward II. it has been a Royal fish, the property of the British Sovereign, wherever it be caught within the limits of his kingdom.

Many of the useful fishes of the world are distinctly "British," such as the trout and the sea trout, the salmon, the charr, the grayling, the smelt, and the white-fish, of which there are three peculiar British species—the gwyniad, the vendace, and the pollan—Celtic fish, as they might be called, since they are almost confined in their distribution to the still Celtic portions of Great Britain and Ireland. Though most of these generic types of the beautiful and comestible salmon family are found in many parts of Europe, Northern Asia, and North America which are not British, yet they are particularly

THE CURIOUS SEA HORSES OF THE BRITISH SHORES
This fish swims with its body in a vertical position

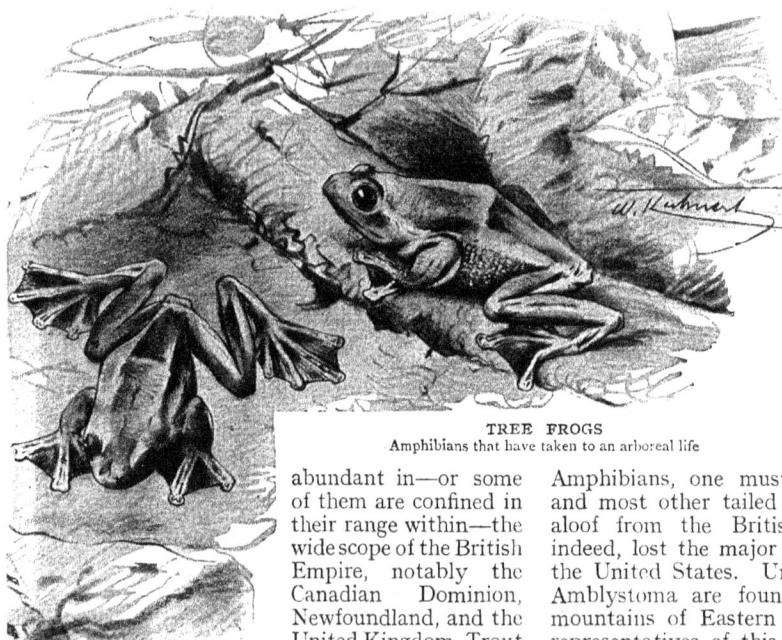

TREE FROGS
Amphibians that have taken to an arboreal life

abundant in—or some of them are confined in their range within—the wide scope of the British Empire, notably the Canadian Dominion, Newfoundland, and the United Kingdom. Trout have also been introduced into New Zealand and British Central Africa, and have become naturalised there. The place of the salmon as a "sporting" fish is taken in India by the mahsir of the Indian rivers and mountain streams, a gigantic barbel or carp, sometimes 5 ft. in length.

Then there are the eels of the British rivers and coasts, the rivers and coasts of Newfoundland, and the West Indies, the pike of the United Kingdom and of Canada, the extraordinary "sea horses" of the British shores, the climbing perch of Ceylon, India, and Malaysia, the cod of the Newfoundland Banks—one of the several fishes or marine forms that have been the cause of national quarrels and the extension of empires. The beautiful sea breams of the West African and Australian coasts frequently assume the most gorgeous colouring of crimson-scarlet or red-gold. The British coasts are well supplied with the invaluable flat fishes—soles, flounders, halibut, dabs, turbots, and brills.

Before leaving the fishes, mention should be made of the lung-fishes, because they descend from an immeasurable age in time, their earliest representatives going back far into the Primary epoch. They attracted great interest when first revealed to science in their living forms in consequence of their breathing air through lungs, but they are now thought to be not so closely connected with the ancestral line of the amphibians as the ganoids. They are noteworthy because of their present curious distribution. One genus, Ceratodus, is *entirely* confined to the fresh waters of Australia. The second genus, Protopterus, is restricted to the lakes and rivers of Tropical Africa—the Congo basin, British

Central Africa from the Zambezi northwards, Victoria Nyanza and the Upper Nile, the Shari, Lake Chad, and British Nigeria, Sierra Leone, the Gambia, and French West Africa. The third genus, Lepidosiren, is not only solely South American in its range, but apparently is confined to the basin of the Amazon and the northern part of Paraguay. Unless, therefore, it is some day discovered in the rivers of British Guiana, it cannot be called a British citizen.

Passing now to the Amphibians, one must confess that salamanders and most other tailed amphibians keep singularly aloof from the British Empire. Great Britain, indeed, lost the major part of them when she lost the United States. Unless one or more species of Amblystoma are found in the Himalaya or the mountains of Eastern Burma, she is without any representatives of this order, with the exception of three species of newt in the British Islands.

As to the tailless amphibia (frogs and toads), on the other hand, the British Empire has much to boast of. There is the remarkable tongueless Surinam toad, chiefly confined in its distribution to Guiana, which develops cavities in the back of the female into which the eggs are pressed by the male, here to develop into tadpoles. The true toads are found all over the world, except in New Zealand, New Guinea, and Madagascar; but the Australian toads are of three genera peculiar to that island-continent. The true tree frogs avoid British dominions somewhat, though they are found in Australia, in Assam, and in portions of Canada. It is curious that they should not have reached England, since they are found all over France and even penetrate to Belgium and Holland. There is a giant tree frog in the forests of British Guiana nearly

THE TUATERA, OR SPHENODON LIZARD OF NEW ZEALAND
The oldest reptile in the world, which may be described as a "living fossil"

5 in. long, reddish brown with a black stripe along the spine, and with very large adhesive discs to its fingers and toes.

Britain occupies a proud position in regard to the Reptiles, for she alone possesses the oldest reptile in the world, the Tuatera, or Sphenodon lizard of New Zealand, not properly a lizard, but a representative of a very ancient and generalised order, the Prosauria, from which all the other reptiles, the mammalia, and even the birds may have ascended. Ancestors or collateral relations of the tuatera once lived in Warwickshire, Shropshire, and Eastern Scotland, and no doubt other parts of Britain and Ireland, as well as Europe, India, etc, in the Triassic period of the Secondary epoch.

It is one of the many surprises in biology that the only survivor of this early, generalised order of true

New Zealand was colonised before her people felt much interest in natural science. Otherwise, New Zealand would have chosen for its national emblem not—whatever it may be—the Southern Cross, or other hackneyed symbol, but the Tuatera lizard ; for, however poor in fauna, New Zealand should be the proudest country in the world as the possessor of a living sphenodon.

With regard to Tortoises and Turtles, the Empire possesses the gigantic land tortoises of the Aldabra, Chagos, and Seychelles archipelagoes of tiny islands in the western Indian Ocean. They were formerly inhabitants of the Amirante Islands, Rodriguez, Bourbon, and Mauritius, as well as of Madagascar and the Comoro Islands, but except in North and South Aldabra, the Chagos, and Seychelles, they are everywhere extinct, having been

CROCODILES AND WATER BIRDS ON LAKE NYASA
From a painting by Sir Harry Johnston

reptiles should live in the very antipodes of its original habitat. The tuatera, as a genus, is without doubt the most ancient survival amongst air-breathing vertebrates. Its existence, more or less in its present form, may go back to a period distant from our time by ten millions of years. During this vast period this order of reptiles became extinct everywhere but in New Zealand, and even here they were nearly wiped out before their importance was fully realised. Bush fires, the multiplication of dogs and cats, and the hatred of "vermin" on the part of the British colonists are the causes which have reduced the existing tuateras to a few hundred specimens on the islets in the Bay of Plenty. Unfortunately,

found to be good for food. Perhaps there would not be a single living gigantic tortoise under the British flag but for the interposition of Mr. Walter Rothschild, who saved the great tortoises of the Aldabras by leasing the northern island from the British Government.

The green or edible turtle, the foundation of turtle soup, is an inhabitant of all the three great oceans within the tropics, but it has to resort to land for breeding purposes, and its most famous places of frequentation are the island of Ascension in the eastern Equatorial Atlantic, the Cayman Islands in the Caribbean Sea, and the coasts of Nicaragua and Honduras. The two first-named are British possessions, and turtle-catching is a very important industry at Ascension and in the British West Indies. Turtles are also captured for commerce off the coasts of British New Guinea, Borneo, and Ceylon; not only the edible form, but the still more important turtle, the hawksbill, which produces "tortoiseshell." The tortoiseshell of commerce from the hawksbill turtle is also exported from the West Indies.

Among the Crocodiles the British Empire possesses examples of every genus except the alligator. Owing to the attention which was given to the alligator by the Spanish discoverers of Florida and Texas, and still more by the British settlers in what is now the United States, this reptile has received a prominence of name and notice wholly disproportionate to its limited geographical distribution. It is still difficult to restrain British sportsmen, settlers, and even officials in Africa or the West Indies from writing about "alligators." There is no alligator in any West Indian island, there are only crocodiles in the West Indies; there is no alligator in India or Burma, but there is a true alligator in the rivers of Eastern China, though not in Hong Kong. The range of the true alligator is limited to the eastern and southern parts of the United States. It scarcely penetrates even into Mexico. Elsewhere in the world it is only found in the eastern part of China. It is somewhat nearly allied to another genus of crocodiles, the caiman. This is a far more remarkable creature than the alligator, grows to a much greater size, and is very ferocious.

Of crocodiles, the large, common, and fierce crocodile of all Tropical Africa extending from the watershed of the Zambezi on the south to Egypt on the north, and the Senegal river on the west, as well as to Madagascar, is Crocodilus niloticus. This, and the great marsh crocodile of India, Ceylon, Burma, and Malaysia, the still larger Crocodilus porosus of Eastern India, Ceylon, and thence across Malaysia to Southern China, Northern Australia, the Solomon Islands and Fiji, are, with the caimans of South America, the only members of this order of reptiles that are really dangerous to man or to large animals. But the most remarkable living representatives of this order are the gavials, or garials, of India, Borneo, and the Malay peninsula. These are far removed from the true crocodiles, though belonging to the same order. They belong to a family that was once of world-wide distribution, having in the Secondaries and early Tertiaries lived in Europe and North America. One genus (Rhamphosuchus) of this family, which inhabited India almost down to the Age of Man, reached a length of fifty feet!

Among the great group of Lizards and Snakes common to the British Empire are the geckoes, or house-haunting lizards, with adhesive feet, of Malta, and of East, West, and South Africa, Southern Asia, Australia, and New Zealand, British Guiana, Honduras, and the West Indies. In West Africa there are gaudy red and blue agama lizards that are noteworthy features in all the native towns, where, owing to their pursuit of flies and other insects, they are unmolested. Among other remarkable agamas under the British flag is the flying dragon of Southern India, the Malay peninsula, and Borneo, which has an extensive ribbed membrane that enables it to take parachute flights. In Ceylon there is an extraordinary agama which imitates chameleons in the most misleading way, so that anyone not learned in biology would mistake it at once for a chameleon, a reptile of quite a different suborder.

The extraordinary iguana lizards, many of which develop fantastic shapes and ornaments like the heraldic basilisk, are citizens of the British Empire, but have a very remarkable and puzzling geographical distribution. They are found in Madagascar, but nowhere in Africa. They are found throughout the West Indies and Tropical America, and also in the Fiji Islands of the Pacific. In the West Indies they are noteworthy for possessing delicious flesh, and are much sought after by the negroes as an article of food. In fact, runaway negroes in the old slavery days largely sustained themselves on iguanas. These lizards sometimes grow to a considerable size, and can bite fiercely and lash with their tails. They are almost exclusively tree-haunting—though there are one marine and one ground genus—but take readily to the water when pursued. For the purposes of self-protection they frequent boughs and trunks of trees that overhang rivers, and on the approach of danger fling themselves into the water. They are thus a source not only of embarrassment, but even of harm to travellers passing in canoes or boats, for if a huge iguana lizard flops on to you from a height of twenty or thirty feet, weighing perhaps fifty to seventy pounds, it might break your neck.

THE GECKO, OR HOUSE-HAUNTING LIZARD
A reptile that can run along a ceiling, back downwards

The monitors, if not the iguanas, are the biggest members of the lizard order, attaining the size of small crocodiles. They are found in India and Malaysia, and in all parts of Tropical Africa right down to Natal, and are persistently misnamed " iguanas " by unlearned Britons in Africa and India. They are celebrated for their long, whip-like tail, with which at close quarters terrible injuries can be inflicted, as well as for their sharp teeth, and their ravages on the poultry-yard, since they are fond of both eggs and fowls.

As regards the great non-poisonous boas and pythons, which kill their prey by crush-ing, the python genus, members of which possibly grow to 30 ft. in length, if not more, is found in British Africa, India, Malaysia, Ceylon, and Australia. Needless to say, with the prevailing pas-sion for misnaming everything, pythons are usually called " boa-constrictors " in the British dominions. The true boa-con-strictor inhabits British Guiana and the rest of Tropical South America, but the largest speci-men known only reaches 11 ft., and most of the stories told of its capacity for swallowing large animals really apply to the anaconda, of quite a different genus. This huge snake, which is also found in British Guiana and elsewhere in Tropical South America, inhabits the water chiefly, and undoubtedly attains an enormous size, perhaps as much as 33 ft. in length. The anaconda, like the python, is no doubt able to swallow animals as large as a small deer, and may thus have killed and eaten human

THE BLACK IGUANA
A lizard of fantastic aspect that haunts trees

beings, though there is no definite record of such a calamity having taken place. The tendency of the python in Africa seems rather to be to " chum " with humanity, and the pythons in the vicinity of native habitations become half tame, and are tolerated for their rat-killing propensities.

With regard to poisonous snakes, these arise from two stems, the colubrine and the viperine, and the last-mentioned divides into two distinct branches, the vipers and the " rattlesnakes." The poisonous colubrines belong to the elapine or cobra group, which is widely distributed over Tropical Africa, Asia, Malaysia, and Australia, with one genus (Elaps) in South America and the Lesser Antilles. The British Empire is the unwilling host of many deadly elapine snakes on its land surface, without counting the poisonous sea snakes which frequent the coasts, breeding among the rocks, of the Indian Ocean from the Cape of Good Hope to the Persian Gulf, India, Ceylon, Malaysia, Hong Kong, Australia, New Zealand, and the Pacific coast of Central America.

In South Africa there is an allied genus of coral snakes (Homorelaps) to that of South America. There is also the genus Naja of true cobras in all Tropical Africa, with at least two species,

and there are several species of tree cobras (Den-draspis) and the South African " ringhals," or African hooded cobra (Sepedon). In India and Malaysia the poisonous elapine snakes are the common cobra, the giant cobra, or hamadryad, the justly dreaded not very large kraits, the resplendent cobras, and the long-glanded cobras.

The true vipers abound chiefly in Africa. Egypt and British Arabia possess the horned vipers and the saw vipers. In British Tropical and South Africa there are the prehensile-tailed tree vipers (Atheris), the egg-laying vipers—the others are viviparous—of the genus Atractaspis, the wide-spread Causus vipers, and the sluggish, short-tailed puff adders.

There are no vipers in America, but their place is amply taken by the rattle-snakes. Yet the rattlesnakes, though so abundant in North America, seem to have origi-nated in Asia. The one genus of aquatic pitted-faced vipers ranges with several species from the shores of the Caspian Sea to the Eastern United States, and is represented in the fauna of India, Ceylon, and Malaysia. The exceedingly venomous genus Lachesis extends its range with numerous species from Eastern India and Ceylon across Malaysia and Southern China, including Hong Kong, to Mexico, Central and South America, and to the Lesser Antilles. Especially noteworthy is the fer-de-lance, found in Guadeloupe, Martinique, St. Lucia, and Trinidad, which, together with the coral snake, is the only example of a venomous snake in the West Indies, and that merely in the Lesser Antilles. There are no poisonous snakes in the rest of the West Indian Islands, large or small. In British Guiana and Trinidad there is the terrible " bushmaster," which strikes, unprovoked, at any passer-by who disturbs its slumbers. Its bite is rapidly and certainly fatal.

The British Empire might well be proud of its range of bird life, were the nation not far more eager to destroy than to preserve. There is the ostrich in the Egyptian Sudan, East Africa, and Northern Nigeria, also in South Africa. Curiously enough, this bird is absent from all intervening British Central Africa and Congoland.

In New Guinea and North-Eastern Australia there are four species of cassowary, and the con-tinent of Australia is the sole habitat of the emeu, which in earlier times also inhabited Tasmania and the larger islands of Bass Strait, besides the large Kangaroo Island off South Australia. In fact, the special form of Kangaroo Island emeu

only became extinct in the nineteenth century. In New Zealand there are still living six species of the "wingless" apteryx; and in these two large islands, which are without any indigenous mammals but a rat and some bats, there existed down to the eighteenth century a few gigantic Moas. This was the name given by the Maori people to an order of Ratite birds whose non-flying representatives ranged in size from a monster half as big and tall again as an ostrich to a bird no larger than a turkey.

The hoatzin of British Guiana, almost restricted to the Guianas in its range, is a remarkable bird, for it seems to unite in itself structural affinities with the cuckoos and plantain-eaters, on the one hand, and with the rails and game birds, on the other. Had the colony owned any imagination when it was founded, it would have chosen either this or the Victoria regia water-lily for its emblem.

Nearly all the genera and species of the penguins are represented on the coasts or islands of the British possessions in the South Atlantic and South Pacific—the Falkland Islands, which are

THE DEADLY RATTLESNAKE OF BRITISH NORTH AMERICA
The "pit vipers" with the rattle are confined to America

almost entitled to be called "Penguin Land," South Georgia, Tristan d'Acunha, Gough Island, Cape Colony, West and South Australia, Tasmania, New Zealand, and British Antarctica.

Among the grebes and divers should be named the picturesque great-crested grebe, which nests in England, the handsome great northern diver, and, of the petrel order, the wandering albatross of the seas around the Cape, with kindred species along the coasts of Tasmania, Australia, New Zealand, and the Falklands. There was actually an albatross frequenting the eastern shores of Britain, say, a million years ago.

Within the British Empire are most of the known species of pelican, gannet—such a striking feature of Scottish cliffs and islets,—tropic bird, frigate bird, and cormorant. Handsomely-coloured species of the last, with green-black and white plumage, scarlet legs, crimson and blue-skinned cheeks, are found within the New Zealand Dominion.

In Africa is the magnificent goliath heron; in British Guiana are found the scarlet ibis and the large

satin-white, black-beaked jabiru stork; in Africa and India the huge marabou stork. In Africa there is the exceedingly handsome saddle-billed stork, whose colouring is glossy black, iridescent green, and snow-white, with a crimson-scarlet beak. The common white stork of Europe used to be a frequent visitor to England until the gunners drove it away. The same agency has extinguished the booming bittern, with its plumage of orange-brown and black, of the East Anglian marshes. The sacred ibis is still found over much of British Africa, as also are two species of flamingo. The rosy flamingo, with its plumage of light vermilion, crimson, and black, is one of the sights of the Bahama Islands in the British West Indies, which it frequents in enormous numbers (see page 55). The horned screamer, a bird in outward semblance and internally like a cross between a goose and a pheasant, comes from British Guiana and other parts of South America. Among notable "British" Anatidæ there are the musk duck of Tasmania and Australia, the eider duck of Northern Britain and Canada, the steamer duck of the Falkland Islands, the spur-winged goose of Africa, the Cereopsis goose of Australia, the black-necked swan and Coscoroba swan of the Falkland Islands, and the celebrated black swan of Australia.

Among "British" birds of prey there are the Turkey buzzards of British Honduras, Jamaica, the West Indian Islands, and Guiana. British Africa in the south and east, in the Sudan, and on the Gambia, can boast of the secretary bird. Amongst extraordinary vultures must be mentioned more especially the king vulture of Eastern India. We must not forget the splendid golden eagle of Scotland and Ireland, and the tiny Microhierax falcon of the Malay Peninsula, scarcely larger than a finch. Among the gallinaceous birds are the brush turkeys and mound-builders of Australia and New Guinea, and the handsome curassows and penelopines of British Guiana, Honduras, Trinidad, Tobago, and the Windward Islands. In Africa, among other noteworthy forms is the vulturine guinea-fowl, with a display of pure ultramarine in much of its plumage. The range of this creature is almost confined to British East Africa and Somaliland. In Asia there are the two species of peacock, mainly British in their range—the peacock pheasants, the argus pheasant (entirely British), and a magnificent array of gorgeously-coloured pheasants and pheasant-like birds in the Himalaya, Northern Burma, and the Shan States. The pheasant of the United Kingdom was probably introduced into England by the

Romans, though it is not absolutely certain that it may not have lingered from prehistoric times; for fossil pheasants are found in France, and true wild pheasants exist in Eastern Europe.

Cranes and bustards, plovers and jaçanas, though world-wide in their range (mostly), are best represented within the limits of the British Empire. The great auk was a British citizen until it was exterminated. As to pigeons, the eccentric dodo was a native of Mauritius, but was killed out before the British took possession of that island. Besides the magnificent crowned pigeons and the gorgeous fruit pigeons, is a very lovely pigeon, fast nearing extinction owing to merciless persecution by plumage-hunters —the "Mountain Witch" of Jamaica. An extraordinary pigeon of Mauritius—soon done to death, of course, because of its plumage—used to be called the "Dutch Ensign," because its

THE SACRED IBIS OF NORTH AFRICA
The moon god Thoth of the ancient Egyptians, whose temples it frequented

plumage was red, white, and blue—the colours of the Dutch flag before the French adopted the tricolour.

Among the cuckoos, besides the well-known harbinger of spring in England, are the remarkable large ground cuckoo of Borneo and the grotesque black ani of the British West Indies. As to turacos, or plantain-eaters, Britain owns the greater number of the species in British Africa, some of them quite peculiar to British regions. Among these is a very handsome plantain-eater discovered by myself in

THE WHITE STORK
These birds were formerly summer residents in England

the westernmost parts of British Uganda, with a large beak exactly the colours of mother-of-pearl, and plumage in which ultramarine blue, emerald green, purple, and rose-colour mingle.

What may not be said about British parrots! The good old grey parrot of Uganda, Southern Nigeria, and the Gold Coast! In true West Africa the light grey parrot with the red tail, which becomes the best talker of all his order, is replaced by a much uglier bird with a purplish-brown tail and grey-brown plumage. British Guiana has splendid macaws, the blue-and-yellow species presenting one of the finest examples of brilliantly contrasted colouring in Tropical South America (see page 53). Several of the smaller British West Indian Islands have parrots of great beauty quite peculiar to themselves. But parrots are the special glory of Australia. Australia's emblem should be the cockatoo. And in this island-continent there are parakeets of unbelievable loveliness and delicacy of plumage. They are not radiant in scarlet, crimson, orange, emerald, blue, and purple, but adopt the exquisite pale tints of æsthetic muslins—pale pink, pale lemon, dove-grey, pale azure, rose, lavender, sea-green, and so forth.

With regard to kingfisher-like birds, I must not fail to enumerate the tody of Jamaica—an exquisite mite, with a long, flat yellow beak, a crimson-scarlet gorget, a white stomach, and an emerald-green body. There is the beautiful little kingfisher of the British Isles, against which a dead set is made by country boys and men because its feathers are useful for making artificial flies. Nor should one overlook the laughing jackass kingfisher of Australia, or the long-tailed kingfishers of New Guinea.

Some of the most extravagant of the hornbills come from India and Malaysia, as well as

West Africa. The great snowy owl is found in Canada, and many species of the eagle owl in parts of the Old and New Worlds under British control. Humming-birds are met with in Southern Canada—especially in British Columbia—and also as far north as Newfoundland, but only during the summer months. There are special and beautiful species in most of the British West Indian Islands and British Guiana. In this last colony toucans are found, and an allied group of birds, the barbets, is well represented in British Africa, one of them, the crimson-breasted barbet, being of gorgeous colouring, and having a large toucan-like bill.

It is impossible to catalogue the passerine birds of the British Empire. This immense order includes thrushes, nightingales, robins, and other warblers, finches, larks, wagtails, titmice, shrikes, starlings, crows, ravens, jays, and paradise birds. These last are peculiar in their distribution to New Guinea and the adjacent islands, and consequently the most beautiful and most wonderful among them are British subjects. So also are the allied bower birds of Australia, the colour arrangements of which are amazing in their subtlety, just as their habits of making ornamental walks or bowers, or museums of shells and seeds, are almost human. One of the most remarkable genera of the passerines is also, perhaps, the largest in size—that of the lyre birds of Eastern Australia. This type is, in a small way, one of the world's wonders, and it is a shame that the three species of this bird, peculiar to the eastern half of Australia, receive no stringent protection from the great Commonwealth.

As to the Beasts of the British Empire, its wealth in this respect is enough to invite an attack from a jealous coalition of other Powers not so favoured. It owns exclusively the most primitive of the egg-laying mammals—the duckbill—in Australia and Tasmania (see page 54). The ordinary five-toed echidna, in several varieties or subspecies, inhabits Tasmania (where it is largest), Australia, and British New Guinea. The Empire possesses nearly all the marsupials of the world in Australia and New Guinea, and some of the American opossums in British Guiana and British Honduras. In certain respects the most remarkable marsupial is the Tasmanian devil—a large, clumsy, and exceedingly fierce dasyure peculiar to that Garden Island of English loveliness.

As regards the edentates, the Empire has all the three ant-eaters in British Guiana, one sloth, and two kinds of armadillos—the giant and the broad-banded. In British India, Ceylon, and Malaysia there are two species of scaly pangolin, or manis, and in Africa four species of this enigmatic animal, which is only called an edentate because it has no teeth, and not because it has any affinities with the American sloths and armadillos. In British Africa also, more than in other states of that continent, there is the equally mysterious, unrelated aard-vark, which likewise lives on ants and termites. Another "British" mammal of great antiquity in type and geological age is the hyrax, or rock-rabbit—the coney of the Bible—which in several species is found in British Africa from Egypt to Cape Colony, and the Gambia to Somaliland.

A COLONY OF AMERICAN FLAMINGOES NESTING IN THE BAHAMA ISLANDS, OFF THE COAST OF FLORIDA
From a reconstructed group in the Natural History Museum, New York

As regards the elephants, the Empire almost monopolises the living Asiatic species, which only exists in a wild state now in Ceylon, Eastern and Southern India, Burma, the Malay Peninsula, Northern Siam, and perhaps Sumatra. The African elephant occurs in British East Africa and Uganda, British Somaliland, Northern Nigeria and Eastern Sierra Leone, British Central Africa and Northern Zambezia.

In British East Africa, and perhaps in British Somaliland, there is Grévy's zebra, the most splendid of all the zebras, besides other handsome varieties. Though there is no wild horse under the British flag, there is the parent of the domestic donkey in the Nubian wild ass of the northern and eastern parts of the Egyptian Sudan.

The Empire has a veritable "corner" in rhinoceroses, as it possesses the remaining specimens of the one-horned or Indian rhinoceros, which continues to exist on the south-eastern slopes of the Himalaya and in Northern Assam. In Eastern Bengal, Assam, Lower Burma, the Malay Peninsula, Borneo, and Java, it is succeeded by the smaller-sized one-horned rhinoceros. · The two-horned Asiatic rhinoceros ranges from Bhutan and Assam down the Malay Peninsula to Sumatra and to Borneo, but not to Java. In Africa there are still living a few white rhinoceroses with the square, unprehensile lip, in Zululand —perhaps literally two or three,—in the northern part of Mashonaland, and in the northernmost parts of the Uganda Protectorate ; and possibly, in addition, a few may yet be found in the southern regions of the Egyptian Sudan. The common two-horned African rhinoceros with the prehensile lip occurs sparsely in British South Africa and the eastern parts of British Central Africa. It is fairly abundant in British East Africa and the northern parts of the Uganda Protectorate. It is also said to exist in the western parts of the Egyptian Sudan, in the countries round Lake Chad, and perhaps elsewhere in Northern Nigeria.

The large hippopotamus, which was once an inhabitant of Britain, and perhaps also of India, is now very rare in the Lower Nile ; has been, indeed, exterminated north of Khartum. But it is very abundant in the Upper Nile and its tributaries, and throughout British East Africa, British Central Africa, the Zambezi and all its affluents, the Niger and Benue rivers, Lake Chad and the Gambia river, together with, perhaps, the western rivers of Sierra Leone. In the extreme east of the Sierra Leone

THE EXTINCT DODO OF MAURITIUS
A large pigeon-like bird that has been exterminated in modern times

Protectorate the vast creature is replaced by the pigmy hippopotamus of Liberia.

As to the wild swine, the wild boar has long been extinct in Britain, but one, at least, of the domestic breeds of pig in Ireland seems to be descended from a wild species. Nevertheless, by far the greater number of modern British pigs are descended really from the Asiatic wild boar.

The handsome red river hog of British West Africa and Nigeria extends its range eastwards till it enters the Uganda Protectorate. In British East Africa and British Central Africa are at least two other species of bush pigs, with bony excrescences above the tusks, and a fourth is found in the eastern parts of British South Africa. Again—only recently discovered—in the few patches of dense forests of British East Africa, and perhaps in the westernmost parts of the Uganda Protectorate, there is the remarkable black forest pig (Hylochoerus). The wart hogs are an extreme development of this type, with enormous tusks, warty excrescences, and huge, broad heads. There are two species at least, if not a third, under the British flag—one in the Egyptian Sudan, another in British East, Central, and South Africa, and perhaps a third in West Africa.

In British West African possessions, from the Gambia to the western frontiers of Uganda, occurs the remarkable water chevrotain (see page 83), which descends almost unaltered from Miocene times, when it inhabited France and Germany. It is very nearly a connecting link between the pig group and that of the ruminants (deer and oxen). Among the antlered deer, the Empire claims the stag, or red deer, yet wild in Scotland, England, and Ireland ; its magnificent relative, the wapiti, in Canada ; and transitional forms between the red deer and the wapiti in Kashmir and North-Western India. In Canada and Newfoundland, also, the elk, or moose, still exists ; and in Eastern and Northern Canada is to be found the handsome big subspecies of reindeer generally known as the caribou. India, Malaysia, and Ceylon possess the large, unspotted sambar deer in various forms, and a large type of Siamese deer, with great development of antlers ; the spotted hog deer, and the singularly beautiful axis, in which the white spots and stripes are carried to a development not known in any other cervine. Here also are found the muntjacs, a primitive type of deer, with large canine tusks and with short antlers starting from very long, bony pedicels.

As to the Giraffe family, the okapi was discovered by the writer within literally a few miles of the western frontier of Uganda, in the adjoining Congo Forest, and in all probability this rare animal is actually a British citizen, as it is reported to exist in the forested regions of the Uganda Protectorate to the west and east of the river Semliki. In British Somaliland is found the remarkable Somali giraffe, a wine-red-coloured giraffe marked with a network of narrow white stripes.

Over the rest of British Africa occur all the sub-species of the other giraffe type, Giraffa camelopardalis, in the hinterland of the Gambia, in the Northern Gold Coast, Northern Nigeria, Egyptian Sudan, British East Africa—where we have the five-horned giraffe, discovered by the present writer,—in the central parts of British Central Africa, and in the central parts of South Africa and the Northern Transvaal.

With regard to the Oxen, the wild cattle of English and Scottish parks are descended from the Roman domestic cattle introduced into the British Islands, though we speak of them as wild white cattle. In Canada are still to be seen several herds, amounting to two or three thousand, of the American bison. In India are found the bison-like yak in the far north, and in the plains and forests of Tropical India and Malaysia the buffalo, or "arni," the magnificent gaur, and the smaller gayal. In British Africa there is the black buffalo with the bossed horns, which in slightly different types ranges from Lake Chad down to Cape Colony, as well as the red buffalo of the Gold Coast and Southern Nigeria and the extreme westernmost parts of Uganda.

Amongst the Capricorns there may perhaps remain a few examples of that aberrant type, the musk-ox—which lived at Bath and Twickenham some sixty thousand years ago,—in Labrador and the north-easternmost parts of Canada along the shores of the Arctic Sea. In the north-easternmost parts of India, on the verge of China and Tibet, occurs the equally peculiar and remarkable takin, and in the Malay Peninsula and the Himalaya there are kindred serows, the southern forms of which are in the Malay Peninsula quite large animals of rather striking colouring—black, bay, and white.

India is celebrated for its magnificent wild sheep, one of which—the Pamir variety—is the biggest wild sheep in the world. The western parts of Egypt contain the African wild sheep, or audad. Cyprus has a wild sheep more akin to the Armenian mouflon ; Western Canada, the Rocky Mountain sheep. North-Western India possesses an ibex and the big markhur goat with its spiral horns. South-Eastern Egypt and the Aden Protectorate still shelter two forms of the handsome ibex.

The British wealth in Antelopes is surpassing. In India alone are found the nilgai and the four-horned antelope. Most of the great and nearly all the lesser kudus are inhabitants of South, Central, and North-East Africa. In the Anglo-Egyptian Sudan is the giant eland, and the handsome Derbian eland is obtained from the hinterland of the Gambia and Sierra Leone, and perhaps exists in Northern Nigeria. And so might be enumerated the handsome bongo and other bushbucks, gnus and hartebeests, oryxes, gazelles, giraffe-gazelles, springbucks, and the Indian blackbuck. The Royal antelope of Sierra Leone and the Gold Coast is the tiniest of horned ruminants. The zebra duiker of Eastern Sierra Leone is among the rarest of small antelopes.

British rodents include the large, handsomely-coloured squirrels of India, Burma, the Malay Peninsula, and Borneo ; the porcupines of Canada, West Africa, and Malaysia ; a hare-like rabbit of Assam ; the Arctic hare in Scotland and Canada ; the large red Siberian hare in England. Egypt has its jerboas, while other jumping rodents are met with in East and South Africa ; and of rats, mice, and voles endless is the story, though be it noted that in the harvest mouse England has the tiniest of all rodents.

Whales and porpoises are necessarily largely international in their range, but a dwarf genus of the whalebone whale occurs in the sea between Australia and New Zealand ; the Ganges and Indus have a dolphin and the Irrawadi a grampus.

Among peculiar Seals and Sea Lions, several species frequent the British shores; the southern sea lion inhabits

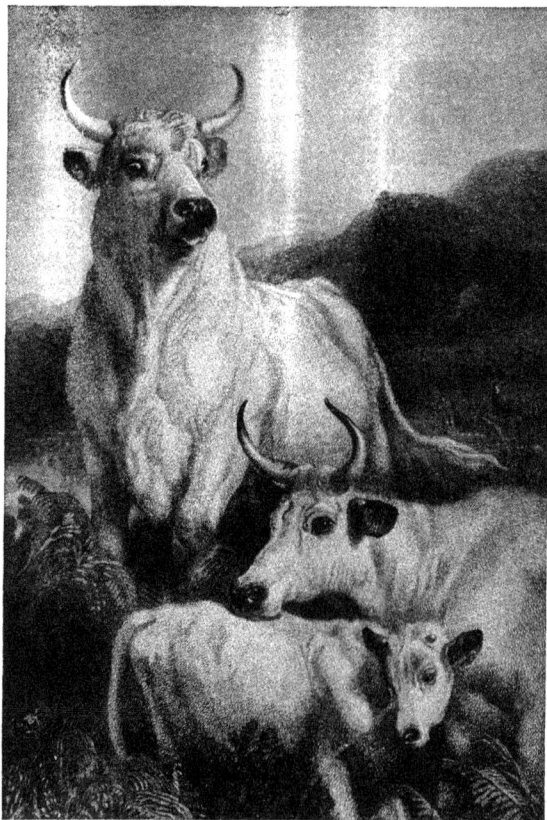

THE WILD CATTLE OF CHILLINGHAM, NORTHUMBERLAND
From the painting by Sir Edwin Landseer, R.A.

SIR EDWIN LANDSEER'S FAMOUS PICTURE OF "THE STAG AT BAY"

the Falklands, Hooker's sea lion the southern shores of New Zealand, and the hair seal the southern coast of Australia; while a fur seal haunts the Falklands and South Shetlands, another the coasts of Cape Colony, and a third those of New Zealand. Some peculiar and large seals are found along the Antarctic shores, such as the crab-eating seal and Ross's seal. Weddell's seal and the large, fierce leopard seals, though mainly Antarctic, range northwards to the Falkland Islands and the extremities of Patagonia and New Zealand. The treatment of the wonderful sea elephant is, from a faunistic point of view, a grave scandal in the history of the British Empire. This amazing seal—the size and bulk of an elephant—at one time resorted to the Falkland Islands, and especially to South Georgia and the South Shetlands. For the sake of their oil we have allowed them to be massacred and massacred till now only very few are left.

British land Carnivores include the lion; the tiger; the leopard; the jaguar; the clouded tiger of Malaysia; the lynxes in India, East and South Africa, Egypt, and Canada; the puma and ocelot in British Honduras, and perhaps Guiana; most of the types of wild cat; and the hunting chita of India, East and South Africa.

Among noteworthy Canines we have the long-eared Otocyon in Somaliland and South Africa; and a very remarkable wild dog in British Guiana (Speothos), which is in appearance much less a dog than (superficially) a weasel, with its short, rounded ears and short legs. In South and East Africa there is the handsomely-marked black-backed jackal, orange, grey, white, and black in colouring. India and Malaysia abound in wild dogs; and Canada, besides possessing wolves of more or less European type, shares with Alaska the wild dog, or wolf, of large size and chocolate-black-coloured hair, the so-called Alaska wolf.

Canada has sables, martens, and most of the fur-bearing weasels. British Africa has handsome black and white weasels (*Pœcilogale*), and India and Africa the large black and grey ratel. The glutton is found in Canada; there are British as well as Indian and Canadian badgers; and skunks are found in Canada as well as in British Honduras. Among the civets there are the extraordinary black binturong of the Malay Peninsula, the large African civet, which supplies the perfume still valued in commerce, and the Indian and Malayan civets, besides many species of genet and allied creatures in South Africa, East Africa, West Africa, India, and the Malay Peninsula. Within British bounds occur all the species of hyæna of which the brown hyæna is monopolised, probably, by the eastern parts of South Africa.

Regarding the Bears, the north-westernmost parts of Canada may perhaps share with Alaska the splendid chocolate-coloured Alaska bear. British Columbia, at any rate, has the grizzly bear and the black bear; and the white Polar bear is found in the northern and north-eastern parts of the Canadian Dominion. In India we have the Syrian bear, the sloth bear, the black bear, and in Assam and Malaysia the small honey-eating sun bear.

The British Empire possesses within its political limits nearly all the genera and a large proportion of

THE MYSTERIOUS AARDVARK OF BRITISH AFRICA
This ungainly earth-pig, of nocturnal habits, lives on ants and termites

the species of Bats. Specially noteworthy are the long-tailed fruit-eating bats of New Guinea and Fiji; the appallingly hideous fruit-eating bat of British West Africa (Hypsignathus); and among insect-eating bats, the horseshoe genus of England and elsewhere, the Anthops ornatus of the Solomon Islands, with an enormous nose-leaf, the brilliantly-coloured orange-brown-cream Kerivoula picta of Ceylon, the still more beautiful orange-white-sepia harlequin bat of the Himalaya, the white-winged, chestnut-brown-bodied Scotophilus albofuscus of the Gambia, the creeping, climbing, narrow-winged Mystacops tuberculatus bat of New Zealand, and lastly the true vampire blood-sucking bat (Desmodus), which is found in British Guiana as well as in other parts of South America.

The order of Apes and Monkeys is well represented in British dominions. In Southern Nigeria are two peculiar forms of lemur; in parts of British Africa are several species of galago; and in Ceylon the loris, and in Malaysia the altogether remarkable and peculiar tarsier. In Guiana and Trinidad some of the American marmoset and monkey types are represented. In British Africa we have an extraordinary array of Old World monkeys — all the species of colobus, nearly all the cercopithecines, and at least one peculiar species of mangabey. On the Rock of Gibraltar exist the only remaining monkeys of Europe, identical, however, with the macaque of the Morocco coast. In India and Malaysia the genus Semnopithecus and its allied type, the proboscis monkey,

are well represented. The British baboons include all the known species of that group, not excepting the mandrill and drill of West Africa, and the extraordinary gelada of Northern Somaliland. The great apes are represented by the gorilla of the south-westernmost portion of the Uganda Protectorate, by several forms of chimpanzee in Uganda, the Egyptian Sudan, Southern Nigeria, the Gold Coast, and Sierra Leone, by the orang-utan of British Borneo, and by the gibbons of Assam and the Malay Peninsula.

Lastly, there is Man. Within the limits of the Empire nearly all human types are, or have been, represented—the Australoid in Australia and Tasmania, and perhaps in Ceylon and Southern India; the Negro and Negroid in the Andaman Islands, the Malay Peninsula, the Solomon Islands, New Guinea, Southern India, and all British Africa. Almost peculiar to British Africa is the Bushman variety of the Negro. Congo Pigmies are found on the western frontiers of Uganda. The Caucasian is represented in the British Isles, Malta, Cyprus, Egypt, India, and Canada; the Mongol in India, Malaysia, Hong Kong, and Borneo; the Eskimo and the Amerindian in Canada, Honduras, Guiana, and the British West Indies; and the Melanesian and Polynesian hybrids in New Guinea, New Zealand, Fiji, and Tonga.   H. H. JOHNSTON

THE ARGALI, A FINE MOUNTAIN SHEEP, OF THE HIMALAYA

ANIMAL LIFE AT THE EQUATOR
STRANGE DENIZENS OF THE TROPICAL REGIONS
By Sir HARRY JOHNSTON.

At the present day the Equatorial regions of the earth's land surface are undoubtedly the most densely populated by all forms of life, excepting man; and the deficiency of man, as compared with the countries farther north, is due in the main to the fact that the land surface of the Equatorial zone—and in this argument " Equator " is equivalent to a belt round the middle of the earth 1,400 miles broad—is far less in area than the vast regions lying beyond this central girdle of heat and moisture.

Yet the fact cannot be evaded that civilised man of the Caucasian subspecies has not taken kindly as yet to the colonisation of Equatoria; although this portion of the earth's land surface, even if it be much smaller in extent than the north-temperate or subtropical regions, is proportionately far more productive, being everywhere well watered, endowed mostly with a rich, not to say rank, soil, and possessing enormous forests, many of them replete with vegetable food-stuffs and a remarkable assortment of wild beasts, birds, reptiles, fishes, and crustaceans suited for human food. It is also highly mineralised; in short, it is the richest region of the earth.

In all probability man originated in the tropics of the Old World, perhaps in India or Further India, not far from the Equatorial region, or in the great Malay Islands. If not in Southern Asia, then the only alternative suggestion, so far without evidence, seems to point to Western Asia or Eastern Europe as the cradle of the human race.

But as soon as Man arrived at full consciousness of his own powers, though he migrated in all directions—south, east, west, and north—he throve most, advanced most towards the mastery of the beasts of the field and of the earth itself in Temperate Europe and Asia, in North and South America. We are entitled to surmise that very early in his human career he shunned the dense forests that combined heat and moisture, and preferred a life on the more open plains and plateaus of cooler regions wherein he could develop his powers as a hunter to the greatest advantage.

Still, though at the Equator man is at his weakest—for in the Antarctic and Arctic, though few in numbers, he is still the master—will it be always thus? Probably not. His only real enemies in this richest

THE WATER CHEVROTAIN
A remarkable primitive ruminant in Equatorial Africa. From a photograph of a living specimen obtained by Sir Harry Johnston

part of the earth are intestinal worms; and those micro-organisms, vegetable or animal, which are introduced into his blood from the already infested human inhabitants indigenous to the Equatorial belt. One means of introduction is now well known to be certain forms of insect or tick. Cure by prophylactic the indigenous peoples of the Tropics; expel from their systems these parasitic worms, bacteria and protozoa; destroy the transmitting agencies—insanitary habits, mosquitoes, flies, fleas, ticks—and you will rid the Equatorial belt of three-fourths of its unhealthiness for the Caucasian and his cattle.

But here, again, the actual configuration of the land surface in this belt of 1,400 miles broad to a considerable extent neutralises the evils attendant on perpetual summer, unremitting rain, never-varying heat. At least a fifth part of Equatorial Africa is at such an altitude above sea-level that either its rainfall is lessened and its sunshine cooled, or, if still moist, it is at any rate cool enough at night-time to admit of sound, refreshing sleep, thus enabling the body of the white man to remain in full vigour. Still more is this the case in Equatorial South America, which is becoming emphatically a white man's country. The lofty snow-topped mountains of New Guinea will certainly some day be the home of a powerful white nation, and white men are even now able to maintain health and vigour in Java, Sumatra, and the Malay Peninsula, in Ceylon and the Nilghiri region of Southern India. Even North Borneo gives us Mount Kina Balu for a sanatorium.

But, it may be objected, why speak of the white man only? May not a high civilisation arise and flourish and be perpetual which is directed by black men and yellow men free from white intermixture? Quite possibly; almost certainly so in Equatorial Africa; but in the other regions of the Equatorial belt in America and Asia the white man has got the start, the indigenous inhabitants of these regions, except in India and Ceylon, being neither sufficiently numerous nor sufficiently powerful to compete with him on their until recently uninhabited mountains. Nor are India and Ceylon quite exceptional, inasmuch as both regions were conquered and permeated racially thousands of years ago by our Aryan congeners from Temperate Asia.

There are many indications in the geological history of life on the earth's surface, and in the present and past distribution of living forms, to suggest that there were two principal areas of fruitful evolution in plants and animals : one—greatest of all—in the Eurasian-American regions grouped round the Arctic Circle, and the other—South America-Australia-South Africa—radiating from Antarctica. We know for a fact that the present vegetation of Florida and of Pliocene Europe existed before the Pleistocene cold as far north as Northern Greenland, within a few hundred miles of the Pole. We also know that vegetation must have been sufficiently abundant at the South Pole for its remains to have become compressed into thick strata of coal. It is easy to suppose that before the Pleistocene or the beginning of the Quaternary epoch there may have been no ice cap at either Pole, but temperatures approximating to those of modern Italy or France. Yet it is difficult to understand how forests of magnolia, cypress, and oak could have endured the six months' darkness of the Arctic Circle in Northern Greenland, or the forests of Antarctica—which made the coal discovered by Sir Ernest Shackleton—have similarly existed despite the absence of the sun for half the year. However that may be, there were evidently once the two great centres of life-swarming, namely, the Northern land area of Eurasia-North America, and the Southern of Antarctica-Patagonia-Australia, conjoined with South-West Africa.

During the second half of the lengthy Secondary epoch—the age in which reptiles were reaching their culmination, but mammals and birds were both increasing and diversifying—the distribution of the earth's land surface seems to have been very different, and this difference to have continued down to early Tertiary times. We are entitled to presume that in these ages of the reptile supremacy and the expansion of the mammalian class there was a good deal more land exposed for the habitation of animals in the southern half of the globe—a continental mass of fanciful outline which united Africa south of the Sahara Desert, the Sudan, and Somaliland with South America mainly by way of a broad isthmus across the Equatorial Atlantic, and again by a great continental expansion in the South Atlantic, of which St. Helena, Ascension, and Tristan d'Acunha are the last vestiges.

At the beginning of the Tertiary epoch Brazil-Guiana was a great peninsula, almost an island, but precariously united—sometimes by isthmuses, sometimes by chains of islands—with Andean South

America, with West Africa, and, to a certain extent, with the Antilles. Much of Central America lay under a broad stretch of sea which separated the Northern from the Southern Continent. These Cretaceous or Eocene waters flowed over a good deal of what is now North America, and parts of the West Indian Islands. And probably from the middle of the Secondary epoch to the Eocene Tertiaries, South America was united, not only with its African neighbour, but also with the Antarctic Continent by way of the South Shetland Islands. Antarctica again extended northwards till it connected with Tasmania and Australia and came very near to New Zealand.

In contrast with this, much of Northern and Central Australia, Malaysia, and India was under water ; the lime-depositing seas of the Secondaries flowed over a good deal of Somaliland, the Nile Valley, the Libyan and Sahara Deserts, Eastern and Southern Europe, the valleys of the Amazon, Orinoco, and Paraná ; so that there was not nearly so much of the Equatorial belt in the occupation of plant and animal life as there is at the present day.

Then as great changes in the earth's surface took place during the Tertiary and Quaternary epochs—low continents subsiding under shallow seas, land emerging above the surface of the ocean, the Alps, the Himalaya and Central Asian mountains rising into great altitudes, the Arctic Ocean retreating from Siberia to the Polar zone ; the Antarctic and South Atlantic waters invaded those lands which once united Africa with South America, and South America with Antarctica and Australia. Meanwhile Africa became the huge, continuous, torrid continent we now behold.

By degrees, as land masses grew greater and more continuous round the Equatorial belt, water and cold prevailed in the far North and the far South. The Glacial periods began which have not yet finished, for the ice still locks up and renders useless an immense area of over five million square miles of land in Antarctica, Greenland, Northern Canada, Northernmost Europe, and Northern Asia.

The advancing frigidity drove the majority of living things towards the Equatorial belt. It was here, most of all, that plant life flourished, and the excessive development of forests in Equatorial America, Africa, and Asia served as a shield to protect ancient types which dropped out from the struggle for existence in the hustle, bustle, and competition prevailing over the open country of the temperate and subtropical regions. The forests of Equatoria were, of all conditions, most hampering to

THE ZEBRA ANTELOPE OF LIBERIA
A rare denizen of Equatorial West Africa

NEWLY-CAUGHT ZEBRAS IN A CORRAL
A scene from Eastern Equatorial Africa

the movements of the great Secondary Age reptiles, to the terrific carnivores of the Tertiaries, and to Man of the perfect human type. No doubt a forest life had been the one condition which enabled the pristine birds and the early mammals to survive, increase, multiply, and prosper, in spite of the presence on the earth of many families of dinosaurs and other monstrous reptiles. The bird itself may have been an early branch of the dinosaurs which took to a life in shallow water, alternately wading and flapping its way with its fin-like upper limbs along the surface of the water. But it soon evaded the attacks of its foes—and yet attached itself more particularly to the pursuit of its favourite insect food—by taking to a tree life, during which its clumsy fin-like upper limbs developed into wings. The mammal, originally, no doubt, a smallish creature about the size of a fox or dog, also took to the forest in pursuit of insects and small reptiles, and further because in the forest no great pterodactyle could swoop, no monstrous dinosaur could penetrate, no crocodile ascend—and pythons and poisonous snakes had not yet arrived. The forest, again, since the Pliocene, has enabled the imperfectly developed anthropoid apes still to linger on the earth's surface out of reach of their enormously successful cousin of the genus Homo. The gorilla, chimpanzee, orang-utan, and the largest of the gibbons (the siamang) are entirely confined in their present-day distribution to the Equatorial forests of West and West Central Africa and Malaysia. The forests of Central and South America and Malaysia hold the three living species of tapir.

Tapirs once inhabited England, France, and, above all, Southern Germany, from the earliest Miocene

period through the Pliocene. But they seemingly never entered the more open lands of the Mediterranean basin—thus they never reached Africa,—but remained in the presumably dense forests of Central or Eastern Europe until the competition with more highly evolved forms became too fierce, when they either died out or migrated eastwards through Central Asia to China, Indo-China, and Malaysia.

In like manner the okapi, a humble relation of the gigantic giraffe, and probably a beast which was evolved in the early Pliocene in Greece or Asia Minor, must have fled southwards through Egypt and the Nile Valley till it found a refuge and a respite in the Equatorial forests of the Congo, amid which even British sportsmen will find it difficult to effect its extermination.

The Equatorial regions of Africa and India—and to a less marked extent Malaysia—have much in common in their fauna and flora. Besides the anthropoid apes, there are buffaloes of a rather primitive type in Western Equatorial Africa and in Easternmost Equatorial Asia. There are the only living representatives of the Tragulina, primitive ruminants almost midway in their structure between the pig and the deer, and once native to Europe and North America. The most remarkable of the living forms is the water chevrotain of Equatorial Africa. Liberia and the Equatorial portion of the Congo basin and of the Cameroons are the home of peculiar and local types of cephalophine antelopes. In the Equatorial forests of Southern Asia there are splendid wild cattle, notably such species as the gaur or seladang, and banteng. In this region also is the curious tarsier-lemur, and there are equally strange lemurs—the potto and the angwantibo—in the Equatorial forests of West Africa. There are the slow lemurs of Malaysia, and the slender lorises of Southern India and Ceylon. The Equatorial belt of West Central Africa has some extraordinary flying rodents—the anomalures, squirrel-like in appearance, and not very distantly related to the squirrel family. Equatorial Eastern Asia has the galeopithecus, another flying mammal placed in an order by itself, though it possesses distant affinities with the lemurs and insectivores. In the same region of Equatorial Asia are the tupaias—squirrel-like insectivores remarkable for their lemurine affinities and strong superficial resemblance to the squirrels. The largest and handsomest of the squirrel race are found in the same countries. West Africa also has large and beautifully-coloured squirrels, such as Stanger's squirrel, the size of a small rabbit, and with a portion of its fur of a brilliant orange, matching closely the orange oil-palm nuts on which it feeds.

Equatorial South America has its own strange and remarkable rodents, the capybara—largest of living gnawing animals which leads a water life—the agutis, and the white-spotted pacas. The Equatorial belt of South America is not so rich in mammalian forms as that of the Old World, but it has developed the peculiar vampire bats, which extend their range into Tropical America beyond the Equatorial belt; the South American monkeys and marmosets, the greater number of the genera and species of which are confined to the Equatorial belt; and a remarkable number of opossums, the principal American marsupial, of the genus Didelphys. Yet the didelphine marsupials do not seem to have been originally Equatorial in their distribution, but to have been numerous in North America, Northern Europe, and Temperate Asia. From the last-named region they probably reached North America, and thence found their way in late Tertiary times to the Southern Continent, where they have multiplied and prospered. No doubt this generalised and rather basic type of marsupial reached Malaysia, New Guinea, and Australia from Asia in the later Secondaries or early Tertiaries, and there gave rise to the extraordinary present development of marsupials of

the two very distinct suborders, the polyprotodont and the diprotodont, distinguished by the many or the few incisor teeth in the lower jaw. Both the dasyurine family of the polyprotodonts and a few of the diprotodonts apparently travelled southwards across Antarctica to South America, where their remains are found fossil in Patagonia. But one genus — Cænolestes—of the South American diprotodonts has survived extinction, and in two species inhabits Equatorial Colombia and Ecuador at the present day.

In Equatorial New Guinea and in the northernmost regions of Australia are the interesting tree kangaroos, and in the Equatorial Malaysian Islands are the handsomely marked cuscus phalangers. Equatorial Asia has its peculiar species of rhinoceros, the rhinocerous of the Sunderfunds, with one horn, and the Sumatran species, with a hairy variety, with two horns. In Equatorial Africa the rhinoceros genus is only found in the eastern and less-forested regions. Both continents possess the elephant within the Equatorial belt, but the elephant of the densely forested portions of Western Equatorial Africa—possibly Liberia, and certainly the Kamerun and North-Western Congo—is a smaller species or subspecies, with ears much reduced in size as compared with the average African elephant; while there is also a smaller type of Asiatic elephant in Easternmost Equatorial Asia.

As regards carnivores, Equatorial South America, Guiana, and Northern Brazil possess a very strange weasel-like dog—Speothos venaticus. Equatorial South America is also the home of peculiar types of the raccoon family, such as the bassaricyon, the kinkajou, and the coatimundis; also of two curious one-coloured cats, the jaguarondi and the eyra. Equatorial Africa has, like Asia, some peculiar civets; and Eastern Asia has the extraordinary - looking binturong and cynogale, the last an otter-like member of the civet family. Equatorial East Africa has the aberrant aard wolf, a dwarf ally of the hyænas, though this creature extends its range into South Africa. Equatorial Malaysia possesses one of the handsomest cats in the world, the beautiful clouded tiger. Easternmost Equatorial Asia also has some surviving marvels—the babirusa pig, with its upper tusks growing vertically through the bones of the face, and the repulsive but interesting Black ape of Celebes, a distant ally of the baboons. Borneo has a monkey with an enormous nose, that procures for it the name of proboscis monkey.

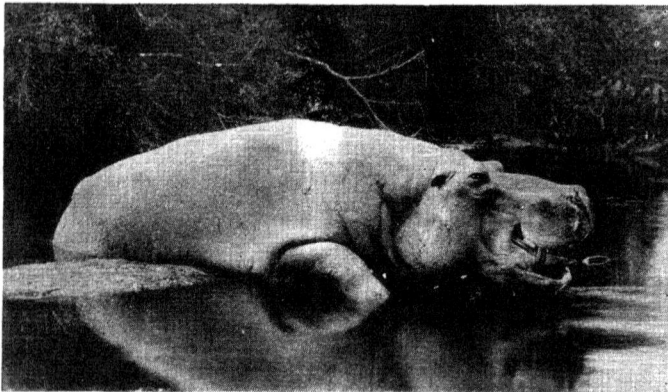

A HIPPOPOTAMUS IN A POOL IN A SOUTH CENTRAL AFRICAN RIVER

Equatorial Africa possesses notably the gorilla and the chimpanzee ; also, in its westernmost forests, the drill ; and in those of the more central regions, the mandrill—two strangely hideous baboons with huge heads and raised ridges of coloured skin on the face. There are also in these Equatorial regions many species of the Cercopithecus and Cercocebus genera ; and all the colobus monkeys are peculiar to Equatorial Africa.

Amongst Equatorial edentates there are the large and small ant-eaters, the sloths and armadillos of Tropical and Equatorial America (especially of Brazil), and the scaly pangolins of Western Equatorial Africa and South-Eastern Asia.

Amongst Equatorial antelopes should be noted the giant eland of the South-Western Egyptian Sudan, and the equally splendid derbian eland of the West African forests ; also the most beautifully coloured of all the antelopes—the bongo tragelaph of Equatorial Africa, its range extending from Sierra Leone and Liberia on the west, to the East African highlands on the east.

A GORILLA SHOT IN GERMAN WEST AFRICA
This anthropoid ape weighed 500 lb., stood 5 ft. 5 in. high, and measured 8 ft. from finger to finger

Equatorial Asia has several peculiar species of deer. Equatorial America has the red brocket deer of Guiana, and the remarkable little pudu deer of Ecuador. Among equatorial bears are the spectacled bear of the Peruvian Andes, the sun bear of Malaysia, and the sloth bear of India and Ceylon.

At least three genera of the African guinea-fowl are restricted to the Equatorial regions. In Equatorial America is the interesting family of tree-dwelling gallinaceous birds, the curassows, whose nearest allies, perhaps, are the megapodes of Equatorial Eastern Asia, New Guinea, and Northern Australia. The grey parrot is an Equatorial bird in Africa ; so are the macaws in America and the lories in Asia and Polynesia. The black cockatoos of New Guinea and Northern Australia are most remarkable bird forms. Still more famous are the birds of paradise of Equatorial New Guinea and the adjacent islands. The hoatzin of Guiana is one of the world's few remaining " missing links," a survivor from a far-distant " generalised " past. Its possible allies, the great blue plantain-eaters, violet and crimson plantain-eaters, green, purple, and crimson-winged turacos of West Africa are notable birds of the Equatorial zone ; so equally are the amazing emerald and crimson trogons of Central America, Colombia, and Peru, and the surpassingly lovely fruit pigeons and extravagantly grotesque hornbills of Equatorial Asia and Papuasia.

Allusion has already been made to the presumed land connection between Africa and South America across the Equatorial belt, and also in the region of the South Atlantic. In the form of a chain of islands this link between the two continents must have remained in the Equatorial zone down to as late a period as Miocene times, to account for the close relationships still existing between African and South American monkeys, sirenians, rodents, bird families and orders, snakes, spiders, fresh - water fishes, land crabs, land shells, etc. But the student of the Equatorial distribution of what might be called Miocene and Pliocene life-forms is frequently tempted — with very little geological or hydrographical evidence, as yet, to support him—to conjure up the existence down to the beginning of the Quaternary epoch of a variable land belt in the region of the Equator across the Pacific Ocean connecting Malaysia and Papuasia with Central and South America ; or at any rate a series of islands separated from one another and from America by narrow straits of sea not impassable for man in a primitive stage, for mammals able to swim or fly, for birds, insects, crustaceans, or land shells.

There is the case of the tapir, for example. The Asiatic tapir exists at present in Tropical or Equatorial Malaysia, its range extending southwards as far as Sumatra and Borneo. In the rest of the world, living tapirs are only found in Tropical or Equatorial America. In Central America —Mexico to Panama—there is a peculiar subgenus, Elasmognathus, which differentiates into two species, Baird's tapir and Dow's tapir. Elsewhere in South America there is the well-known South American tapir, with a mountain species or variety, Roulin's tapir, in the Equatorial Andes. But it is remarkable that the common species of Brazil is less different in some of its structural features from the Malay tapir than it is from the Central American tapirs of the elasmognathine subgenus.

We know that the Asiatic tapir extended its range into Southern China down to a few thousand years ago, and that on the American side there were tapirs, apparently of the existing genus, living as far north as California and South Carolina. Was there some vanished land connection of Pliocene or Pleistocene times which united Eastern Equatorial Asia with Western Tropical America ? Or did the tapir genus enter America by way of Kamchatka, the Aleutian Islands, and British Columbia before the beginning of the Pleistocene cold ? Of course, there is another solution : the tapir family probably originated, like the rhinoceroses, in Eocene or Miocene North America, and spread thence to Asia and Europe by the northern land connections, ages before there was any menace of glacial conditions. But the actual genus Tapirus seems to have been evolved early in Pliocene Europe, and to have emigrated thence to Asia. Its nearest known relation in North America may have been Tapiravus, which has not been found in European formations; but unless it descended directly from that form, and is a case of parallel, converging development, the true tapir must either have had Alaska for its seat of origin, and thence have migrated southwards to its present homes in Equatorial Asia and America, or have found its way from Asia to America at a later date and by a more southern land route.

Man himself offers an addition to this puzzle in the geographically far-removed affinities between certain living types of Western Equatorial Asia and America. Despite the general similarity of the American aborigines, the "Amerindians," one cannot fail to remark that, after accounting for much intermingling, they fall naturally into two types: (1) a slender, rather short-legged, flat-faced, small-chinned, small-nosed people, whose black-brown hair is not without a slight tendency towards waviness; and (2) a robust, big-nosed type, of which the North-American Indian and the more successful peoples of Western South America are good examples. This type of robust build and big nose would seem to have come from the north and north-west, and may be the result of repeated Proto-Caucasian (Ainu-like) invasions from North-Eastern Asia,— migrations of the white race in ancient and modern times. But as regards type I, who can refuse to admit its extraordinary physical resemblance to the indigenes of Borneo, to the Proto-Mongols of the Nicobar Islands, the Malay Peninsula, Indo-China and much of Polynesia ? Place photographs— especially of unclothed types—of Amerindians from the Equatorial zone of South America alongside those of the Eastern Asiatic peoples already mentioned,

THE HEAD OF THE ELATE HORNBILL
From a drawing by Sir Harry Johnston

especially the Borneans, and the physical affinities are obvious. Not only that, but the resemblances in weapons, arts, industries, and customs of the Equatorial Amerindians and the Eastern Equatorial Asiatics are equally striking. Weapons and designs from the Amazon Valley match to a nicety those from Central Borneo, and frequently reappear among the Polynesians of the Pacific archipelagoes.

In the botany of Equatorial America and South-Eastern Asia, the affinities are equally striking. So also are the relationships between the most highly developed Tertiary groups of snakes, such as the blunt-headed snakes—non-poisonous, but allied to the vipers—and also the rattlesnake sub-family. In the case of these crotaline vipers there is one genus, Lachesis, which is common to Equatorial Asia and Equatorial America. So also is the species of sea snakes, Hydrus platyurus, which ranges from the coasts of the Indian Ocean, Western Pacific and Pacific archipelagoes, to the west coast of Central America and Panama. These sea snakes, despite their aquatic life, resort to rocky coasts to breed, and otherwise frequent the neighbourhood of the land for their fish food. Therefore they are of littorine habitat, and are rarely met with at any great distance from land, though, no doubt, oceanic currents have something to do with their distribution.

The caiman-like alligator of mid-China—Yang-tse-kiang river—and the alligator and caimans of Tropical and Sub-tropical America are among the other affinities in the two faunas, and an additional connection may be traced in the Vejovidæ family of scorpions and the Thelyphonidæ family of "whip-scorpions." But the relationships of the birds characteristic of the Equatorial belt in America are, besides their recent alliance with Sub-tropical North America, divided pretty equally between Malaysia, Papuasia, Polynesia, on the one hand, and West Africa on the other.

It is not easy to account for some Equatorial affinities, among forms obviously of late Tertiary origin, by obliging their ancestors—in theory—to journey for thousands of miles to the far north in order to find land communication across the Atlantic or the Pacific into the New World, of which they now only inhabit the Equatorial zone. Neither the Aleutian Isthmus, which connected Kamchatka with Alaska, nor the union of North-West Europe with Greenland by way of Spitsbergen, will suffice altogether to explain the intimate relationships in certain features between the Equatorial fauna and flora of America and those of West Africa, and also, in a lesser degree, of Polynesia, Papuasia, Malaysia, and Indo-China.   H. H. JOHNSTON

ANIMAL LIFE IN THE POLAR REGIONS
THE BIRDS AND BEASTS OF FROZEN LANDS
By AUBYN TREVOR-BATTYE

WHEN the mind ranges over all that has been recorded of adventure and enterprise in the Arctic and Antarctic regions it is apt to remember and dwell upon the sadder side of the picture, upon the barrenness, the desolation, the dreariness, upon the barely sustained fight of men's lives against the cruel forces of Nature—a fight constantly carried to all but the snapping-point of human endurance, against the fellness of the cold, the paralysing influence of the long Polar night. By many, therefore, these regions have come to be regarded as the land of death. It is apt to be forgotten that it is only civilised man, the intruder into a world of conditions opposed to himself, who is at this hopeless variance ; that he is then really in a world of organic existences quite self-contained within their own limits, adapted by long ages of gradual change to their frigid and rigorous environment; that the white lands at the end of the earth are, in a phrase, not lands of death, but lands of life.

Let us take the Arctic regions first. It is impossible to consider the organic life of this area without touching upon its human life, upon those human beings of whom this region is the chosen home. Of these the best known are the Eskimo. Other human beings dwell in that Far North—in Greenland, Danes, but they are historically recent introductions—European settlers. The Eskimo are part and parcel of the native fauna; by immemorial influence of their surroundings as completely modified from the average type, and as completely adapted to their environment as are the plants and lower animals among which they move. Their range extends from Labrador nearly to the 80th parallel, and all but half-way round the world—as far, that is, as Bering Sea. Through all this immense distance they are able to communicate with one another in a common language, a fact that has no equal among any other people in any part of the globe. Whence they came and why they came are lost in the dimness of past ages. They

THE MUSK-OX OF GREENLAND
This animal is considered by naturalists to be on the borderland between the sheep and the oxen

have a culture that is all their own and exactly fitted to their needs, a language that is elaborately constructed, and they have very great inventive power. Beyond this, they have song and saga, and a distinct attainment in decorative and pictorial art.

So much has been said about them, not only by the earlier adventurers from the North-West Passage, but of late years by Dr. Henry Rink and others, by the great explorers Nansen and Peary, that we are able to picture every detail of their daily life. We see these little people, their skin garments thrown aside, working in their curious dwellings all the winter through, shaping bows and arrows, working up horns of deer or tusks of walrus into weapons for the chase, and generally engaged in manifold preparations against the coming of the sun. Then spring breaks upon them, and the dog teams are straining at their collars, and with the opening of the leads in the ice the kayaks begin paddling in the sea.

But though the Eskimo dwell and wander further north by hundreds of miles than any other human beings, the Arctic lands of the Eastern hemisphere also have their own inhabitants. These are the races of the mainland of the Arctic littoral. Along that vast European and Siberian coastline roam the Samoyedes from Barents Sea to the Urals, the Ostiaks of the Ob, the Chuckchis of the Lena and Bering Sea.

These people, because land fails them, cannot range so far to the north as the Eskimo range ; on the other hand, they are able to lead less isolated lives, for they, like the reindeer Lapps, are in touch with civilisation—such as it is—and trade with the Russians for small necessaries of life. Like the Eskimo, they are wanderers, but in their case it is from a different cause. For while some live chiefly by fishing in the great rivers, others must follow the reindeer, upon which their prosperity and even their existence depend. As the reindeer must shift from one pasturage to another, so their owners must go with them.

The tribes best known are the Samoyedes, and on the islands off the coast, only visited at rare intervals by anyone from the mainland, they are seen in their most primitive condition. They are not, like the Eskimo, dependent upon blubber, horns, and bone, for the drift from the great rivers which is carried by currents out to sea and laid upon their coasts supplies them with material for many purposes. From this material—unpromising, and of the roughest description—cast upon their solitary shores they fashion sleighs, bows, arrows, traps for foxes, stockades for moulting geese, and many domestic articles. They live in wigwams,

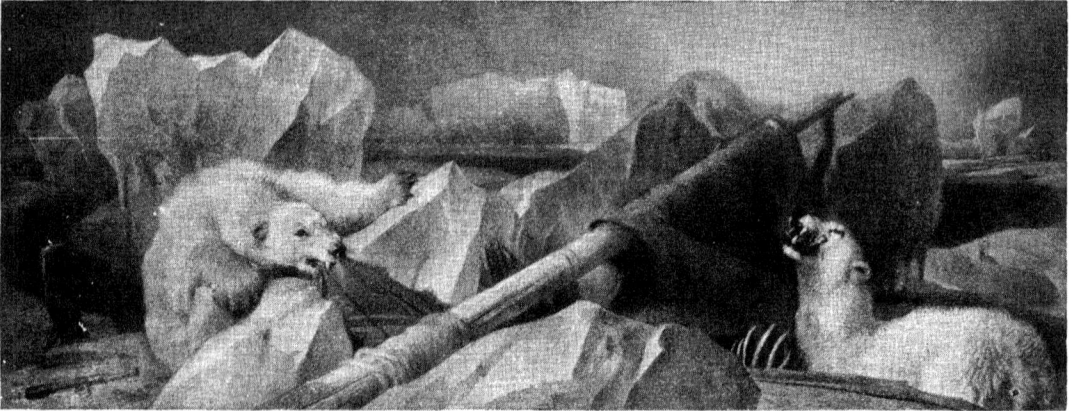

"MAN PROPOSES, GOD DISPOSES." SIR EDWIN LANDSEER'S FAMOUS PICTURE OF POLAR BEARS
This scene represents Polar bears sporting among the relics of Sir John Franklin's ill-fated expedition

or "chooms," covered with birch bark or with skins. Skins also form their clothing. They even use their sleighs in summer when the snow has melted, for these are so constructed that they can run over grass and pass easily across the summer bogs and swamps. These people are hospitable and courteous. Though they have not a written language, they keep records by means of symbols, and some of them, like the African Bushmen, can draw with quite recognisable accuracy pictures of animals and of hunting scenes. They have a system of religion akin to Shamanism, and the rites connected with birth and death are curious and interesting. Certain hills are held as holy, and dedicated to their deity, Nûm. When a man dies, his sleigh, his spear, his bow, or other object used in his daily life, is taken to the top of such a hill, and there is broken and left as evidence of his death and in propitiation of the god. The dead man is laid out on the moss of the *tundra*, or treeless plain, and deposited there under a rude covering.

The oral traditions of the Samoyedes have been carried down through very many generations; they regard their neighbours, the Ostiaks, almost with terror, as "very bad and murderous" people. Nothing, however, is known of any warfare, and scarcely of any contact between these two tribes within historic times; and it is therefore possible that this fear points to incidents of long, long ago, perhaps even as far off as the original migration.

Such, then, is a brief sketch of those human beings who, in the Arctic regions, stand at the head of the animal kingdom.

We will turn to some of the "lower" animals. Even at the northernmost extension of Greenland to the 84th parallel, wherever the borderland between the inland ice and the frozen ocean supports a scanty vegetation, there we find herbivorous mammals. It is the home of the musk-ox, one of the most singular animals in the whole zoological scale. It has affinities both with the oxen and the sheep. It is covered with woolly hair, has heavy horns with great frontal development, and lives in herds usually varying from twenty to thirty individuals. Like all Arctic animals, they are exceedingly confident in the presence of man. There was a time when they lived in Europe, and a skull has been found in the Drift gravel of Norfolk.

The wild reindeer of the Arctic regions are also astonishingly fearless; the writer has known them come up quite close to him as he sat and sketched. And on one occasion they actually sniffed at the biscuit-boxes just outside the door of a traveller's tent, so that the occupant, Sir Martin Conway, was able to take a snapshot photograph of them at close quarters. Of course, in districts where they have been much hunted, this confidence has, to a great extent, been lost.

Terrestrial and amphibious carnivorous animals are represented in these regions by the Polar bear, Arctic wolf, Arctic fox, ermine, walrus, and various species of seal; ungulates, by the musk-ox and reindeer; rodents, by the hare and lemming; while in the sea live a remnant of the right whale, the narwhal, and certain species of rorqual. Were one to choose one animal rather than another as characteristic of the North Polar area, one would certainly choose the Polar bear. The "ice-bear," as the Scandinavians appropriately call it, seems indeed the very genius of the Arctic world. Without those structural modifications that characterise the seals, it is yet all but, if not quite, as amphibious as they are. The swimming powers of Polar bears are probably determined by questions of food; because, unlike the seals, they are not capable of swallowing their prey under water. Though these bears, therefore, are not, like the seals, subject to periodic migrations through great stretches of sea, they are capable of travelling wide distances in the water, and their hunting is largely done in that element. When a bear sees a seal lying near the edge of the ice, it slips into the water and dives, rising at intervals for breath. Nearer and nearer it draws, slipping along under water by the edge of the pack or the berg, only when it rises showing its face, and again silently sinking until within striking distance of the unconscious seal, when with a blow of its ponderous paw it kills its quarry (see page 56).

Male Polar bears roam about through the winter, and do not hesitate to cross stretches of land. I have, for instance, seen a considerable and well-marked track made by these creatures when passing backwards and forwards across the neck of the inhospitable island of Spitzbergen.

The white colour of the Polar bear is of obvious advantage as a means of concealment from its prey; and, indeed, it would seem that this white dress is given to some creatures as a blind for purposes of attack, to others for purposes of escape. The Polar bear, however, is the only Arctic animal which retains this colouring throughout the whole year. Others assume the white coat only on the approach of winter, and lose it again more or less in the spring. Thus the Arctic wolf is whiter in winter than in summer; the Arctic fox, like the Alpine hare, is blue or bluish brown in summer, but quite white in winter. In changing the coat this fox assumes the white first on the sides, and does not become white all over by a general and simultaneous change, so that during October and November it is a parti-coloured animal. The ermine, again, of the Arctic and sub-Arctic area, which is bright red in summer, assumes, by a gradual change, its white winter dress. Even in England, as is well known, the animal there called the stoat occasionally towards winter changes to pure white, and often to a pied condition. The exact influences which bring about this change are not understood; that it is not directly due to the action of cold on the nerves and pigment cells is evident, since the change may take place while the weather is still far from wintry. Among Arctic birds, while there is perhaps only one—the ivory gull—white all the year round, there are some which assume in winter plumage more or less entirely white. Other examples are the rock ptarmigan, and, in the sub-Arctic area, the willow grouse and the common ptarmigan. It is easy to dispose of these instances under the convenient and alluring phrase "protective resemblance," but the exceptions are the puzzle, as exceptions often are. The musk-ox is not white in winter; the caribou and reindeer, though whiter in their winter coats, are nowhere nearly white enough to be invisible in snow and safe from the sharp eyes of their inveterate foe, the wolf. It may be argued that in either case whiteness would be no protection from an animal that hunts by scent. What, then, of the

hare and, among birds, of the ptarmigan? To add to the puzzle, the cock snow bunting assumes his white plumage in the summer.

Is it not possible that phenomena such as these must have a very ancient origin and should be referred to the slow and gradual influences of great climatic periods? To those who incline to this belief these facts would then seem to indicate a state of gradual recovery from glacial conditions of unknown duration, but in any case prolonged.

Another very interesting fact about Arctic animals is the general absence of hibernation. The female Polar bear lies up in the snow at certain times, but this is connected with the birth of her young and is probably not an instance of true hibernation. We do not even know for a fact that she remains without food during this period; indeed, it appears exceedingly probable that she does not.

The lemming quite possibly goes to sleep at intervals, though this has never, so far as we are aware, been actually proved, but it is certain that a great deal of activity goes on beneath and in the snow, for as spring sets in their runs are seen everywhere. The creatures that seem thoroughly to hibernate are those on the border-line—the brown and the black bears, the marmot, the dormouse, for examples. If, therefore, we regard these animals as a migration from the East—and there seems to be some evidence that the dormouse is a wanderer from Asia—then we shall recognise that winter sleep overtakes them *as they approach* the direction of greater cold. Once over the line, it might seem that varieties have become species able to maintain

THE WALRUS, A CARNIVOROUS MONSTER OF THE ICE-BOUND COASTS
Its huge tusks, the canine teeth of the upper jaw, may measure 30 inches in length

THE NARWHAL, THE UNICORN OF THE POLAR SEAS
Its horn is probably used for fighting its enemies and as a means of catching its prey

their activity during the greatest cold, and hiber-nation has been dropped. There is, however, another theory of Arctic life, which is briefly touched upon on the next page.

Very few birds—perhaps not more than five species—remain in the Arctic regions throughout the winter. The rock ptarmigan remains there, and also the ivory and Ross's gulls. Ross's or the wedge-tailed gull, also called the rosy gull, because of a beautiful rosy blush diffused over its breast feathers when in breeding plumage, is one of the loveliest of birds. It will always have a pathetic interest for Arctic voyagers, for when the dead company of the ill-fated Jeanette were discovered, the devoted naturalist Newcomb still had beneath his shirt three skins of this rare bird, placed there for safety, and so carried through all that terrible struggle on foot from the wreck to the Lena delta. There is some evidence to show that the raven also remains in these high latitudes throughout the winter season ; at any rate, this bird has been recorded as appearing in these regions with the very first light of the returning sun.

Spring brings back a large company of birds to the North Polar regions. While some—chiefly perching birds—arrest their migration in the Lower Arctic, and some at intermediate intervals between there and the Extreme North, others press on into the High Arctic, and a few almost to its farthest limits. Yes, when spring returns, a wonderful tide, so to call it, of bird life sets in towards the North Polar regions. The groups chiefly represented are waterfowl and wading birds—the "waders," as they are generally called.

Most of the Arctic birds are familiar in the United Kingdom in autumn and winter, and many nest there, though the majority of those that do so stay nest and breed high up in the British Isles ; but the breeding range of the waders varies very much, some nesting in the High Arctic, others barely reach-ing the more rigorous regions of the Arctic world. The common snipe is a good example of the latter class. It is doubt-ful whether in the Western hemisphere it has bred north of latitude 70° (Greenland), but in the Eastern hemisphere it breeds in the Siberian swamps above the Arctic Circle. On the American side, wading birds are naturally referable to the American and Canadian mainlands, comparatively few crossing Bering Sea to the Asiatic side. As examples we may men-tion the pectoral sand-piper, Buonaparte's sand-piper, and the American stint. Of those waders which are familiar on the English coasts in autumn and spring, but do not stay to breed, good instances are the knot and sanderling. The knot is one of those birds whose breeding-place was a mystery to ornithologists until a nest of young were found in Grinnell Land, 82° 33′ N., by the British Arctic Expedition of 1875-6. The winter plumage of this bird is made of shades of inconspicuous grey, but before they leave British shores many have begun to put on their beautiful red nesting dress.

The list of waterfowl that nest in the Polar area is a long one. The pink-footed goose—so numerous in winter on the Norfolk coast—nests in Spitsbergen, the white-fronted and the bean goose along the European littoral. The brent goose, essentially a sea goose and very familiar to all punt-gunners, possibly nests as far north as any migratory bird. One of the most noticeable of the ducks is the long-tailed duck, which is indeed a striking bird in its beautiful summer plumage, and although it is also found in lochs, it is, perhaps, more of a true sea duck than even the eider, and seems indifferent to the roughest sea, diving in the breakers as if in calm water. It is generally believed that seals feed exclusively on fishes, but this is not so, for the writer once saw one of these birds taken by a seal. He was watching a party of long-tailed ducks in open water among the ice, when suddenly the head of a seal appeared. The ducks settled further off, whereupon the seal vanished, but rose once more among the ducks, which again flew off. The seal raised itself high out of the water, looking round as a terrier does which has lost a rabbit, again sank, and again came up, this time successfully, for it caught a duck and took it under.

There are certain birds to which every Arctic voyager refers, because they are never long out of sight, and, indeed, come to be regarded as his familiar companions. These are the fulmar, the skuas, the guillemots, and the little auk.

The fulmar is the northern representative of a group of birds which is represented by species not only in British seas, but as far as the Antarctic regions ; they are called petrels and shearwaters, and are characterised by having nostrils enclosed in a tube which extends along the bill. The fulmar, known to whalers as the "mollymoke" or "mollymauk," seldom, probably never, visits the land except at breeding time ; it rests only, and very seldom, on the sea. To Arctic voyagers it seems to have an evil eye and never to rest, merely settling for a very short time to feed, and flying off again. Backwards and forwards it goes with wonderful untiring flight, as if looking for something it can never find. The skuas are gull-robbers ; as soon as they see a gull which has food in its beak they set to work to chase it, and at last, by persistence of attack, succeed in making it drop its food, which, swooping down, they catch in mid-air. No bird is more clever than is either a Buffon's or an Arctic skua in its efforts to draw away an intruder from its nests ; the helplessness one of these birds will then simulate is beyond belief.

The British common guillemot is represented in the Arctic region by a rather larger form known as Brünnich's guillemot. These birds are colonial cliff-builders, and each pair incubates a single egg on narrow ledges of the most precipitous places. A visit to such a "loomery," with its sounds from thousands of birds' throats and its bewildering confusion of myriad wings, is an experience which the explorer is never likely to forget.

Of the Arctic sea-birds two are especially beloved by all who have visited those waters—namely, the black guillemot, or "dovekie," and the little auk, or "rotje." The dovekie is a very beautiful little bird ; its plumage is black with a white spot on the wing, and its legs are bright vermilion red. It nests in clefts in stacks of rocks. It has a pretty habit of constantly dipping its bill into the water. The little auk—a "dumpy" little bird, only 8½ in. long—also nests in holes, but high up on the cliffs. One of these birds flew round the writer's head at an altitude of 3,000 ft., on the highest mountain in Spitsbergen. One of the most striking pictures of beauty in the memory of every Arctic explorer is that of the Polar calm : the sea quiet as a mirror, the lovely snow-white ivory gulls sitting upon the bergs of blue and green ice, and the dovekies and little auks floating in the water spaces among the ice. But so placid a picture may have a discordant note ; as a shadow passes over the birds they dive for safety, for is not the shadow that of the ruthless pillager, the burgomaster, or glaucous gull ?

Two questions naturally arise out of what has gone before—namely these : first, "How did all these animals get there ?" and, secondly, "Why do birds go there to nest ?"

The first question is a very big one. To begin with, we must remember that the Arctic area was not always the frozen, inhospitable region that it is at present. To-day those seas swarm with pteropods and other lowly forms of animal life, but they were warm enough once to support even the frail life of the coral, while on the land the magnolia flourished. This is fact. Geology, which has shown us this, also points to another condition held as probability by many eminent scientific men. This condition is that in the Tertiary Period (*our* period) a great land causeway joined Europe with America. This would give continuity of organised life, and would explain the occurrence of musk-ox remains in the continent of Europe and the river drift in Norfolk.

Does not this also seem partly to supply an answer to the second question, that about the birds ? Such movements become more intelligible to us when we realise that in them the birds are but obeying an old memory, a memory of the species that lies away back in the dimness of unrecorded ages —that in thus going up into the Polar summer to rear their young, they are really only *going home*. One of the most remarkable of all the facts of migration is that the young birds require no leading. In the very great majority of wading species it is the young birds which first arrive on British shores, while the old birds follow later, so unchangeably is the memory of the route to travel stamped upon the race.

This sketch reveals some of the points connected with Arctic life. Let us now turn to the Antarctic.

It is popularly supposed that the two Poles are very much alike in character, but scientifically they are really very different. But in fact it is

THE SEA ELEPHANT, WHICH IS LARGER THAN THE LAND ELEPHANT
The nose of this colossal seal is furnished with a proboscis, which the animal, when enraged, is able to inflate. The sea elephant is of vast proportions, measuring some 20 feet in length and from 15 to 18 feet round. Photograph by A. Rudland

scarcely an exaggeration to say that they have little in common except cold and ice.

To begin with, their geographical conditions are quite distinct, for one is water, the other is land. In the North Pole we have an ice-laden sea, studded with islands, but the South Pole is itself set in a great continent, which is contained in, and almost entirely fills the space included in, the Antarctic Circle. Outside of this lies the surrounding ocean, the nearest point of any continental mainland being Cape Horn, which, far from being within the Antarctic Circle, is some ten degrees distant from it.

Instead of the placid Arctic seas, the ocean that surrounds the Antarctic continent is constantly stormy, its immense rollers, unstayed by any land, having no parallel near the other Pole. No human beings live within this area, and no evidence has so far been discovered to show that any human beings ever did. This absence of human life forms one of the most striking differences between the areas we are comparing, and it is accentuated by the additional fact that no land mammals live there either. The only mammals are seals and whales. Readers of "The Heart of the Antarctic" will remember how its author describes the Bay of Whales in Victoria Land as teeming with life. He tells us that hundreds of whales, finners, killers, and h u m p b a c k s , were blowing all around, while on the ice were numerous groups of Weddell's seals.

One of the most remarkable of the Antarctic seals is the sea elephant. This animal has a loose, tubular sack of skin placed over the nostrils, which can be inflated and erected, and then resembles a short trunk. The bladder-nosed seal of the Northern seas has—as its name implies—a feature which can be distended in the same way. Another Antarctic seal is the southern fur seal. Unfortunately, it is only in high southern latitudes that these animals can now be found in any numbers, to such an extent have they been persecuted and killed off.

The bird life is also very limited, consisting only of penguins, petrels, and gulls, the first-named being among the most remarkable birds in existence. No living bird is more astonishingly adapted in form to suit the conditions of its life, its feathers modified to almost a scale-like covering, its wings like flippers, its legs situated far back—every feature has relationship to rapid progress in the water. On land it walks in an upright position, and bears a quaint resemblance to a human being. The largest of the penguins is known as the emperor penguin, and these and the adelie penguin nest in immense colonies. These penguin "rookeries" have been often described. It is not easy to walk among the nesting birds, for they peck savagely at an intruder—indeed, a dog belonging to the Challenger was lost in one of these bird colonies. A most interesting fact is mentioned by Dr. James Murray, biologist to the Shackleton Expedition. He tells us that when the nesting is over, the emperor and adelie penguins deliberately feed up, and then set off to see the country. They appear to keep walking on for days and even for weeks without further food. Although several of the petrels are observed as far as the Antarctic Circle and down to the edge of the pack-ice, only one or two species reach the Antarctic continent itself. The snowy petrel, the sooty albatross, and even the Dominican gull have been recorded by each expedition as being observed about the edge of the pack, but only the giant petrel and Wilson's petrel appear to reach the land. These, with the penguins and the Antarctic skua, alone among birds seem to redeem that inhospitable region from the impression of absolute lifelessness.

That the Antarctic region has had, like the Arctic, its period of warm seas is shown by the discovery in Victoria Land of a perfect specimen of a solitary coral. The Swedish Expedition under Nordenskiöld found on Seymour Island remains of trees, showing that during the Tertiary Period this region had supported great forests. The Antarctic continent is now separated from Tierra del Fuego by a sea of 2,000 to 3,000 fathoms in depth, yet Nordenskiöld points out that these trees were in some instances similar to Tertiary forms he discovered in Tierra del Fuego, and that in Brazil allied forms are in existence.

In conclusion we may cite two very remarkable facts. The first is, that of five flowering plants found by Moseley on Heard Island, one (*Callitriche verna*) was a British plant. The other fact is that of the discovery by the Shackleton Expedition at Cape Royds of a rotifer (*Macrobiotus arcticus*), which, to quote Dr. James Murray, "is an Arctic species only previously known in Spitsbergen and Franz Josef Land." The day may come when we shall understand such problems ; meantime each expedition brings back new material for study and new contributions to knowledge of the life of the Polar regions. AUBYN TREVOR-BATTYE.

A GROUP OF PENGUINS IN THEIR ARCTIC HOME
The wings of these strange birds have become flippers or paddles. Thus the penguins cannot fly, and they walk clumsily on webbed feet. From a photograph by Lewis Medland

THE INTELLIGENCE OF ANIMALS

THEIR POWERS OF REASON AND CAPACITY FOR EDUCATION

By Professor C. LLOYD MORGAN, F.R.S.

THE more fully the structure of animals is studied, the more extensive is the evidence in support of the conclusion that delicate adaptation to the conditions of life prevails throughout organic Nature. But the adaptation is in each case that which is practically serviceable, not that which meets the demands of theoretical perfection. Were this the appropriate occasion, abundant illustrations of this fact could be adduced. One must here suffice. There is perhaps no organ which is more admirably adapted to the needs of daily existence than the eye. And yet Helmholtz, the well-known German physicist, after prolonged and careful study, assured us that the eye has every possible defect that can be found in an optical instrument, and even some which are peculiar to itself. This may at first seem to be a somewhat sweeping condemnation. None the less, Helmholtz can speak of the adaptation of the organ of vision to its function as "most complete," and he urges that the completeness of adaptation is shown by the very limits which are set to its defects. The perfection is practical, not absolute. It is that which meets the demands of the conditions of its normal use, not those which the theorist may formulate. And whenever, taking a wider survey, we scrutinise the construction of physiological organs, we find the same character of serviceable adaptation to the needs of the organism.

On the theory of natural selection, this is exactly what we should be led to expect. In accordance with that theory, every animal has to submit to the most searching of competitive examinations in the struggle for existence. Only on condition that the organism, and its ancestors before it, have succeeded in passing this examination does the type which they represent persist. But the severity of the struggle is subject to some variation. As the examination becomes more searchingly stringent, the level of adaptation requisite for a pass—to keep up the metaphor—is raised. The relative perfection increases, but it still remains relative. Practical success suffices; theoretical excellence is not required. So long as the adaptation is sufficient, with a margin of safety, to secure the pass-mark of survival, all the biological conditions for the persistence of the type are fulfilled. Nay, further, the existence of organs, if such there be, whose perfection is far in advance of the practical needs of their possessors would be evidence of the co-operation of some factor in organic evolution other than natural selection.

The elementary and, no doubt, familiar considerations which have just been advanced with regard to organic structure bear with equal cogency on the study of animal behaviour. The instincts and the intelligence of animals afford a fascinating but very difficult field of investigation—fascinating from its intrinsic interest and from its bearing on evolutionary theory; difficult because of its subtly inferential character. Unfortunately it has, until lately, been left too much in the hands of well-meaning but untrained amateurs, who have appealed for corroboration of their views to the common-sense of brother amateurs. In no department of the study of animal life has the amateur been given a freer hand; in scarcely any department are naïve, uncritical opinions of less weight and value. If the inquiry into the nature and psychological backing of animal behaviour is to be raised to something approximating to the level of scientific investigation, it must be prosecuted in accordance with accredited methods of careful observation and experiment by those who have undergone a suitable preparatory discipline; and conclusions must be drawn with due regard to scientific caution.

TWO TURKEYS "COURTING" A HEN
From a photograph by Professor B. H. Bentley

If, then, we attempt to bring the study of behaviour into line with that of zoological structure, and if we regard animal intelligence as of functional value for survival in the struggle for existence, we may do well to ask at the outset what characteristics we should expect to find in the mental processes of animals. Accepting evolution through natural selection as a valid induction, we ask in effect what preliminary deductions we may draw with respect to the psychological status of the higher animals, assuming their mental processes to be practically effective to the end of survival.

Now, in order to be practically effective to this end, the behaviour of an animal must be of such a character as to enable it to secure an adequate amount of food of a kind appropriate to its needs. It must be of such a kind as to enable it to escape from its natural enemies. Neither of these objects can be attained by the higher animals unless they have niceness and keenness of perception, to which sense organs of well-developed practical perfection afford the requisite data of experience. Nor can they be attained unless the organs of locomotion are fitted to their task, and unless all the motor activities are under complete control, with that delicacy of co-ordination which is essential to prompt and effective action. Nor, again, can they be attained unless the niceness of sensory perception and the delicacy of co-ordinated activity be brought into the closest and most accurate accord—unless, in a word, that which answers in the animal to the "training of hand and eye" in the child has reached a high pitch of serviceable perfection. Thus, in predatory mammals and birds, quickness to perceive their prey and to detect the subtlest indications of its presence or whereabouts, skill in its capture, like ability through keen perception and ready response to escape from enemies, enough wit to avoid or get out of the occasional difficulties incidental to life in a complex and changing environment—these are essential in order that intelligence should be a factor in survival. The presence of these, on the doctrine of evolution, we should be led to expect, and these assuredly we find. Add to these such instinct and intelligence as are necessary to enable the male to woo and win a mate—perhaps to woo, but in any case to win her—such instincts and intelligence, diverse in many ways, in accordance with the conditions of life, as are necessary to enable the parents to shelter and provide for the young till they are able to fend for themselves; such instincts and intelligence as further the common life of those animals which live in flocks, herds, packs, or other communities; such instinct and intelligence as enable the young to fall into the ways of their fellows among whom they are reared; such instinct and intelligence as fit them to range freely over their habitat. Add these and their like, for the list makes no pretence to be exhaustive, and we have before us the mental equipment which is practically serviceable to beast or bird in the struggle for existence. Keenness of perception, readiness of response, the ability to profit surely and swiftly by the teachings of direct experience—such are the saving graces of animal intelligence.

These saving graces are founded in heredity. There is, first, the hereditary ability to respond in certain adaptive ways to the surrounding conditions,

irrespective of any process of learning. This is the instinctive basis of behaviour which is always present, but is relatively predominant in the less intelligent animals. There is, secondly, the hereditary ability to profit by experience and to learn its lessons—what, following Sir E. Ray Lankester, we may term educability. This is most marked in the more intelligent animals. And there is, thirdly, in these more educable animals, an exuberance of vitality and varied response finding expression in a restless output of activity which is of the highest value both for the training of skill and for the acquisition of experience, since the animal, especially the young animal, is thus brought into relation with the greatest possible number of varied situations.

No one has done more to help us to realise the evolutionary importance of this third element than Dr. Karl Groos, through his able discussion of the so-called play of animals. Such play affords to the animal not only physical, but mental training; it is the necessary means to the full development of bodily power and of intelligent behaviour, and resourcefulness in emergency. In animal life the maxim "practice makes perfect" finds abundant illustration, and the biological value of animal play is that it is the field for the ready attainment of serviceable perfection through practice under conditions where the results of failure are not fatal.

This necessarily brief discussion of some of the essential characteristics of animal intelligence will no doubt be regarded by lovers of anecdotes as unsatisfactory, in that it lays stress on what they will regard as a lower order of intelligence, and altogether ignores the higher mental faculties. The aim, however, has been to determine on general grounds what is essential to survival—what kind of intelligence is practically effective in fitting the animal to obtain a pass in the stringent examination of life's struggle. It has, indeed, been suggested that just as, under natural selection, an organ does not reach a higher level of perfection than that which is practically useful to the organism, so, too, on the theory of natural selection, we have no valid ground for expecting a higher level of intelligence in the animal than that which suffices for its daily needs and the continued preservation of its race. It will, however, be urged, and justly urged, that the question is not what we may have valid grounds for expecting on merely general

A MILKMAN'S HORSE KNOCKING AT A CUSTOMER'S DOOR

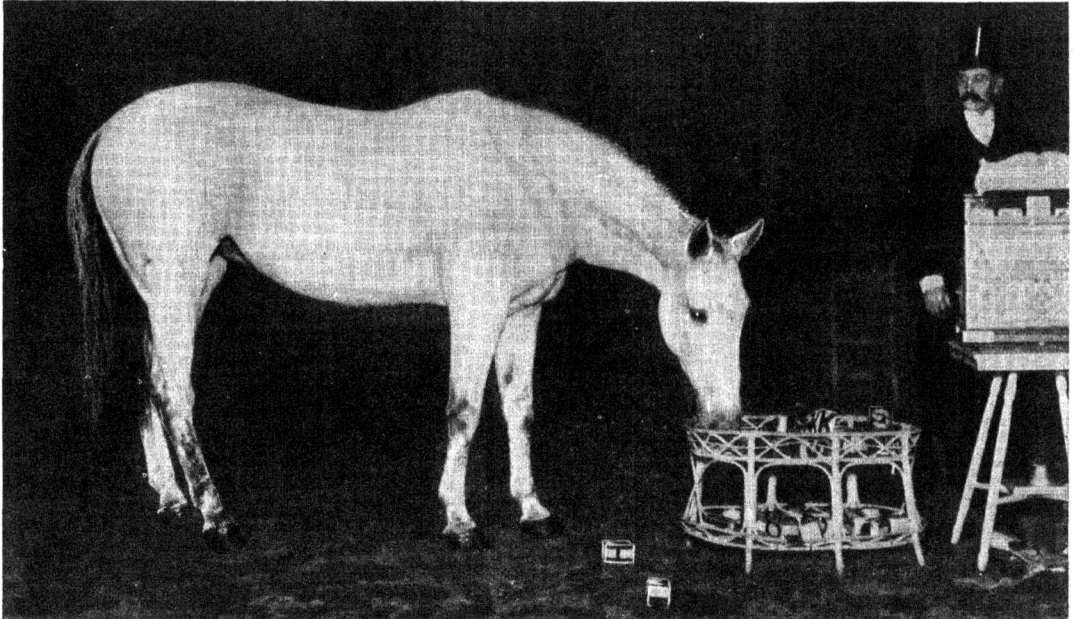

AN INGENIOUSLY TRAINED HORSE SPELLING OUT A WORD FROM LETTERED BRICKS

principles, but rather what we are bound to accept on the evidence of properly accredited observers.

We may, first, consider the evidence afforded by the remarkably clever tricks of performing animals. P. G. Hamerton, many years ago (1873), described how, in his own house, a cleverly trained dog would fetch in their right order the letters which spelt the English or German equivalents of common French words, and do other wonderful things. But the owner of the dog (M. du Rouil) admitted that there was a means of *rapport* between them which he was not prepared to divulge. A few years ago we read in the newspapers of the remarkable arithmetical powers of Kluger Hans, the calculating horse. The case is probably typical. Clever Hans gave, by tapping on the ground with his hoof, correct answers to the questions which were put to him. When told that the day was Tuesday and asked which number in the series of week-days this represented, he would tap thrice. Similarly, he would tell not only the hour, but the number of minutes indicated by a watch, the number of men and women in a particular row of visitors, the tallest and the shortest members of the party. In " Nature " (September 22, 1904), we are told that a representative committee, which included the director of the Berlin Zoological Gardens, a veterinary surgeon, and a professor of the Physiological Institute of the Berlin University, witnessed these performances with the view of ascertaining whether they were the result of trick, or whether they were due to the mental powers of the animal. Their verdict, it is reported, was unanimous in favour of the latter view. But in a letter to the same journal (October 20, 1904), the Rev. J. Meehan criticised this verdict and stated that the equally clever horse, Mahomet, had been taught, with

infinite pains by his trainer, to begin pawing the ground when his master looked straight at him, and to cease pawing when his trainer turned his gaze to the floor. Similarly, he had been drilled into bowing his head at one tone of his trainer's voice, and shaking it on hearing another. The same trainer showed Mr. Meehan how a collie would spell his own name and words sent up by the audience by bringing the right letters from an alphabet. The collie went along the letters, picked out the one he needed, and brought and laid it before the footlights. But the trick was really simple. His master carried his gloves in his hand. A little twitch of the gloves as the dog passed the particular letter wanted was the cue. The well-trained animal took in the slightest stir of the gloves with the corner of his eye. This dog even played a game of cards—and won. " I wonder," says Mr. Meehan, " what the German professors, good, easy men, would have said had they seen, as I did, Mahomet figure out a sum with his tail to the board ! "

To return now to Kluger Hans. His wonderful performances were made the subject of further inquiry by a commission of psychological experts, headed by Professor Stumpf, of Berlin University. The conclusion they reached was that the horse was not capable of independent thought. I quote again from " Nature " (December 15, 1904) : " When asked a question, Hans knows that he has to beat with his hoof in reply ; but he does not know when to cease beating until he detects some movement on the part of the person questioning him. The commission expresses the opinion that, so far as Herr von Osten, the owner, is concerned, these movements are given involuntarily, and are sometimes of so imperceptible a nature as to be undetected save by highly trained human observers. There has been no trickery, says Professor Stumpf, but, on the other hand, there have been no reasoning powers on the horse's part. The whole secret is in Von Osten's skill, patience, and judicious reward, and on Hans's part in keen powers of observation."

But what were the visual clues which appealed to these powers of perception ? To this question the following reply is given by Herr Pfungst, who has published a memoir on the inimitable Hans (Berlin, 1905). "After Herr von Osten had stated the problem, he tended, involuntarily, always to bend the head and trunk slightly forward, whereupon Hans would extend his right foot and begin to tap without putting his foot back after each successive tap. When the desired number of taps was reached, Herr von Osten would, involuntarily, give a slight upward jerk of the head. At this second signal, the horse would retract the foot to its normal position." This completes the story of Kluger Hans, which has been given in some detail because, as we have said, it may be regarded as probably a typical example of the cleverness of which adroitly and patiently trained animals are capable.

The really noteworthy facts in all such cases are the extraordinary delicacy of perception of which they afford evidence and the sure linking of the perception with some particular mode of behaviour. As I have elsewhere said, those who have seen a shepherd's dog working sheep on a moorland fell, and have taken the trouble to ascertain how the results have been reached, will appreciate, on the one hand, how well the dog responds to the signals of his master, and, on the other, how completely all initiation is due to the shepherd who has trained him. Those who merely witness such a performance without inquiry or investigation will probably misunderstand the whole matter. In the north of England competitions are not uncommon where, say, three sheep have to be driven over a definite course, between certain posts and round others, through narrow passages and into a small pen—the whole within a definite time-limit.

The dog has been trained to respond to some six or eight whistle signals, sometimes accompanied by gestures and movements of a stick. The signals, given in different whistle tones and inflections, have each a particular response in behaviour ; on hearing one the dog drives the sheep straight ahead ; on hearing others he rounds them from the right or from the left ; yet others make

him at once stop, lie down, or creep slowly forwards. The dog's whole business is to obey these signals ; and nothing can be more admirable than the instant response of a well-trained animal.

It being now generally admitted that casual observation supplemented by amateurish interpretation is of little or no value, it will be more profitable to give some account of the systematic experiments conducted by trained psychologists, and to indicate some of the more salient conclusions to which, according to experts, they point, than to furnish a *réchauffé* of anecdotes of a type that is already sufficiently familiar. I must assume, however, that readers have sufficient interest in the psychological aspect of the inquiry to be at the pains to grasp what are the problems which trained observers are trying to solve. This will necessitate some further preliminary discussion.

It will be well, first, to devote a short space to a consideration of the manner in which animal skill in effective behaviour is developed. But what does one mean by development ? Always, whether we are dealing with animal structure or animal behaviour, one means by development the continuous passage to a further stage of organisation from a foregoing stage of less complete organisation. Now, animal skill is always based on instinctive foundations. There is little doubt but that the flight of a swallow or a seagull is at the outset as truly instinctive as the swimming of a duckling or a waterhen. A bird does not have to learn to fly, or a mammal to run or leap, without any inherited co-ordination of limb movements. The hereditary co-ordination, sometimes more, sometimes less perfect to start with, is the organised foundation on which the far more complete organisation of perfected locomotion is based. Consummate skill is the outcome of continued practice ; and through that practice, in the flight of the swallow, for example, the data afforded to the sense of sight, those afforded by the varied pressures of wind currents on the wings, breast, and tail, and those afforded by the execution of the appropriate activities, are all brought into perfect harmony. Those redundant or aberrant movements of the wings or body which conduce to failure and awkwardness are suppressed ; only the movements essential to the real business of perfected flight in closest touch with the situation of the moment are retained. Unless we are prepared to regard the organism capable of such wonderful improvement through

A CAT THAT KNEW HOW TO RING A BELL

PHOTOGRAPHS FROM LIFE, SHOWING A CAT USING HER PAW TO GET CREAM FROM A JUG

practice as an unconscious automaton, we must admit that the perfected skill is the outcome of experience, and is therefore every whit as much a mental product as a bodily process. Indeed, such skill development affords an admirable example through which to illustrate the distinction drawn by many psychologists between naïve " experience of " situations with behaviour in appropriate relation to them, and " knowledge about " either the situations or the behaviour. No doubt those of my readers who ride a bicycle have both naïve " experience of " the procedure appropriate to a given situation, and " knowledge about " the principles involved. They know how the principle of inertia is utilised in correcting innumerable incipient tumbles, now to right and now to left. Yet there are many boys of ten who can ride as well as they do, but who neither know nor care to know what principles are involved. I think we may fairly presume that the bird, with its marvellously developed practical experience of all that conduces to effective flying, has no systematic knowledge of mechanical principles. In what, then, does the process of learning consist ? In bringing the series of contributory items of experience into line, so that the fitting sequence $a b c d e f$, etc., runs smoothly, and f, the swallowing of an insect, duly follows upon a, the sight of it in the air. The initial awkward or redundant series might be $a n q b c m p d r e$— the important f never being reached owing to failure in skill. The redundancies n, q, m, etc., are eliminated, and each part of the business series $a b c$, etc., is improved and linked with its neighbours.

It should be noticed that the development of skill is always in relation to a situation within which the skilled behaviour is appropriate. More, therefore, is involved than the mere motor activity ; there is involved also a delicate perception of what is presented to the senses in the surrounding world. Suppose, then, that an animal has learnt a good number of varied modes of behaviour suitable to frequently recurring circumstances in its daily life. Its restless activities are constantly bringing it into new situations resembling, with some difference of detail, those with which it is already familiar. Fresh sequences of experience which may still be symbolised by $a b c d e f$, etc., are again and again

presented. A good deal of experimental work has been directed to the ascertainment of the manner in which such sequences come to be established. There are, however, some questions of interpretation which may here be briefly considered.

It is clear that if the experience of to-day is to be of any value for behaviour to-morrow, there must be some sort of retention of to-day's series, and some sort of revival of that series on the fitting occasion to-morrow. It is clear, too, that the items of experience symbolised by a, b, c, etc., must be in some way connected, so as to form a series. Let us take a simple illustration, which may help us to follow up the matter a little further. When I lived at the Cape, I used to take my two dogs on my walks up a spur of Table Mountain. On one occasion, at a certain spot, they put up a coney, or rock-rabbit, and gave chase down a side track. They had the experience $a b c d b e f$, etc., where b is the coney. No doubt b was continuously present until the coney, unfortunately, at e popped out of the range of experience. But we cannot here unduly complicate our alphabetical symbolism. Whenever in the course of a subsequent walk they came to that particular spot a, they became excited and raced down the side track. They had, we may presume, the series $a c d e f$. The important item b, the rock-rabbit, was not on the scene of action. For more than three years, every time they came to that spot, the series $a c d e f$ was repeated. To race down the side track became a settled habit. But what about b ? We naturally, as amateurs, say that the dogs " remembered " having seen and hunted a coney at this spot. The psychologist must ask just what is to be understood by the word " remember." He urges that it is not in any way necessary to the interpretation of the facts that there should be called up in the mind of the dog a memory picture of the previous occurrence with b present, as contrasted with the existing situation with b absent. He urges that the primary business of revival, through pre-established connections in experience, is not a backward-looking review of the past, but rather a feeling of present expectancy. To meet the case, many psychologists have adopted the word " meaning " in a rather specialised technical sense. They would say that, after the first occasion, the spot *a* had for the dog the meaning "rock-rabbit hunt"—just as the sight of its bottle has meaning for the infant. We may term it "expectancy meaning," or, in more technical phrase, "pre-perceptive meaning"—a feeling of what is coming based on the experience of what has on previous occasions been found to come. But a further question here arises. Does the expectancy implied when we say that *a* means *b c*, etc., necessarily involve the formation of a definite image of *b* —that is to say, a mental picture of the coney? The generally accepted view is that it need not revive any definite image.

Put the matter to the test of your own experience. Suppose that, while you are writing in your room, you hear a sound in the garden that "means" startled blackbird. Does a definite image of a blackbird arise before the mind's eye? Remember that the question is, not whether you can, but whether you *do* see it.

For myself, though I visualise freely, I am pretty confident that in ordinary circumstances I do not. Or take another case. I am one of those who are quite incapable of forming definite images of taste. Of course, I know what the taste of coffee is like, and can recognise it when it comes. The sight of my cup of coffee after dinner has pre-perceptive meaning. But I have no anticipatory taste image. Some years ago my cook, by a stupid blunder, sent up some soup, which was being kept hot for my wife, in my coffee-cup. I added sugar and milk, and then had a striking example of erroneous pre-perception. But though I remember that the stuff was exceedingly nasty, I can form no image of its taste on the mental palate in any degree analogous to a visual image or an auditory image. Many people have no visual images, but, none the less, these people have visual expectancy. It is therefore held that there is no necessity, for the satisfactory interpretation of the facts, that there should have been formed in the consciousness of my dog any definite visual image of the rock-rabbit we encountered in our walk. For many of us, no doubt, such images play an important part in retrospective memory—in the picturing of past scenes.

But here another distinction comes into view. There are two kinds of images (revivals of sensory impressions), or, if not two kinds of images, two modes in which they may occur. There is the "anticipatory image," foreshadowing the impression which is just coming; such is the auditory image of the expected report when the flash of a distant gun is seen or the lightning streaks the sky. It is held by many psychologists that this is the form or mode in which images have their genesis, arising out of and being, so to speak, floated off from pre-perceptive meaning. Their value is that they afford more definite and clearcut warning of what is just coming

into our experience. The second form of image, or mode of its occurrence, is the "memory image." The immediately serviceable value has gone; it is more of a luxury. It is not anticipatory, but retrospective, and, when fully developed, carries with it definite retrospective reference to the occasion on which the sensory impression thus revived had its place in vivid experience. One of the much-debated questions in comparative psychology is whether, and if so in what degree, animals are capable of forming definite images, and whether they are anticipatory images or free memory images, or both. It is not easy to devise experiments and observations which shall enable us to give decisive answers; and, as we shall see, different interpretations of the observed facts may be given. Another question is whether animals afford evidence of framing general and abstract ideas. Out of his varied experience of things good to eat, does the dog form an abstract and general idea of "good for eating"? That a number of things that the dog is accustomed to eat carry pre-perceptive meanings which have certain characteristics in common, which are none the less differentiated for his practical experience, and which call forth different modes of behaviour, is likely enough; but has a dog the capacity of attaining to that strange inversion which is a distinctive step towards thought—the topsy-turvydom of fixing attention on the meaning as exemplified by the thing?

Let us consider the manner in which we aid little children to reach the latter point of view. A simple case must suffice. We take a number of objects

TERRIERS ACCOMPANYING A WOUNDED COLLIE TO A HOSPITAL
From the painting by Y. Carrington at King's College Hospital, by permission of Messrs. A. & F. Pears

which are (*a*) red, yellow, or blue ; which are also (*b*) round, square, or triangular ; and, further, (*c*) made of paper, wood, or flannel. We get the child to arrange them first in colour groups, then in shape groups, and then in stuff groups. In the first case, colour is that which matters for the purpose in hand ; shape and stuff are present to experience, but may, nay, must be neglected. In the second case colour, which before mattered so much, does not matter ; shape does. And so on. The child makes a number of mistakes at first, but gradually learns this new game. The fixing of attention on what counts for the end in view, to the temporary neglect of much that is present to experience but does not count, this is the essential feature in the formation of abstract and general ideas. It should be observed that these ideas are not in the least up in the clouds and divorced from the things which the child can see, touch, and handle ; on the contrary, their vitality lies in their being embodied in the things ; but the objects are now regarded as examples of the embodiment of the idea, and one thing can be freely substituted for another. The round bit of wood can be substituted for the square bit of flannel, if both are examples which embody the idea blue.

When the child has reached a somewhat higher stage of what is termed ideational process—since he regroups experience in terms of ideas—he arranges these ideas in sequence, and searches for objects, facts, and modes of behaviour which shall give to the trains of ideas concrete practical shape. He forms plans of action which shall enable him to reach a wished-for but hitherto unattained end. Here he is at *c*, having passed through *a* and *b* ; he badly wants to get to *f*, but how is it to be managed ? An ideational plan which he frames in thought, in advance of practical experience in this particular connection, suggests that *d* and *e* afford the appropriate means by which *f* may be reached. He has had, indeed, no previous experience of particular examples of *d* and *e* in this connection ; but such examples have, maybe, proved serviceable in another connection. Why not try them here ? Perhaps seeing someone else pass through *d* and *e* may suggest the right clue. In any case, he does the thing in idea before he does it in fact, the doing of it in fact being only a translation of the idea into action. As Dr. Stout has well said, it is one of the characteristics of thought-process that you can cross a bridge before you reach it ; you can cross it ideally before you traverse it actually.

Much careful experimental work—that is, observation of animal behaviour under controlled conditions—has been carried out with a view to ascertaining how far an animal is capable of supplying the *d e* in the series of actions necessary to reach *f*, and the conditions under which the requisite

A FOX-TERRIER LIFTING THE LATCH OF A GARDEN GATE

series is completed. To a consideration of this experimental work we may now address ourselves.

I have elsewhere described how my fox-terrier learnt to open the iron gate outside my house by lifting the latch with the back of his head. The series, when the trick was established, was : *a b c*, running out of the door and down the steps ; *d*, jumping on to a low parapet ; *e*, looking out with his head beneath the latch ; *f*, running out into the road when the gate had swung open. But in the early stages of behaviour the series was *a b c d m n o d p q d e f*.

In plain English, he jumped up on to the parapet, gazed out through a number of openings between the vertical rails, jumped down again, and up again, looked out elsewhere, and only after some minutes chanced to stumble upon the finally effective sequence *d e f*. There was in this case no evidence of a plan of action of which *d e* was the fitting expression. Only gradually were the useless and redundant activities *m*, *n*, *o*, etc., eliminated, until eventually the effective series in due order of sequence was established. The type of learning was that of trial and error—trial along the lines marked out by normal dog behaviour, error in the sense of failure to get out into the road that way—of the stumbling by chance on the successful behaviour, and of the gradual linkage of *d, e, f* by repetition of this behaviour. It may be mentioned that the gate took a little time to swing open by its own weight when the catch was released. Hence, in the early trials, the dog had passed on to look out elsewhere before he saw that the gate was open, and this probably delayed to some extent the linkage of *e* to *f*.

Let us, however, pass to more definitely systematic experimental work. Dr. Small was, I believe, the first to institute an inquiry into the manner in which an animal learns to thread its way through a labyrinth of passages. He experimented with rats, which are suitable subjects, since under natural conditions they grow accustomed to narrow and branching passages. Professor J. B. Watson has carried the investigation a step further. The maze is built on the model of that at Hampton Court, and involves a 40-ft. or 50-ft. run from the entrance to a central chamber in which lies food, and in which the animals have been previously fed.

Two kinds of error are possible: (1) To take a wrong turning which will lead into a cul de sac; or (2) to follow a longer and more roundabout course when a shorter passage is open. It takes a little time—two or three traverses—to establish in the experience of the rat that such a traverse "means" food. In the early trials of the maze all possible errors are made and the time occupied is variable and long. With successive traverses the errors gradually diminish, the right turnings are more often taken, and the times occupied in the complete traverse are progressively shorter, till after, say, thirty runs, the course has been learnt with accuracy, and the times are nearly constant, about two feet being covered per second. For example, the average time in minutes, based on experiments with nineteen rats, in successive trials, here arranged for convenience in groups of three, are as follows:

Trials	1-3	4-6	7-9	10-12	13-15	16-18	19-21	22-24	25-27	28-30
Times	10·14	2·81	1·66	·95	·71	·52	·60	·46	·42	·36

In the first three trials we may surmise that the "meaning" had not yet been well established, and that the animals were timid amid new surroundings. After that, save in the seventh group, there is a fairly regular decrease in the times. From specially devised experiments, Professor Watson is confident that the guiding sensation cues are those derived from running a certain distance. So much trotting along is the experience *a b c*, which is followed by *d*, the appropriate turn to right

or left. There is little or no guidance from sight or smell. In his latest experiments he has used a maze in which the lengths of the runs can be increased or diminished without disturbing the number and sequence of turnings; and he finds that the animals which had learnt the lengthened form of the maze and were then suddenly introduced into the shortened form in nearly all cases ran squarely into the ends of the alleys affected by the change. On the other hand, the animals which had become habituated to the shortened form attempted to round corners at the old distances, regardless of the fact that the alley into which they then should have turned was farther along in the course.

Like observations on the manner in which a maze is learnt have been made on other animals—the dancing-mouse, the guinea-pig, the sparrow, the pigeon—and, with simpler forms of the maze, on the frog, the turtle, the perch, and the crawfish. Dr. Kinnaman's results with two monkeys show the same method of learning as in other cases; they, too, attain to about the same rapidity of progress—some two feet per second; but Professor Watson, after comparing these results with those for rats, says that, taking the two sets of records at their face value, the monkey isn't "in it" when it comes to a race in learning the maze. In all cases the method of learning seems to be the same—that of trial and error, with the gradual elimination of the false turns.

PERFORMING SEAL BALANCING A LAMP ON ITS NOSE

A different kind of experiment, first brought into prominence by Professor Thorndike, is that with puzzle boxes. An animal, say, a cat, is placed in a cage, with a tempting piece of fish outside; the door of the cage opens when a latch is lifted, and this may be effected by pulling a string passing along the top of, and inside, the cage, by clawing a button, by pressing a lever, and so forth. In some cases the door was opened by the experimenter (hidden from view) when the cat licked itself or scratched itself. Here, in the terms of our letter formula, the wished-for fish is *f*, the fit and proper series leading up to it *a b c d e*. Suppose that *a, b, c* are the initial experiences in terms of perception and behaviour, by what methods are the essential *d e* supplied? Observation seems to show that all possible letters of the alphabet—all sorts of scratchings and pullings and tearings—are tried, without any apparent plan of action based on a scheme of general ideas. As the range of possible modes of procedure is not large, it is natural that *d* should ere long be hit upon by chance ; then there follows the gradual linkage of the whole correct series through repetition, and the elimination of random and useless actions. There is no clear evidence of the perception of the relation between the means and the result. The reason why scratching itself should have the effect of opening the door cannot well be obvious to the cat ; and yet this is established as the d in the series in just the same manner as pulling a string.

In conducting experiments with monkeys it is convenient to place the piece of banana or small bunch of grapes, which functions as f, inside the cage. They learn the trick of the fastenings much more rapidly than cats ; they can deal with more difficult and more delicate catches, and with combinations of modes of releasing the door-latch, six or seven acts of behaviour having to be carried out in serial order. This is to be expected, first, because the monkey is the more intelligent animal ; secondly, because its ability to grasp with the "hand" is a distinct advantage ; and, thirdly, because its natural curiosity and its tendency to pick and push at anything which is projecting and loose are well-marked characteristics of its kind.

But unquestionably the learning of the trick of dealing more rapidly and surely with more complex puzzle boxes may be due to the formation in the monkey mind of an ideal plan of action with free memory images, to the thinking of $d e$ as the appropriate means of attaining to f, and the execution of requisite behaviour, because it is seen to be related to the wished-for end.

Here there is diversity of opinion. Dr. Thorndike sees no evidence to lead him to believe that ideas and free memory images are formed, and Professor Watson has come to the same conclusion. On the other hand, Mr. L. T. Hobhouse is satisfied that such ideas are formed ; and Dr. Kinnaman sees evidence of the framing of memory images.

CONSUL, THE TRAINED CHIMPANZEE, AT BREAKFAST
From a photograph by G. Meggy

It is not clear, however, what kind of evidence would satisfy all parties to the discussion. But there is one question the answer to which may throw definite light on the subject. Do animals supply the $d e$ through imitation ? We must first be clear as to what we desire to ascertain through experiment and observation. There are two kinds or stages of imitation. There is the natural tendency, shown by many animals, to do what they see others doing, just because they see others doing it and are instinctively provided with follow-my-leader proclivities. Few would be so bold as to doubt the existence of this kind of imitation in the monkey's inherited make-up. There is, secondly, imitation which is subservient to a purpose—the imitation which enables the imitator to insert missing links in the means to the attainment of an end. In the terms of our alphabet scheme, it is the supplying of $d e$ because it is seen that $d e$ affords just what is wanted to complete a plan of action, though one did not happen oneself, unaided, to think of this particular means of attaining one's object. One would certainly suppose that if a monkey were trying to think of the right means to get a piece of banana that was beyond his reach, the means being at hand, and if, having failed to utilise his opportunities, he saw the means effective in the hands of his master, he would imitate the successful procedure. Mr. Hobhouse reports that this was the case with his monkeys. Professor Watson finds that, with his monkeys, it is not the case. The difference may lie in the monkeys, or in the skill and caution of the observer. I will give a short account of the negative results obtained by Professor Watson, for the most part in his own words.

A TRIUMPH OF ANIMAL TRAINING : TWO MONKEYS RIDING BICYCLES
Photograph by Bert, Paris

It should be stated that the monkeys concerned in the following tests were very intelligent, one of them, a rhesus, the quickest animal in learning mechanisms Professor Watson has ever observed. The animal was tethered in an open floor-space; a grape was placed out of reach; a light 10-in. stick with a 2-in. T-piece fastened to one end was left near. In order to get the grape the monkey would have to hook the T-piece behind the grape and pull it in. Three monkeys were given several hundred trials; never once was the slightest effort made to use the rake in the right way. Then Professor Watson tried to show them how to draw in the food. He would wait patiently until he apparently had their attention, and then slowly hook the T-piece round the grape and slowly draw in the food till the monkey could grab it. Though this was often done, never once did any of the animals subsequently push out the rake, hook the blade round the grape, and then pull it in. Again, a piece of banana was placed in the bottom of a tall bottle, and a sharpened stick was stuck into it. The animals on their first trial immediately grasped the stick and jerked out the banana. The experiment was then repeated, but the stick was not stuck into the banana. The monkeys, as before, jerked out the stick, threw it down, and tried to reach the food with the paw. Professor Watson tried to teach them the use of the stick, but failed signally. Yet again, the animal had to push a piece of banana from the middle of a 15-in. glass cylinder fastened to a table. A light 16-in. stick was placed near. The monkeys clawed and bit at the cylinder near the banana; they then tried to reach the food from one end with the paw. Day by day Professor Watson took the light stick, put it into the cylinder, and then slowly forced the banana out. The monkeys watched its passage and grabbed it when it came out at the end of the cylinder. They did not use the stick themselves. Meeting no success with this procedure, he put the stick inside the cylinder just in contact with the food; the monkey would simply have to push the food out. Invariably, after oft-repeated lessons, the monkey first jerked out the stick and then began its efforts to extract the food by means of its paw.

Professor Watson tells us that the anecdotal material he has collected while observing his monkeys would compare favourably with that presented by Romanes and others; but he says that close examination of such acts, especially during the period of their genesis, does not lead him to think that the higher forms of imitation are present, or that the monkeys ever resort to learning by the perception of the relation between means employed and results attained. But he wishes it to be understood that his present conviction stands ready to be changed as soon as the evidence calls for it. Such is a sample, and, of course, it is only a sample, of the kind of experimental work which is being carried out.

Although the evidence is, to my mind, not quite convincing, it is not improbable that we shall learn that animals have at least the earlier and simpler forms of ideas, and shall learn, too, more of the nature of these ideas and the manner in which they are used. When the research has been carried somewhat further, we shall need to correlate its results with the more open-air results of such observations as were so admirably recorded by Le Roy. Meanwhile, we must keep an open mind, duly cautious and critical, but wholly without prejudice. If language is the chief tool of thought, we cannot expect, on general grounds, the animal to get far without the free use of the potent symbolism of words. The facts, as they seem to be gradually shaping, are what the evolutionist might well anticipate. After all, the trial and error method, though it may seem to us rather hopelessly haphazard, is likely to be successful in at least 99 cases out of 100 in the natural life of the animal. One thing, at any rate, comes out with increasing clearness—that the perceptions of the higher anima's are wonderfully delicate and keen, probably in their own fields as keen as, or keener than, our own, and that these dumb brutes are marvellously quick—perhaps as quick as, or even quicker than, most of us—to profit by experience.　　　　C. LLOYD MORGAN

THE PATHOS OF THE WILD: A YOUNG ELEPHANT FINDING THE DEAD BODY OF HER MOTHER AND TRYING TO ROUSE HER

THE CHAMÆLEON

A species of lizard which possesses the power of "rapidly adjusting its colours to those of any of its normal environments.

Prof. Poulton

LION AND LIONESS SIGHTING PREY IN THE DISTANCE

These denizens of the open desert "are often nearly devoid of markings, being sufficiently concealed by their obliteratively shaded colour." *Prof. Poulton*

THE CONCEALING COLOURS OF ANIMALS

NATURE'S STRANGE DISGUISES FOR PROTECTION AND AGGRESSION

By EDWARD B. POULTON, F.R.S.

Hope Professor of Zoology in Oxford University

THE fact that animals are commonly well concealed in their natural surroundings must have been known to the primitive hunter, taught by the difficulties of his craft. The amount as well as the perfection of this concealment is liable to be greatly under-estimated by those who are chiefly familiar with animals in captivity, with specimens in museums, or illustrations in books.

An animal may be concealed by its likeness to a particular object, such as a leaf or twig, the lichen on the bark of a tree, a stone, or a lump of earth. It may also be hidden by such a general likeness to its surroundings that it becomes more or less invisible. Form is of great importance in the first kind of resemblance ; of less importance, and sometimes of no importance, in the second ; thus, the concealment of a stick-like caterpillar or a leaf-like butterfly is largely brought about by form, while, on the other hand, the shape of a transparent bluish jellyfish, floating on the surface of the ocean, is of little importance for the purpose we are considering.

Looking at the earth as a whole, we find two principal classes of surface from the point of view of the concealment of living forms. There are, first, the uniform expanses of the deserts and the sea at a distance from the shore ; and there are, secondly, the expanses of land diversified by the myriad forms of plant life, and the coast lines and neighbouring shallow waters diversified by animal and vegetable growths of the most varied kinds. It is common for animals in these latter parts of the earth's surface to exhibit detailed resemblances to particular objects, while the general resemblances are more characteristic of the uniform areas.

The separation between the two classes of surface is not by any means complete ; for example, parts of the ocean remote from land are permanently covered with floating seaweed, while in close proximity to diversified surfaces of land or shallow water are the broad stretches of uniform sand. Furthermore, large tracts of Africa are desert-like during the dry season, but support an abundant and varied vegetation during the wet. So also the distinction between the two kinds of resemblance is not sharply marked ; for both are often to be found on the same animal. Thus, its general background of colouring may resemble a surface of earth or sand, while its markings convey the effect of particular objects, such as dead leaves or stems, found upon that surface (see, however, page 112). Again, the stones that are commonly scattered over the desert areas of the earth's surface may be resembled in shape as well as colour.

Thus certain plants of the Karroo, and other parts of Africa which are desert during the dry season, present the most beautiful and detailed likeness to stones (see figure on page 110), and the same is true of certain grasshoppers (Acridians). This observation was made in 1812 by W. J. Burchell.

It must be remembered that there are also general resemblances, not only to uniform areas and the surfaces of the diversified parts of the earth, but also to particular objects. This will be made clear by an illustration. When we are shown a stick-like caterpillar or a leaf-like butterfly, the meaning of the form and colouring is clear and the concealment obvious. When, however, we look at the large caterpillar (figure, page 111) of the Privet-hawk moth apart from its surroundings, we find it difficult to realise that such a large green object, with its distinct oblique white stripes bordered with purple, can be anything but highly conspicuous among the leaves of the privet or lilac on which it is most commonly found. But, as a matter of fact, the caterpillar is well concealed on these plants, and it is often a task of some difficulty for even an experienced naturalist to find an individual whose presence is indicated by freshly eaten leaves.

There is little doubt that the purple-margined white stripes add greatly to the resemblance by breaking up the green surface of the large caterpillar, and perhaps by giving the effect of leaf-shadows tinged with complementary colours. It is to be noted that the purple margins do not appear while the caterpillar is in its younger stages and comparatively small, and that they become a marked feature only in the last stage. There is, further, strong evidence that this caterpillar is specialised for concealment, inasmuch as it possesses the power of developing a different and appropriate shade of green, according as its food-plant is privet or lilac. I have fed Privet-hawk caterpillars, hatched from eggs laid by the same moth, on these two bushes, and found that those on the privet became a brighter, yellower green than those on the lilac. The same difference is also to be found in the wild state. The concealment of the Privet-hawk caterpillar—which is an undoubted fact, however difficult it may be to realise it apart from the surroundings—is a general resemblance to leaves, while that of a leaf butterfly (such as Kallima ; figure, page 111) is a special resemblance. The one produces the effect by colour almost unaided ; in the other the resemblance in shape is as perfect as are the details of the markings.

It has been necessary to introduce in the preceding paragraph an example from the insect world instead

N

K

of considering it in its place on a later page, in order to illustrate a class of resemblances which is probably very large, but as yet imperfectly appreciated and understood. The concealment brought about by general protective resemblance can only be appreciated by the study of living Nature, and can only be completely understood by the insight of the artist. The principle itself was brought forward by the present writer in a paper which was read before the Entomological Society of London on November 7, 1883, and published in the "Transactions" of that body in 1884 (pages 48-50); and it was afterwards analysed and advanced in a remarkable way by the discoveries and interpretations of the great American artist-naturalist, Abbott H. Thayer. These have grown out of the important fundamental discovery which I now proceed to describe.

We are here considering a general principle of colouring, which holds more or less extensively throughout all the animal groups that depend on concealment. The white under sides of animals were formerly explained on the same principle as the white colour so commonly observed in birds' eggs laid in concealed nests, or the pale tints of subterranean forms, or those found in caverns. In all these cases it was assumed that the colour, being useless and no longer sustained by natural selection, had disappeared, or was on the way towards disappearance. Mr. Thayer, however, showed in 1896 that the gradation usually seen in animals, from dark on the back, through increasing paleness on the sides, into white underneath, is of the utmost value in promoting concealment, and bears the evident stamp of adaptation. He showed that the colouring is brightened so as exactly to compensate for the gradually decreasing illumination of the animal's body at lower and lower levels. The back, which is darkest of all, is exposed to the full illumination of the open sky, the sides, becoming lighter and lighter, are illuminated less and less, finally passing into the white of the belly, where the illumination is least. And the gradation in tint is accurately adjusted to the varying degree of illumination from the normal source. The remarkable degree of accuracy may be inferred from the fact that Mr. Thayer's models were painted by him for the special illumination of the position where they may now be seen, in the Natural History museums of London, Oxford (figure, page 112), and Cambridge.

PLANTS THAT RESEMBLE STONES

This plant, Mesembryanthemum bolusii, grows in the African Karroo, and resembles the stones around it. From an illustration by Sir W. Thiselton-Dyer

The wonderful effect which he has produced could only have been obtained in this way, and the balance between illumination and depth of colour is so delicate that it is only at its best at the time of the year when Mr. Thayer did his work, and is disturbed by the fine film of dust which gradually collects on the surface of the models. The appearance of solidity and relief is conferred by shadow, and the neutralisation of shadow on the surface of animals and of Mr. Thayer's models removes this appearance, and produces a ghost-like, translucent effect. When to the neutralisation of shadow is added a colour which harmonises with the surroundings, the animal seems to melt into and become one with its background. Natural selection, operating upon variations which tend in the direction of concealment, has produced what all artists in all ages have, until now, failed to discover.

The artist, by *painting in* the shadow, can give objects on the flat surface of his canvas the most convincing appearance of relief; but until Mr. Thayer's discovery he never realised that by *painting out* the shadows of a solid object he can destroy the appearance of relief and solidity. I once asked Mr. Thayer how it was that he found what so many have missed. He replied that he found it in Nature. He observed that wild animals in their natural surroundings were ghostlike and inconspicuous — so much so indeed that in order to render them clear and distinct on the canvas it was necessary to paint them silhouetted against the light on the sky-line, or else as they would appear in an unnatural illumination. He was led by observation to examine their colours carefully, and thus discovered the principle of the neutralisation of shadow, which he then confirmed by painting models so that they became well-nigh invisible, or, in the most favourable circumstances, entirely invisible.

The colour harmony between such an animal as the hare and its background is produced in the following manner. The dark, earth-coloured back, bathed in the cold blue-white of the sky, resembles the earth, which is, of course, itself equally illuminated from the same source; and a similarly close resemblance to the ground is presented by the cold blue-white of the belly steeped in shadow and yellow earth reflections. Passing from the back towards the belly the illumination becomes less and less, and the colours correspondingly brighter. In fact, the animal appears to possess a uniform earth colour owing to the fact that its colours are not uniform, but always related to the amount of light received from the open sky. Mr. Thayer has painted models (placed beside those I have mentioned, as shown in the figure on page 112) and the feathers of stuffed birds so that they are of a uniform depth of colour, and these appear to be darker and darker as the illumination decreases in passing from the back to the belly. Such a specimen is cut off from the ground and rendered prominent by a deep black shadow.

The principle of shadow neutralisation was recognised by the present writer in 1887 and 1888, but its far-reaching importance was not grasped until Mr. Thayer published his great discovery in 1896. The earlier examples are of interest, for they show the same principle of concealment at work among insects. The first was a stick-like caterpillar (allied to the one figured on this page), which holds on to the branch by its two posterior pairs of claspers and stands out like a side twig.

A CATERPILLAR THAT RESEMBLES A TWIG
The caterpillar of the waved umber moth clings to a stem with its hind claspers and stands out like a twig

The body of the caterpillar between the last pair of claspers and the last but one is cut off from the branch by a distinct groove or chink, and here the shadow would show that the two objects are really different and not continuous, were it not for the neutralising effect of a bright colouring. The principle is the same as that by which there is elimination of shadow in the chink between the earth and the rounded body of a mammal or bird. The second instance was the green chrysalis of the Purple Emperor butterfly (figure, page 112), which, in spite of its thickness, closely resembles one of the flat leaves of the sallow on which the caterpillar feeds. The appearance of flatness was shown to be produced by a white stippling, which neutralised the shadow that would have betrayed the rounded surface, and hence the thickness. The amount of stippling was adjusted to the steepness of the slope. "By this beautiful and simple method a pupa, which is 8·5 mm. from side to side in its thickest part, appears flat, and offers the most remarkable resemblance to a leaf which is a small fraction of 1 mm. in thickness."

A complete and detailed account of Mr. Abbott H. Thayer's discoveries has been published in a

THE DRY-SEASON KALLIMA, OR LEAF BUTTERFLY
The figure shows a hanging position. In wet-season specimens the bent leaf-tip is barely represented

THE PRIVET-HAWK MOTH CATERPILLAR
The purple-edged white stripes of the caterpillar help to make it inconspicuous among the privet leaves

finely printed and illustrated work by his son, Mr. Gerald H. Thayer, under the title of "Concealing Coloration in the Animal Kingdom: An Exposition of the Laws of Disguise through Colour and Pattern." In this volume there are embodied the facts and conclusions of all Mr. A. H. Thayer's papers dealing with the subject, as well as a large amount of new matter and great numbers of novel and very beautiful illustrations. Some of these—to be mentioned in the later pages of this article—illustrate ideas which naturalists will probably consider erroneous; but nothing in the work ought to be allowed to diminish or obscure the importance of the great fundamental discovery of the neutralisation of shadow and its significance. This must stand for all time as a landmark in the progress of investigation into the colours of animals.

After showing that *light and shade* on solid objects are the means by which "the eye is made aware of their *existence*, their *main form*, their *position*, and all their *minor modellings*," the author points out "the fallacy of the statement . . . that a protectively coloured animal of the type described above escapes attention because, being of a dull-brown colour like the ground and the bushes, it looks when it sits motionless like a clod or a stump—or some such inanimate thing. For clods and stumps are solid objects of a uniform tint, and manifest to the eye, by the laws of light and shade, not only their solidity, but all their smaller modellings. They are not inconspicuous, except in so far as their great abundance makes the eye inattentive to individual ones. The protectively coloured animal, on the other hand, is, as it were, *obliterated* by his counter-gradation of shades, and in the cases where he escapes notice, it is by virtue, not of the eye's perceiving his solid form, and taking it for that of an inanimate object, but of its failure to recognise it as a solid object of any kind, seeming, if it rests on it at all, to see *through it to what is beyond*" (Thayer: page 19). The effect produced by Mr. A. H. Thayer's models is of this kind—if the eye rests on them at all it seems to see through them. The soundness of these conclusions may be demonstrated, as the author suggests, by holding an obliteratively shaded creature "*upside down*, in its normal lighting, and against its normal background. It will be seen not merely that its ghostly dimness has vanished, but that it is extraordinarily conspicuous—just doubly as conspicuous, in fact, as any stick or clod placed in the same position would be. For an inverted animal not only lacks counter-shading, as a stick or clod does, but is even fully shaded the *wrong* way—brightest where it catches most light, and darkest where it catches least. No other conceivable arrangement of colours could make an object as conspicuous as this. Yet an animal held thus inverted is, materially, as truly 'coloured like his surroundings' as he ever was" (Thayer : page 20).

Other convincing evidence that obliterative shading is by far the most important element is offered in the same work. Thus Mr. Thayer reproduces (in his figure 6) a photograph of the Plymouth Rock hen posed against a background of skins of the same breed. In this fowl, as in so many artificially selected races, the natural obliterative shading has been lost, and the hen stands out conspicuously in spite of a pattern which is, of course, precisely the same as that of its background.

The markings on an obliteratively shaded animal are "not an exact reproduction of the actual background pattern, but a picture of that pattern as it looks when more or less altered and refined by distance." The animal "must bear *a picture of such background as would be seen through it if it were transparent*" (Thayer : page 31). This picture, Mr. Thayer maintains, is made up of a smaller pattern on the higher parts of the animal, which are seen from the side against a more distant background. The dorsal pattern looked down upon from above is coarse, to match the details of the ground below, which are only reduced by a distance equal to the height of the animal's back.

The following passage describes these concealing markings on the obliteratively shaded body of an animal as Mr. Thayer sees them : " Only an artist, perhaps, can rightly appreciate the profound and perfect realism of these background pictures worn by birds and other animals. Just as a good caricature drawing of a man looks in one sense more like the

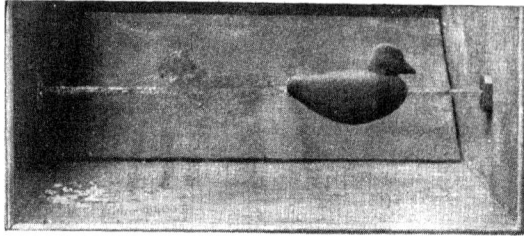

MR. THAYER'S MODEL IN THE OXFORD MUSEUM

The lefthand duck, being counter-shaded, is almost invisible ; the righthand duck, coloured uniformly like its background, is made conspicuous by shadow

CHRYSALIS OF THE PURPLE EMPEROR

The chrysalis, as it hangs by its tail, gives the appearance of a leaf in form and thickness, though it is much thicker than one. Photo by A. E. Tonge

man than the man himself, so, in a far more high and wonderful degree, do these pictures on animals' coats exceed the verisimilitude of the actual scenes they imitate. They have been compounded and epitomised and clarified till only pure, essential typicality remains. . . . Just as in great human art, but far more essentially and surely, the trivialities and chance individual abnormalities have been eliminated, or subordinated to the scheme of ultimate, impartial typicality. To learn, then, the purely characteristic colours and light-and-shade effects of leaves and sticks and stones and other parts and types of natural scenery, we should look not at the scenes themselves, but at the animals whose patterns picture them. The essential realism of these pictures is such as the keenest artist among men could never hope to match. Nay, for Nature herself [The words " Nature herself " are intended by the author and his father to express " natural selection, pure and simple and omnipotent "] has made them—Nature herself has discovered and applied, to a point utterly beyond human emulation, the art of painting pictures" (Thayer : pages 39, 40). The naturalist who, not being also an artist, is therefore unable to see all this in the concealing colours of animals, would naturally wish to learn from the man who has thrown so much new light upon the whole subject ; and the naturalist would probably be right in accepting the above conclusions, at any rate in large part. It cannot be doubted, however, that his faith is likely to be severely shaken by some of Mr. Thayer's illustrations, to be considered in later pages, and that a scientific instinct would rightly make him very suspicious of backgrounds like that of the wood duck in Mr. Thayer's plate 3, which was " copied, colour-note for colour-note, from the bird himself."

The " Obliterative Colouring " of A. H. Thayer, of which a brief account has been given above, is a further and far higher development of the principle of " General Protective Resemblance," which I briefly described and illustrated in 1883. The immense advance is, of course, due to the discovery, in 1896, of the great and far-reaching law of obliterative shading. Mr. Thayer uses the term *mimicry* to express what I called *special protective resemblance*, and he discriminates between *mimicry* and *obliterative coloration* just as I discriminated between *special* and *general protective resemblance*, always allowing, however, for his classical discovery of shadow obliteration. The parallelism between our lines of thought may be judged by comparing the following passage from page 25 of Mr. Thayer's book with the words on page 109 of the present article: "Protective or disguising coloration, then, as we define it, falls into two main divisions; the one including *concealing* colours mainly based on counter-shading, and the other including mimicry, in almost all its branches. As has already been explained, the goal of the former principle is the rendering animals *invisible* in their normal haunts. [Compare also the description of the concealment of the Privet-hawk caterpillar, on page 109, with Mr. Thayer's words on page 36 of his book on the patterns of the forest *Caprimulgidæ* (Goatsuckers): "The fact that none of these detail-picturings is so patently realistic as to be appreciable to everyone when the bird is seen away from its natural environment is part of the very marvel of the thing."] Mimicry, on the other hand, aims at *deceptive visibility;* it makes an animal *look like something else than what it really is.* It will be seen that the latter principle is open to unlimited variations of method and result, whereas the former, as we have proved, is in its main essentials strictly limited. There are innumerable kinds of solid objects for animals to simulate in appearance, but there is only *one* way to make a solid object in a natural lighting cease to appear to exist. Both these are principles of disguising costume, and both are protective, yet they are fundamentally unlike. . . . We have, then, *obliterative coloration* and *mimicry* as the two main principles of 'protective coloration.' The same ideas were expressed in the description of the ruffed grouse, beautifully represented in its forest environment in Thayer's plate 2 : "Nature has, as it were, used the bird's visually unsubstantialised body as a canvas on which to paint a forest vista. In this there is nothing of mimicry, as we define it. Mimicry uses the solid aspect of an animal's body, modified in form and colour, to simulate some other solid object. But vista- or background-picturing, based on the complete obliteration of the animal's solid aspect, which causes its actual form to pass for an *empty* space, is a widely different principle."

A FLOWER-LIKE MANTIS

We have here an example of aggressive coloration, other insects being attracted to the flower-like mantis and so caught

THE COTTON-TAIL RABBIT

An example of protective coloration among mammals
Photograph by H. C. White Co.

It is certainly true that the word "mimicry" was originally used in Mr. Thayer's sense by H. W. Bates to include the above resemblances as well as the deceptive likeness of one animal form to another. It is, nevertheless, a very inconvenient practice to employ the same term for the resemblance of a caterpillar to a stick and of a moth to a wasp. Concealment, pure and simple, is the object of the one; the suggestion of efficient defence is the object of the other. Although deprecating the use of "*Mimicry*" for this purpose, I recognise the desirability of possessing new terms to express more precisely the differences which I formerly discriminated by the use of (1) *Special* and (2) *General Protective Resemblance;* and I suggest that "*Imitative*" (with "*Imitation*") should be employed for the first, and Mr. Thayer's "*Obliterative*" (with "*Obliteration*") for the second. The more accurate technical term "Eikonic" (with "Eikon") may be employed as alternative to "Imitative," and "Aphanistic" (with "Aphaneia") as alternative to "Obliterative."

Adjustable protective resemblance is the highest and most complex of all the methods by which concealment is attained. An animal which possesses this power can adjust its colours to those of any of its normal environments. The adjustment may be *rapid*, in forms which are apt to wander from one environment to another, such as fish (figure, page 119), frogs, the chamæleon, certain crustacea, and cuttlefish. Probably this rapid adjustment is always initiated by the influence of reflected light upon the eye and a resulting nervous impulse conveyed by the optic nerve to the brain. Thus stimulated, the brain-centre discharges impulses by the outgoing nervous paths to the pigment-cells, whose partial or complete contraction or expansion determines the colour. *Slow adjustable protective resemblance* is found, on the other hand, in comparatively sedentary forms passing their lives in a single environment, which, however, is only one out of many different environments, any other of which might have been prearranged for the individual. Thus, the Peppered moth (*Amphidasys betularia*) lays its eggs on trees or bushes with twigs which may be dark brown, light brown, green, or whitish: the Lappet moth (*Gastropacha quercifolia*) may lay them on old lichen-covered hedges or on hawthorn of younger growth. In all such cases the power of adjustment is of the utmost importance; but it only requires to be exercised once or twice—and generally once only—in a lifetime. So far as the subject has been investigated, it appears that slow adjustment is effected through the stimulus of light upon nerves of the skin, and not upon the eye.

Adjustable obliterative shading is better considered under the reptiles (page 117), for up to the present this principle has been observed in no other class.

Changes corresponding with the summer and winter are well known in northern mammals and birds, etc., and corresponding with the wet and dry seasons in tropical butterflies. These will be more usefully dealt with in detail later.

A plausible interpretation of animal colouring is well refuted by Mr. Thayer: "This indisputable fact, that animals tend to be *dark* in thickets and dusky forests, and *pale* on the glaring desert and on ocean beaches, is a complete refutation of the theory that the counter-shading is due to the *tanning* effect of light. On the other hand, the idea that the paleness of desert creatures is due to *bleaching* is equally well answered by the fact that their *shadowed under sides* are still the lightest, as in the case of almost all other animals " (Thayer: pages 27, 28).

Either kind of concealing coloration — obliterative or imitative—may be employed for the purpose of seizing prey or defence from enemies. Anticryptic, or aggressive, resemblances for the purpose of attack may thus be divided into (*a*) Aggressive Obliterative Coloration, and (*b*) Aggressive Imitative Resemblance. The far more abundant examples of procryptic, or protective, resemblance may be similarly classified as (*a*) Protective Obliterative Coloration, and (*b*) Protective Imitative Resemblance. (The corresponding technical terms would be Antaphanistic, Anteikonic, Proaphanistic, and Proeikonic.)

CRYPTIC RESEMBLANCES IN THE MAMMALIA

Obliterative coloration is by far the commonest mode of concealment in mammals, and it is often employed for the purpose of attack as well as defence. The great majority of quadrupeds are, as Mr. Thayer states, "darkest on the back and lightest on the belly, usually with connecting intermediate shades. *White* is by far the commonest colour for the middles of their under sides, while the dark of the upper sides very often culminates in a black or dusky median line, a sort of painted 'ridge pole,' laid along the centre of the back, over the

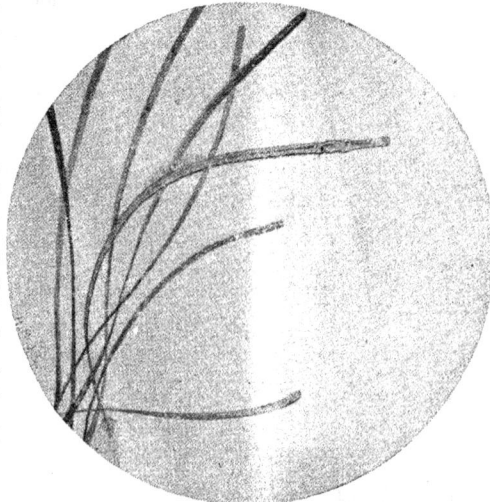

THE ENGLISH PIPE FISH

This fish resembles the green strap-like leaves of the zostera. Painted from life in the Plymouth Laboratory by Mr. H. M. J. Underhill

tips of the dorsal vertebræ. With or without such extreme accentuation, complete obliterative shading characterises most of the species of almost all the mammalian orders " (Thayer: page 119).

A good example of protective obliteration is represented in the cotton-tail rabbit (figure, page 113) while aggressive obliteration is illustrated by the jaguar. Mr. Thayer justly claims concerning the former, as well as certain other coloured illustrations of his work: " These paintings of ours . . . are, we believe, the first ever published which rightly illustrate and in some respects do justice to the wonderful effects of obliterative coloration, based on the great law of *obliterative shading* . . . The world has had enough, or must soon have had enough, of pictures of birds and beasts with their light and shade falsified to make them show. Outdoor Nature as it really is, in the matter of marvellous and exquisite visual correlations between animal and environment, offers to art, in this late age, an almost boundless virgin field " (Thayer: pages 127–8).

The value of obliterative colouring in the normal position of a mammal is well seen by placing an animal such as a hare on its back, so that the effect of its colour gradation is reversed.

Looking at the Mammalia as a whole, the absence of bright colour is remarkable; and when it is seen, as in certain monkeys, its seat is the skin, and not the hair. Strong contrasts of dark and light are, on the other hand, well known, and may be strongly obliterative, as in the zebras, and probably so on the legs of the okapi. Mr. Thayer truly points out that mammals of the open country, such as lions (see page 108) and many species of hares, are often nearly devoid of markings, being sufficiently concealed by their obliteratively shaded colour. In the concealment of forest forms, on the other hand, pattern plays a very important part. Darwin did not accept the usually received interpretation of the markings of the tiger as an aid to concealment, believing that they were first developed in the male by sexual selection, and later inherited by both sexes. In this case Darwin's wonderful scientific judgment was probably at fault, for it is far more likely that the pattern is an aggressive resemblance to vistas of upright stems and their shadows. But, as Mr. Thayer points out, even the tiger's inconspicuousness is primarily due to obliterative shading, and only secondarily to the pattern.

Imitative resemblance is very scarce in mammals as compared with obliteration, and when present is usually observed in arboreal forms. It is said that the manis, or pangolin, clings to a trunk with the anterior part of its body standing out like a broken branch. Some bats also resemble the roughnesses of bark, while sloths look like masses of a vegetable nature. Sloths are sometimes green, owing to the presence in their hair of an alga—perhaps the food of the larvæ of small moths which breed in the fur of these remarkable animals.

The seasonal changes of Arctic mammals have been insufficiently studied. Some species appear to become white by a moult, which still occurs when the individual is removed to a more southern land with a mild climate. In other species there is some evidence of changes in the hair caused by the stimulus of cold. It is quite likely that the change is produced in more than one way; but the whole subject requires careful experimental investigation.

The artificial production of low temperatures is now so easy that there should be no difficulty in conducting a long series of well-considered experiments on this interesting and important subject. It is to be observed that seasonal whitening in the north is aggressive (as in the Arctic fox and ermine), no less than protective (as in the Alpine hare and lemming). Obliterative shading is impossible in northern and Alpine forms which match the snow; for, their backs being white, no lighter shade is possible on any other part of their bodies. "But," as Mr. Thayer states, "the upward reflection from the snow itself goes far towards concealing the shadow on such animals."

ARCTIC AND ALPINE ANIMALS IN THEIR SUMMER COATS
In the summer the Arctic fox, the Alpine hare, and the ptarmigan, shown here, are coloured like the heather and rocks among which they live. Natural History Museum, South Kensington

Mr. H. Lyster Jameson obtained evidence in 1895 that the great majority of individuals of the common house mouse living wild on the sand-hills of North Bull, Dublin Bay, are more or less pale and sand-coloured. Only five specimens out of thirty-six were as dark as the ordinary mouse. North Bull is, at the lowest tides, separated from the shore by a strip of water 20 yards wide, which, however, is crossed by a bridge. Mr. Jameson suggests the reasonable hypothesis that the local change of colour has been due to selection by hawks and owls, which may often be seen hunting for prey on the island. These birds, guided by sight, would most readily capture the darker individuals. The North Bull is a formation of recent date, and Mr. Jameson's study of ancient maps allows "about 100 years for the evolution of the pale race, with a maximum limit of 120 years."

Before concluding the mammalia, it is necessary to say a few words about the part played by stripes and patches of white or pale colour. Mr. Thayer calls these "secant" and "ruptive" markings respectively, because he believes that their function is to "cut up" or "break up" the pattern, so that, for example, the silhouette of a dark animal against the sky loses its revealing form, and is cut or broken up by the light markings into irregular dark masses. Although this interpretation is no doubt often correct, Mr. Thayer adopts it for slow-moving, specially defended mammals, such as the skunk, whose markings, whatever secondary significance they may possess, primarily serve to warn their enemies.

ARCTIC AND ALPINE ANIMALS IN THEIR WINTER COATS
During the snow season the Arctic fox, Alpine hare, and ptarmigan adapt their coats to the pure white of their environment. Natural History Museum, South Kensington

CRYPTIC RESEMBLANCES IN BIRDS

Mr. Thayer gives a far more complete account of obliteratively shaded birds than of any other group of animals. He classifies the examples under the following heads, according to the character of the background picturing which he recognises upon the obliteratively shaded body : (1) a picturing of the larger details of the nearer ground, as in snipes and woodcocks (not including the European woodcock) ; (2) an intensely elaborate picturing of the minute details of the near ground, as in forest goatsuckers ; (3) a picturing of the more distant background, as in certain forest grouse, owls, and the European woodcock ; (4) grass and heather patterns, as in sparrows, larks, pipits, quails, ptarmigans, and many other birds ; (5) bark and forest vista picturing of climbing birds, such as woodpeckers, wrynecks, nuthatches, and creepers ; (6) beach-sand and pebble patterns of shore birds ; (7) reed picturing of bitterns and certain herons ; (8) water markings and colours of certain aquatic and swamp-haunting birds, such as duck, teal, etc. ; (9) the coloration of ocean birds.

To each of these types of colouring Mr. Thayer and his father have devoted much careful observation. The illustrations are numerous and very beautiful, the Rocky Mountain white-tailed ptarmigan on its nest (Thayer : Fig. 41) being the finest photograph of a bird I have ever seen. This magnificent representation of the obliterative effect of grass picturing was obtained by Mr. Evan Lewis. I do not doubt that Mr. Thayer's conclusions are, as a whole, correct, and that he has given us by far the best introduction to the study of the cryptic colouring of birds. The creation of a background out of the bird's own pattern has already been criticised (see page 112), and I must also express the belief that the effect of the author's admirable work will be weakened by his poetic but fantastic interpretation of the tints of spoonbills and flamingoes. No less than three out of the sixteen coloured plates illustrate the hypothesis that the rosy, red or white and red colours of these birds represent the sky at sunset or dawn. Not only is there the obvious difficulty that such tints are fleeting, but there is the further objection—fully recognised by the author—that against sunset or dawn itself the birds, whatever their colour, could only appear black ; and that, therefore, they could only represent the tints reflected in some other quarter of the sky. In deprecating the introduction of these plates I feel bound to allude to their remarkable beauty. They are magnificent ; but in such a place they are an injury, and not an aid to science. It is also extremely improbable that the author is correct in his opinion that the colours and patterns of male birds when far more brilliant than the females are still essentially cryptic. In the absence of incontrovertible positive evidence furnished by careful observations in the natural habitats of the species, it would be unjustifiable to accept the hypothesis illustrated by the beautiful representation of the peacock in Mr. Thayer's plate 1 ; of the male wood ducks in his plates 3 and 4 ; and the male birds of paradise in his plate 6.

Birds' eggs are commonly white when laid in inaccessible places or hidden from sight in holes or covered nests ; they are commonly coloured when exposed on the ground or in open nests. This antithesis suggests that the colour of eggs is protective — a conclusion obviously true of many birds, such as the plover. Certain species which nest on open ground even possess the instinct of scattering fragments of shell, etc., over the surrounding earth, thus creating a background which promotes concealment. And even when eggs are bright blue, and at first sight appear to be conspicuous, it by no means follows that they are really so in Nature. The whole subject requires investigation in the spirit of Mr. Thayer's work on obliteration. When such an inquiry is undertaken, it will probably be found that birds' eggs which stand out conspicuously against their cotton-wool backgrounds in the drawers of museums are very far from conspicuous in Nature.

Imitative resemblance is probably even rarer among birds than mammals. Northern birds in their white winter plumage, without obliterative shading (see page 115), may resemble lumps of snow, and Mr. Thayer also considers that the North American screech owl may sometimes resemble a bit of a branch. The ostrich in certain attitudes is said to bear a marked likeness to a bush. The concealing coloration of birds is more commonly protective than aggressive, but predaceous species with anticryptic colours and patterns are numerous.

The mode in which the beautiful winter plumage is gained by Arctic and Alpine birds (figure, page 115)

THE COMMON WILD DUCK ON ITS NEST
In colour and markings the bird resembles the undergrowth in which it has its habitat. Photo by W. Bickerton

requires experimental investigation, and offers at least as promising a field of inquiry as the corresponding changes in mammals.

CRYPTIC RESEMBLANCES IN REPTILES AND AMPHIBIA

The concealment of these two groups of animals is chiefly obliterative, and often due to a patternless single colour, harmonising with the tint of the surroundings, as in many brownish-green and bright green snakes, lizards, and frogs. Ordinarily, however, a background-picturing pattern is added to the counter-shaded body. Concealment is more commonly aggressive than in the higher classes we have been considering hitherto. Imitative resemblance is rare, as compared with obliterative, but many examples will at once occur to the naturalist, such as the protective likeness of the little green tree-frog (*Hyla*) to a bud or folded leaf, and the aggressive resemblance of the huge python, hanging like a broken branch from a forest tree, as it lies in wait for prey. We also find in the amphibia an interesting modification of the ordinary methods of concealment, assistance being rendered by the use of foreign objects. Thus, the horned frog of South America (*Ceratophrys*) sits motionless in a hole scooped in the ground, and also covers a large part of its back with earth. In part concealed by this means and in part by its dark greenish colour, the frog waits patiently for prey. Concealment thus afforded may be called allocryptic or adventitious cryptic resemblance. It is found in many insects, especially during their larval and pupal stages. In these, however, it is nearly always alloprocryptic or employed for the purpose of defence, whereas in the horned frog it is allanticryptic.

We now meet, for the first time, and in both reptiles and amphibia, the remarkable power possessed by the individual of adjusting its colour to that of any ordinary environment. It is not only possessed by chamæleons (see page 105), but also by other kinds of lizard and frog. In all cases the change is brought about by the stimulus of light upon the eye. It was shown in 1905 that the little South African chamæleon (*Chamæleon pumilus*) possesses the power of adjustable neutralisation of shadow, the illuminated side being dark, and the shaded side bright, and the compensation so accurate that the two sides of the animal appear to be of exactly the same shade of green. The observation was made much earlier by J. S. Beuttler, but the meaning of the adjustment was hidden until Mr. A. H. Thayer's great discovery of the neutralisation of shadow.

CRYPTIC RESEMBLANCES IN FISHES

Mr. Thayer points out that the fishes of the open sea possess a constant and simple obliterative coloration. "The gradation of shades and tints— from silvery-white bellies to dusky backs on almost all free-swimming pelagic and freshwater fishes, is true and delicate and exquisite beyond description. Among fishes of the same general form and habits, however distantly related, this obliterative shading varies scarcely at all from species to species. Even their colour-tones differ but little. They almost all wear exquisite, soft, water-colours—green, olive, grey, pearl-blue, silvery, deep olive-brown, etc. These tints are all components in the obliterative shading, and its resultant uniform, soft, water-tone, and have, in general, little independent pattern-effect" (Thayer: pages 161, 162). Exceptions are afforded by the glassy transparent pelagic larvæ of fishes, and notably by *Leptocephalus*, now known to be a stage in the life-history of the eels. All these immature fish possess the same method of concealment as the hydrozoa, or "jellyfishes." In *Leptocephalus* even the blood is transparent, a very rare phenomenon in the vertebrata ; for in nearly all of them respiration is only rendered possible by the red colouring matter of blood, hæmoglobin, the carrier of oxygen to the tissues.

Certain African fishes (of the Silurid genus *Synodontis*) " are much in the habit of floating or swimming leisurely on the surface with the belly in the air, as was well known to the ancient Egyptians, who have frequently depicted the fish in this anomalous position." Now, the two species in which this reversal of the usual attitude is best known exhibit a reversal of the usual system of colouring, the ventral surface being dark brown or black,

THE EGGS OF THE OYSTER-CATCHER LAID AMONG PEBBLES SIMILAR IN FORM AND COLOUR
The lefthand picture shows a view of three eggs, to be found exactly in the centre of the picture. The righthand picture shows a close view of these same eggs. From photographs by W. Bickerton

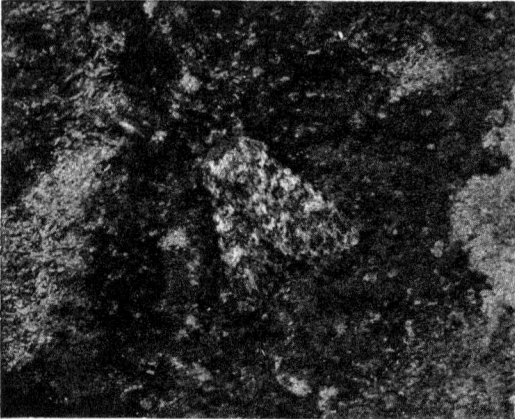

A BRITISH MOTH AT REST ON LICHEN-GROWN WALL
Polia flavicincta. From photograph by Mr. A. H. Hamm

and the dorsal a pale silvery grey. It is impossible to imagine a more beautiful confirmation of Mr. Thayer's principle.

In the shallow waters near the shore the resemblance of fishes is often imitative, a very good example being the long, green, bandlike pipe fish, *Siphonostoma typhle* (figure, page 114), found in the zostera beds around the English coasts. The apparent flatness of this fish is doubtless due to counter-shading, and its appearance to this extent is obliterative ; but taken as a whole it is imitative, the contour, attitude, and colour (except for the counter-shading) being all suggestive of the green, strap-like leaf of the zostera, floating upright in the water. The rounded profile of the face resembles the end of the leaf, while the fins, which would interfere with the cryptic likeness, are obliterated, partly by transparency, partly by their rapid vibratory motion. Among the commonest and most beautiful of the cryptic resemblances of fishes are the sand-and gravel-picturing patterns of those that haunt the well-lighted bottom of shallow water. Mr. Thayer has pointed out that these patterns are accompanied and rendered efficient by obliterative shading.

The chamæleonic power of colour-adjustment is very highly developed, and very commonly present in fishes. Mr. Charles H. Townsend, who has published many beautiful illustrations of these colour-changes, states that in the northern fishes, which he has studied in the New York Aquarium, " the changes are *slow*, requiring days or even weeks ; but the colour-changes of tropical species are *sudden*, or actually *instantaneous*" (figure, page 119). The following label is attached to the Aquarium tanks containing fishes liable to such rapid changes :

" This species *may change colour* at any moment.

" A few minutes' observation of the fish is usually enough to reveal a change either in colour or in pattern of marking.

" The colour-cells of the inner skin are under the instant control of the fish.

" Under natural conditions the changes of colour are made chiefly for the purpose of

concealment from enemies. They are also used for the capture of prey, for signalling, warning, mimicry, courtship, and other purposes."

The ordinary cryptic changes in the colours of fishes are due to the stimulus of light upon the eye, and blind fish lose the power of colour adjustment. But the condition of the nervous system may become the cause of rapid changes, such as those which follow excitement of various kinds.

CRYPTIC RESEMBLANCES IN INSECTS

Until the appearance of Mr. Thayer's book, obliterative coloration would have been held to play but a small part in the insect world. It must be admitted, however, that this conclusion will probably need revision in the light of the American naturalist's observations on caterpillars which rest with their most brightly coloured surfaces downwards and in the strongest shadow. And even as I am writing these pages there arrives confirmation of Mr. Thayer's views in a letter from Mr. Joseph Neale, of Bournemouth, who finds that the large caterpillars which have come under his observation assume a resting position in which shadow is obliterated. In this very position, namely, the caterpillar horizontal, with its back downwards, Mr. Neale points out that the oblique stripes of sphinx larvæ (figure, page 111) resemble the high light on projecting oblique leaf-veins, while the darker borders of these stripes resemble the shadows of the veins. In the ordinary position in which the larvæ are figured, the shadows are represented by dark borders *above instead of below* the stripes which stand for the veins.

Although it is probable, in the light of the observations alluded to in the last paragraph, that obliterative counter-shading plays a far more important part in insects than has been supposed hitherto, there can be no doubt that imitative resemblance is their chief method of concealment. Furthermore, the counter-shading here co-operates with the rival principle of imitation instead of replacing it. By its means, for example, the appearance of flatness is given to a chrysalis which in all other respects bears the most beautiful imitative resemblance to a leaf. (See page 111, and figure, page 112.)

Even a brief account of the imitative resemblances of insects would occupy the whole space

COMMON WHITE BUTTERFLY ON A WHITE DAHLIA
An example of an insect seeking an appropriately coloured resting-place
From photograph by Mr. A. H. Hamm

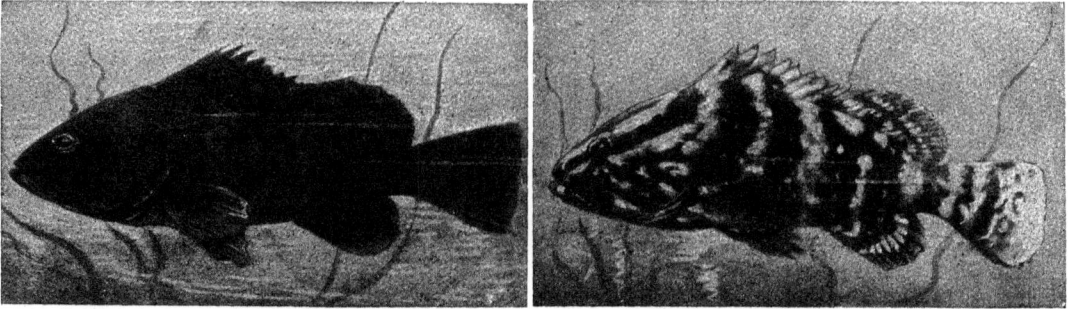

TWO STATES OF THE NASSAU GROUPER, A FISH THAT CAN CHANGE ITS COLOUR IN AN INSTANT

Many tropical fishes have the power of rapidly varying their colours when excited, feeding, or when alarmed. In the latter case they make for a hiding-place, assuming a dark colour. From an illustration by Mr. Charles H. Townsend

allotted to this article. Every characteristic feature of the vegetable kingdom becomes the model for protective likeness—leaves of all colours and conditions of decay and injury, and of many shapes (figures, pages 106, 107, 111, 112, 118) ; twigs and branches of all textures and tints ; the rough bark of the trunk, sometimes brown and at others greyish green with lichen or bright green with alga ; buds and the brilliant colours of flowers (figure, page 113) ; fruits and seeds in utmost variety. Rather than attempt to describe examples, which may be found by every observer of our British insects, I will refer to certain important aspects of cryptic resemblance in this section of the animal kingdom.

It is sometimes argued that the protective resemblances of insects are due to superficial characters, but it must be remembered that the colour and pattern would be useless without the co-operation of instinct leading the species to select an appropriate background and adopt an appropriate attitude. The wonderfully precise and accurately adjusted instincts of insects have been clearly produced by natural selection acting upon the central nervous system, and cannot be explained as the inherited memory of intelligent actions. Many of the instinctive actions, e.g., cocoon-making, are performed but once in a lifetime, and are adapted to meet the dangers of the future rather than the present.

The power of adjusting colours to those of a particular normal environment is widely possessed by caterpillars and by the exposed chrysalises of butterflies, but the changes are slow, and have never been shown to depend on the eye. In all chrysalises the power can only be exercised once in a lifetime, in caterpillars not more than two or three times. This is sufficient ; for the first cannot wander at all, while the second, during their period of growth, wander but little. The most striking result that has been obtained hitherto is the production of greenish or grey lichen-like marks upon stick- or bark-like caterpillars—the effect of the presence of lichen-covered sticks in the surroundings. When sticks without lichen were present, caterpillars—the offspring of the same parents as the others—only produced traces of the marks, or did not produce them at all. The species employed were the Lappet (*Gastropacha quercifolia*) and the Scalloped Hazel (*Odontopera bidentata*). It is probable that effects such as these are wrought by the stimulus of reflected light upon the nerve-endings in the skin, and are never produced directly as on the photographic plate.

No insect in the perfect state has been shown to possess the power of adapting its colours to any particular environment, but it is probable that such an adjustment can be effected in many species. In 1901 I observed that a very abundant grasshopper was invariably reddish brown, like the earth, upon the island of Heligoland, but that the same species was always sand-coloured or green on the flat, sandy Düne, separated by three-quarters of a mile of sea.

Similarly, my assistant, Mr. W. Holland, has observed that a well-known weevil (*Cleonus sulcirostris*) is reddish brown on the red sand of Boar's Hill, Oxford, but dark grey on the neighbouring Shotover Hill. In all such cases we have to decide between two probable interpretations. The local resemblance may be due to an individual susceptibility like that of many caterpillars, or may be due to the local operation of natural selection. Experimental investigations of a very laborious kind are required in order to reach a safe decision. The only experiment of the kind as yet undertaken was made upon a common British moth, *Gnophos obscurata*, well known to be dark in peaty and pale in chalky districts. The experiment showed that, at least from the mature caterpillar stage onwards, there was no susceptibility, and I therefore concluded

TWO DRY-SEASON OFFSPRING (LEFT) AND A WET-SEASON PARENT (RIGHT)

The African *Precis antilope*, showing the under sides of the wings. The dry-season offspring are dead-leaf-like and beautifully concealed. The wet-season parent is conspicuous, and perhaps a very rough mimic of an *Acraea*

that the local colours are the result of natural selection, the pale forms becoming exterminated on peat, the dark on chalk. There is little doubt that this is the interpretation of the recent progressive darkening of many moths in the smoke-producing districts of Lancashire and Yorkshire and the adjacent area to which the smoke is carried.

Over this large part of the North of England, bark-frequenting lichens and algæ are killed, and the trunks of trees present a uniform dark, sooty appearance, forming a background against which a greyish varie-gated moth would stand out conspicuously. These local develop-ments of a dark colour in the moths of smoky districts—observed on the Continent as well as in England—may be compared with an interesting Irish ex-ample recorded by Mr. W. F. de V. Kane. On the south - west coast of Ireland, at places where the rocks are dark-coloured, me-lanic examples of a c o m m o n geometrid moth (*Camptogramma bilineata*) were found intermixed with the ordinary yellow indi-viduals, while on a small island off the coast selection had been carried much further, and only black forms (the var. *isolata*) occurred. The end was tragic, for Mr. Kane informed me that the last time he visited the spot it had been swept by storms and the whole island race exter-minated.

It is impossible to do more than allude to the large body of ob-servations upon the seasonal changes of insects. In northern latitudes, the differences between the early and late broods of the same species sometimes corre-spond to differences in the surroundings, and thus pro-mote concealment. In tropical countries the dry-season forms are often better concealed than those of the wet season, when the struggle for existence is less severe.

The vast majority of insect resemblances are protective, but examples of Aggressive Imitative Resemblance are to be found in the predaceous leaf and flower-like *Mantidæ* (figure, page 113), but even in these the likeness is also of value as a protection against insect-eating enemies.

In the above outline of the vast subject of insect concealment I have attempted to select for consideration those aspects which, it is hoped, will induce the reader to think and make observations for himself. A more de-tailed treatment will be found in the author's "Essays on Evolution" and " Dar-win and the Origin."

Concealment for de-fence or attack is by far the most important of all the uses to which the superficial colours of animals are put. Mr. Thayer argues that it is well-nigh the only use, and that all com-binations of colours and all patterns serve the purpose of con-cealment. With this conclusion I cannot agree, fully believing that special modes of defence are associated with special types of appearance, easily seen and remembered by e n e m i e s (Warning Colours), and that such types often advan-tageously r e s e m b l e each other (Common Warning Colours, or Müllerian Mimicry), and become the models for mimicry by animals without special modes of defence (False Warning Colours, or Batesian Mimicry). I believe, furthermore, that the brilliant colours of the males of many animals are displayed in courtship (Epigamic), and have been gradually developed by the choice of the female. But all these purposes, although they are undoubtedly of the highest interest, are insignificant when compared with the great purpose which has been briefly sketched in the present article.   EDWARD B. POULTON

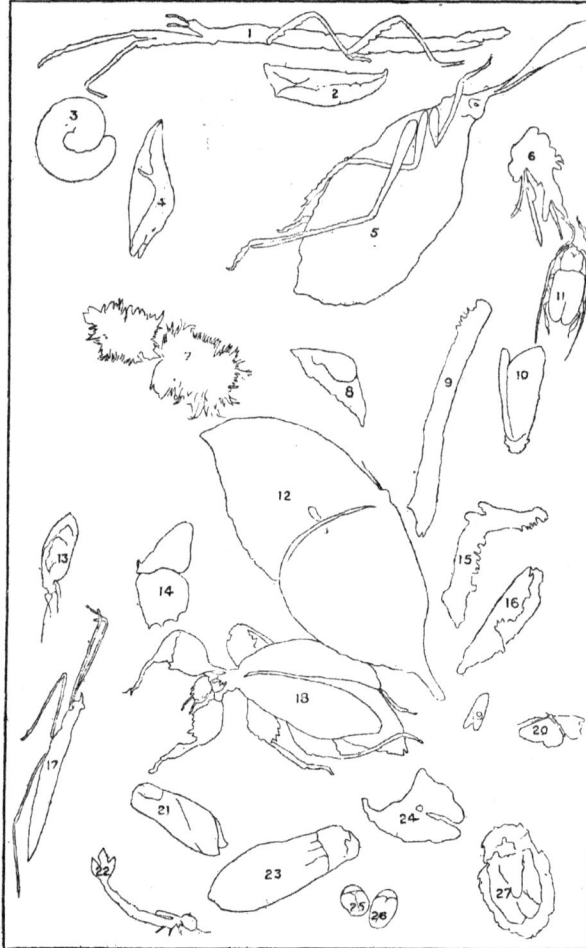

KEY TO COLOURED PLATE, PAGE 106

1, Stick insect, Bacillus rossii ; 2, Chrysalis of Papilio podalirius ; 3, Larva of Cimbex Letulæ ; 4, Chrysalis of Apatura ilia ; 5, Leaf-like locust, Pterochroza colorata ; 6, Psychid caterpillar ; 7, Pemphigus xyloste ; 8, Chrysalis of Rhodocera rhamni ; 9, Caterpillar of Arapteryx sambucaria ; 10, Scarce Merveil du Jour moth, Moma orion ; 11, Mesosa cerculionoides ; 12, Dead-leaf butterfly, Kallima philarchus ; 13, Dead-leaf mayfly, Drepanopteryx phalœnoides ; 14, Green hairstreak butterfly, Thecla rubi ; 15, Caterpillar of Notodonta ziczac ; 16, Chrysalis of Papilio evander ; 17, Water-bug, Ranatra linearis ; 18, Leaf insect, Phyllium siccifolium ; 19, Tortrix ocellaria ; 20, Clouded border moth, Abraxas marginata ; 21, Green silver-lines moth, Halias prasinana ; 22, Young dragonfly ; 23, Bufftip moth, Phalera bucephala ; 24, Cocoon of Aides amanda ; 25, Histerid beetle ; 26, Byrrhid beetle ; 27, Phloea corticata

THE CLASSIFICATION OF ANIMALS

By R. LYDEKKER, F.R.S.

IN dealing with any large assemblage of objects of different kinds, whether natural or artificial, it is absolutely essential to have some definite plan of arranging, or classifying, them in order to be able to refer to each type individually, and also to understand the relationship of one to the other. Such arrangements, or classifications, may rest on different bases. We may arrange the various objects according to the countries from which they come ; this being what is called a geographical or distributional classification. On the other hand, when we have to do with extinct as well as living animals, we may arrange them according to their respective ages. This is a chronological classification. Now, either classification affords much information when applied to animals, including extinct as well as living types. But a still more valuable classification of the animal kingdom is one based on the mutual relationships, and the grade of development of its various members, so far as these can be ascertained ; this being termed a systematic, or natural, classification.

Even the naturalists of antiquity, such as Aristotle, attempted a natural classification of animals, but signally failed, owing to their taking one structural character, or a common feature in the mode of life, as a basis for grouping together or sundering the various species. A new light dawned, however, upon systematic natural history, or rather zoology — for natural history in its fuller ·sense comprises the study of plants and minerals as well as animals—during the eighteenth century when the great Swedish naturalist, Carl Linné, or Linnæus, as his name, according to the fashion of the time, was Latinised, invented a system which has become the foundation stone of all modern methods of classification.

Linnæus had the merit of seeing that animals are divisible in the first place into species—we will discuss races later—and that species are capable of being arranged in groups, the members of which exhibit a greater or lesser degree of resemblance to the one taken as the type of such group. Thus, all the members of the animal we know as the British wild cat constitute a *species*. But, although the name " cat " properly belongs only to the wild animal and its domesticated relatives—or possibly only to the latter—yet there is a number of other carnivorous animals which present a more or less marked resemblance to that species, such as lions, tigers, leopards, pumas, and tiger-cats.

Taking the wild or the domesticated cat as the starting-point, Linnæus grouped with that species a number of other more or less cat-like animals into what he called a *genus*. Moreover, as the English language does not lend itself nicely to classifications of this nature, he gave to the cat a species as well as a genus name in Latin. Thus the wild cat became *Felis catus ; Felis* being the genus, or generic name, and *catus* the species name. Within the same genus are included the lion, tiger, leopard, puma, and tiger-cat under the respective names of *Felis leo, F. tigris, F. pardus, F. concolor,* and *F. tigrina ;* with a number of other closely allied species.

There is, however, another animal, the so-called hunting leopard, or chita, of India and Africa, which, while agreeing in a number of characters with the cats, yet differs in certain features, notably in the absence of the power of completely retracting the claws, from all the animals which can well be included under that name. When such a different type is met with it is referred to a different genus ; the chita being called *Cynælurus jubatus.*

But although a chita differs much more from all the more typical cats than does any one of the latter from any other, yet it is obviously much more nearly related to these animals than it is, for instance, to a wolf or a dog. This affords grounds for making a division of higher rank than a genus, namely, a *family.* Consequently we have the wild cat (*Felis catus*) and the chita (*Cynælurus jubatus*) brigaded together as members of the cat tribe, or cat family, under the name of *Felidæ ;* while, on the other hand, the wolf (*Canis lupus*) and its African relative the hunting dog (*Lycaon pictus*), which belong to distinct but allied genera, collectively constitute a second family group—the *Canidæ,* or dog tribe. The members of the cat tribe agree, however, with the dog group in their carnivorous habits as well as in the general type of structure of their teeth, and many other features in their organisation. Consequently, the *Felidæ* and the *Canidæ,* together with a number of other families, are combined in a group of still higher rank, known as an *order,* in this case termed the Carnivora.

But this is not all, as the members of a number of orders of animals agree with one another in the common character of nourishing their young by means of milk secreted by the female parent, and sucked or otherwise imbibed by the young. All such animals, whether they be four-legged terrestrial creatures like dogs and cats, or marine animals with the front limbs modified into paddles and the hind limbs absent, like whales and sea cows, are grouped in one great *class,* known as sucking animals, mammals, or, technically, Mammalia.

Going a step farther, we find that mammals, birds, reptiles, salamanders, fishes, and lampreys agree with one another in possessing a jointed rod, or axis, popularly called the backbone or spine, but scientifically the vertebral column, running along the middle line of the back. To express this community of character the classes furnished with a vertebral column, of which the individual units are the vertebræ, constitute a *subkingdom,* known as the Vertebrata.

There are, however, certain other groups of animals

possessing what may be called a kind of rudimentary backbone, the notochord, and these are brigaded with the Vertebrata under the designation of Chordata.

All other animals are without this structure, and in consequence are often collectively termed invertebrates, or Invertebrata; although this term is usually employed in a general rather than in a strictly systematic sense. These so-called invertebrates are divisible into a number of subkingdoms, such as the Echinoidea, or sea urchins and starfishes, and the Arthropoda, or jointed animals, such as lobsters and insects, of equal rank with the Vertebrata.

Finally, all these subkingdoms form sections of the animal *kingdom*, or Animalia, as distinct from the vegetable kingdom, or Plantæ.

In addition to the partition into subkingdoms, the animal kingdom is split into two primary *divisions*, namely, the Protozoa, or animals whose whole body consists of but one cell, or primary organic unit, and the Metazoa, or those whose bodies are formed by the aggregation of a number of such cells.

Thus we have the following grades of classification:
Subkingdom: VERTEBRATA.
Class: MAMMALIA.
Order: CARNIVORA.
Family: *Felidæ.*
Genus: *Felis.*
Species (*a*): *Felis catus*—cat.
Species (*b*): *Felis leo*—lion.

Linnæus used generic terms in a comprehensive sense, but the general tendency of modern naturalists has been to reduce the limits of these groups, in many cases to a very unnecessary extent. The older naturalists, when reducing the extent of a genus, considered it permissible to allow the species-name of an animal to be made the type of a new genus in a generic sense, and accordingly the fox—the *Canis vulpes* of Linnæus—became, in the view of those who considered it entitled to represent a genus by itself, *Vulpes communis.* Not so, now say the purists; a species-name must remain a species-name for ever, and accordingly they have rechristened the fox *Vulpes vulpes.* A more absurd instance of pedantry and slavish adherence to a self-made rule could scarcely be conceived.

This, however, is by the way; and it remains to give the following list of subkingdoms and of classes into which the existing representatives of the animal kingdom may be divided.

SYSTEMATIC ARRANGEMENT OF THE LEADING GROUPS OF THE EXISTING MEMBERS OF THE ANIMAL KINGDOM.

KINGDOM ANIMALIA

DIVISION A. COMPOUND ANIMALS..METAZOA
Subkingdom I. Backboned Animals   ..Vertebrata
CLASS

1. Mammals ..		..Mammalia
2. Birds	..	..Aves
3. Reptiles	..	..Reptilia
4. Salamanders and Frogs..		..Amphibia
5. Fishes	..	..Pisces
6. Lampreys and Hagfishes		..Cyclostomata

Subkingdom II. Lancelets   ..   ..Cephalochordata
Subkingdom III. Balanogiossus ..   ..Hemichordata
Subkingdom IV. Sea Squirts, or Ascidians ..Urochordata

Subkingdom V. Sea Urchins and Starfishes ..Echinodermata
CLASS

1. Starfishes ..	..	..Asteroidea
2. Brittle Stars	..	..Ophiuroidea
3. Sea Urchins	..	..Echinoidea
4. Sea Cucumbers	..	..Holothuroidea
5. Sea Lilies ..	..	..Crinoidea

Subkingdom VI. Jointed Animals   ..Arthropoda
CLASS

1. Lobsters and Barnacles..		..Crustacea
2. Centipedes	..	..Myriopoda
3. Insects ..	..	..Insecta
4. Spiders and Scorpions	..	..Arachnida

Subkingdom VII. Lamp Shells   ..Brachiopoda
Subkingdom VIII. Sea Mats   ..Polyzoa
Subkingdom IX. Worms and Leeches   ..Annelida
CLASS

1. Worms and Serpulas	..	..Chætopoda
2. Leeches ..	..	..Hirudinea
3. Gephyrean Worms		..Echiuroidea
Of uncertain rank	..	{ Sipunculoidea / Priapuloidea / Phoronidea }

Subkingdom X. Shelled Animals ..   ..Mollusca
CLASS

1. Nautilus and Cuttlefishes		..Cephalopoda
2. Whelks and Limpets	..	..Gastropoda
3. Chitons	..	..Placophora
4. Tooth Shells	..	..Scaphopoda
5. Bivalves	..	..Lamellibranchiata

Subkingdom XI. Wheel Animalcules   ..Rotifera
Subkingdom XII. Thread Worms   ..Nemathelminthes
CLASS

1. Round Worms ..	..	..Nematoda
2. Thread-shaped Worms ..		..Nematomorpha
3. Spiny-headed Worms ..		..Acanthocephala

Subkingdom XIII. Nemertean Worms ..Nemertea
Subkingdom XIV. Flat Worms and Flukes..Platyhelminthes
CLASS

1. Turbellarians	..	..Turbellaria
2. Trematodes	..	..Trematoda
3. Liver Flukes, etc...		..Cestoda

Subkingdom XV. Corals and Sea Anemones..Cœlenterata
CLASS

1. Polyps ..	..	..Hydromedusæ
2. Jellyfishes ..		..Acalephæ
3. Corals and Sea Anemones		..Actinozoa
4. Beroë, etc.	..	..Ctenophora

Subkingdom XVI. Sponges   ..Porifera
CLASS

1. Lime Sponges	..	..Calcarea
2. Glass Sponges	..	..Triaxonia
3. Ordinary Sponges		..Demospongiæ

DIVISION B. ONE-CELLED ANIMALS..PROTOZOA
CLASS

1. Amœba and Foraminifera		..Gymnomyxa
2. Ordinary Animalcules ..		..Infusoria
3. Blood Parasites, etc.	..	..Sporozoa

There is no consensus of opinion as to what constitutes a genus or a species, and it has been proposed, in certain cases, to split up genera into subgenera, and species into local races, or subspecies. There exist, for instance, certain more or less well-defined local forms or varieties of the tiger. If these are treated as distinct species, all connection with the typical tiger is completely lost. If, on the other hand, they are regarded in the light of local races of a single variable species, the connection is maintained, so that a species may be split up into local races or subspecies as represented in the case of the tiger by the following:

The Tiger: *Felis tigris.*
The Indian race, *F. tigris typica.*
The Manchurian race, *F. tigris longipilis.*
The Persian race, *F. tigris virgata.*
The Malay race, *F. tigris sondaica.*

CHARACTERISTICS OF THE VERTEBRATA

VERTEBRATES, which, in their higher phases, are the most perfectly organised of all animals, take their name from the presence of a jointed rod extending from the back of the head to the hind extremity of the body or the tip of the tail—when this appendage is developed—and known as the backbone, or vertebral column.

The joints, or segments, of which this generally flexible column is composed are known as vertebræ; and although usually formed of bone in the adult, in some of the lower vertebrates these vertebræ may permanently retain the cartilaginous condition in which they are first developed. Each fully-developed vertebra consists of two distinct portions, a basal portion, forming a disc-like or cylindrical solid segment, known as the centrum, and above this—or, in the case of the human species, at the back of this—an arch supported on two pillars, and terminating in a more or less well-developed spine. Perhaps from the presence of these spines, the whole backbone is commonly called the spine.

In the canal formed by the arches of the vertebræ, the so-called spinal canal, is contained what is known as the spinal cord, or spinal marrow, a rope-like structure formed of grey and white nervous tissue protected by a strong fibrous sheath; this cord running backwards—or, in the case of man when in the upright posture, downwards—from the brain, of which it is, in fact, a continuation, to the hind extremity of the body.

The spinal cord is thus seen to lie in a tube situated, in the ordinary position of quadrupeds, above, or, as naturalists say, on the dorsal side—from the Latin *dorsum*, " the back "—of the bony or cartilaginous column formed by the centra, or bodies, of the vertebræ. On the opposite side of the backbone to this nervous, or neural, tube is a much larger tube or cavity containing the viscera, such as the heart, lungs, stomach, and intestines. Accordingly, if a transverse section be made across the body of a vertebrate, it will be found to comprise two tubes—a smaller one on the dorsal side, containing the spinal cord; and a much larger one on the lower, or ventral—from the Latin *venter*, " the stomach "—side of the centra of the vertebral column, containing the viscera.

Another noteworthy peculiarity of vertebrates is that the limbs, which never exceed four in number, are always directed away from that aspect of the body which contains the spinal cord, and towards that enclosing the viscera; whereas in all the invertebrates, the reverse is the case. Vertebrates are likewise distinguished by the fact that the two jaws work in a vertical plane, or, in other words, are upper and lower, instead of being right and left, as they are in insects.

In this connection it is of the highest importance to notice—although we are getting on difficult ground—that the vertebral column is developed from a primitive structure known as the notochord, which persists throughout life in the groups classed in the table on page 122 as the Cephalochordata, Hemichordata, and Urochordata, but in the Vertebrata is found only in the embryo. The occurrence of this structure in these subkingdoms has been regarded as the most important feature of the whole assemblage, the members of which are consequently brigaded together under the common designation of Chordata, or animals with a notochord, in contradistinction to the Invertebrata, in which it is absent.

The Chordata are likewise distinguished by certain other structural features of importance, and more especially by the existence of lateral outgrowths from the pharynx, or throat, which unite with the skin of the neck to form a series of perforations leading to the exterior. These perforations persist in fishes as the gill-slits; but, with the assumption of a life on land, have been lost in the higher vertebrates as functional organs, respiration being performed by lungs in place of gills. Even in the higher vertebrates, however, the gills are apparent in the embryo, and remains of one pair can usually be detected in the adult.

Here it may be mentioned that the fact of whales breathing by means of lungs is a conclusive proof of their descent from land animals, for had they been directly derived from fishes they would certainly have retained gills, which are in many respects better adapted for an aquatic existence, although it is doubtful if they could act with sufficient energy to maintain the blood at the high temperature characteristic of mammals.

The Vertebrata—or, as they are often termed, Craniata—are distinguished from the lower Chordata by the formation of a distinct head, combined with the internal structure known as the skull for the reception of the brain.

Although the lowest members of the group—not to speak of the lower Chordata, which are probably, in some degree, at any rate, degraded animals— are decidedly inferior to many of the higher forms of invertebrates, yet vertebrates as a whole are clearly the supreme and highest development of the animal kingdom, their most advanced type, the mammals, being, in truth, the " winners in life's race," and culminating in man himself, who, we are told, is but " little lower than the angels." The development of brain-power has been the dominating agency in the evolution of the higher vertebrates, and it is the big-brained mammals which have gradually and completely ousted the reptiles which in former ages were the dominating animals of the land.

In the sea conditions naturally are different, and the lowly-organised fishes have maintained their supremacy alongside of the more highly organised whales and dolphins; and, indeed, are likely to persist long after all the larger representatives of the cetaceans have, owing to the intervention of man, been exterminated.   R. LYDEKKER

LION AND LIONESS

TIGRESS

From photographs by Gambier Bolton, F Z.S., by permission of the Autotype Company, 74, New Oxford Street, W.

THE KINGDOM OF MAMMALS
A SKETCH OF THE VARIOUS ORDERS AND THEIR CHARACTERISTICS
By J. ARTHUR THOMSON, M.A.
Regius Professor of Natural History, Aberdeen University

To the naturalist, every class of animal is interesting, but a special human interest attaches to the warm-blooded, hair-covered, milk-giving quadrupeds which Linnæus called mammals. For it is to this class that man belongs. Apart from " the all-pervading similitude of structure " between man and other mammals, which Sir Richard Owen long ago spoke of, the instinctive and intelligent behaviour of mammals often approaches human conduct in a curiously suggestive way. In illustration, we need only refer to the behaviour of mammals in courtship and in bringing up their young, to the activities of the play-period, and to the mutual aid among gregarious and social species. There is a special interest also in the educability of many mammals, for it has led in a number of cases to their intimate domestic association with man, who has acted as a directive factor in their evolution, replacing Nature's automatic sifting by a deliberate artificial selection.

There are said to be about three thousand different kinds of living mammals, and an equal number of extinct forms. Many of the latter are without direct descendants, for a distinction must always be drawn between extinct forms which are represented to-day by lineal or collateral descendants, and extinct forms which are in the strict sense lost races. Without taking the fossils into account in the meantime, we may state one of the first impressions of a visit to a museum-gallery of mammals, or to a good zoological garden—that the class is a most motley assemblage. What a gamut of size from the harvest mouse, that can run up a stalk of wheat, to the gigantic whale; what a variety of habitat—the earth and under the earth, the moor and the loch, the shore and the open sea, the mountains and the air itself; what differences in locomotion, in diet, in intelligence, in disposition, in number of offspring; and, implied in all this, what diversity of structure between the duckmole and the monkey, the whale and the bat, the sea cow and the horse, the mole and the elephant! In short, we find among mammals a remarkable variety of habit and habitat, and of form and structure adapted thereto. They have possessed the whole earth, they exploit it in a multiplicity of ways, and they show a correlated counterpart of bodily adaptation.

There are three grades of mammals—monotremes, marsupials, and placentals.

1. The monotremes—otherwise known as prototheria and ornithodelphia—include three genera, the duckmole and two spiny ant-eaters, which differ markedly from all other mammals in being oviparous. In other words, they lay eggs which are hatched outside the body. In addition to this singular feature, it may be noted that the monotremes are, in certain respects, primitive and reptilian-like.

2. The marsupials—otherwise known as metatheria and didelphia—include a large number of genera, such as kangaroo and bandicoot, phalanger and wombat, all confined nowadays to the Australian region, except the opossums (in North and South America) and the selvas (in South America). Most of the females have an external pouch, or marsupium, to which the tender young are transferred, and within which they are nourished and protected for a considerable time. This strange arrangement is in correlation with the fact that the organic connection between the unborn young and the mother is much less developed than in higher mammals, and that the young are born after a short gestation, as it were prematurely, unable even to suck. The mother has to inject the milk down the throats of her offspring.

3. The placentals—otherwise known as monodelphia and eutheria—include numerous orders, such as carnivores, ungulates, rodents, insectivores, bats, and monkeys. As compared with marsupials, they show a well-developed organic union (allantoic placenta) binding the unborn young to the mother's womb, and the brain displays in most cases a considerable advance.

There are so many different orders of placental mammals that we are apt not to see the wood for the trees. Therefore it may be of service to notice that there have been three main lines of evolution, each leading to great success. There is the herbivorous line of advance, best represented by the very successful order of hoofed mammals, or ungulates; there is the carnivorous line of advance, best represented by the very successful order of beasts of prey, or carnivores; and there is the series which leads through various grades of monkeys to a climax in man. On the herbivore line a premium was on hoofs and horns; on the carnivore line it was on teeth and claws; on the simian line it was on brains.

This is not to be regarded as a classification, though it is in some ways the result of one. It indicates the significance of the great evolutionary movements. It must, however, be noted (1) that there are several archaic orders—the sirenians, the sloths, the ant-eaters; (2) that the herbivorous rodents, the insectivora, and the lemurs, correspond in a general way to the ungulates, the carnivores, and the monkeys; (3) that there are several very divergent orders, notably, the cetaceans and the bats, to which may be added the divergent family of men; and (4) that some of the entirely extinct orders are of a generalised character, and unite several of the now widely separated modern orders. To take a single type, as an instance, it is believed by some that the extinct phenacodus, which figures as an early member of the horse lineage, has affinities with the ungulates, the carnivores, and the lemurs

THE EGG-LAYING MAMMALS

The oviparous monotremes are represented by the duckmole in Australia and Tasmania, the spiny echidna in Australia, Tasmania, and New Guinea, and another spiny ant-eater, also from New Guinea, which is often referred to as a distinct genus, Proechidna. They are exceedingly interesting mammals in many ways, because they exhibit certain reptilian affinities, because they are primitive compared with higher forms, and because they have numerous peculiarities. Let us illustrate each of these points in turn.

The monotremes show several reptilian features. The pectoral girdle resembles that of lizards—the coracoids meeting the breastbone, and so forth—the temperature of the body is unusually variable, the creatures being, in fact, imperfectly warm-blooded ; the eggs are relatively large, with abundant yolk, and hatch outside the body. There are many more technical evidences of reptilian affiliation.

The monotremes are primitive in many respects. Thus the milk is exuded on the skin, without there being any teats, on a bare flat patch with numerous openings in the duckmole, in a depression within a temporary pouch in the spiny ant-eaters ; and the young ones lick it, instead of sucking. Another instance may be found in the state of the brain, which is primitive, especially as regards the commissures passing from one side to the other, when compared with ordinary mammals. The rudimentary ribs attached to the vertebrae of the neck remain distinct for a time at least—and this is another primitive feature.

The monotremes have many peculiar features which cannot be called either reptilian or primitive. The skull is smooth and polished as in birds, bone fusing to bone at an early date. In the echidna there is no trace of teeth even in the embryo ; in the duckmole there are true mammalian teeth, twelve in all, but only in the young, for after the first year or so they are lost, and replaced by horny

THE FORE AND HIND FOOT OF THE DUCKMOLE

The web of the fore foot is folded back when digging. The hind foot has not such an extensive web, and bears a spur, the use of which is uncertain

plates. On the side of the ankle in the duckmole there is a spur perforated by the duct of a gland, and there is a similar smaller one in the spiny ant-eater. These peculiar structures may be used by the males in fighting ; they are rudimentary or absent in the females. As other examples of peculiarities we may note the soft, sensitive collar of the duckmole where the flat bill joins the rest of the skull, or the long, tubular snout and long, worm-like tongue of the spiny ant-eater. Thus we see that the monotremes are in some respects reptilian, in others primitive, in others quite peculiar.

The duckmole, or platypus, is a shy aquatic animal, swimming and diving well. It grubs in the mud for insect larvæ, crustaceans, and worms, collects them in its cheek-pouches, and chews them at leisure when it ascends to breathe. The webbed fingers are obviously adapted for swimming. But they are clawed as well, and, along with the hind feet, are used in making long burrows in the banks. In further adaptation to aquatic life, we may note the short fur, the very inconspicuous ear-flap, the short, flat tail, and the soft collar, which seems to have considerable tactile sensitiveness. Like the hedgehog and some other animals, the duckmole rolls itself into a ball when about to sleep. In the recesses of the burrow, which has often two openings, one above and one under water, the female lays two eggs, each enclosed in a flexible white shell, through which the young animal has to break its way.

The spiny ant-eaters live in dry, rocky regions, and are nocturnal in their habits. Their spines afford ample protection, and the animals burrow very rapidly when molested, sinking vertically into the ground so that when they have got a little way down they are unattackable. Their limbs are very strong, and when they get a grip among the scrub it is very difficult to dislodge them. The long snout and viscid, worm-like tongue, which is jerked out and drawn in with great rapidity, are adapted to catching ants. A single

DUCKMOLES IN THEIR HABITAT

This monotreme, sometimes called the duck-billed platypus, makes a long burrow which has one opening in the bank and another under the water, as shown in the above sectional drawing

egg is laid, and it is at once placed in a temporarily developed external pocket, where it is hatched—a hint of the state of affairs in marsupials.

THE MARSUPIALS

The pouch-bearing mammals, or marsupials, include a great variety of types : (1) the arboreal opossums (*Didelphys*), the aquatic yapock (*Chironectes*) which catches fish ; (2) the dog-like Tasmanian wolf (*Thylacinus*), the civet-like dasyure (*Dasyurus*), the banded ant-eater (*Myrmecobius*) ; (3) the unique notoryctes, with mole-like habits ; (4) the burrowing bandicoots (*Perameles*) ; (5) the divergent South American selvas (*Cænolestes*) ; (6) the large, somewhat bear-like wombat (*Phascolomys*), with rodent-like dentition ; (7) the woolly arboreal phalangers (*Phalanger*), the small, mouse-like honey-eating tarsipes, the large koala, or "native bear" (*Phascolarctos*) ; (8) the terrestrial herbivorous kangaroos (*Macropus*).

This list is necessary to show that the marsupials constitute more than an order of mammals ; they form a subclass by themselves. The figures indicate the eight families composing the subclass : Opossums, dasyures, notoryctids, bandicoots, selvas, wombats, phalangers, and kangaroos, and the point of special interest is that there are many suggestions of placental orders. The dasyure prefigures a carnivore, notoryctes a mole, the wombat a rodent, the kangaroo a herbivore, and so on—not that the resemblances in any case mean more than similar adaptations to similar conditions of life, or habits.

The facts in regard to the geographical distribution of marsupials are of peculiar interest. First, there is the palæontological fact that marsupial types once had a wide range in Europe and America. Their fossil remains prove this. Second, there is the fact that, with the exception of the American opossums and the South American selvas, all the living marsupials are natives of the Australian region. Third, there is the fact that no mammals, except marsupials, are indisputably indigenous to Australia. Fourth, there is the geological fact that the great island-continent of Australia was in former ages connected with Asia by a land bridge across the Java Sea. Reconstructing the past, then, we may say that marsupials were exterminated in many parts of their wide range as the better-brained placental mammals gained a footing, that they reached Australia before their placental competitors had arrived, that the insulation of Australia saved them, and gave them a new territory in which to evolve, relatively unhindered, for long ages in many different directions. And it is this that gives point and interest to the number of varied types to which we have called attention.

THE SPINY ANT-EATER, A CURIOUS MONOTREME
The long snout, with the worm-like tongue, is adapted for catching ants

For an alliance so heterogeneous as that of the marsupials it is difficult to find general characters. Those that are most readily intelligible have to do with the production of offspring. In the first place, the marsupials do not carry their young before birth so long as is usual in placental mammals. The gestation. as it is called, lasts only a fortnight in the opossum, about five weeks in the kangaroo. Furthermore, the organic connection between the embryo and the mother is of a simple type, except in the bandicoot, where the ordinary condition seen in the placentals is exhibited ; the young are born very helpless, unable even to suck ; they are transferred at once to the external pouch, or marsupium (which is absent in some opossums), and there the milk is forced down their throats by the contraction of a muscle around the mammary gland.

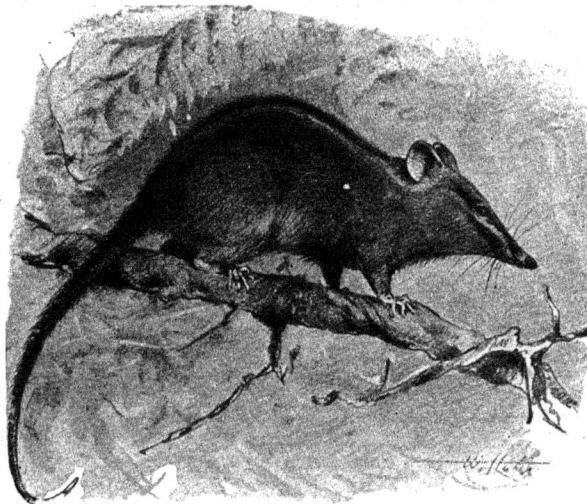

THE LONG-SNOUTED PHALANGER, A TINY MARSUPIAL
This little creature is very similar in form and size to the common shrewmouse. The female is provided with a well-developed pouch for carrying its young

Another intelligible general character relates to the brain, which is less developed than in the average placental mammal. But, when we go further, we are arrested by the fact that the most general characters do not seem to have much direct

THE BEAR-LIKE WOMBAT OF TASMANIA
A marsupial that lives in burrows or clefts in the rocks

meaning. In all marsupials, except tarsipes, the angle of the lower jaw is turned inwards towards the similarly inflected angle on the other side ; the cheekbone (molar or jugal) reaches far back to share in making the socket (glenoid cavity) in which the lower jaw works ; there are more incisor teeth in the upper jaw than in the lower except in the wombat ; and so forth. But these general characters, it should be added, are not in themselves significant, they are only indices of something deeper.

THE EDENTATES

Under the title edentates is included a number of very distinct types which cannot be compelled within the rubric of a single order. They may be appropriately called " living fossils." More precisely, they are the specialised survivors of waning and probably very primitive stocks whose golden age is long since past. They are not very nearly related to one another, and there seems little warrant for connecting them with any of the other present-day orders of mammals.

In the New World there are the arboreal sloths (Bradypodidæ) of South and Central America, the shaggy, toothless South American ant-eaters (Myrmecophagidæ), and the armadillos (Dasypodidæ), which range from Texas to Patagonia, unique in their encasement of body skin-plates. Along with these may be ranked the extinct ground sloths, or megatheriidæ, one of which—a very fortress—exceeded the rhinoceros in size, and the likewise extinct glyptodons, whose body was often huge and covered by armour of great strength.

Seemingly built not for a day but for all time, these glyptodons and similar types, highly specialised as regards defence, have passed entirely from the lists of life, while less evolved types persist and prosper. But it must be remembered that the day of the glyptodons was of no mean duration.

In the Old World there are the Oriental and Ethiopian pangolins, with strong, horny scales covering their back, like slates on a roof, and the shy, nocturnal aardvarks of Africa, but these two types have not much in common.

If these edentates are living anachronisms, it is useful to consider them from that point of view, to inquire into the protective features in structure and in habit, which secure their survival in the modern world. The shaggy hides of the sloths—the most arboreal of all mammals—harmonise with the mosses and lichens on the branches, and the protective resemblance is increased by a remarkable greenish tint on the hairs, due to a unicellular plant like that which makes tree-stems green in wet weather. The armadillos are covered with bony shields and rings, they can roll themselves up into unopenable balls, they run quickly and burrow quickly, and they come out at night—a very remarkable combination of protective adaptations. The African aardvarks that burgle the hills of the white ants are destitute of armour, but they make deep burrows and are shy nocturnal creatures. Just as simple dwarfish peoples among mankind have in different countries and ages taken refuge in forests and underground—becoming, in fact, fairies and gnomes —so most of these quaint edentates, survivals of a bygone age, linger on in burrows and among the branches, living a quiet, unobtrusive life, avoiding the glare of day unless the possession of strong armature lends them the welcome assurance of temporary security in the wilder parts of the earth.

THE SIRENIANS, OR SEA COWS

Another peculiar order is that of the sirenia, represented nowadays by two sluggish aquatic vegetarian types, the dugong and the manatee. A third type, Steller's sea cow, from the North Pacific, belonging to the genus Rhytina, seems to have been exterminated in the eighteenth century.

We may take these quaint creatures as an illustration of what is meant by evolutionary convergence—the occurrence of similar adaptations in types which are not closely related to each other. It is evident that the sirenians are in several ways like cetaceans : in both we notice the plump, rounded body, the horizontally flattened tail, the absence of hind limbs, the flipper-like form of the fore limbs, the great reduction of hairs on the skin, and so on. Even inside the body there are many resemblances,

THE RABBIT BANDICOOT, AN OMNIVOROUS MARSUPIAL
This creature derives its popular name from its large ears

such as the backward extension of the lungs, the very oblique midriff, or diaphragm, and the occurrence of wonderful networks of blood-vessels (retia mirabilia). But when we go further, we find that in regard to other organs the structure is fundamentally different—*e.g.*, as regards brain, dentition, skull, backbone, larynx, and stomach. The two orders are not nearly related, but they have undergone similar adaptations to similar conditions of life. From the structure of halitherium and other extinct relatives of the living Sirenia, it seems that the unknown ancestors must have been related to the ungulates, or hoofed mammals, and probably to a stock from which the elephants also diverged.

CETACEANS

The baleen whales, such as the right whale (*Balæna*), the humpback (*Megaptera*), and the rorqual (*Balænoptera*), differ in so many ways from the toothed cetaceans, such as the sperm whale (*Physeter*), dolphin (*Delphinus*), porpoise (*Phocæna*), and narwhal (*Monodon*), that it is an open question whether there should not be two orders of "cetaceans." Waiving this, however, let us consider the cetaceans as illustrating what is meant by adaptations. That their ancestors were quadrupeds, there can be no doubt, but how far is the body of modern cetaceans from being supportable on four limbs! Yet the spindle lies comfortably in the water and is well suited for cleaving it. The hind limbs are represented at most by internal vestiges, the fore limbs have become flippers, the tail is horizontally flattened into flukes, the skin has lost almost all its hair, a thick layer of blubber retains the animal heat and makes the animal buoyant, the ear has no trumpet and the sound must travel inwards through the bones of the skull, the lungs are greatly elongated backwards, helping the animal to lie horizontally in the water, there is a neat contrivance for enabling the creature to breathe and swallow at the same time, and so on through a long list of characteristic modifications, for, as Weismann has well said, "*Take away the adaptations from the whale and how much is left?*" The whale's body is a huge bundle of adaptations.

PÈRE DAVID'S DEER, AN UNGULATE OF MANCHURIA

HOOFED MAMMALS, OR UNGULATES

One of the most successful lines of mammalian evolution has been that leading to the modern ungulates—a terrestrial, herbivorous series, in which the bones of the palm and the sole stand high off the ground and the nails usually grow round the ends of the digits to form hoofs. It is a rather heterogeneous order, including unwieldy forms like the rhinoceros and hippopotamus, graceful forms like the gazelles and roedeer, quaint old-fashioned types like the giraffe and the okapi, highly specialised modern types like the horse and the elephant. More precisely, the ungulate alliance includes : (1) the even-toed pigs and peccaries, hippopotamuses, camels and llamas, chevrotains, and the true ruminants—cattle, sheep, deer, okapi, and giraffes ; (2) the odd-toed rhinoceros, tapir, zebras, asses, and horses ; (3) the small rodent-like coneys, or hyraxes ; and (4) the elephants. Besides the living ungulates there are many extinct types, some of them of very large size, and often "generalised"—that is to say, exhibiting characters which connect what are now separate orders. Better, perhaps, than any other types do the fossil ungulates—*e.g.*, those in the lineage of the horse and the elephant — furnish proof of the reality of gradual evolutionary advance from the general to the special.

RODENTS

There are more members of this order than of any other among mammals. They succeed not because they are strong, certainly not because they are clever, but because they are many. They are represented in all parts of the world, having gained access, through man's heedlessness, even into Australia, the "preserve" of marsupials. Most are terrestrial and small and vegetarian, but the water vole is very largely aquatic, the capybara is four feet long, and rats are omnivorous. Quite characteristic are the chisel-edged incisor teeth—an almost mechanical adaptation, for the hard enamel is either restricted to the front of the teeth, or is extremely thin posteriorly, so that as the lower teeth work against the upper, the back part wears away more rapidly, and thus a chisel-edge is formed. The lower jaw works

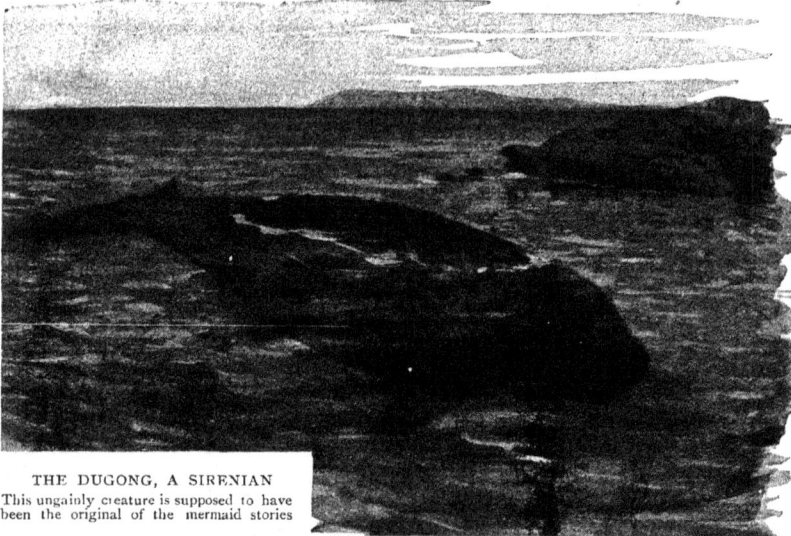

THE DUGONG, A SIRENIAN
This ungainly creature is supposed to have been the original of the mermaid stories

we see the strong crests for the better fixing of the muscles, the deep socket for the lower jaw, which gives the animal a strong grip and permits only of up-and-down movement, the protruding cheekbones bounding a large space for the muscles of the lower jaw, the strong killing canines and a specialised cutting blade among the back teeth. If we watch the domestic cat, we see the tiptoe tread, the neat contrivance for retracting and protruding the claws, the sensitive hairs on the face, and curious details of habit which appear as reminiscences of the arboreal wild stock.

INSECTIVORES

This is an order of small mammals in some superficial respects like rodents, but on a quite different line of evolution. Shrewmice, moles, hedgehogs are familiar types, but there is so much diversity that brief characterisation is very difficult. Most of the insectivores run about on the ground, but the mole has sought safety and earthworms underground, a few are aquatic, and a few arboreal. As regards brains, they represent a low level of organisation, and many of them are otherwise primitive. Some, however, like the mole, are highly specialised. It is interesting to notice how many rodent types have their counterparts among the insectivores—the mouse balances the shrewmouse, the jerboa the jumping shrew, the water vole the water shrew, the porcupine the hedgehog, and so on ; but, as has been said, there is no close relationship. According to

backwards and forwards with considerable play, and a characteristic " munching " results—very familiar in the rabbit. Some rodents, such as squirrels and rats, make affectionate pets ; viscachas, prairie dogs, jerboas, and other forms have attractive habits ; and one hears tales of the beaver's ratiocination. Nevertheless, on the whole, the rodents, or gnawing animals, form a distinctly " lower order " —that is, a relatively unintelligent one.

CARNIVORES

As successful as the ungulates, but in a very different direction, are the carnivores, which include four great tribes : (1) the cats, large and small, the civets and mongooses, the aard-wolf, and the hyæna ; (2) the dogs, wolves, jackals, and foxes ; (3) the bears, raccoons, otters, skunks, badgers, martens, sables, stoats and weasels ; and (4) the seals, the sea lions, and the walrus—adapted for marine life. Most of them prey upon other animals and devour their warm flesh, and in adaptation to this habit there are well-developed claws, sharp teeth, keen senses, and agile movements. Most are bold and fierce animals, but there is a well-marked play-period in youth, which is always a mark of great educability. We see the climax of this educability in the dog.

As in other cases, it is most useful to consider the general characteristics in relation to the dominant habits and conditions of life. Thus, if we examine a lion's skull,

PRAIRIE DOGS, OR PRAIRIE MARMOTS
Rodents that dig burrows and share them with the ground owl and rattlesnake, all living amicably together

many authorities, a separate order is required for the remarkable flying foxes (*Galeopithecus*) of the Malayan Archipelago and the Philippines, which have the fore and hind limbs connected by a parachute, and are able to glide for half a hundred yards, or more, from one tree to another.

BATS

Undoubtedly related to the insectivora—though their popular name is flitter-mice—the bats are the most specialised of mammals. Like the birds and the extinct flying dragons, they have discovered the secret of flight, and their whole structure is adapted thereto. The wing is formed by a fold of skin, which usually begins at the shoulder, extends along the upper margin of the arm to the base of the thumb, thence from one long finger to another, and eventually down the sides of the body to the hind legs, or even to the tail. Among the bat's adaptations to flight we may notice the following: The bones are slender, and the long bones have large medullary cavities; the backbone has little mobility, but serves as a stiff fulcrum; the breastbone has a slight keel for the insertion of muscles; and so on. Even the fact that there is usually only one offspring at a time may be correlated with the flying habit. In many ways bats are peculiar. Thus the sense of touch is extraordinarily acute, and is located in the hot skin of the wing, the large mobile external ears, the whisker hairs of the snout, and in strange plaited " nose-leaves " around the nostrils. This tactility is, of course, to be associated with their crepuscular habits, for they have to catch small prey in scanty light. Some of the larger bats are fruit-eaters, but the great majority are strictly insectivorous. In cold countries bats hibernate, hanging head downwards from the roofs of caves, wrapped up in their wings.

LEMURS

Leading on to the monkeys, but remaining at a much lower level, are the lemurs. Their name means ghosts, and they are ghost-like in their silent movements and in their love of darkness. Most of them are nocturnal, and all are arboreal. They are small, soft-furred mammals, with fox-like faces, and with considerable resemblance to monkeys. In early

THE VISCACHA, A FAMILIAR RODENT OF THE ARGENTINE
These animals are related to the chinchilla which yields the lovely fur

Tertiary times they had a much wider range than now, occurring, for instance, in Europe and North America. In the latter continent they became extinct ; in the Old World they migrated to Ethiopian and Oriental regions. Of the fifty or so living species there are thirty-six in Madagascar. It seems that they must have reached that region while it was still connected with Africa, and that insulation took place before any of the larger carnivora had got across. Where they have not been insulated, they linger in forests—a feeble folk, without the brains or agility of monkeys, saved only by their quiet, shy ways. Although there are some highly specialised lemurs, such as the aye-aye of Madagascar, famous for its extraordinarily long skeleton-like third finger, and the Indo-Malayan tarsier, of weird appearance, the lemurs, as a whole, are lowly organised creatures, and link the monkeys to the lower mammals.

PRIMATES

The title of this order indicates its pre-eminence — the pre-eminence of including man. Man is the supreme mammal, and his lineage, though we do not know its details, is simian. We began our survey by pointing out that a special interest attaches to the class Mammalia because we who study it are mammals—deny it as men often do in their conduct and policy—and we end our survey by pointing out that the interest is intensified in the case of the primates, for it is from within this order that the ascent of man began. It must be carefully noticed, however, that within the order there are many grades, each of which represents a big step in evolution. The New World broad-nosed monkeys, usually called platyrrhines, are at a level anatomically far below the Old World, narrow-nosed catarrhines, which include the four anthropoid apes ; but they are also very different, and it is not improbable that the order has developed from two distinct roots, one leading to the marmosets and American monkeys, the other leading to the more dog-like baboons and macaques, and to the more man-like gibbons, orangs, chimpanzees, and gorillas.

DIET OF MAMMALS

It has been well said that " animals, from their own point of view, have two, and only two, occupations in the world. These are to care for themselves and to care for their offspring." It is interesting to consider the general life of mammals in this light, beginning with the securing of food, which is the first and fundamental chapter in the never-ending story of caring for self. What a variety of food-getting we find—the duckmole collecting small bivalves on the floor of the pond, the Cape ant-eater breaking into the hills of the termites, the sea cows browsing on seaweeds and estuarine plants, the giraffe reaching up to the leaves on the high branches, the reindeer scraping off the snow to get at the lichens, the right whale rushing through the ocean with enormous gape, engulfing myriads of sea butterflies and other pelagic fry, the sperm whale doing battle with huge squids, the lion devouring antelope, the bear berries, the seals fish, the mole pursuing the earthworms underground, the hedgehog searching for the grubs of beetles, the bats feeding on insects, the monkeys plundering the orchard, and how many more!

SELF-PROTECTION AMONG MAMMALS

An adequate picture of a class of animals must include some illustration of the give and take between its members and their enemies. There are some that stand confident in their strength, such as lions and tigers, elephants and rhinoceros; there are others that avoid controversy, preferring to lie low and say nothing. Some are more or less secure in their armour—e.g., porcupines and hedgehogs, armadillos and echidnas; others are well equipped with weapons —teeth and claws, horns and hoofs. The greenish sloth is inconspicuous among the branches, often shaggy with epiphytic plants, and some of the antelopes in their khaki suits are hard to see in the veldt. Some, like the armadillo and hedgehog, roll themselves into balls; others burrow quickly and are gone; the flying foxes, the flying squirrels, and the flying phalangers show similar extensions of skin that enable them to save themselves by taking great swoops from tree to tree; the jerboas, the squirrels, and some lemurs and monkeys have a surprising swiftness of leap; the bats and many others have become creatures of the night. But there is no end to the variety of self-protection.

THE INSECTIVOROUS HEDGEHOG AND YOUNG
When the hedgehog is alarmed, it rolls itself into a ball, presenting a formidable surface of spikes to its assailant

ADJUSTMENT TO PHYSICAL SURROUNDINGS

It is plain that the structure of the mammal's body is often well adapted to its particular environment: the spindle-shaped porpoise to the open sea, the snake-like weasel for creeping through holes, the monkey with its five hands for gymnastics among the branches. Similarly, the white Polar bear is suited in its colour to the snow and ice of the Far North, and the tiger to the tropical jungle. The more penetrating our observation, the more of this fitness of structure do we discover: the mole's hand is a marvellously effective spade, the elephant's nasal dexterity passes belief, the horse's legs are suited for galloping on tiptoe. The adaptations are found in the inmost recesses of the body: the sheep's four-chambered stomach is fitted for rumination, the whale's lungs are very large and long, the heart of the swift ungulates is specially suited to stand the strain. But we follow the quest farther, and find that in many cases the mammal is able to adjust itself in an effective way to changes in its surroundings. Thus the stoat, mountain hare, and Arctic fox turn white in winter, becoming at once safer and physiologically more comfortable; the hedgehog, dormouse, marmot, and bats fall into a strange coma when the cold of winter sets in; there are also interesting migrations—reminding us of those among birds—for instance, in the reindeer and some antelopes, in seals and also among cetaceans.

MATING AMONG THE MAMMALS

It is interesting to compare birds and mammals in regard to their wooing. In many birds the males display their beautiful plumage, their graceful flight, their powers of song, apparently working up excitement in their desired mates.

THE MOLE, AN INSECTIVORE WHICH LIVES UNDERGROUND
The mole's hand is peculiarly adapted for burrowing

Among mammals there is also some display both of beauty and of vocal powers, but the wooing is on the whole much more forcible. Rival males—e.g., stags and sea lions—fight fiercely with one another, and the females follow the conqueror. The antlers of stags, the horns of antelopes, the great tusks of the narwhal, and the large canines of boars are used as weapons in these battles of the males; while manes and beards, bright colours and strongly scented secretions, seem sometimes to play a part in exciting the females and overcoming their coyness. It is interesting to notice how profoundly the nature of the animal may be modified at the breeding season: minute alterations are effected in remote corners of the body, and the whole temper may be changed. The timid hare becomes a fierce and heedless combatant, and love is often stronger than hunger. But if the courtship of mammals is like a storm in violence, it is also soon over, for while there are exceptional pairs that marry, such as beavers and some antelopes, remaining constant year after

internal parasite, within the mother, sharing the details of her life. In a complicated way—a *chef d'œuvre* of contrivance—the unborn offspring, enwrapped in ante-natal vestments, obtains food and oxygen, and gets rid of its waste products by what may be called vital diffusion between its blood and that of the mother. There is a passage of fluids and gases from the mother to the offspring and conversely, and, though no ordinary solid particle can pass the filter, living microbes sometimes get through from the mother, bringing about an infection before birth. But the point that we wish to emphasise is that the connection between mother and offspring is very thorough and very subtle. There is much

TWO SEA LIONS AT PLAY ENGAGED IN A SHAM FIGHT
From a photograph by Mr. F. Martin Duncan

year, this is not the way with the majority. After the pairing season is past the males usually return to their ordinary, often solitary, life; and, in the great majority, the family life is distinctly matriarchal.

ANTE-NATAL LIFE

No one can understand mammals, even a very little, without taking account of the way in which the young are carried within the womb of their mother. The monotremes, as we have seen, lay eggs; the marsupials bring forth their young prematurely, and stow them away—in most cases—in an external pocket; but in placentals there is a period of gestation in which the young creature lives, like an

nurture before birth; there is a "mysterious wireless telegraphy of ante-natal life." Very interesting also are the special adaptations to the life before birth; thus, to take a detail, the claws sometimes show—as in reptiles and birds—an embryonic "glove finger," which disappears soon after birth, but is probably of use in saving the delicate vestments from being torn by movements of the embryo.

In his well-known pre-Darwinian essay on evolution, "The Vestiges of Creation," Dr. Robert Chambers suggested that the prolongation of the period of ante-natal nurture would give the brain time to grow more complex, so that there came about, in the course of ages, a gradual improvement in

STAGS FIGHTING IN THE MATING SEASON

It is scarcely surprising that great biological interest should attach to the play of mammals, which is familiar to everyone in the case of kittens, puppies, lambs. kids. foals, and the like.

THE PLAY OF MAMMALS

As may be supposed, the simplest forms of play are gambols and frolics. Also fundamental is the game of experiment, in which the animal, without serious purpose, tests things, itself, or its fellows ; and from these roots arise more complex forms of play —the sham hunt, the race, the sham fight, and so on. The subject has been thoroughly studied by Prof. K. Groos, who shows that play is more than an expression of overflowing energy whose forms are defined by imitation, that it is rather the outcrop of definite instincts, which have been evolved like other instincts, arising by germinal variation, and fostered in virtue of their utility. But what can be the utility of play, which by definition has no serious purpose ? To this Groos answers that play is the young form of work— a rehearsal without responsibilities, a sham fight before the battle of life begins, a preliminary canter before the real race. Animals play because they are young, but it is perhaps as true that they have a period of youth in order that they may play. A second justification of play is found in the opportunity it affords for exercising and perfecting instinctive activities, which, therefore, do not require to be so definitely engrained in the cerebral constitution. Thus play is a device for lightening the burden of inheritance, which leaves the brain more educable. Thirdly, it is a very suggestive idea that the play-period affords opportunities for initiatives, new departures, and idiosyncrasies, before the struggle for existence has become keen. The playing animals are notably plastic and docile, and, apart from a few cases—like sheep—they are also notably intelligent.

THE STRUGGLE FOR EXISTENCE

The " balance of Nature "—the established order of things—is a *moving* equilibrium, which requires to be continually readjusted. Whenever there is a disturbance or dislocation of the established inter-relations, whenever there is a conflict of interests, whenever the lusty, self-assertive living creature is confronted with difficulties and limitations, we have the struggle for existence. For, as Darwin clearly

mammalian intelligence. Whether this theory be right or not, there is no small interest attaching to the length of the gestation in different kinds of mammals. We give a few figures from a very valuable study by Dr. John Beard (" The Span of Gestation and the Cause of Birth." Fischer, Jena, 1897)—horse, 336 days ; cow, 280-287 ; man, 276-280 ; sheep, 145-150 ; pig, 112-120 ; guinea-pig, 63-66 ; dog, 58-64 ; cat, 56 ; kangaroo, 38 ; rabbit, $30\frac{1}{2}$-31 ; white mouse, $19\frac{1}{2}$-$19\frac{2}{3}$; opossum, $7\frac{5}{8}$. These periods are doubtless wrapped up along with other periodicities in the mammal, but three interesting points may be noted. The period of ante-natal life is, on the whole, longer in the higher types ; longer in the larger than in the smaller forms at the same level of organisation ; and longer when the number of offspring born at a time is small.

In regard to the condition of the young mammals at birth, we may quote a suggestive statement made by Sir W. H. Flower : " In those forms which habitually live in holes, like many rodents, the young are always very helpless at birth ; and the same is also true of many of the carnivora, which are well able to defend their young from attack. In the great order of ungulates, or hoofed mammals, where, in the majority of cases, defence from foes depends upon fleetness of foot, or upon huge corporeal bulk, the young are born in a very highly developed condition, and are able almost at once to run by the side of the parent. This state of relative maturity at birth reaches its highest development in the cetacea, where it is associated with the conditions under which these animals pass their existence."

recognised, this includes much more than an internecine struggle around the platter at the very margin of subsistence, it includes every thrust and parry in face of difficulties, every reaction against environing limitations, every endeavour after well-being which makes for the survival of the stock. In face of difficulties or limitations one alternative is, of course, increased competition, but another is increased combination ; and Nature has, so to speak, approved of both lines of solution, as is particularly well seen in the case of mammals.

Let us first of all, however, illustrate from the natural history of mammals the *reasons* for the struggle for existence. The first reason is to be found in the tendency to over-population, for struggle is the safety-valve against the internal pressure of rapidly increasing population. The slowest breeder among mammals is the elephant ; it is supposed to rear a single young one every ten years, but, as it lives to more than a hundred, it can have a fairly large family before the end of its days. And Darwin calculated that in 750 years each pair of elephants would, if all their offspring lived and bred, be the ancestors of nineteen millions. It is well known that the rabbits introduced into Australia about 1860 soon became calamitous in their prolific multiplication ; and every few years there is over-population among the Scandinavian lemmings, or among the common field voles. The river of life continually tends to overflow its banks.

A second reason for the struggle for existence is to be found in the pattern of the web of life—there are nutritive chains, one link depending on another for support. Thus the carnivores struggle with the herbivores and the herbivores with the carnivores, in natural antipathy deeply rooted in the past and justified by the conditions of the present. A third reason is the irregular changefulness of the physical environment, which is quite careless of life. Mammals have to struggle with the vicissitudes of Nature—with *weather* in the widest sense. There is a fourth reason for the struggle for existence which is often overlooked—namely, the self-assertiveness of the vigorous animal, which elbows its way through the crowd, jostling its neighbours.

We find among mammals vivid illustrations of the three chief forms of the struggle for existence.

1. The struggle between nearly related fellows is well seen in the deadly combats between the brown rat and the black rat. Here it is war to the death. The black rat, which seems to have migrated from the East, was in possession of many European towns when the stronger and fiercer brown rat—perhaps a native of Mongolia—crossed the Volga about 1727 ; but wherever the brown rat arrived the black rat had practically to go to the wall. Thus, at the present day, there are almost no black rats in Britain. It may be admitted that the arrival of the brown rat, which is much more obtrusive than the black rat, roused man to more strenuous methods of extirpation, which were naturally more effective against the weaker species, but there can be no doubt as to the keen competition between the two kinds, and the result that the weaker goes to the wall. Similarly, in South Africa to-day one of the fiercest and most aggressive of the native rats (*Mus microdon zuluensis*) seems quite unable to compete with imported European brown rats and black rats, and has consequently disappeared in the larger towns when the immigrants have established themselves. The next step is that the brown immigrant ousts the black immigrant, forcing it from town to country.

A STAG FOLLOWED BY ITS MATES, FOR WHICH IT HAS FOUGHT ANOTHER STAG

The struggle between fellows of the same or of nearly related species need not be for the necessaries of daily life—for food, foothold, or breathing-room ; it may be for such luxuries as a fourth or fifth wife, as in the case of the sea lions, which fight fiercely for the increase of their harem. It is difficult to exaggerate the fury of the combats between rival stags or antelopes. Or, to take a quite different instance of struggle for a luxury, the wild stampede of the reindeer when the longing to visit the salt seashore becomes irresistible. Many are overthrown and trampled in the mad rush.

2. The struggle between foes which are not nearly related to one another is seen in the manifold thrust and parry between herbivores and carnivores, between the small rodents and insectivores, on the one hand, and the birds and beasts of prey on the other.

3. The struggle with Fate—between the living creature and the changeful, merciless physical conditions—is seen in the endeavours many mammals make when the winter is more than usually severe.

The results of the struggle for existence are in the main always three, and mammals illustrate them well. (1) There may be a reduction in numbers which relieves the pressure of population without directly making for progress. The lemmings, obeying the instinct of their smooth brains to move on, an instinct that sometimes leads them to a promised land, may swim out in thousands into the North Sea, which solves their population problem, but without obviously improving the wits of the race. This is indiscriminate elimination, which never

TIGER CUBS AT PLAY   Photo H. Irving
The most intelligent mammals have a play-period in their youth

makes for progress. (2) In many cases, the exigencies of the struggle drive the animals to seek out new habitats, to explore new territories, to adopt new habits. When the beaver village gets overcrowded there is a flitting, and whether it be true or not—probably not—that the old experienced hands leave the huts to the youngsters, and explore for themselves, there is certainly a recurrent coercive expansion of range. It must have been thus that the whales returned to the freedom of the sea, that moles invaded the Eldorado of the earthworms, that the bats made the great venture of flight. (3) In the third place, the result of the struggle may be discriminate elimination of the relatively less fit to the given conditions of life, and it is this kind of result that has meant most in the evolution of the races. The black rat yields before the brown rat, and there is a survival of the fittest—*i.e.*, the fittest in respect to the given conditions. It is very important to realise that the struggle for existence will make for progress without *rapidly* killing off the less fit. For if the struggle mean that the less fit have a more difficult life, and do not live so long or so lustily, if it mean that they have smaller and less vigorous families, then it will in the long run work out to the same result as if the less fit had come to a rapid, violent end.

This struggle is going on now, and has been going on since early Triassic mammals began to emerge from their ancestral reptilian stock. Given a continuous crop of variations, and Nature's sifting, the result is Evolution.   J. ARTHUR THOMSON

A COLONY OF BEAVERS AT WORK BUILDING A DAM

CHARACTERISTICS OF MAMMALS

AN ACCOUNT OF THEIR STRUCTURE, ANATOMY, & OTHER FEATURES

By R. LYDEKKER

WITH the exception of the word " beasts," there is no true English term for this group of animals, which constitutes the highest class in the vertebrate division. The term " quadrupeds " was, indeed, long in popular use, but since it is inapplicable to whales, while it would also include most reptiles, it is now largely superseded by the term " mammals," the animals in question being so named from the fact that their young are nourished by milk sucked from the mammæ of their mother.

In addition to the presence in the females of mammary glands secreting the milk, mammals differ from other living groups of the higher vertebrates in the mode in which the lower jaw is articulated to the skull. In other vertebrates this articulation is effected by the intervention of a separate squared bone known as the quadrate, upon the lower end of which the articular hollow of the lower jaw plays, while its upper end is wedged into the skull proper. In mammals, on the other hand, this intermediate bone is absent, and the lower jaw consequently articulates by means of a convex surface, or condyle, directly with the walls of the skull itself. Moreover, in all mammals each half of the lower jawbones consists of only a single bone, instead of several distinct bones joined together. Thus an isolated jawbone is always sufficient to prove whether its owner was a mammal or some other vertebrate. Another very important feature in mammals is that they always have hair, although it may be only a few bristles near the mouth, on some portions of their bodies during a certain period of their existence. Again, that part of the large cavity of the body which contains the heart and lungs is completely separated by a horizontal partition, known as the midriff, or diaphragm, from the cavity containing the stomach and intestines. Further, at least in all living members of the class, the brain of mammals is much more highly organised than that of other animals ; one of its distinctive features being the presence of a transverse band on its lower surface, by means of which its two lateral halves are intimately connected, this being called the *corpus callosum.*

The foregoing are some of the chief features distinguishing mammals from all other vertebrates, and reference may now be made to a few characteristics in which mammals differ from certain of the lower classes, although agreeing with others. One of the most important of these differences is that the skull of mammals is jointed to the first vertebra by means of a pair of transversely disposed knobs, or condyles,

THE BLACK RAT
The tail is covered with scales with their edges in apposition

as they are technically called. In this respect mammals are broadly distinguished from birds and reptiles, in which there is only a single knob, or condyle, placed in the middle line of the skull. Frogs and newts, constituting the class of amphibians, agree, however, with mammals in the mode by which the skull is jointed to the backbone, although they differ very widely in other parts of their organisation.

On the other hand, mammals differ from fishes, amphibians, and reptiles in having warm blood, which is propelled from a four-chambered heart through a double circulatory system ; one part of this system causing the blood to pass through the lungs for the purpose of taking in a fresh supply of oxygen from the air, and the other being subservient to the supply of freshly oxygenated blood to the various organs and members of the body. This circulatory system also differs from that of birds and reptiles in that the blood for the nourishment of the body is propelled from the heart by a single vessel, known as the aorta, which passes over the left branch of the windpipe ; whereas in birds and reptiles the aorta crosses either the right branch or both branches of the windpipe.

All mammals, whether they live on the land or in the water, breathe air by means of lungs suspended in the chest ; and during no period of their life do they ever develop gills ; neither do they ever undergo a metamorphosis analogous to that presented by the change of a tadpole into a frog. By these last two negative characters they are, therefore, sharply distinguished from the amphibians, with which, as we have seen, they agree in the mode by which the skull is articulated to the first joint of the backbone.

With the sole exception of the few species of egg-laying mammals, or monotremes, of Australia and New Guinea, which are the lowest members of the class, the young of mammals are invariably born in a living condition.

A remarkable feature in mammals is the circumstance that, with only three constant exceptions, the number of joints, or vertebræ, in the neck is seven ; this number being equally constant in the immensely elongated neck of the giraffe, or in the extremely shortened one of the whale, in which the vertebræ are reduced to thin plates of bone.

As a rule, mammals have the two pairs of limbs characteristic of vertebrates, but occasionally, as in whales, the hind pair may be wanting. In a large proportion of species the hind and fore limbs are of approximately equal length. In some cases, however, the hind limbs may be enormously elongated at the expense of the front pair, as we see in the kangaroos and jumping-mice ; and progression is then effected by means of leaps and bounds from these strong hind limbs. The opposite extreme of limb - structure is shown among the bats, where, while the hind pair retain their normal structure, the fore limbs are greatly elongated to afford support to a leathery winglike structure, by means of which these strangely modified creatures are enabled to fly in the air with the same ease and swiftness as birds. In the whales and dolphins, which lead a purely aquatic life, we find the fore limbs modified into paddles for swimming, while the hind ones are, as we have said, totally wanting. Similar conditions obtain in the dugongs and manatis; but in the true seals, which are less completely aquatic, the hind limbs are still well developed, although directed backwards to form, in connection with the tail, a kind of rudder. The bats are the only mammals which are wholly adapted for flight, but we meet with certain forms in other groups, such as the flying - squirrels among the rodents, and the flying-opossums among the pouched mammals, which are enabled to take long leap-like flights from tree to tree by means of a kind of parachute formed of folds of skin running along the sides of the body from limb to limb. The limbs themselves are not, however, specially modified; and true flight, in the sense of propulsion caused by up-and-down strokes of the fore limbs, is not performed by these mammals. Something more on the subject of limbs is mentioned in the paragraphs which are devoted to the skeleton (see page 140).

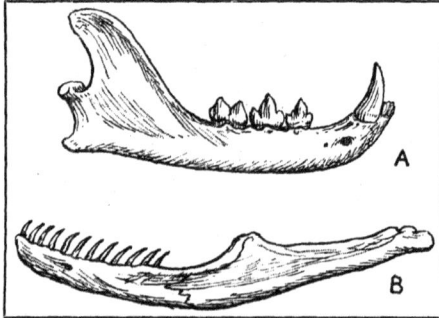

LOWER JAWBONES OF CAT (A) AND SNAKE (B)
By the lower jawbone alone a mammal's remains can be distinguished from those of any other vertebrate

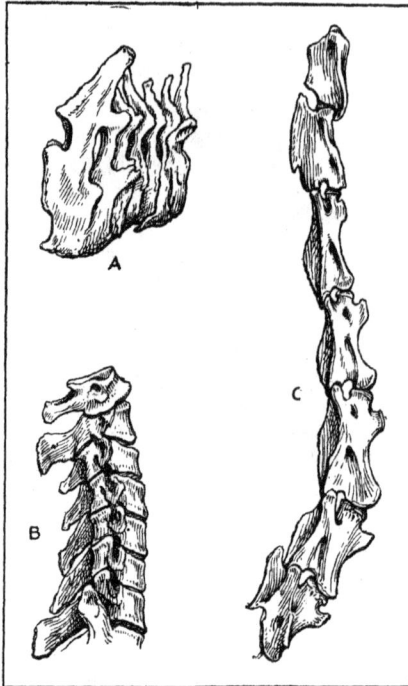

THE NECK VERTEBRÆ OF GIRAFFE (C), MAN (B), AND WHALE (A)
In almost every mammal the number of vertebræ in the neck is seven, whether they are elongated as in the giraffe or reduced to thin plates of bone as in the whale. A, whale ; B, man ; C, giraffe

of the tail is greatest is a small species from Madagascar belonging to the insectivorous order, and known as the long-tailed tenrec, in which the tail is nearly three times as long as the body. In some of the apes and monkeys the tail is absent ; and it is very short in the bears among the Carnivora, and in many deer among the hoofed mammals, or ungulates. In many ungulates, however, such as cattle, it is of great length ; and in that group its extremity is furnished with a tuft of hair, and thus forms an effectual instrument for brushing away flies from the body.

In the spider-monkeys of South America, as well as in the opossums and phalangers, in certain porcupines, and other forms, the tail is prehensile, and serves as an important aid in climbing, or in enabling its owner to hang head-downwards. In the beaver the tail is expanded into a flattened oar-like form, which probably acts as a rudder in swimming. But the most remarkable modification of this useful organ occurs in the whales and dolphins, in which it is expanded into a large forked structure, termed by whalers "flukes," and is the main agent in propelling the body through the water.

In regard to the external covering, hairs are always present on some portion of the body during some period of life. Mammals never develop that modified kind of hair-structure known as feathers, which are peculiar to birds ; but the body may be covered with overlapping scales, like those so common in reptiles, although this now occurs only in the pangolins, or scaly ant-eaters of India and Africa. The tail of the common rat is an example of a part of the body covered with scales having their edges in apposition ; but in both these instances hairs are mingled with the scales. Still rarer than scales are bony plates, developed in the true skin. These structures are only met with among the armadillos, which are furnished with bucklers and transverse bands of these bony plates, and are in some cases able to roll themselves up into a ball, presenting

Almost as great variations are displayed in the modifications and uses of the tail of mammals. In the majority of cases the tail is present and forms a tapering axis, often clothed with long hair, which may considerably exceed the total length of the body. The mammal in which the relative length

on all sides an impenetrable coat of mail. In the Pleistocene, or latest geological period before the present, however, South America produced a number of mammals allied to the armadillos, and known as glyptodonts, which were covered with a continuous cuirass of bony plates, reaching in some cases more than an inch in thickness. That these huge and well-armoured creatures, which might well be regarded as typical examples of animals fitted to withstand all enemies, have perished, while their smaller and less completely defended allies have lived on, indicates that there are other causes at work in the destruction of animals besides the attacks of foes. Between the plates of the armour of the armadillos hairs are always developed, and in one species these are so abundant as completely to hide the plates themselves, and render the general appearance that of an ordinary hairy mammal.

The chief use of hair is to protect the body from cold, and thus to aid in the maintenance of a uniform high temperature; and when hairs are absent, we find this function performed by a more or less thick fatty layer beneath the skin, the layer, when excessively developed, as in whales, being known as blubber. To make up for the difference between the temperature of winter and summer, many of the mammals which inhabit the colder regions of the globe develop a much thicker coat of hair in the former than in the latter season, of which we have an excellent example in the horse. In some mammals, such as the hare and cat, the body is covered with only one kind of hair; but in other cases, as in the fur-seals, there is one kind of long and somewhat coarse hair, which appears at the surface, and another of a softer and finer nature, which forms the thick and warm under-fur. This under-fur is greatly developed in mammals of all groups inhabiting Tibet, where it is locally known as " pashm "; and it is this pashm of the goat of these regions which affords the materials for the celebrated Kashmir shawls. Curiously enough,

too, animals which usually do not develop pashm almost immediately tend to its production when taken to the Tibetan region, as is notably the case in the dog. Less frequently the hair of the body takes the form of stiff bristles, as in the pig; and still more rarely this thickening is carried to such an extent as to produce spines, of which we have the best instances in the porcupine and hedgehog, belonging, it should be borne in mind, to distinct orders, the former being a rodent, while the hedgehog is one of the insectivora.

The solid horns of the rhinoceroses, and the hollow horny sheaths covering bony processes of the skull in cattle and antelopes, are very similar in their nature to hairs, and may indeed be compared to masses of hair welded into solid structures.

Although a fair idea of mammals as a whole may be gained without investigation into the nature of their soft internal parts, yet anyone who desires to obtain really accurate knowledge of them must make up his mind to acquire at least some slight idea of the general structure of the bony skeleton, and also of the form and nature of the teeth, since these parts are of the highest importance in classification.

It has already been incidentally mentioned that the skull consists of two portions—the skull proper, which contains the brain, and the lower jaw. It will suffice to mention, in addition, that the hind part of the skull is known as the occiput, and that on the front surface the pair of bones roofing over the cavity of the nose are known as the nasals, while those behind them, forming the region of the forehead, are termed frontals. Further, the bones in the upper jaw which carry the hind, or cheek, teeth are known as the maxillæ, while those in which the front cutting teeth are implanted are termed the premaxillæ. All the other numerous bones of the skull have received distinct names; but the reader desirous of becoming acquainted with these must refer to one of the many excellent textbooks on anatomy.

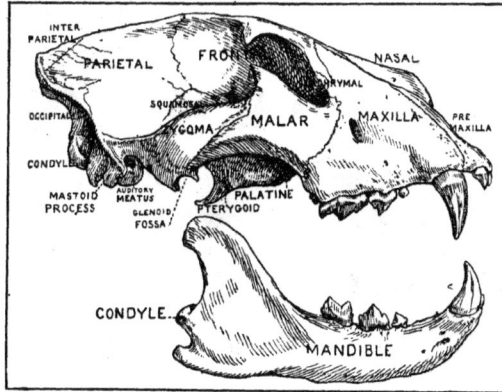

SKULL OF THE LION

The mammalian skull, of which that of the lion is taken as a type, differs from that of the lower vertebrates, especially in having its parts more firmly welded together

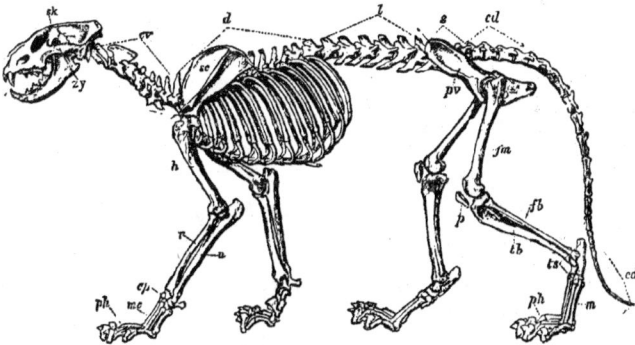

SKELETON OF THE LION

sk. skull; zy. cheekbone (zygomatic arch); cv. vertebræ of the neck; d. vertebræ of the back; l. vertebræ of the loins; s. sacrum; cd. vertebræ of the tail; sc. shoulderblade (scapula); h. armbone (humerus); r. u. bones of forearm (radius and ulna); cp. wrist (carpus); mc. metacarpus; ph. toebones; pv. haunchbone (pelvis); fm. thighbone (femur); p. kneecap (patella); tb. fb. bones of lower leg (tibia and fibula); ts. ankle (tarsus); m. metatarsus

In the backbone, or vertebral column, the first vertebra, or the one which articulates with the skull, is known as the atlas; following this is the axis vertebra, which is remarkable for having attached to it the body or basal portion of the atlas vertebra, known as the odontoid process. This separation of the body of the atlas vertebra from its proper segment is constant throughout the greater part of the vertebrate division. The remaining five of the cervical, or neck, vertebræ, are distinguished from the dorsal, or vertebræ of the region of the chest, by the absence of ribs. The ribs of most of the dorsal vertebræ articulate in the middle line of the inferior aspect of the body with the breastbone, or sternum, which is itself composed of several segments. The dorsal vertebræ are succeeded posteriorly by a smaller number (forming the region of the loins) which have no ribs and are termed lumbars. Behind the latter are several coalesced vertebræ forming the so-called sacrum, to which the haunch-bones, collectively constituting the pelvis, articulate; and these are again succeeded by the tail, or caudal, vertebræ, of which the number varies according to the length of the tail itself.

In the great majority of mammals the fore limb is connected on each side with the trunk simply by the blade-bone, or scapula, which lies on the back surface of the anterior ribs;

four outermost of which is succeeded by the three phalangeal bones of the fingers or digits. The thumb, or first digit, which lies on the same side as the radius, has, however, only two of these phalangeals.

The hind limb differs from the fore limb in that the innominate, or haunchbones, which together form the pelvis, are usually connected by an immovable bony union with the sacral region of the vertebral column. The thighbone, or femur, corresponding to the humerus of the arm, articulates with a cavity in the innominate termed the acetabulum. The leg has two parallel bones articulating with the lower end of the thighbone or femur, of which the larger, or tibia, occupying the inner

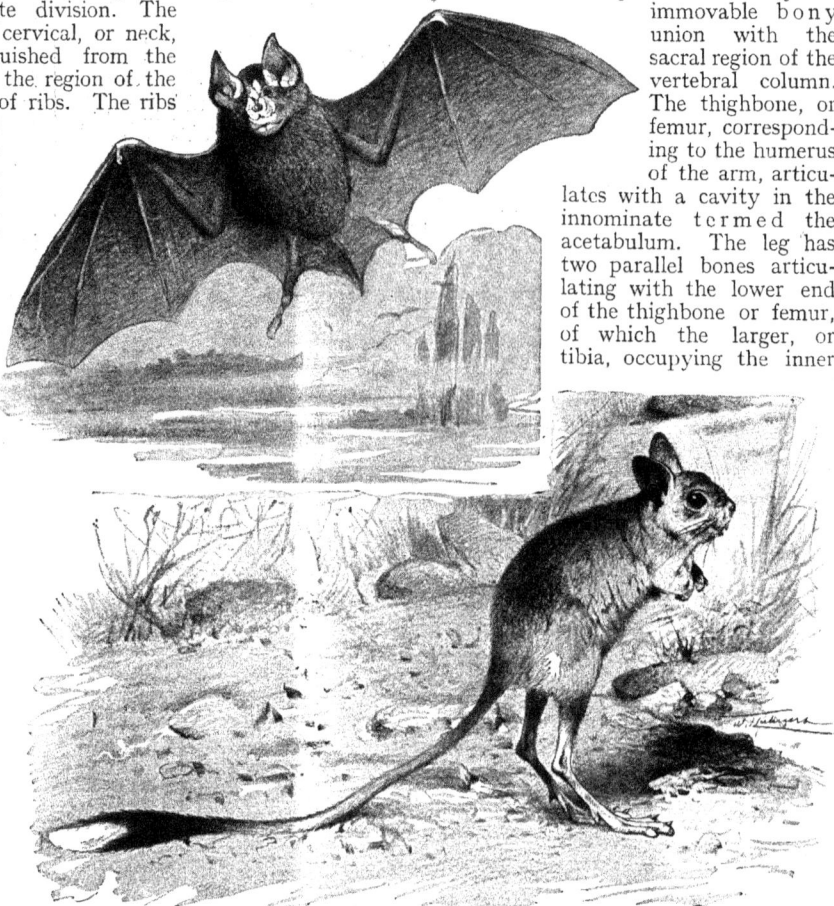

BAT AND JERBOA, ILLUSTRATING TWO EXTREMES IN THE DEVELOPMENT OF LIMBS

and, in front, by the collarbone, or clavicle, which connects the scapula with the sternum. The bones of the fore limb are, firstly, the armbone or humerus, which has condyles at its lower end; and, secondly, the two parallel bones of the fore arm, of which the outermost—when the palm of the hand is turned forwards—is the radius, and the other the ulna. The radius is always present, but in many hoofed mammals only the upper end of the ulna remains, which is fused with the radius. The radius articulates below with the upper of the two transverse rows of small solid bones forming the wrist or carpus; beyond these, we have in man and monkeys, as well as in certain other groups, five elongated bones, termed metacarpals, each of the

side of the limb, corresponds to the radius of the fore arm, while the smaller outer bone, or fibula, represents the ulna. The ankle, or tarsus, corresponds to the carpus in the fore limb, and likewise consists of two transverse rows of small bones. Two bones of the uppermost row, namely, the calcaneum or heelbone, and the astragalus or hucklebone (the uppermost bone of the ankle), are specially modified. In the foot proper the bones correspond with those of the hand; those representing the metacarpals being, however, termed metatarsals. In the course of this work it will frequently be found convenient to speak of the extremity of the fore limb, or hand, as *manus*; while the hind foot may be termed *pes*.

In the foregoing summary we have spoken of the hand and foot as consisting of five fingers and toes, or digits ; and this is the case in most monkeys, many carnivora, rodents, etc. In other cases, however, and especially among the hoofed mammals or ungulates, there is a tendency to the reduction of the number of digits. Thus, in cattle and deer, commonly known as ruminants, the number of functional digits is reduced to two, corresponding to the third and fourth of the typical series of five, while in the horse only a single digit remains, which in the fore limb corresponds to the middle or third finger of the human hand, and in the hind limb to the middle toe.

THE MUSK-OX, WHOSE HAIRY COAT IS ADAPTED TO THE ARCTIC REGIONS

Almost all mammals when adult have both jaws provided with a series of teeth, varying greatly in number and structure in the different groups. These teeth are almost invariably fixed in separate sockets, and, while each of the front teeth has only a single root or fang, the side or cheek teeth very generally have two or more such roots, each of which occupies a separate division of the socket. In all cases the teeth are fixed in their sockets merely by the aid of soft tissues connected with the gum, and are never welded to the jaws by a deposit of bone. Very generally there is a sharply-marked line of division, termed the neck, between the root, or portion of the tooth implanted in the jaw, and the crown or exposed portion.

In most of those mammals in which the teeth of different parts of the jaw differ in structure from one another, there are two distinct sets of teeth developed during life. The first of the two includes the milk or baby teeth, which are generally shed at a comparatively early age, are of small size and few in number, and are finally succeeded by the larger and more numerous set which remains during the rest of life, unless previously worn out.

In those mammals in which the permanent teeth differ from one another in form in different regions of the jaw, we are enabled from their position, and also from their relations to the temporary series of milk teeth, to divide them into four distinct groups. Taking one side of the upper jaw, as that

THE PORCUPINE, IN WHICH SOME OF THE HAIRS HAVE BECOME QUILLS

of the wolf, of which the teeth are shown in the annexed figure, we find the front bone, or premaxilla, carrying a small number (in this instance three) of simple cutting teeth, termed incisors. Behind these teeth, from which it is generally separated by a longer or shorter gap, there is a tooth with a simple and often conical crown, which, like the incisors, is inserted in the jaw by a single root. This tooth, which is usually larger than the incisors, is termed the tusk, or canine tooth, and in the wild boar and most carnivorous mammals attains a very large size. It can always be distinguished from the incisors by the fact that it is implanted in the maxilla, or second bone of the jaw, or at least, on the line of junction between that bone and the premaxilla. Behind the canine comes a series of teeth, which may be as many as seven, although only six in the wolf, with more complicated crowns, and, except the first, inserted in the jaw by two or more roots. This series is collectively known as the cheek teeth, but they are divided into two minor groups according as they are preceded by milk teeth or not. In the wolf the four teeth immediately behind the canine, with the exception of the first, are the vertical successors of milk teeth, and are known as premolars ; while the two hindmost teeth, which have no such temporary predecessors, are termed true molars, or molars. In the lower jaw the tooth, usually larger than the others, which bites in front of the upper canine is the lower canine. In advance of this tooth are the incisors, and behind it the premolars and molars, distinguished from one another in the same manner as are the corresponding teeth of the upper jaw. With the exception of the pouched mammals (marsupials), there are, in practically all the mammals with teeth of different kinds, never more than three incisors, one canine, four premolars, and three molars on each side of each jaw, so that the total number of teeth on both sides of the two jaws is not more than forty-four. In the figured upper jaw of the wolf the number falls short of this full complement, owing to the circumstances that there are only two in place of three molars.

As it is inconvenient always to have to describe the number of teeth in any given mammal by writing them down at length, a graphic formula has been invented by which the number of teeth of each species can be shortly and clearly expressed. Thus, taking only one side of each jaw, and indicating the incisors by the letter i, the canines or tusks by c, the premolars by p, and the molars by m, and taking the number above the line as representing the teeth of the upper, and those below the teeth of

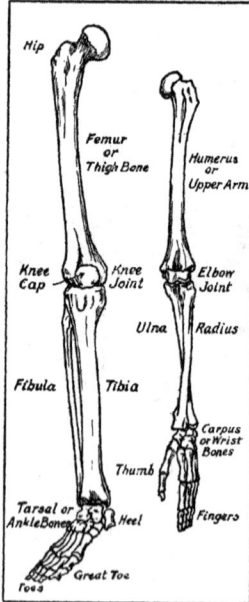

HUMAN LEG AND ARM BONES COMPARED

the lower jaw, we may state the number and kinds of the teeth of the wolf by the formula : $i\frac{3}{3}$, $c\frac{1}{1}$, $p\frac{4}{4}$, $m\frac{2}{3}$. The total thus given is 21, and double this shows the number of teeth on both sides of the two jaws, which in this case is 42.

A few words must be said regarding the internal structure of teeth, as without some knowledge of this it is impossible to understand the modifications which they undergo in different groups of mammals. Taking a simple more or less conical tooth like the tusk of a lion or tiger, or any tooth of a sperm-whale, it may be observed that when such a tooth first appears above the gum it is open at the base, where it forms a hollow cone. And in teeth like the tusks of the elephant, which grow throughout the whole life of their owner, such a condition remains permanent. Usually, however, a tooth ceases to grow after a certain period, and the base of the root or roots then becomes completely closed, and assumes a pointed shape. A tooth of this simple conical type is composed internally of a substance known as the ivory, or dentine, generally coated externally with a thin layer of a much harder nature and having a highly polished appearance, which is termed the enamel. Moreover, outside the base of the crown there may be patches of a coarser substance, called the cement. In order that its structure may be readily understood, a model of such a tooth may be made by taking the finger of a kid glove, filling it with beeswax, and putting some smears of sealing wax at the base of the outer surface, when the beeswax will represent the ivory, the kid the enamel, and the sealing wax the cement. If we then cut off the summit of the finger we shall have a central disc of beeswax (ivory) surrounded by a circle of kid (enamel), which will represent the condition of such a type of tooth when its summit has been worn away by friction against the opposing tooth of the opposite jaw. If, however, before cutting off the end of our model, we indent the summit with several deep pits, and also mark the sides with one or more grooves, and fill up such pits and grooves with sealing wax, it will be obvious that we shall have a much more complex type of structure. This complex model will serve to explain the type of tooth-structure found in many of the hoofed mammals ; and it will be obvious that if we now cut off the summit of our model we shall find a series of irregular discs of beeswax (ivory), each surrounded by a sinuous border of kid (enamel), in the folds of which will be masses of sealing wax (cement). R. LYDEKKER

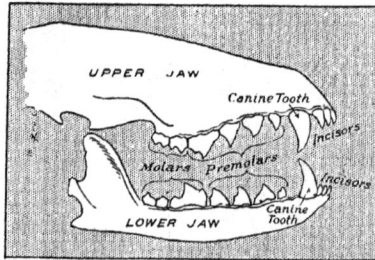

THE DENTITION OF A WOLF

THE EVOLUTION OF MAMMALS

WITH A TABLE OF THEIR CLASSIFICATION

By R. LYDEKKER

As regards the origin of mammals, it is now practically certain that the class is descended from a group of extinct reptiles known as the Anomodontia, or Theromorpha, of which the remains occur in great abundance in the Triassic or lowest Secondary strata of South Africa, Russia, and certain other countries. The members of one section of these anomodonts were carnivorous reptiles, with teeth divided into incisors, canines, and cheek teeth, after the mammalian pattern, and also implanted in distinct sockets in the lower jaw. Some of these theriodonts, as they are called, were of great size, rivalling in this respect wolves and leopards ; but it must have been from the species of the size of opossums that mammals are descended. The resemblance between these early theriodont reptiles and mammals is not confined to the teeth, the skull exhibiting a number of mammalian features, especially as regards the small size of the quadrate, by which, as mentioned already (page 137), the lower jaw of reptiles is articulated to the skull itself. The

THE LEFT HALF OF THE LOWER JAW OF AN EXTINCT POUCHED MAMMAL

quadrate bone of the theriodonts is, in fact, on the point of vanishing, and there may some day be discovered one of these reptiles in which it has actually disappeared. When such a specimen turns up, it will be almost impossible to say whether its owner was a reptile or a mammal ; and, indeed, we are already acquainted with one skull from South Africa which has been referred by some naturalists to a reptile, and by others to mammals.

The shoulder-girdle and the pelvis of all the anomodont reptiles are strikingly like those of the egg-laying mammals, or monotremes ; so that there can be no doubt as to the direct descent of the latter from the former. Unfortunately, the existing representatives of the egg-layers have either completely lost their teeth, or develop only a few of a highly specialised character while young ; and it is therefore impossible to say whether any member of the group had a dentition like that of the theriodonts. If not, it may be that the egg-laying mammals were derived from anomodonts independently of other mammals.

In all anomodonts the skeleton of the limbs is of an essentially mammalian type.

In all likelihood the evolution of mammals from

THE GOLDEN MOLE

the theriodont reptiles took place in Southern Africa, which is known to have existed as dry land since a very early period of the earth's history. And, apart from the egg-laying group, there is a probability that the early Tertiary group of extinct Carnivora known as creodonts is nearly related to the ancestral stock of ordinary mammals.

Many of these creodonts have the crowns of the cheek teeth surmounted by three cusps arranged in a triangle, as shown in the lower jaw figured on this page ; this tritubercular type of molar, as it is called, which persists in its simplest form among recent mammals only in certain groups of insectivora, such as the African golden moles, being common to the early members of many of the mammalian order.

Whether or no the aforesaid creodont carnivora are or are not the direct descendants of the theriodont reptiles, they are certainly a very ancient type. Their dentition presents many resemblances to that of the carnivorous marsupials ; but it seems probable that this is not indicative of near relationship.

The creodonts are, however, known to have given rise to certain extinct whale-like mammals termed zeuglodonts ; and if, as seems on the whole probable, these zeuglodonts are related to whales, the latter must also have originated from primitive Carnivora.

The other group of exclusively aquatic mammals, namely, the Sirenia, as represented by the existing manati and dugong as well as by a number of extinct kinds, are now known to be intimately related to the ancestors of the elephant group, or Proboscidea, so that both are evidently divergent branches of a single stock.

As to elephants themselves, discoveries in the early Tertiary strata of the Fayum district of Egypt have shown them to be descended from animals which depart comparatively little from the early type of hoofed mammals.

Indeed, as we descend in geological time, we find the ancestral forms of nearly all the orders of land mammals approximating more or less markedly to a common type, so that there can be no doubt as to their having originated from one stock, although it is at present impossible to point out their precise relationships to one another.

Although commonly placed high up in the series,

the insectivora are evidently mammals of a very low type, with a possible relationship to marsupials.

Numerous as are the species of mammals now living, it must never be forgotten that they form but a small portion of those which flourished at earlier periods of the history of our earth. The mammals of the present day may, indeed, be com-

THE SKELETON OF A MAMMAL-LIKE REPTILE, THE PAREIASAURUS
A gigantic land-reptile in which many mammalian features are to be found

pared to the topmost branches and twigs of a giant forest tree, of which the larger limbs and trunk are concealed from our view ; and it will accordingly be manifest that anyone who confines his studies to the existing species will have but a very imperfect idea of the whole array of mammalian life, and of the mutual connection of its various branches. The study of fossil mammals is, however, very difficult, and one requiring an extensive knowledge of comparative anatomy. All, therefore, that can be done is to direct attention, as occasion arises, to some of those extinct mammals which are of especial importance and interest as showing the manner in which groups now widely separated from one another were formerly more or less completely connected.

Although the number of extinct mammals is very large, yet the greater proportion of these belong to the latest of the three great epochs into which the geological history of our globe has been divided. During the long-past epoch known as the Secondary period, during which the Chalk and Oolites were deposited, the earth was tenanted by gigantic reptiles of strange form, and it is not till we come to the rocks above the Chalk—such as the London Clay and the overlying strata—that we find mammals taking an important place among the inhabitants of the earth. It was, indeed, during this Tertiary period that these animals attained the dominant position which they now occupy ; and the present stage of the earth's history may be truly called the Age of

Mammals and Birds. We are not, however, to suppose from this that mammals were unknown before the Tertiary period, a considerable number of species, mostly of small size, having been already discovered in various strata of the Secondary period.

An additional importance attaches to the study of extinct mammals, since it is by their means alone that we are able to explain several apparent anomalies in the geographical distribution of living groups. It would, for instance, be impossible to explain the present existence of tapirs only in such widely remote areas as the Malay Peninsula and Islands, and South America, unless we had learnt by geological explorations that these animals formerly roamed over large portions of Europe and Asia, whence their descendants gradually migrated to the regions where they now remain.

The former occurrence of an epoch of great cold in the Northern hemisphere, known as the Glacial period, explains how nearly related animals are now confined to isolated mountain chains ; their ancestors having been enabled, during the prevalence of the cold, to spread over the plains of the temperate regions, whence they retreated with the advent of warmer conditions to seek a congenial climate in the nearest mountain region.

The mutual relationships of the different groups of mammals are so complex that they cannot be

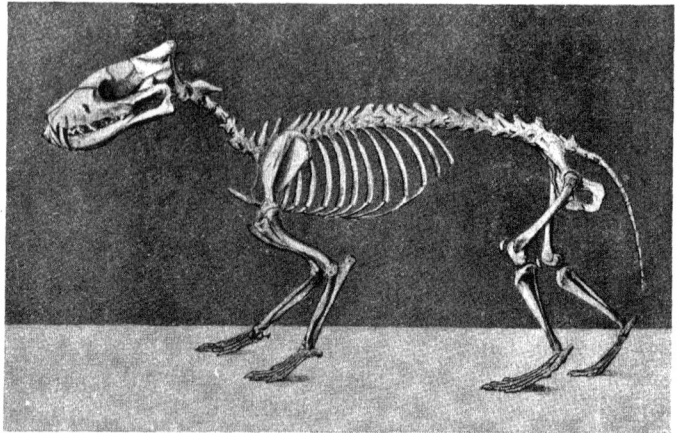

THE SKELETON OF A HYÆNODON, AN EARLY CARNIVORE
The relatively large and heavy skull is a notable feature

represented by any system of linear classification, and a tree-like diagram would be the only method of properly displaying these relationships. But since it is impossible to give such a genealogical tree with any certainty, we are driven to the linear arrangement, as in the following table of the existing orders and families.   R. LYDEKKER

CLASS I.—MAMMALIA

SUBCLASS I.
Viviparous Mammals (Eutheria, or Vivipara)
SECTION A.
Placentals (Placentalia)

ORDER I.

Man, Apes, Monkeys, Lemurs	..	**Primates**
SUBORDER I.—Man, Apes, Monkeys	..	**Anthropoidea**
FAMILY		
1. Man	..	Hominidæ
2. Man-like Apes	..	Simiidæ
3. Old World Monkeys and Baboons..		Cercopithecidæ
4. American Monkeys	..	Cebidæ
5. Marmosets	..	Hapalidæ
SUBORDER II.—Lemuroids ..	..	**Lemuroidea**
FAMILY		
1. Lemurs	..	Lemuridæ
2. Aye-aye	..	Chiromyidæ
3. Galagos and Lorises ..	..	Lorisidæ
4. Tarsier	..	Tarsiidæ

ORDER II.

Bats	..	**Chiroptera**
SUBORDER I.—Fruit-eating Bats ..	..	**Megachiroptera**
FAMILY		
Flying Foxes	..	Pteropodidæ
SUBORDER II.—Insect-eating Bats	..	**Microchiroptera**
FAMILY		
1. Horseshoe Bats	..	Rhinolophidæ
2. False Vampires	..	Nycteridæ
3. Typical Bats.. ..	..	Vespertilionidæ
4. Free-tailed Bats ..	..	Emballonuridæ
5. Vampires	..	Phyllostomatidæ

ORDER III.

Insect-eating Mammals ..	..	**Insectivora**
SUBORDER I.—Flying Lemurs ..	..	**Dermoptera**
FAMILY		
Cobegos, or Kaguans	..	Galeopithecidæ
SUBORDER II.—Tupais and Jumping Shrews ..		**Typhlomena**
FAMILY		
1. Tupais	..	Tupaiidæ
2. Jumping Shrews	..	Macroscelididæ
SUBORDER III.—Shrewmice, etc...	..	**Insectivora vera**
FAMILY		
1. Hedgehogs, etc.	..	Erinaceidæ
2. Shrewmice	..	Soricidæ
3. Moles	..	Talpidæ
4. Tenrecs	..	Centetidæ
5. Solenodons	..	Solenodontidæ
6. Otter-Shrews	..	Potamogalidæ
7. Golden Moles	..	Chrysochloridæ

ORDER IV.

Flesh-eating Mammals ..	..	**Carnivora**
SUBORDER I.—Typical Carnivora..	..	**Fissipedia**
FAMILY		
1. Cat tribe	..	Felidæ
2. Civets and Mungooses ..	..	Viverridæ
3. Hyænas	..	Hyænidæ
4. Aard-Wolf	..	Proteleidæ
5. Wolves, Dogs, Foxes ..	..	Canidæ
6. Bears	..	Ursidæ
7. Raccoons and Pandas ..	..	Procyonidæ
8. Weasel tribe	..	Mustelidæ
SUBORDER II.—Eared Seals, Walrus, Seals		**Pinnipedia**
FAMILY		
1. Sea Lions and Sea Bears ..	..	Otariidæ
2. Walrus	..	Odobænidæ
3. Typical Seals	..	Phocidæ

ORDER V.

Rodents, or Gnawing Mammals	..	**Rodentia**
SUBORDER I.—Squirrels, Mice, etc.	..	**Simplicidentata**
FAMILY		
1. Scaly-tails	..	Anomaluridæ
2. Squirrels and Marmots ..	..	Sciuridæ
3. Beavers	..	Castoridæ
4. Sewellels	..	Haplodontidæ
5. Dormice	..	Myoxidæ
6. Jerboas	..	Dipodidæ
7. Mice and Rats	..	Muridæ
8. Gundi, Degu, etc. ..	..	Octodontidæ
9. Porcupines	..	Hystricidæ
10. Chinchillas and Viscachas	..	Chinchillidæ
11. Agutis and Pacas ..	..	Dasyproctidæ
12. Cavies	..	Caviidæ

SUBORDER II.—Hares, Rabbits, Picas

SUBORDER II.—Hares, Rabbits, Picas ..	..	**Duplicidentata**
FAMILY		
1. Picas	..	Lagomyidæ
2. Hares and Rabbits	..	Leporidæ

ORDER VI.

Hoofed Animals	..	**Ungulata**
SUBORDER I.—Even-toed Group ..	..	**Artiodactyla**
A. Cattle, Sheep, Antelopes, Deer, and Giraffes	..	**Pecora**
FAMILY		
1. Cattle, Sheep, and Antelopes	..	Bovidæ
2. Prongbuck	..	Antilocapridæ
3. Giraffes and Okapi ..	..	Giraffidæ
B. Camels and Llamas ..	..	**Tylopoda**
FAMILY		
Camels and Llamas ..	..	Camelidæ
C. Chevrotains	..	**Tragulina**
FAMILY		
Chevrotains	..	Tragulidæ
D. Pigs and Hippopotamuses..	..	**Suina**
FAMILY		
1. Pigs	..	Suidæ
2. Peccaries	..	Dicotylidæ
3. Hippopotamuses ..	..	Hippopotamidæ
SUBORDER II.—Odd-toed Group ..	..	**Perissodactyla**
FAMILY		
1. Tapirs	..	Tapiridæ
2. Rhinoceroses	..	Rhinocerotidæ
3. Horse tribe	..	Equidæ
SUBORDER III.—Hyraxes ..	..	**Hyracoidea**
FAMILY		
Hyraxes	..	Hyracidæ
SUBORDER IV.—Elephants ..	..	**Proboscidea**
FAMILY		
Elephants	..	Elephantidæ

ORDER VII.

Manatis and Dugongs	..	**Sirenia**
FAMILY		
Manatis and Dugongs	..	Halicoridæ

ORDER VIII.

Whales and Dolphins.. ..	..	**Cetacea**
SUBORDER I.—Whalebone Whales	..	**Mystacoceti**
FAMILY		
Whalebone Whales	..	Balænidæ
SUBORDER II.—Sperm Whales and Dolphins		**Odontoceti**
FAMILY		
1. Sperm Whales	..	Physeteridæ
2. Beaked Whales	..	Ziphiidæ
3. Susu	..	Platanistidæ
4. American River Dolphins ..	..	Iniidæ
5. Porpoises and Dolphins ..	..	Delphinidæ

ORDER IX.

Sloths, Armadillos, etc. ..	..	**Edentata**
FAMILY		
1. Sloths	..	Bradypodidæ
2. Ant-eaters	..	Myrmecophagidæ
3. Armadillos	..	Dasypodidæ
4. Pangolins	..	Manidæ
5. Ant-Bears	..	Orycteropodidæ

SECTION B.
Pouched Mammals (Implacentalia)

ORDER X.

Marsupials	..	**Marsupialia**
FAMILY		
1. Kangaroos and Wallabies ..	..	Macropodidæ
2. Cuscuses and Phalangers ..	..	Phalangeridæ
3. Wombats	..	Phascolomyidæ
4. Selvas	..	Epanorthidæ
5. Bandicoots	..	Peramelidæ
6. Tasmanian Wolf, etc. ..	..	Dasyuridæ
7. Opossums	..	Didelphyidæ

SUBCLASS II.
Egg-laying Mammals (Hypotheria, or Ovipara).

ORDER XI.

Platypus and Echidnas ..	..	**Monotremata**
FAMILY		
1. Spiny Ant-eaters	..	Echidnidæ
2. Platypus, or Duckbill ..	..	Ornithorhynchidæ

The Negroid races shown in this picture are : 1, Bushman ; 3, Negro ; 5, Zulu. The Mongolian races are : 2, Australian Aboriginal ; 4, Maori ; 6, Red Indian ; 7, Japanese ; 9, Chinese ; 10, Eskimo. The Caucasian races are : 8, Hindu ; 11, Turk ; 12, Persian ; 13, Greek ; 14, Russian ; 15, Jew

MAN THE SUPREME MAMMAL
HOW HE BECAME THE HEAD OF THE ANIMAL KINGDOM
By J. ARTHUR THOMSON, M.A.

Regius Professor of Natural History, Aberdeen University

ZOOLOGICALLY considered, man is a mammal, bound to other mammals by an "all-pervading similitude," to use Sir Richard Owen's phrase, in structure, in function, and in development. At the same time, he is worthy of being called the supreme mammal, so highly is he specialised in body and brain; and he shows a number of distinctive features, or what we may call human characteristics.

Man alone, after his infancy is past, walks quite erect—a thorough-going biped. Though his head is weighted with a heavy brain, it does not droop forwards on the chest. Dr. Robert Munro and others have given strong reasons for correlating the upright attitude with the increased command of the vocal mechanism. Playful reference is often made to the absence of a tail in man, but in this reduction the anthropoid apes have gone even farther. In the human skeleton there are four or five obvious caudal vertebræ, which fuse together to form what is called the coccyx. In young embryos there is a short projecting portion surrounded by skin, and this sometimes persists after birth as a distinct tail. The sparseness of well-developed hairs over the body is also distinctive, but there is great variability in this respect in different individuals and races. Darwin was inclined to interpret the relative nakedness of man as due to some extent to a warm climate, but in greater part to sexual selection, the less hairy men and women having selected one another for æsthetic reasons.

Both the fore limbs and hind limbs have undergone modification in adjustment to man's erect walk. It is evident that when the bipedal position was attained, and the hand liberated from any share in locomotion, the human foot was bound to develop a steady but flexible sole. Thus there is something distinctive even in the way man plants his foot on the ground, with the whole sole flat. The great toe is often longer, never shorter, than the others, and lies in the same line with them. There is a better heel than in any monkey. It used to be the custom to contrast man and monkeys as Bimana and Quadrumana respectively, but there is very little in this distinction. For although all monkeys have an "opposable" great toe, or hallux, which they use against the other toes in grasping things in a way that is not possible to man, the value of the contrast diminishes when we note that some wild peoples are able to grip with their toes, and that babies show a marked power of "opposability" that is afterwards lost.

When we compare the human skull with an ape's, we see that man has more of a true forehead rising above the eyes, a less protrusive face—not less, however, than some American monkeys—smaller

cheekbones and supraorbital ridges, no crests for the attachment of muscles on the top or back of the head, projecting nasal bones above the nostril, an early disappearance of the junction between the premaxillary bones (that bear the four front teeth, or incisors) and the maxillary bones (that bear all the other teeth), a true chin or bony process at the anterior junction of the two halves of the lower jaw, and much more uniform teeth, forming an uninterrupted horseshoe-shaped series without very conspicuous canines. At the same time, it must be remembered that there is considerable variability among men in regard to these characters; thus the ridges over the eyes are much more strongly marked in some people than in others, the cheekbones differ greatly in their degree of prominence, and so on.

Compared with most monkeys, man has longer legs and shorter arms, and there are other minor differences that might be mentioned. Of enormously greater importance are the differences in cranial capacity and weight of brain. "The cranial capacity is never less than 55 cubic inches in any normal human subject, while in the orang-utan and chimpanzee it is but 26 and $27\frac{1}{2}$ cubic inches respectively." The brain of a healthy human adult never weighs less than 31 or 32 ounces; the average human brain weighs 48 or 49 ounces; the heaviest gorilla brain does not exceed 20 ounces. A man *may* have a brain three times as heavy as that of a gorilla.

Like many other mammals, man shows a noteworthy degree of sex-dimorphism—that is to say, there is a marked difference in the detailed characters of man and woman. The typical female form, compared with the typical male form, shows a relatively longer trunk, shorter arms, legs, hands, and feet—"relatively to short upper arms, still shorter forearms, and relatively to short thighs, still shorter lower legs, and relatively to the whole short upper extremity a still shorter lower extremity"—all pointing to a less actively locomotor type. Man is, on an average, heavier, taller, more muscular, shorter-lived, and so on; but the important fact from the biological point of view is that the fundamental sex difference saturates, as it were, through the organism, expressing itself in large features, such as the proportions of different parts of the body, and in minute details, such as the number of red blood corpuscles.

The large and complex brain of man is correlated with those differences which everyone admits to be most distinctive: (1) That he has a power of working with abstract ideas, making "conceptual inferences," exercising *reason*; (2) that he has the habit of guiding his conduct in reference to certain ideals, exerting, when he will, a power of ethical choice, "*thinking the ought*"; and (3) that he has a language in the true sense, a power of expressing his judgments in a manner intelligible to others, which is something more than having words, being, in short, a *logos*.

Darwin was not sanguine of discovering a cause for the intellectual stride which marked the emergence of man; he gave more attention to showing, by his observations on dogs and the like, that, even in regard to reason — or conceptual inference—there are no sharply defined limits between the mental powers of man and animals. In regard to morality, he showed that the raw materials were present in animal behaviour, especially among social and gregarious species. The same fundamental springs of conduct, such as family affection and kinsympathy are there, and in some cases there is a deliberation which is a foretaste of conscience. But the power of speech was required to develop public opinion, which has so much to do with morals.

In regard to language, Darwin pointed out that various mammals and birds have particular sounds with distinctive significance, and that sheepdogs can understand a complex order which their master gives. "I cannot doubt," he said, "that language owes its origin to the imitation and modification of various natural sounds, the voices of other animals, and man's own instinctive cries, aided by signs and gestures." But its development went hand in hand with that of intelligence, and "the fact of the higher apes not using their vocal organs for speech no doubt depends on their intelligence not having been sufficiently advanced"—a view which agrees with what a modern psychologist has said, that animals would speak if they had anything to say.

MAN'S AFFILIATION

Darwin's "Descent of Man" (1871) is an expansion of a chapter in his "Origin of Species" (1859), for the argument which leads to the conclusion that man arose from an ancestral stock common to him and to the higher apes is logically the same as that which leads to the general doctrine of descent—that the present is the child of the past and the parent of the future. The heads of the familiar argument may be indicated:

(1) *As to Structure.*—Bone for bone, organ for organ, man is very like one of the anthropoid apes. Even in such a trivial detail as the arrangement of the hair in different parts of the body, man is dis-

tinctly simian. Some of the anatomical differences have been already indicated, but the only very important one is the large and heavy brain. As Huxley proved in his famous book, "Man's Place in Nature" (1863), the structural differences between one of the lower monkeys, such as the marmoset, and the chimpanzee are much greater than those between the chimpanzee and man. Furthermore, as we shall see later, man is a walking museum of relics—vestigial structures — in his muscular, skeletal, and other systems, which are quite enigmatical except when interpreted as dwindling items in a complex mammalian heritage. In a very literal sense, he is heir of all the ages. Many of the occasional abnormalities of structure in man have a very interesting historical interest, throwing light on his pedigree. Thus, Testut, one of the greatest authorities on such matters, says of the muscles that "the gap which usually separates the muscular system of man from that of the anthropoids appears to be completely bridged over" by the abnormalities.

(2) *As to Functions and Habits.*—The bodily life of man is very like that of his presumed relations. Some peculiar "idiosyncrasies" that now and then assert themselves seem like outcrops of the pre-human. Various features of physiognomy and trivial details of gesture are paralleled among the apes. Men and monkeys are subject to similar diseases, such as rheumatism and phthisis.

One of the most interesting additions to the list of resemblances has to do with the blood. Let us take it first in its simplest form. Friedenthal has shown that when the blood of a cat is transfused into a rabbit it exerts a destructive influence, dissolving the red blood corpuscles. But when the blood of a horse is mixed with that of an ass, or a dog's with a wolf's, there is no destruction of corpuscles, and this is a sign of near relationship. Friedenthal found that this held good for man and chimpanzee.

This matter has been worked out by a number of investigators, Tschistowitsch, Uhlenhuth, Nuttall, and Friedenthal. If a rabbit, for instance, has repeated injections of blood serum from another *unrelated* kind of animal, it reacts to the foreign material by forming and accumulating in its blood a substance which, when added to a solution of the serum injected, gives rise to a precipitate. When human blood has been injected into a rabbit, the serum of this rabbit forms a precipitate when added to human blood. When added to the blood of an anthropoid ape, like the orang-utan, it gives a

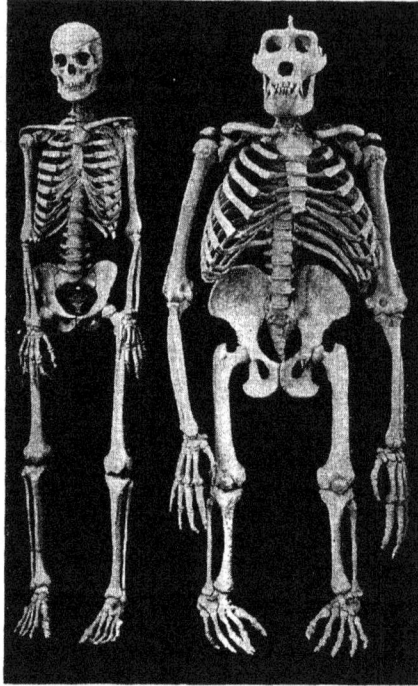

SKELETONS OF MAN AND THE GORILLA
These skeletons show how man differs from the gorilla in the length of arms and legs and form of the skull

similar precipitate, but this is weak in the case of the more distantly related Old World monkeys, and practically absent in the very remotely allied New World monkeys. When added to the blood of a horse or the like, there is no reaction. There is a common property in the blood of related forms ; in short, *blood-relationship* is a literally accurate term. " We have in this," as Prof. Schwalbe says, " not only a proof of the literal blood-relationship between man and apes, but the degree of relationship with the different main groups of apes can be determined beyond possibility of mistake." There are some anomalies, but most of the results correspond closely with the conclusions of comparative anatomy. Thus, according to the blood tests, the ass is considerably nearer to the horse than the orang-utan is to man.

(3) *As to Racial History and Individual Development.*—Remains of primitive man are few and fragmentary, and though it must be granted that some of the early skulls are nearer the simian type than those normal to-day, we have more security in dealing with individual development than with racial history. There is a close resemblance between the unborn human child and the corresponding stage in the development of one of the anthropoid apes.

THE APOLLO BELVEDERE, IN THE VATICAN
An ideal type of manly beauty. Photo. by Anderson, Rome

Not only is there a general likeness, but there are several trivial resemblances that are even more convincing. Thus, the human embryo, about the middle of the ante-natal period, has a pointed ear, somewhat similar to that of a macaque monkey, and this animal ear-point is sometimes quite clearly seen in adult men. Children born in untoward conditions, after a siege or a famine, are sometimes below the ordinary human level ; and apart from such cases of arrested development there are a few instances of what seems most reasonably interpreted as a structural harking-back to a pre-human stage.

In man, as in other types of animal life, we find that the development of the individual recapitulates in a general way the historical evolution of the race. This is best illustrated in the parallelism sometimes observable in regard to particular organs between stages in embryonic development and stages in presumed ancestral evolution. Thus, it is noteworthy that the unborn human offspring shows a hairy coat, or lanugo, a short projecting tail, arms as long as legs, and other characters which point to a quadrupedal, tailed, hairy ancestor. Similarly, the newly-born baby shows peculiarities which seem to be more than the results of its adaptation to the conditions of the ante-natal life. Thus, we may refer to the mobility of the feet and of the individual toes, and to the extraordinary power of gripping and hanging on with the fingers.

THE VENUS OF MILO, IN THE LOUVRE
The highest ideal type of woman. Photo. by W. E. Mansell & Co.

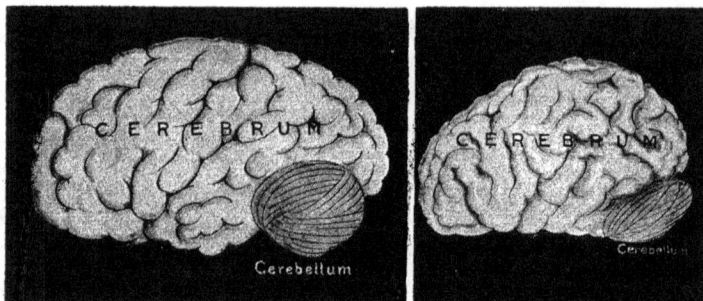

THE BRAIN OF MAN AND THE GORILLA COMPARED

It is in the development of the brain that man's superiority over the ape is most characterised, as the size and number of the convolutions show

There are four genera of anthropoid apes—gibbon, orang-utan, gorilla, and chimpanzee—and, as Huxley pointed out, each has its particular points of resemblance with man, while each is specialised along a line of its own. In this connection it is important to notice how the generally accepted conclusion that man has arisen from a stock common to him and to the anthropoid apes is borne out by embryology. We may illustrate this with reference to the gibbons, which approach man in many ways, not only in sociality and vocal powers, but in numerous structural details ; for instance, in regard to the teeth. Yet the gibbons are highly specialised for climbing, and have disproportionately long arms almost reaching the ground. There is at once marked resemblance and marked difference, but the particularly instructive fact is that worked out by Selenka, that up to a certain point there is a very striking resemblance between the gibbon embryo and the human embryo, and for a considerable time the gibbon embryo has its fore limbs and hind limbs in their usual proportions. Human embryo and gibbon embryo travel for a time side by side on parallel paths of development, but eventually each takes a way of its own.

Thus, whether we consider finished structure, or functions and habits, or the facts of development, we find so many strong resemblances between man and the anthropoid apes that it seems impossible to doubt that there is a real blood-relationship. They and he must be regarded as offshoots from a common stock. The structural affinities are so numerous, again, that we must reject the view that man has evolved " parallel with the monkeys, but without relation to them, from very low primitive forms." The zoological evidence clearly points to the conclusion that the divergence of man from the anthropoids occurred after the divergence of the common stock from the Old World monkeys (Catarrhini) in general. As Darwin said : " The Simiidæ branched off into two great stems, the New World and the Old World monkeys ; and from the latter, at a remote period, man, the wonder and glory of the universe, proceeded."

Darwin compared rudimentary (or, better, vestigial) organs to the unsounded letters in words, such as the "o" in leopard and the " b " in doubt, which have only a historical interest, having ceased to be functional. We see the same sort of "survival" every day in our clothing ; for instance, in various functionless buttons and unopenable buttonholes, and illustrations abound in social institutions. The general idea is that the vestigial structures—whether bones or buttons, letters or offices —are dwindling residues of what were once of functional importance, and man's body is a veritable storehouse of such relics.

(I.) It is useful to distinguish, first of all, those antique structures which are present only in the embryo, and do not normally come to anything in the adult. Thus, to take one of the most familiar cases, there are on each side of the neck of the human embryo four clefts from the pharynx to the exterior—the visceral clefts or branchial clefts. They are present in the embryos of all reptiles, birds, and mammals, but only the first of them has any functional import. It becomes the Eustachian tube passing from the ear-opening to the back of the mouth. The others normally disappear without leaving a trace. There is no doubt as to the historical significance of these visceral clefts ; they represent the gill clefts which persist throughout life in fishes, and are associated with gills. Amphibians represent a transition stage, for there the first gill cleft becomes for the first time the Eustachian tube in the service of the ear ; and the other clefts, though provided with gills and used in respiration in tadpoles and other larval stages, often disappear entirely in adult life. To pursue the inquiry one step farther, it is interesting to find that a not very infrequent abnormality in man is the presence of a " cervical auricle " on the side of the neck—a persisting vestige of a posterior gill cleft which has not wholly disappeared in embryonic life. It may take the form of a minute aperture, or it may show a small flap like a hint of an external ear. Similar cervical auricles are occasionally seen in horses, goats, and other mammals. In further illustration of embryonic vestigial organs, we may

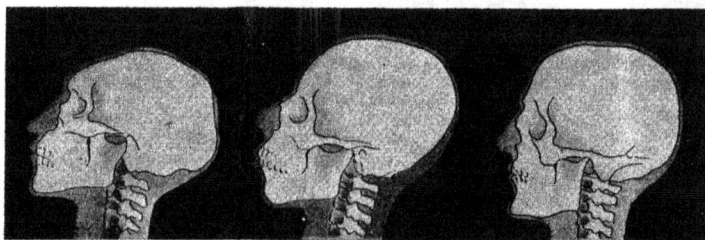

THE DIFFERENCES BETWEEN THE SKULLS OF LIVING RACES OF MAN

These pictures show how the skull of man has evolved. The first is that of an Australian native, a low type ; the second that of a Negro ; and the third that of a European, the highest type

notice (1) the transitory notochord which precedes the backbone, and is reminiscent of the permanent dorsal axis in old-fashioned primitive vertebrates like the lancelet and the lamprey; (2) the primitive, soft woolly covering, or lanugo, on the body of the unborn child, which occasionally persists in adult life in one of the kinds of exuberant hairiness—to be distinguished from other kinds where the hairs are not of the lanugo type; (3) the occurrence of ribs in connection with vertebræ which have none in the adult, the vertebræ supporting the hip girdle, the vertebræ of the loins, and the sixth and seventh neck vertebræ, the last occasionally persisting in the adult. Besides these vestiges there are many others, especially in connection with the muscular, skeletal, and vascular systems, a careful account of which will be found in Wiedersheim's remarkable treatise, "The Structure of Man." One caution must be emphasised: that it is beyond our knowledge to deny utility to all of these vestigial embryonic structures. Some of them, such as the transient notochord, may have some architectural significance, serving,

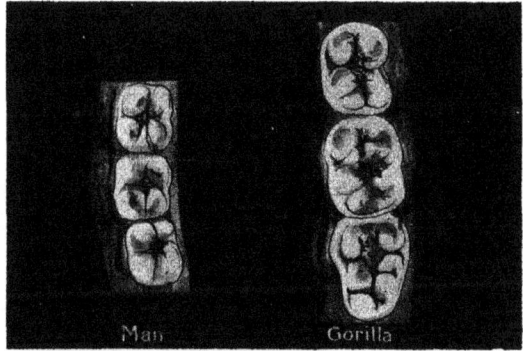

THE MOLAR TEETH OF MAN AND THE GORILLA

There is the same number of teeth in man and the man-like apes, and, excepting the great canines, or tusks, in the male apes, there is also similarity in structure

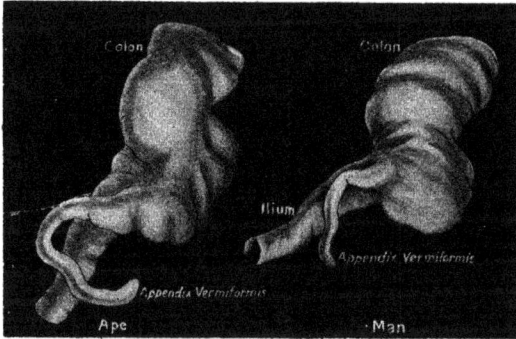

THE VERMIFORM APPENDIX IN MAN AND THE GORILLA

The difference between the appendix in man and the ape is well shown in the diagram. The organ has become atrophied in the human intestine, and its inflammation is a well-known source of mischief

perhaps, as a preliminary scaffolding for its more effective substitute—the backbone.

(II.) In a second grade we may place those structures which persist in adult life, but in a much disguised form. The intergrade between the two sets of cases is the first gill cleft, which becomes the Eustachian tube and tympanic cavity. In the same way, the gill arches, whose primary significance was to support gills, are in part persistent in the skeletal support of the mammalian tongue (the hyoid) and in the framework of the larynx (Adam's apple). Very striking, but more difficult to state in a few words, is the extraordinary way in which antique arrangements persist in disguised form in the vascular system, now and again revealing their historical significance by some tell-tale variation, such as the occurrence of a double aortic arch. The arteries leaving the heart of the human embryo are disposed as paired aortic arches going to the visceral clefts, as in a fish or a tadpole; and in a very remarkable way the dorsal aorta (to the body) is derived from what corresponds to the second branchial

artery, and the pulmonary artery (to the lungs) from what corresponds to the fourth branchial artery.

(III.) In the third place, we rank vestigial structures in the stricter sense—dwindling residues persistent in adult life, but either functionless or relatively unimportant. A good instance is afforded by the third eyelid—a minute fold which lies in the median angle of the eye. It is larger in some races than in others—these vestigial structures are notably variable—but it is never of use. It is, however, a relic of a functional third eyelid, or "nictitating membrane," seen in most mammals, birds, and reptiles. In these it has a special muscular arrangement which flicks it across the eyeball, and it serves to clean the surface of the eye. It is absent in whales and their allies, and we may connect this with the continual washing of the eye with water. It is vestigial in monkeys and mankind, and we may connect this with the mobility of the upper lid, and also with the better development and greater mobility of the lower lid. It is instructive to compare the rudimentary third eyelid in man with the so-called "epicanthus" which is conspicuous in Mongolians, giving rise to the slit-like appearance. This is a prolongation of the upper lid over the inner angle of the eye, and is apparently due to the flatness of the bridge of the nose which leaves the

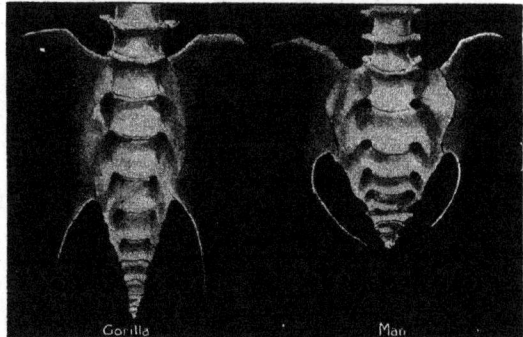

THE SO-CALLED TAILS IN MAN AND THE GORILLA

The caudal, or tail, vertebræ of the spinal column in man being rudimentary, the tail is practically absent, but is distinctly evident in the gorilla

HARMSWORTH NATURAL HISTORY

superfluous skin to form the fold in question. " According to Ranke, about 6 per cent. of Caucasian children exhibit a markedly Mongolian type of eye during the first six months of their lives."

Darwin gave a number of good examples of vestigial muscles in man, such as those which are occasionally strong enough to move the trumpet, or pinna, of the ear, or the thin sheet below the skin of the neck and some other regions—a residue of the *panniculosus carnosus* which spreads like a mantle over the back, head, neck, and flanks of many mammals, and is of use in twitching off flies and the like. There is a memorable passage in "The Descent of Man": "He who rejects with scorn the belief that the shape of his own canines, and their occasional great development in other men, are due to our early forefathers having been provided with these formidable weapons, will probably reveal by sneering the line of his descent. For though he no longer intends, nor has the power, to use these teeth as weapons, he will unconsciously retract his ' snarling muscles ' (thus named by Sir Charles Bell) so as to expose them ready for action, like a dog prepared to fight."

The list of retrogressive muscles is very long, and the degree of their retrogression is diverse and variable. Two of the best illustrations are afforded by the palmar muscle of the fore limb and the plantar muscle of the hind limb, which used to have to do with bending the digits, but have become surperfluous by the differentiation of other muscles, and through the specialisation of the foot as a supporting and ambulatory organ. The two muscles in question vary considerably in the degree of their degeneracy. They are sometimes still functional, but they are often completely reduced to tendinous tissue. As Professor H. F. Osborn says, these two muscles illustrate " organs so far on the down grade that they are mere pensioners of the body, drawing pay—*i.e.*, nutrition—for past honourable services without performing any corresponding work."

As another instance of a dwindling muscle, we may mention the small subclavius which underlies our collar-bone, or clavicle, and steadies the latter during the movement of the arm. It is an insignificant tag of flesh ; it is sometimes represented by a non-contractile band of fibrous tissue ; it is occasionally absent ; it is certainly dwindling.

It is of some importance to distinguish vestiges which have dwindled into insignificance from vestiges which, though in process of reduction, are still very far from a vanishing point, and are often large enough to be mischievous. It is a noteworthy fact that these larger vestigial organs are particularly apt to be seats of disease, as is well known to all in connection with the most familiar instance, the vermiform appendix on the human intestine. Structures which are not essential to the economy of the body, on which selection has lost its grip, are apt to cross the border-line between degenerative change and disease.

To understand the vermiform appendix we have to recall the fact that in many vegetarian animals there is a large cul-de-sac, or cæcum, growing out at the junction of the small and large intestine, which serves to increase the digestive absorptive surface. In the rabbit, for instance, the cæcum is the largest part of the whole intestine, and it bears at its blind end the vermiform appendix. In man and anthropoid apes there is a very short cæcum, and at the end of it there is the vermiform appendix, a dwindling vestige, but still from two to eight inches in length.

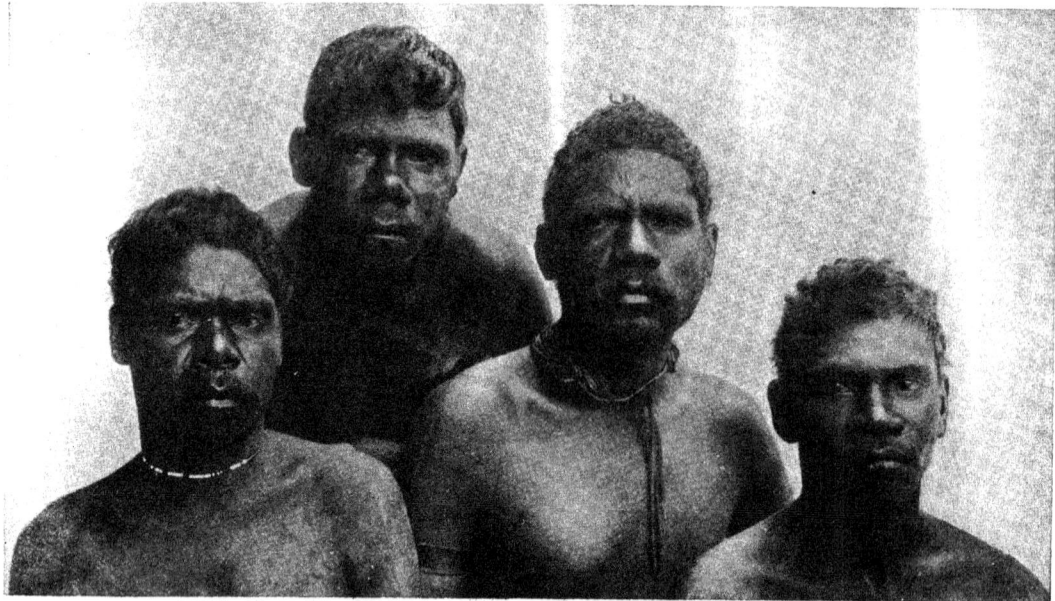

A GROUP OF AUSTRALIAN ABORIGINES, THE LOWEST TYPE OF THE HUMAN RACE

It has the same general structure as the intestine, though with a large proportion of adenoid or lymphatic tissue, but everything points to its retrogressive character. It is *relatively* longer in the embryo than in the adult; it is very variable; it is often closed either partially or wholly—apart from artificial blocking or pathological obliteration; it often shows degenerative changes in its tissues. Everyone knows that this remnant that has outlived its usefulness, this anachronism surviving from a remote past, often costs a man his life. It is "like an idle person in a community," peculiarly liable to go wrong and give rise to serious mischief.

It is important to get away, once for all, from the static conception of a living creature as something fixed and finished, for careful measurement always shows that organs fluctuate continually from generation to generation. Most of them are kept relatively stable by the averaging implied in the conditions of inheritance and by the pruning action of natural selection, but some are on the up grade, others on the down grade, and vestigial structures are among the latter. When we consider the matter in this broad way, we see why it is impossible to give a strict definition of a vestigial organ. For some persist in adult life, and others are purely embryonic; some have a slight secondary use, others an occasional use, and others apparently no use at all. We understand, too, why vestigial organs should be very eminently variable. "This is intelligible," Darwin said, "as they are useless, or nearly useless, and consequently no longer subjected to natural selection. They often become wholly suppressed. When this occurs they are nevertheless liable to occasional reappearance through reversion." What has lain latent for generations may suddenly find re-expression.

There is something very fine in the closing words of "The Descent of Man": "We must, however, acknowledge, as it seems to me, that man, with all his noble qualities, with sympathy which feels for the most debased, with benevolence which extends not only to other men, but to the humblest living creature, with his God-like intellect which has penetrated into the movements and constitution of the solar system—with all these exalted powers—man still bears in his bodily frame the indelible stamp of his lowly origin."

RECONSTRUCTION OF PRIMITIVE MAN

"By considering the embryological structure of man—the homologies which he presents with the lower animals, the rudiments which he retains, and the reversions to which he is liable—we can partly recall in imagination the former condition of our early progenitors. We thus learn that man is descended from a hairy, tailed quadruped, probably arboreal in its habits, and an inhabitant of the Old World" (Darwin: "Descent of Man," p. 930). It is very difficult to advance from this to an imaginative reconstruction of the appearance of those who first called themselves men, who first began to recognise the year—with its suggestive object-lesson of recurrent sequences—who first became aware that their race had a history. Two general statements may be made, (*a*) that the stock had diverged considerably from the ancestral stem common to it and to the anthropoid apes, in acquiring a bigger brain, a better forehead, a free hand, an upright gait, and so on; (*b*) that we must not attach too much importance to the lowest races of men living to-day, like the Bushmen of South Africa, since these peoples appear

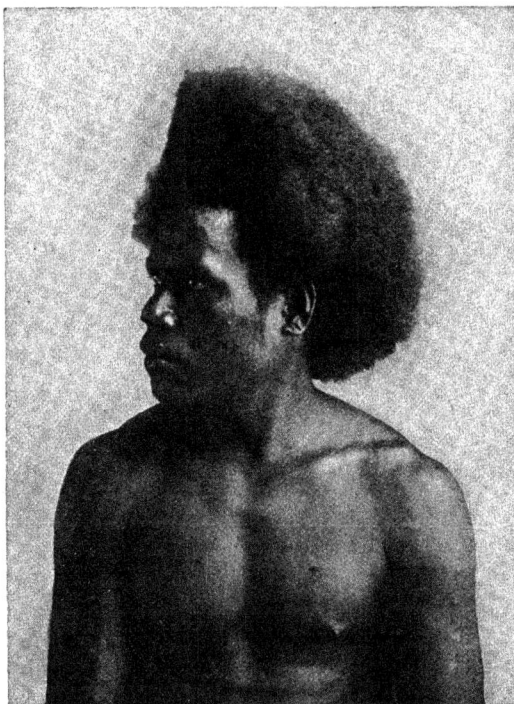

A NATIVE OF THE FIJI ISLANDS

to be in some respects actually degenerate.

In regard to many points, we can only guess, which is not science; but one of the safest lines is to study the dwindling structures in man, and work from these backwards to a reconstruction of the primitive type. Thus Professor Wiedersheim writes ("The Structure of Man," Trans., p. 216): "There would appear to have been a time when our ancestors were protected against the inclemencies of the weather by a natural covering of hair, and against insects and other injurious influences by an extensive tegumental musculature; when the pinna of the ear, more advantageously disposed than at present, and moved by numerous and powerful muscles, collected the sounds of approaching danger incomparably better than at the present day, and when the sense of smell, probably intensified by Jacobson's organ, was more highly developed than now." . . . "The intestinal tube may have been longer, and thus better suited than at the present day for vegetable diet, the ancestor of man enjoying at any rate more favourable conditions of existence as a vegetarian than his successor now does (compare also the former greater number of cheek teeth). He may also have had the further advantage of not possessing a vermiform process of the cæcum which predisposes to disease, and causes the destruction of a considerable percentage of his fellows."

It is probable that Primitive Man was more muscular, especially about the shoulders, neck, and jaws ; that he had a stronger development of skin-twitching and ear-moving muscles ; that he had a moderate amount of pigment in his skin, and that he was very variable in this respect ; that he had a retreating forehead, strong ridges above the eyes, prominent cheekbones, a flat nose, a highly developed sense of smell, and larger, less regular teeth. It is probable that some features were only incipient, such as the chin, the mobile lips, and the woman's breasts.

As to the mental endowment of our early ancestors, we cannot, of course, do more than speculate. But perhaps there is some degree of truth in that extraordinarily vivid picture which Æschylus gave of them—living in caves without fire, without woodwork, without system, without seasons, without foresight ; a dream-life without judgment :

And, let me tell you, not as taunting men,
But teaching you the intention of my gifts,
How, first, beholding they beheld in vain,
And hearing, heard not, but, like shapes in dreams,
Mixed all things wildly down the tedious time,
Nor knew to build a house against the sun
With wicketed sides, nor any woodwork knew,
But lived like silly ants, beneath the ground,
In hollow caves unsunned. There came to them
No steadfast sign of winter, nor of spring
Flower-perfumed, nor of summer full of fruit,
But blindly and lawlessly they did all things,
Until I taught them how the stars do rise
And set in mystery, and devised for them
Number, the inducer of philosophies,
The synthesis of letters, and, besides,
The artificer of all things, Memory,
That sweet muse-mother.

Compared with his ancient ancestors, man has probably lost much in the way of brute strength, but he has profitably bartered this for a more alert and educable brain, a greater manual dexterity, an increased power of articulate speech, and a growing gentleness. We can well imagine that what seems to have been true at other great epochs in the history of animate Nature was true of the ascent of man—that the loss of certain advantages was at once a condition and a consequence of gaining others.

The Origin of Man

We cannot at present get beyond the general conclusion that man sprang from a stock common to him and to the higher apes. It is quite certain that he is not a scion of any existing simian stock, though his lineage has been collateral with theirs. Evidence is accumulating to show that discontinuous variations, large in amount and qualitatively novel, sometimes occur in Nature, and it may be that man arose as a *mutation* of this sort, as an anthropoid genius, taking a big step all

at once. Not, indeed, that he can be thought of as rising suddenly to the height of his dignity, but rather that some important cerebral variation occurred brusquely and led on to consequences which were very gradually disclosed as in any childhood to-day. The origin of man, whether by the slow accumulation of minute cerebral variations or by a brusque mutation, was a step so wonderful that it is on *a priori* grounds difficult to believe that it was taken more than once. There are, however, accomplished anthropologists who firmly believe that the human race made more than one start.

As to the factors which led to the emergence of the human type in an ancestral simian stock, we

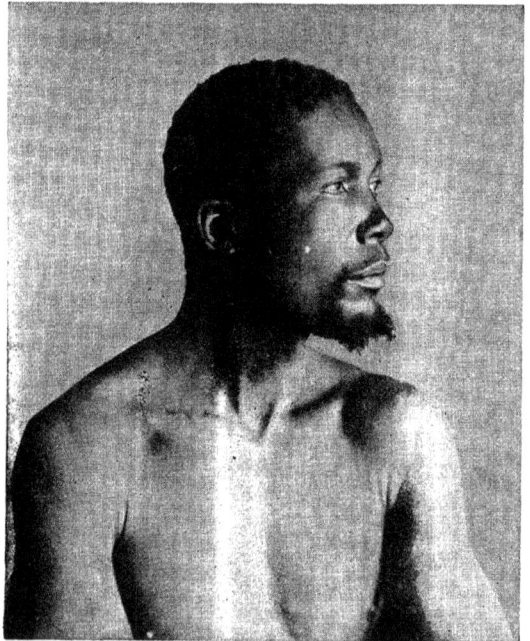

A ZULU, OR KAFFIR, OF CAPE COLONY
Photograph by G. W. Wilson & Co.

can only speculate. We must remember that the order of Primates is marked by great intelligence ; that we find amongst them habits of walking semi-erect, of using sticks and stones, of building shelters, of living in families, of co-operating in bands, and of talking a great deal.

It is noteworthy, also, that we find among apes the prolonged ante-natal life which is characteristic of very intelligent mammals, and may have had something directly to do with the improvement of wits, just as the prolonged human infancy has had to do with the growth of gentleness. Darwin fully recognised that the survival of the human pioneers must have been due, not merely to their cleverness, but to their social instincts.

It is often said that an alleviation of the struggle for existence must have been a condition of the ascent of man, but, while we do not in the strict sense know anything about it, we may perhaps state the case more accurately in the proposition that the ascent of man was associated with a form of struggle in which brains meant more than muscle, and co-operation more than competition.

Although the anthropoid apes are not social creatures, there are many social monkeys. Families combine for protection, and the sociality promotes emotional and intellectual strength. When we realise how largely the human mind is still a social product, we see the force of the epigram : " Man did not make society ; society made man."

THE ANTIQUITY OF MAN

A false standard of time and a misunderstanding of the Holy Scriptures led pre-Darwinian naturalists to ignore the immense antiquity of the human race. It is true that in 1836 the French antiquary, Boucher de Perthes, discovered flint axes lying along with the bones of extinct mammals in un-disturbed gravel-beds from twenty to thirty feet below the surface, but he had to wait almost twenty years for a fair hearing—even Darwin looked askance—and yet longer for decisive corroboration. Both were gained, however, when Sir Charles Lyell summarised the existing evidence in 1863 in his " Antiquity of Man."

It is certain that men lived in Europe at a time when mammoth and rhinoceros, hyæna and lion, were still abundant there, and that he used stone weapons against them. From the situations in which these palæolithic implements have been found, it is inferred, more or less precariously, that these must have dropped from their makers' hands at least 150,000 years ago. But these stone implements are not the work of novices, they are well-finished articles compared with the earlier eoliths, which probably take us back for another 150,000 years, or farther. It must be noted, however, that these eoliths are often very debatable, for it is difficult to be certain that their marks are those of workmanship and use.

No fossil remains of man have been found except in Post Tertiary (Diluvial) deposits, but there are some vague indications of his presence in the pre-ceding Pliocene Age. An indirect argument leads us to put the origin of the human stock even earlier ; for as it is certain that man could not have arisen from any of the known anthropoid apes, but from a stock common to them and to him, it seems likely that the human stock had diverged before the time when the anthropoid apes are known to have been established as a distinct family. This takes us back to Miocene ages, for there are remains of extinct genera of gorilla and chimpanzee in Miocene strata, and that means hundreds of thousands of years ago.

Whether man had emerged as man in Tertiary times or not, his antiquity is to be measured not in centuries, but in millennia, and his migrations had begun while land bridges existed that are now broken, for instance, while Asia was still linked to North America.

THE PREHISTORIC RECORD

The study of prehistoric man is quite young, practically beginning about the middle of the nineteenth century in the study of caves and lake-dwellings, " kitchen-middens," stone inplements, and actual human remains, but it has already been rewarded by discoveries of the highest interest. It is generally recognised that man had made considerable progress by the time of the later Glacial periods, and that he was a contemporary in Europe of the mammoth and its predecessor *Elephas antiquus*, of the cave lion, and the cave bear, and other extinct mammals. There was a long-drawn-out Stone Age antecedent to the use of metals, and it is divided into a Palæolithic period, when the weapons and instruments were rough and unpolished, and a Neolithic period, when they were polished, neatly finished, and often bored. Here and there—in England, in the valley of the Meuse, in the Neanderthal, at Heidelberg, in North America—remains of early man have been unearthed, which disclose interesting primitive features. " The salient features of these ancient men," writes Professor G. B. Howes in his preface to the translation of Wieder-sheim's book already named, "are a low, retreating and contracted forehead, and an inwardly shelving occiput—indicative of a primitive type of brain and of powerful neck muscles ; a high temporal ridge, and an expanded palate—indicative of powerful jaws and jaw muscles ; and, further, the presence of ape-like brow ridges—for which the famous Neanderthal calvaria is so notorious—appears also to have been a racial character."

The older Diluvium deposits in the Neanderthal, near Düsseldorf, yielded in 1856 the roof of a skull and some limb bones, and authorities such as Professor Schwalbe regard the skull as belonging to a form different from any of the races of man now living, and as type of a more primitive species, which Schwalbe named *Homo primigenius*.

In 1886 Fraipont described two skulls and skele-tons from Spy, in Belgium, found along with remains of the cave bear, mammoth, and rhinoceros ; in 1899 and subsequent years Gorjanovíc-Kramberger obtained parts of about ten individuals in a cave near Krapina, in Croatia, along with bones of cave bear and rhinoceros ; an important skull from Gibraltar has been carefully studied by Schwalbe and Sollas ; various remains have been found at Mentone, at Heidelberg, and elsewhere. The net result, according to many anthropologists, is to suggest a still somewhat dimly described species, *Homo primigenius*, occupying an intermediate position between the highest apes and the lowest living human races. It is mainly as regards the skull that this " Neanderthal race " differs from the *Homo sapiens* of historic times ; the forehead was low and retreating, the supraorbital ridges were strong, the cranial capacity was distinctly less, the teeth were stronger, and the chin was only hinted at.

It is the opinion of many anthropologists that the species *Homo primigenius* was not gradually transformed into *Homo sapiens*, but was supplanted by the latter, a scion of the same stock, possibly originating in Southern Asia.

" PITHECANTHROPUS "

In 1891 Dr. Eugène Dubois discovered in alluvial volcanic tufa on the banks of the Bengawan river, near Trinil, in Java, some very interesting, genuinely fossilised remains, a skull-cap, a thighbone, and two back teeth—which he regarded as belonging to a type intermediate between man and anthropoid apes. The water-borne deposit in which they were found probably dates from the Older Diluvium, but some would say Pliocene. Dubois called his discovery *Pithecanthropus erectus*, and, although there was not very much to come and go on, it may be noted that species, which subsequent discoveries have proved to be very distinctive, have often been based on a vertebra, a tooth, or even a scale. But there is not as yet unanimity in regard to Pithecanthropus. Some critical investigators, such as Pettit, Cunningham, and Turner, maintain that the remains are strictly human, while others, such as Schwalbe, regard them as transitional—a link no longer missing—between apes and man. On the one side, it is confidently stated that the skull-cap is human, but indicative of a type rather more primitive than the man of Neanderthal ; on the other side, it as confidently stated that Pithecanthropus, though intermediate between apes and man, is more closely allied to the apes. Some regard this rather shadowy form as one of the direct ancestors of man, and others as a side-track failure.

An experienced zoologist, Mr. F. E. Beddard, in his volume on the " Mammalia " in the " Cambridge Natural History," reconstructs *Pithecanthropus erectus* in the following terms. " The animal when erect must have stood 5 ft. 6 in. high. The contents of the cranium must have been 1,000 cm.—that is to say, 400 cm. more than the cranial capacity of any anthropoid ape, and quite as great or even a trifle greater than the cranial capacity of some female Australians and Veddahs.

" But as these latter are not 5 ft. in height, the ape-like man had really a less capacious cerebral cavity. The skull in its profile outline stands roughly midway between that of a young chimpanzee—young in order to do away with the secondary modifications caused by the crest—and the lowest human skull, that of Neanderthal man. This creature is truly, as Professor Haeckel put it, ' the long-searched-for missing link '—in other words, represents ' the commencement of humanity.' "

As the greatest experts differ markedly in the interpretation of the sparse remains of Pithecanthropus, a suspension of judgment seems the most scientific frame of mind, but the divergence of opinion is interesting in showing the close relationship of the simian and the human.

RACES OF MANKIND

No secure conclusion has been reached as to the most natural classification of the numerous human races. So far as is known, they are all mutually fertile, and there are often intergrades between strongly contrasted types. Some characters seem to blend readily, while others, more firmly established, tend to persist without blending when grafted on to another race. Crosses between remotely related races do not usually prove a success, yielding unstable or even reversionary half-breeds. It must be remembered that in the breeding of animals steady advance has always rewarded mating within a narrow range of relationship.

The following scheme of classification is very widely accepted as at least convenient.

I. NEGROID RACES. Usually with frizzly hair, dark skin, broad nose, thick lips, prominent eyes, large teeth, narrow pelvis, and dolichocephalic or long-headed skulls—*e.g.*, Sudanese Negroes, such as Hausas ; Bantus, such as Zulus, Bechuanas ; Hottentots ; Bushmen (practically extinct).

II. MONGOLIAN RACES. Usually with black, straight hair, yellowish skin, broad face, prominent cheekbones, small nose, sunken narrow eyes, teeth of moderate size, and a variable skull—*e.g.*, Chinese, Japanese, Burmese, Tibetans, Lapps, Finns, Eskimos, Malays.

III. CAUCASIAN RACES. Usually with soft straight hair, well-developed beard, variable skull, retreating cheekbones, narrow and prominent nose, small teeth, and broad pelvis—*e.g.*, Europeans ; Hamites, such as Berbers ; Semites, such as Jews and Arabs ; Hindus.

STRUCTURAL PROGRESS IN MAN. While some parts of the human frame are dwindling, there are others which the anatomists declare to be on the up grade. In other words, there is evidence of progress in fitness. Professor Wiedersheim gives the following illustrations among others: higher differentiation and more subtle development of the muscles of the thumb ; increase in the physiological efficiency of the hand in general ; strengthening of the arch of the foot, the ankle, and the great toe ; perfecting of the whole lower limb for support and locomotion ; widening of the hip girdle in women ; more subtle differentiation of the facial muscles as distinguished from those of the scalp and of the ear ; more subtle differentiation of the muscles of the larynx ; higher development of the brain, especially of the cerebral hemispheres. It should be noted that the careful collection of data in regard to variations in man is of the greatest significance, for in this way alone can a basis be found for judgment as to whether a structure is progressive, retrogressive, or practically stationary. It is also necessary to distinguish from these inborn *variations*, which arise from within, all the acquired *modifications*, which are superimposed from without. For, while the former are transmissible in inheritance, we cannot at present say that this is ever true of the latter.

While there is definite evidence of progressive variation in man, there is nothing to lead us to believe that any marked change will be wrought out in man's highly finished bodily frame. Everything appears to point, rather, to the conclusion that the line of human evolution will be that of increasing utilisation and command of the brain, with which there will be doubtless correlated an improvement in the social framework which forms man's external heritage.　　　　J. ARTHUR THOMSON

THE GORILLA

"Although somewhat unwieldy creatures, they are active and indefatigable climbers, and are said to ascend to the very tops of the forest trees."

N

THE ORANG-UTAN

This "Man of the Woods" is supposed to need its shaggy coat of reddish brown hair to protect it from the tropical rains of its native forests and the attacks of mosquitoes and other pests.

158

THE GREEN MONKEY

"Although gay and gentle during youth, this monkey becomes morose and vicious when old.

THE EVOLUTION OF THE LEMURS, MONKEYS AND APES

A DISCUSSION OF THEIR ORIGIN AND DESCENT

By R. I. POCOCK

THE positive characters absolutely distinctive of the Primates as a whole are comparatively few, yet every existing member of that order may be distinguished from every other mammal by a complex of structural features ; so that if in one or more characters a species fails to conform to the plan typical of the rest, appeal may be made to others to establish its right to be classified as a primate. The dentition, for instance, of the aye-aye (*Daubentonia*) deviates widely from that of other genera of the order in its analogy to the dentition of rodents ; nevertheless, the structure of the hands and feet and of other organs shows clearly that the animal is an aberrant lemur. Similarly, the great toe of marmosets and men is not opposable to the other toes, yet the sum total of the characters of these animals attests unmistakable kinship with monkeys.

When the characters of existing primates are passed in comparative review, the order seems to be derivable from a generalised type, the hypothetical ancestral form, presenting the following features, mainly structural. A skull with prominent jaws armed with teeth of a tolerably primitive pattern, but reduced in numbers to $i.\frac{3}{3}$, $c.\frac{1}{1}$, $p.m.\frac{3}{3}$, $m.\frac{3}{3}$, the inner incisors of the upper jaw being large and in contact ; an orbit encircled by a bony ring, but communicating freely with the temporal fossa ; a comparatively small cranial portion containing a brain, with the cerebellum not overlapped behind by the cerebrum ; a long non-prehensile tail ; a pair of collar bones ; limbs plantigrade, nearly equally developed, freely movable from the hip and shoulder girdles, the two long bones (radius and ulna, tibia and fibula) separated from each other, the feet five-toed, the first toe (hallux) of the hind foot being flat-nailed, large, and freely opposable, while the corresponding toe (pollex) of the front foot, although relatively smaller, was also opposable, but to a lesser degree ; an alimentary canal with a simple stomach and a large cæcum, and other features concerned with the internal anatomy.

In these conclusions is found complete justification for the statement that the primates, as a whole, are a group of lowly organisation and humble origin. Except, indeed, for the reduction in the number of the teeth, the annular orbit, and the opposable first digits, they do not differ greatly from the hypothetical primitive eutherian mammal ; and when some fossil forms, closely related apparently to the lemurs, are discussed, it will be found that one form (Hyopsodus) has been discovered with the complete dentition and incomplete orbits regarded as present in the generalised type. We may even go still further, and say that the presence of collar bones, of five toes on each foot, and the distinction of the radius from the ulna, and of the tibia from the fibula are features the primates have inherited unchanged from some very early five-toed type which preceded the reptiles in vertebrate evolution ; and it is highly probable that the freedom of the limbs from the shoulders and the hip must also be assigned to this same remote and lowly origin. One has only to note the undeniable likeness between the limbs of human beings and of frogs or salamanders to find ample support for this supposition.

Coming now to a consideration of the subdivisions of the order, the Prosimiæ, which constitute the first suborder, are shown by a complex of characters not necessarily found in any one individual to be less highly organised than the anthropoidea—that is to say, they have departed, as a group, less widely from the generalised eutherian type from which the hypothetical ancestral primate was not far removed. In the skull the jaws, as a rule, are larger and the braincase smaller, and the brain itself less highly developed than in the anthropoidea ; and only in one case is there a bony partition separating the orbit from the temporal fossa. The dentition is more varied than in the anthropoidea ; but one family, the Lemuridæ, retains typically the primitive number of teeth believed to have been possessed by the immediate ancestor of existing primates. Moreover, by means of certain fossil forms the existing Prosimiæ are connected with other extinct species presenting a still earlier and more generalised type of skull and teeth.

These extinct lemur-like primates, represented principally by skulls or pieces of skulls and teeth, have been discovered in Eocene beds of Europe, North America, and South America. Their classification is a matter of very great difficulty, and this is in no sense overcome or concealed by referring them to a distinct order or suborder, the Mesodonta or Pseudolemuroidea, as has been done by some naturalists.

One of the most interesting of these early types is the North American genus Hyopsodus, which was about the size of a lemur, with a somewhat dog-shaped skull, the jaws being strong and prominent, and the cranium depressed. The word hyopsodus means " with pig-like teeth," and the name alludes to the fact that the animal had forty-four teeth, like a pig, the dental formula being : $i.\frac{3}{3}$, $c.\frac{1}{1}$, $p.m.\frac{4}{4}$, $m.\frac{3}{3}$, that is, the same as in primitive members of other orders of eutherian mammals. In this respect Hyopsodus differed from all other primates.

Another of its primitive and peculiar features was the absence of the bony bar surrounding the orbit behind, which is found in all other primates, whether living or extinct, but is commonly absent in other orders of mammals.

It is interesting to trace in these extinct lemuroids a gradual lessening in the number of the teeth, until the number found in existing lemurs and American monkeys is reached. The European Adapis and the South American Notopithecus, for example, while retaining the same number of cheek teeth and canines as Hyopsodus, had lost one of the incisors above and below on both sides, the formula being $i.\frac{2}{2}$, $c.\frac{1}{1}$, $p.m.\frac{4}{4}$, $m.\frac{3}{3}$. Again, in Microchœrus, also a European genus, the formula was $i.\frac{2}{1}$, $c.\frac{1}{1}$, $p.m.\frac{3}{3}$, $m.\frac{3}{3}$, as in the Cebidæ, except that an additional lower incisor on each side was lost ; while the number of teeth in Plesiadapis was nearly the same as in the indrisine lemurs, except for the suppression of one more lower incisor ; the formula is $i.\frac{2}{1}$, $c.\frac{1}{1}$, $p.m.\frac{2}{2}$, $m.\frac{3}{3}$. It is not to be inferred, however, that the genera just mentioned differed only in the numbers of their teeth ; nor that those that approached existing primates in these particulars were like them in other respects. They exhibited great differences in the number, size, and shape of the tubercles of their grinding teeth, in the shape of their skulls, the size of their jaws, and in other cranial features. So great, indeed, are the differences that each one of the genera mentioned has been made the type of a special family ; but not one of these families appears to be closely related to or exactly in the line of descent of any of the existing primates. Such forms as Hyopsodus and Adapis, however, do supply evidence of the descent of the primates from some more generalised mammalian stock akin to the insectivora.

The first tribe of Prosimiæ, the lemuroidea—which includes the lemurs, galagos, lorises, pottos, and sifakas—contains far the greater number of existing and extinct species of the group. The variations in structure and habits they present are almost as great as those exhibited by the anthropoidea. The lowest position in the tribe must be given to the Lemuridæ, or true lemurs, most of which retain the dental formula $i.\frac{2}{2}$, $c.\frac{1}{1}$, $p.m.\frac{3}{3}$, $m.\frac{3}{3}$; but, apart from this, the teeth are not all of a primitive type, the front teeth in both jaws being modified in a way that attests considerable departure from the ancestral

COQUEREL'S MOUSE LEMUR

type. For example, the canines and incisors of the lower jaw are greatly reduced in size, and project so as to form a small comb-like structure, the function and appearance of the lower canine being assumed by the first premolar. The canine of the upper jaw is normal and sharp, but the incisors are very small, and show a tendency to partial or total suppression in the adult. In the African galagos and in the Madagascar lemurs, of the genus Lemur, the four incisors are retained with a distinct gap between the inner pair ; but in the hattock (*Mixocebus*) one pair is lost ; and in the sportive lemur (*Lepilemur*) both pairs are absent. The sportive lemurs, unlike the true lemurs, are nocturnal, and have large orbits and naked membranous ears. They also have the ankle joint somewhat elongated. In these particulars they approach the so-called mouse lemurs (*Chirogaleus, Microcebus*) of Madagascar, and the galagos of Africa, in which the ankles are still more elongated. This structural peculiarity reaches an extreme in some members of the Galago genus, where it is clearly connected with the astounding leaping powers of these arboreal animals. Another structural peculiarity connected with their activity is the extraordinary gripping power of the hands and feet, the fingers of which are furnished with expanded terminal adhesive discs so that when springing from bough to bough they clutch with a strength which makes a fall impossible. In these respects the galagos are more highly organised than the true lemurs, though in others connected with the ear bones of the skull they are more primitive.

All the lemurs hitherto mentioned have long tails and are active leapers, with the hind legs longer and stronger than the front legs. Contrasted with them in the matter of habits are the slow lemurs (page 163), which may be described as specialised crawlers, just as the galagos are specialised jumpers. The arms of the slow lemurs are relatively longer, and they appear to be quite unable to leap. They have, however, the same tenacity of clutch as the galagos, and, to give the hands a wider span, the digit corresponding to the forefinger of the human hand is greatly reduced in size or quite vestigial. The tail also is shortened or reduced to a small knob. They furnish, indeed, an admirable illustration of the law

that the tail undergoes degeneration in arboreal mammals that lose activity, and especially jumping power. The slow lemurs are found in Tropical Southern and Eastern Asia (*Loris* and *Nycticebus*), and in Tropical West Africa (*Perodicticus* and *Arctocebus*) ; and it is probable that they were evolved from some early lemuroid form, which was also the ancestor of the galagos, in a continental area embracing Africa and Southern Asia, but not including Madagascar.

Although agreeing with the Lemuridæ in most points, the group of existing lemurs called the Indrisidæ, and containing the endrinas and avahis, differs from them essentially in dentition, the formula being $i.\frac{2}{2}$, $c.\frac{1}{0}$, $p.m.\frac{2}{2}$, $m.\frac{3}{3}$. They have thus lost the lower canine tooth and one premolar tooth above and below on each side. The lower incisors are procumbent, as in the typical lemurs ; but the median upper incisors are sometimes, as in the sifakas (*Propithecus*), of comparatively large size and closer together, approaching in these particulars the condition of the ancestral form. The brain also is better developed than in the Lemuridæ. In habits the endrinas are less quadrupedal than are the true lemurs, and are said to walk upright among the branches of trees, and to progress on the ground by leaping, holding the arms above their heads. It will thus be seen that both in structure and habits the Indrisidæ are, on the whole, a more highly specialised group than the Lemuridæ. Nevertheless, they are not derivable from any known member of the latter family.

Researches in Madagascar, however, have brought to light the remains of many remarkable lemurs, some of which prove that this group formerly attained a size and bestiality of appearance unsuggested by existing species ; while others reveal affinities not only with the Lemuridæ and Indrisidæ, but also with the South American monkeys, or Cebidæ. The most interesting forms belonging to the first category are Megaladapis, related to the Lemuridæ ; and Palæopropithecus, akin to the Indrisidæ. Judging from the structure and dimensions of the skull, which sometimes exceeds twelve inches in length, and is low, flattened, and narrow, with long, projecting, powerful jaws and small, oblique orbits, Megaladapis was an animal of small intelligence but of great size and strength. The teeth, broadly speaking, are like those of the Lemuridæ, except that the elongation of the jaw leaves a long, toothless space between the long cutting canine of the upper jaw and the caniniform premolar of the lower jaw and the upper and lower cheek teeth respectively. This gigantic lemur was nearly equalled in size by the huge indrisoid genus Palæopropithecus, the skull of which measured about seven or eight inches in length. The braincase was relatively larger and the jaws were shorter than in Megaladapis. The jaws, nevertheless, were considerably produced, very massive, and somewhat upturned at the extremity. The small orbits, too, looked forwards and upwards. The conformation of the skull, indeed, recalling somewhat that of a hippopotamus in the elevation of the orbits and nose, suggested to Dr. Standing the possibility of

this lemur being aquatic, and adapted for lying under water with only the eyes and nostrils projecting above the surface. It is not unlikely, however, that the upwardly inclined nose and orbits, associated with a massive lower jaw in Palæopropithecus, were accompanied by a greatly inflated hyoid bone giving resonance to the voice, as in the case of the howler monkey (*Alouatta*) of South America, to the skull of which that of the great extinct lemur in question shows many curious resemblances in profile view.

These two highly specialised lemurs throw no light upon the descent of existing forms of the Prosimiæ or of the Anthropoidea. Much more interesting from this point of view is the smaller extinct form Archæolemur, which should, perhaps, be made the type of a distinct family, intermediate in certain respects between the Lemuridæ and Indrisidæ, and

THE SLOW LORIS

approaching nearer than either to the American monkeys, or Cebidæ. The dental formula was $i.\frac{2}{2}$, $c.\frac{1}{0}$, $p.m.\frac{3}{3}$, $m.\frac{3}{3}$. Thus in the numbers of the premolars Archæolemur resembled the Lemuridæ and the Cebidæ, but differed from them, and approached the Indrisidæ in the loss of the lower canine. Especially monkey-like were the incisor teeth, of large size, the inner pair of the upper jaw being exceptionally large and contiguous, while those of the lower jaw were hardly at all procumbent. The jaws were comparatively short, and the profile of the face ran obliquely upwards to the postorbital frontal region, and the orbits were small and directed forwards. The braincase, however, was low and receding, and was deeply constricted behind the orbits, as in other extinct lemurs, and there was no postorbital bony plate such as all anthropoids possess.

It was probably from some lemurs akin to Archæolemur that the existing Lemuridæ and Indrisidæ have been evolved, and possibly some extinct form combining some of the characters of Archæolemur and of Tarsius was the ancestral type of the anthropoidea. In any case, Archæolemur and Tarsius—to be described presently—are the two genera of the Prosimiæ which in their characters overlap the Prosimiæ, on the one hand, and the Anthropoidea, on the other.

The second tribe of prosimian primates, the Chiromyoidea, contains the Mascarene nocturnal lemur known from its cry as the Aye-aye. This genus differs widely in dentition from all other primates, the formula of the adult being $i.\frac{1}{1}$, $c.\frac{0}{0}$, $p.m.\frac{1}{0}$, $m.\frac{3}{3}$. Not only has there been suppression in the number of incisors, canines, and premolars, but the incisor teeth that remain are enlarged, laterally compressed, in contact, with enamel only on their front surfaces, and grow from persistent pulps like those of a rodent. The aye-aye is sufficiently peculiar in other ways, such as the high, compressed muzzle, the cleft upper lip, and claw-like nails, to be placed in a separate family (Daubentoniidæ) from the true lemurs. The immediate ancestor of the aye-aye is unknown, but its structure supplies evidence that it traces its descent from some extinct form of the Indrisidæ. The formula of the milk teeth, for example, is $i.\frac{2}{2}$, $c.\frac{0}{0}$, $p.m.\frac{2}{1}$, which, making allowance for the absence of the molars, is the same as that of existing Indrisidæ, except that one additional premolar has been suppressed. Moreover, as has also been remarked, the large size and juxtaposition of the incisors of the permanent set suggest that the ancestor of this animal branched off from some form of Indrisidæ in which the inner incisors, both above and below, were also large and in contact. Nevertheless, this lemur cannot logically be placed in the last-mentioned family.

The third prosimian tribe, the Tarsioidea, contains the single genus, known as the Spectral Tarsier (*Tarsius*), which, with a close superficial resemblance to some galagos, combines structural characters some of which are peculiar to itself, while others affiliate it with the galagos, and yet others with the anthropoid primates. The teeth are of a more primitive type than in any existing lemur,

the dental formula being $i.\frac{2}{1}$, $c.\frac{1}{1}$, $p.m.\frac{3}{3}$, $m.\frac{3}{3}$, and coinciding closely with that of the Lemuridæ, except for the suppression of one lower incisor on each side. The remaining incisor, however, is not so procumbent as in that family, and the canine is more vertical, and the upper inner incisors are moderately large and in contact, as in some Indrisidæ. A marked anthropoid feature is the closing of the orbit behind by a bony wall, the possession of which and of the metadiscoidal and deciduate placenta—which is a later type than the diffuse and non-deciduate placenta of other Prosimiæ, and shows that the tarsier has reached a higher grade of development than the latter have—suggests that tarsier is descended from some extinct lemuroid stock, also with those attributes, whence the Anthropoidea are also descended. And the dentition of tarsier confirms the evidence from other sources that this stock had a dental formula of $i.\frac{2}{2}$, $c.\frac{1}{1}$, $p.m.\frac{3}{3}$, $m.\frac{3}{3}$, with canines and incisors subvertical and the inner incisors of the upper jaw larger than the outer and in contact in the middle line. For teeth of this character amongst the Prosimiæ we may look to a form that preceded Archæolemur, and differed from that genus in retaining the lower canine teeth.

The second suborder of the primates—variously called Anthropoidea (man-like) and Pithecoidea (monkey-like)—stands at a higher grade of development than the Prosimiæ. The brain is of a more advanced type, on the whole, as is especially attested by the backward extension of the cerebral hemispheres over the cerebellum, though the howlers amongst the South American monkeys, and the siamang, one of the gibbons, are scarcely, if at all, more advanced in this respect than lemurs of the family Indrisidæ.

The anthropoid primates are divisible into two tribes : the Platyrrhini, containing the South American monkeys and marmosets, which are confined to Central and South America, and are unrepresented by fossil forms outside that continent ; and the Catarrhini, including the monkeys and apes of Europe, Asia and Africa, and man, a group which appears to have been evolved in the Old World, and, with the exception of man, is restricted thereto, both as living and fossil forms.

The Platyrrhini, or American monkeys and marmosets, are more akin to the Prosimiæ than are the Catarrhini. They differ from the latter, and resemble the Tarsiidæ and the true lemurs of the

SPECTRAL TARSIER

family Lemuridæ in retaining three premolar teeth above and below on each side, and in having no definite bony tube developed in connection with the external orifice of the ear in the skull. They further differ from the Catarrhini in having a broad cartilaginous partition between the nostrils, so that in the living animals these orifices are widely and not narrowly separated. The country of their origin is unknown. Fossil species, referable to the two existing families (Cebidæ and Callitrichidæ) occur in recent (Pleistocene) deposits in Brazil; and others (Homunculus), akin to the Cebidæ, in Eocene beds in Patagonia. Unfortunately, these Eocene forms throw no more light upon the descent of the Cebidæ than is supplied by existing species. As has been explained, however, there formerly existed in Madagascar a lemur (Archæolemur) which had many monkey-like characters, showing special resemblance to the Cebidæ; and on the strength of this resemblance it has been suggested that the Platyrrhine monkeys are of African origin—descended, that is, from a lemuroid type evolved in Africa before the subsidence of the Transatlantic bridge which once existed between Africa and South America. Along this bridge, it is supposed by Standing, the ancestors of existing Platyrrhini may have entered South America.

Judging by their teeth, represented by the formula $i.\frac{2}{2}, c.\frac{1}{1}, p.m.\frac{3}{3}, m.\frac{3}{3}$, as well as by other characters, the Cebidæ—of which the best-known examples are the squirrel monkeys, howlers, spider monkeys, and capuchins—stand nearer the lemurs than do the marmosets and tamarins. As compared with the monkeys of the Old World, as well as with the lemurs, the Cebidæ are feeble and inactive. Although essentially arboreal, their leaping powers are poorly developed, and specialisation within the family has taken the line of proficiency in slow, safe climbing among the high branches of the trees they frequent. The squirrel monkeys (Saimiris) are, perhaps, the most agile of all, and these resemble the typical lemurs in having a well-developed non-prehensile tail, a character in which they and genera akin to them, like the night apes (Nyctipithecus), are more primitive than the rest. Like some extinct lemurs, too, they have vertical lower incisors. Cebidæ with these characters are referred to a sub-family, Nyctipithecinæ, to contrast them with the Pitheciinæ, the uakaris (Uakaria) and the sakis (Pithecia), which have procumbent lower incisors recalling somewhat those of existing members of the lemuroidea. This likeness, however, may be merely adaptive. Both these genera have non-prehensile tails; but while this organ retains its primitive length in the sakis, it is greatly reduced in the uakaris, which, in this respect, depart from the normal type. In the howlers, the capuchins, and the spider monkeys, on the contrary, the tail is retained in a fully developed condition, and has become modified to form a grasping organ of great utility in climbing, exhibiting a deviation from its primitive use as an aid to leaping not found elsewhere in the primates.

The howlers (Alouatta) show retrogressive metamorphosis in the structure of their skulls and brains, their intelligence is low, and their disposition inert as compared with the larger-brained South American monkeys. On the other hand, their vocal organs are highly specialised. Nevertheless, in this particular, as well as in others, they show indications of affinity with some of the non-prehensile-tailed American monkeys, of which they are probably an offshoot, degenerate in some respects, elaborated in others. The capuchins (Cebus) more nearly resemble the monkeys of the Old World in the structure of their brain and in intelligence than do any other members of the

THE RED, OR GOLDEN, HOWLER

Platyrrhini. Their limbs are also comparatively short, the hind pair being longer than the front, as in most of the Old World monkeys. The tail, too, although truly prehensile, is normally hairy at the end, whereas in the woolly monkeys (Lagothrix) and the spider monkeys (Ateles, Brachyteles) the tail is naked beneath at the end, and is so sensitive and perfected an organ of prehension that it can be used for picking up objects. In general proportions, and in possessing a well-developed, though not highly opposable thumb, the woolly monkeys have not departed from the more primitive capuchin type ; but in the spider monkeys, the most adept climbers amongst the Platyrrhini, the limbs are very long, the arms exceeding the legs in length, while the great elongation of the hand as a climbing hook has rendered the thumb useless for grasping purposes, so that it has almost or quite disappeared. So far as specialisation is concerned, the spider monkeys must be regarded as the highest type of the Cebidæ.

It seems that the marmosets (page 167) and tamarins, constituting the family Callitrichidæ, may be regarded as a specialised offshoot of some extinct group of the Cebidæ stock. They differ from them in the loss of one molar tooth above and below on each side, the dental formula being $i.\frac{2}{2}$, $c.\frac{1}{1}$, $p.m.\frac{3}{3}$, $m.\frac{2}{2}$. Their other differences appear to be adaptive to a special mode of life. Unlike the Cebidæ, they are extraordinarily active, and possess great leaping powers. Their limbs are short, and their tails long and hairy. In these particulars, as well as in size, they are not unlike squirrels, which they also recall in the structure of their hands and feet, these extremities being more paw-like than in any other primate, the nails resembling claws, and neither the thumb nor the great toe being opposable. Now, the smallness of the hands and feet in these pigmy, probably dwarfed, monkeys makes them incapable of grasping branches of any size, and renders an opposable thumb and great toe of little use for maintaining a secure clutch. Sharp claws, on the contrary, as in the case of squirrels, are excellently adapted for climbing over the rough bark of trees.

In these facts probably is to be found the explanation of the conversion of " nails " into " claws," and of the loss of opposability of the first digit of both front and hind leg in the marmosets. The one method of securing a firm hold has been changed for another in these small arboreal primates. The mouse lemurs, galagos and tarsiers, too, with the same arboreal habits as marmosets, have at the end of their digits specially well-developed discs, with which they tightly grip a branch when alighting on it, a circumstance which suggests that an opposable finger and toe are in themselves insufficient to prevent a fall in leaping animals with hands and feet too small to span stout branches. Now, many Cebidæ have such strongly compressed nails that very little modification would be required to turn them into " claws." The thumb, too, is only slightly opposable, and the change required to convert it into the unopposable first digit seen in marmosets would be insignificant. Hardly more elaborate would be the structural change needed to transform the large opposable great toe of the Cebidæ into the degenerate unopposable great toe of the marmosets, which still, however, retains its primitive lateral direction and its flat nail.

The Catarrhini, or monkeys and apes of the Old World, and man, differ from the Platyrrhini in many characters, the principal being the loss of one premolar above and below on both sides, giving the dental formula $i.\frac{2}{2}$, $c.\frac{1}{1}$, $p.m.\frac{2}{2}$, $m.\frac{3}{3}$, and the presence

in the skull of a bony tube to the orifice of the ear. Of the two tribes into which the Catarrhini are divided, the Cynomorpha stand nearest to the Cebidæ. They typically possess a well-developed tail. The cæcal dilatation of the intestine is not reduced in size at its free end to form an " appendix," and the sternum is laterally compressed.

In these characters they resemble the Cebidæ, and differ from apes and man. But they further differ from the Platyrrhine type in certain characters, one of which, the presence of ischial callosities, or hairless sitting-pads on the rump, is found in both the families, serving to link them together ; while of the remaining characters, one, the presence of pouches in the cheek for storing food, is peculiar to the Cercopithecidæ ; the other, the presence of saccular folds in the wall of the stomach—to give increased digestive surface and storage-room for the diet of leaves on which these monkeys subsist—is found only in the Semnopithecidæ (Presbytidæ). These three special characteristics suggest that the Cynomorpha are descended from some extinct type of monkey resembling the Cebidæ and the lemurs in having no ischial callosities, no cheek pouches, and a simple stomach. But since there is no proof as yet of the former existence of Platyrrhine monkeys in the Old World, where the Cynomorpha were beyond doubt evolved, there is no reason to think that the latter are modified descendants of the former. All that can be said is that possibly the two groups trace their descent from some extinct Old World primate combining some of the characters of both and linking them to the lemurs.

Palæontology supplies no data to make further discussion of this question profitable. Nevertheless it proves that the Cercopithecidæ were formerly much more widely distributed than at present. Some of the fossil forms belong to the extinct, some to existing genera. Of the extinct forms, the most important are Mesopithecus, of which the typical species (M. pentelici) inhabited European forests in Pliocene times, and is said to have had a skull like that of the langurs (Presbytes), and the remainder of the skeleton like that of the macaque (Macacus) ; Dolichopithecus, which lived in France at the same period, was also akin to the Semnopithecidæ ; while the older Oreopithecus, from the Middle Miocene of Italy, was a large, powerful monkey, combining to a certain extent the skull characters of some Cercopithecidæ and Simiidæ. Referable to existing genera are fossil remains of baboons (Papio), showing that this genus inhabited India in Lower Pliocene and Pleistocene times, and also Algeria at the latter period. Abundant remains of Macacus attest the existence of this monkey in Central and South Europe and India as far back as the Pliocene. Langurs (Presbytes) lived in the forests of the Upper Pliocene in France and Italy, as well as in North India, where they occur at the present day ; while the essentially African guereza (Colobus) is represented by an extinct species in Miocene beds in Bavaria.

The Cynomorpha, like lemurs and Cebidæ, are essentially quadrupedal primates, adapted for climbing trees or rocks. Their hind limbs are usually longer, and, apparently, never actually shorter than the fore limbs. They are able to stand erect, but very seldom walk on their hind legs, though they can be taught to do so. When on the ground, they habitually walk and run like ordinary four-footed mammals. All the species, including even the heavily-built baboons, can leap to a greater or less extent, some of the more lightly-built forms quite equalling typical lemurs in this respect; and all the strong leapers possess long tails. The thumb is shorter, more opposable, and less like the other digits than it is in the Platyrrhini and lemurs, but it is never a very proficient organ of prehension. The nails, especially of the feet, are, with the exception of that of the great toe, often compressed and claw-like.

The Presbytidæ, resembling the hypothetical common ancestor of the family in the absence of cheek pouches, but differing from it in possessing a complicated stomach, are represented by genera in South-eastern Asia and Tropical Africa. All are active climbers and powerful jumpers, furnished with long tails. The langurs (*Presbytes*) of India and Malaysia may be regarded as a generalised type of the subfamily. The most aberrant and specialised types are the Bornean proboscis monkey (*Nasalis*), which has a larger nose even than man; and the guerezas, which have lost the first digit, or thumb, of the hand, exhibiting in this respect analogy with the spider monkeys among the Cebidæ.

Much more varied in appearance and structure than the Presbytidæ are the Cercopithecidæ, which differ from the hypothetical ancestral form in possessing cheek pouches, but resemble it in the simple structure of the stomach. Broadly speaking, these monkeys are divisible into four groups, known as guenons (*Cercopithecus*), mangabeys (*Cercocebus*), macaques (*Macacus*), and baboons (*Papio*).

The first two genera, confined to Africa, contain species differing from each other in characters of comparatively minor importance, all being long-tailed, active monkeys, principally of arboreal habits. Much more diversified in accordance with different conditions of existence are the macaques and baboons. Both of these groups contain species

THE SILKY MARMOSET

which are largely terrestrial and have the tail shortened, or sometimes reduced to a stump-like vestige, exhibiting in this respect marked deviation from the primitive primate stock. Among the macaques the almost tailless Barbary ape (*M. innuus*), which inhabits the rocks of Gibraltar and Morocco, is a good illustration of this phenomenon; and similar suppression of the tail is exhibited by the two West African baboons, the mandrill (*P. maimon*) and the drill (*P. leucophæus*). The baboons, as a whole, also show marked specialisation in the immense development of their jaws, the arched carriage of the tail, when present, their long fore limbs and well-developed thumbs. These facts, coupled with their great intelligence, give them, perhaps, the right to be regarded as the "highest" of the Old World monkeys, though in the opinion of some authors they are the "lowest."

The second group of the Catarrhini, the Anthropomorpha, containing the single family the Simiidæ, comprising the anthropoid apes, or gibbons, orang-utans, gorillas, chimpanzees, and man, has many points in common with the Cynomorpha. But although they stand on a higher evolutionary plane in the sense of being further from the Platyrrhine type, it is quite certain they are not descended from any existing cynomorphous monkey nor from any extinct form which, if living, could be classified either with the Cercopithecidæ or Presbytidæ. They have neither the sacculated stomach of the latter nor the cheek pouches of the former, and the cæcum of the intestine is reduced in bulk terminally to form the narrow coiled tube called the vermiform appendix. The breastbone is flattened from before backwards; there is no external tail, and, except in man, the fore limbs are much longer than the hind limbs.

On the difficult question of the descent of the Simiidæ, palæontological records are too incomplete to throw much light. Some of the extinct forms, which are based, for the most part, upon teeth or other fragments of the skeleton show affinity with existing apes; others are less closely related to them, and even approach the human species in certain particulars. For example, Pliohylobates, from the Miocene of Europe, was related to the gibbons, but exceeded any existing species in size. Pliopithecus, also from the European

Miocene, was likewise akin to the gibbons. Palæopithecus, from Pliocene beds of North India, on the contrary, was very closely allied to the chimpanzees (*Anthropopithecus*); and a molar tooth from the same strata has been regarded as belonging to an orang-utan (*Simia*). On the other hand, Dryopithecus, which inhabited the forests of South and Central Europe, was a strongly-built, perhaps gorilla-like, ape, with jaws recalling those of the Cercopithecidæ in structure, and with other characters showing more affinity to man than was exhibited by any other Tertiary ape hitherto discovered. Griphopithecus, based upon some teeth, was of doubtful affinities; while Anthropodus showed points of resemblance both to Dryopithecus and to man in the structure of its teeth.

Thus, some of the fossil Simiidæ exhibit remote and perplexing affinities to the Cynomorpha, on the one hand, and to man on the other; and since there are no strong reasons for believing that any of them is on the direct line of descent of these types, the points of resemblance in question must be probably regarded as due to inheritance from a common Catarrhine stock, whence the Cynomorpha and Anthropomorpha have been evolved along independent lines.

From the habits and characters of existing apes it is possible to form a conception of the general features of their ancestral type. On comparing the living animals with the Cynomorpha, the first fact to arrest attention is the great difference they exhibit in capacity or method of climbing. It would be wrong to say that the Simiidæ, as a whole, are inactive climbers, because few primates excel the gibbons (*Hylobates*) in arboreal agility. Nevertheless, their method of passing from branch to branch is widely different from that of the Cynomorpha. The latter resemble the lemurs and marmosets in having the propelling power in their hind quarters. In true quadrupedal fashion, they leap from one branch, stiffen and extend the tail in the air, and alight on the upper side of another, first with their hands, then with their feet.

A gibbon, on the contrary, has leaping powers of the feeblest kind, and when he passes from branch to branch, launches into the air with a weak " take off," and with long arm upstretched catches the bough aimed at, swinging his body beneath it. If he wishes to alight on it, he maintains the hold of his hand, and the weight of his body carries him round and up the other side of the branch, so that he comes down upon it in a sitting or standing posture. If, on the contrary, he wishes to pass on from the second branch, he quits his hold as his body passes beneath and is carried by its momentum and a jerk of the arm towards the third branch, which he grasps with the other hand. Now, this "swinging" method of the gibbon and the "leaping" method of the cynomorphous monkey are equally efficient means of speedily traversing a forest; and it is impossible to believe that one method has been changed directly for the other. A gibbon can no more leap with its hind legs like a mangabey or

langur than either of the latter can swing with its arms like a gibbon. The gibbon's method is unique and not derivable from the other; but the monkey's method has been inherited unchanged from some long-tailed lemuroid ancestor. And since the gibbons are descended remotely from the same stock, the conclusions seem unavoidable that their immediate ancestor had lost the primitive leaping power and that the peculiar acrobatic style of the gibbons has been specially learnt by them.

Turning to the other apes, they are good, but, on account of their weight, slow climbers when pitted against the Old World monkeys. The orang-utan is, of the three, the best adapted for arboreal life, and the least fitted for life on the ground. His thumb and great toe are shorter than in the gorilla and chimpanzee; but his hands and feet are of great length with powerful, hook-like fingers and toes, the great toe being set far back on the foot so as to grasp branches big and strong enough to support the animal's weight. The more human hand and foot of the chimpanzee and gorilla confer greater facility in traversing flat ground. When moving over the ground these three great apes either walk on all four limbs, after the manner of monkeys, or swing their bodies forward between their long arms, using them as crutches and resting on the knuckles of the hand. This method of walking is never apparently practised by ordinary monkeys in good health.

Orang-utans seldom, if ever, stand erect without support from the arms. Chimpanzees occasionally do so (page 169), but very rarely walk in that attitude. Gorillas, on the contrary, can both stand and walk upright, and do so when angered or excited; but, like chimpanzees and orang-utans, they resort to the primitive quadrupedal method when retreating from danger. Terrestrial progress in the gibbons is very different. They never actually rest on the hands at all, but run or walk upright on their legs, using their arms as balancers, and resting on the flat of the foot, supported mainly by the heel and large laterally projecting great toe. They move along horizontal branches in the same way. Thus, in their method of swinging from branch to branch, as well as in their method of traversing flat ground, the gibbons differ more from the monkeys and ordinary lemurs than do the other apes. The latter, however, it must be added, are practically incapable of leaping from bough to bough as the monkeys do, the weakness of the hind legs, the great weight of the body, and the absence of the tail making it almost impossible for them to steady themselves and balance on the upper side of a branch when alighting upon it. They therefore climb by reaching from branch to branch with their long arms and strong hands, swinging from one to another with the body suspended, when necessary, in mid-air. Their method of climbing, in fact, is much the same as that of the gibbons, due allowance being made for the fact that they never apparently venture to let go the grasp of one hand until the other is securely placed. These are useful facts to remember in discussing the question of the origin of apes and man.

THE EVOLUTION OF THE LEMURS, MONKEYS, AND APES

In the primates, reduction of the tail either accompanies the adoption of a terrestrial mode of life by species whose ancestors were arboreal—like the drills, mandrills, and rock macaques—or it accompanies alteration in the mode of climbing, the swift-leaping method being changed for the slow-creeping method as in the slow lemurs (lorises and pottos), or for the upright method of climbing about trees exhibited by the endrinas. Now, in the case of the endrinas and slow lemurs, the reduction of the tail has accompanied the adoption of a more complete arboreal mode of life than is exhibited by their more generalised ancestors, of which the true lemurs and the galagos are living representatives. But the evidence supplied by the structure and habits of anthropoid apes, the total absence of the external tail, their peculiar method of progression on the ground, and of movement through forests, suggest their descent from ancestors which for long ages had lost more or less the characteristic activity, manifested by running on all-fours and of leaping from branch to branch, of the partly arboreal, partly terrestrial, but essentially quadrupedal cynomorphous monkeys, but which, while adopting to a great extent a terrestrial mode of life, had retained the power of ascending trees for safety from enemies, and for the purpose of feeding.

From a primitive Catarrhine stock of this nature, tailless, but with ischial callosities, with opposable great toe, and with the limbs probably subequal in length, it is not unlikely—perhaps, indeed, it is probable—that the anthropoid apes and man are descended, each genus having departed from it in varying degrees along its own line of evolution. Moreover, since both in gibbons and in man, the two types of this group which differ most widely from each other, we find bipedal progression with an upright attitude the sole or habitual method of traversing the ground, and since the same habit is practised, though in a less complete and perfected form, by the gorilla—which structurally lies midway between the two—it is not unreasonable to assume that erect walking was originated by this hypothetical ancestor. No doubt, however, it retained at the same time the ordinary quadrupedal walk of monkeys and baboons.

Great increase in the length of the arms is one marked feature in which all these anthropoid apes have deviated from this type. It is owing to the length of these limbs, as compared with the body and legs, that they can be used as crutches by chimpanzees, gorillas, and orang-utans in terrestrial progression. In the gibbons they are not so used, because the better-developed legs are capable of supporting with comparative ease the lightly-built body. It is improbable, however, that their usefulness as crutches has been the guiding factor in the development of the arms. For this we must look probably to their proficiency as climbing organs. It is quite clear that they are admirably adapted for enabling heavily-built animals, which cannot leap, to pass with speed and safety from branch to branch, or tree to tree, without quitting their hold. Let it be remembered, too, in this connection, that the best gymnasts amongst men are those who have relatively long and strong arms, and short, lightly-built legs, men who approach the anthropoid apes in these respects.

The apes which depart most from our hypothetical ancestral type in arboreal activity are the gibbons and orang-utans, both in their own way being exceedingly skilful climbers. They also depart most widely from it in terrestrial activity; but whereas in the orang-utan the deviation has taken the direction of loss of activity, in the gibbons the incipient power of standing and walking erect, postulated for that type, has been greatly developed, very likely as an accompaniment of loss of bodily bulk and weight; for the unique power of swinging from tree to tree possessed by these lightly-built apes suggests its origin from the slower, more careful method exhibited by their bulkier relatives. If this be so, the comparatively small size of the gibbons is due to dwarfing, and, like the habit in question, is a secondarily acquired attribute. Structural adaptation to special arboreal habits has carried the gibbons and orang-utans further from the primitive Catarrhine stock whence apes and men originated than the chimpanzees and gorillas have carried. Hence it comes about that the two former apes, and especially the gibbons, are structurally less like man than the two latter. Of these, the chimpanzee, which has the arms shorter in proportion to the legs than is the case in any other anthropoid ape, and presents other characters in which it is more akin to man, moves with greater ease and speed over the ground in the ordinary quadrupedal walk of monkeys and baboons—though using, as a rule, the knuckles instead of the palm of its hand—than any other ape. Thus, man's nearest relation among the apes retains in a marked manner the mode of walking on all-fours, believed to have been practised by the hypothetical primitive stock whence the genera of this group emanated; and the gorilla, with nearly as many human characters as the chimpanzee, perhaps shows in an almost unaltered form the powers of standing and walking erect exhibited by that type.   R. I. Pocock

CHIMPANZEE IN AN UPRIGHT ATTITUDE
Photograph, Carl Hagenbeck

169

CLASSIFICATION OF THE PRIMATES

ALTHOUGH every naturalist has his own ideas about the classification of the animal kingdom, and no single scheme can be said to command universal assent, yet differences arise on questions of detail rather than on matters of principle or on the broad lines of demarcation. In the HARMSWORTH NATURAL HISTORY the plan has been adopted of presenting a good working schedule which shall show the place of those animals, whether high or low in the kingdom, which have called for mention in the text. A complete classification, however, is not practicable, since it would involve neither more nor less than a catalogue of every species or race of living creature, and occupy the whole of a large volume. As a helpful compromise, it has been arranged to give, at the beginning of each order of the vertebrate division, a classification thereof, coming as far down as the species described, but avoiding those refinements of grouping that are frequently irritating in themselves, and, in any case, of value only to the expert. So far as they have been ascertained, both the popular and scientific names of animals are cited. We begin with the classification of the first order, that of the Primates.

ORDER I.
MAN, APES, MONKEYS, AND LEMURS—PRIMATES
SUBORDER 1
Man, Apes, and Monkeys—Anthropoidea
FAMILY 1
Man—Hominidæ
GENUS

Man (fossil)—Homo primigenius
Man—Homo sapiens

FAMILY 2
Man-like Apes—Simiidæ
GENUS 1
Chimpanzee and Gorilla—Anthropithecus
SPECIES

Chimpanzee ...... Anthropithecus troglodytes
Gorilla .... a. gorilla

GENUS 2
Orang-utan—Simia
SPECIES

Orang-utan .... Simia satyrus

GENUS 3
Gibbons—Hylobates
SPECIES

Siamang .... Hylobates syndactylus
White-handed gibbon.................h. lar
Hulock .... h. hulock
Hainan gibbon....................h. hainanus
Agile gibbon .... h. agilis
Malay gibbon .... h. rafflesi
Crowned gibbon .... h. pileatus
Wou-wou, or silvery gibbon ...... h. leuciscus

FOSSIL APES

Man-ape .... Pithecanthropus erectus
Miocene ape .... Pliopithecus
Tree ape............... Dryopithecus
Oligocene ape .... Propliopithecus

FAMILY 3
Old World Monkeys and Baboons—Cercopithecidæ
GENUS 1
Langurs—Semnopithecus
SPECIES

Hanumán, or true langur..Semnopithecus entellus
Himalayan langur .... s. schistaceus
Madras langur.................s. priamus
Banded langur .... s. femoralis
Negro langur .... s. maurus
Crested langur .... s. cristatus
Nilgiri langur .... s. johni
Purple-faced langur .... s. cephalopterus
Capped langur .... s. pileatus
Red-bellied langur............ s. chrysogaster
Dusky langur .... s. obscurus
Phayre's langur .... s. phayrei
Hose's langur .... s. hosei
Douc..........................s. nemæus

GENUS 2
Snub-nosed Monkeys—Rhinopithecus
SPECIES

Snub-nosed monkey..Rhinopithecus roxellanæ
Slaty snub-nosed monkey .... r. bieti

GENUS 3
Proboscis Monkeys—Nasalis
SPECIES

Proboscis monkey .... Nasalis larvatus

GENUS 4
Guereza Monkeys—Colobus
SPECIES

Black-and-white guereza .. Colobus guereza
Black guereza .... c. satanas
King guereza .... c. polycomus
Ursine guereza .... c. ursinus
Angola guereza .... c. angolensis
White-thighed guereza .... c. villerosus
Bay guereza .... c. ferrugineus
Crested guereza .... c. cristatus

GENUS 5
Guenon Monkeys—Cercopithecus
SPECIES

Malbrouck monkey .. Cercopithecus cynosurus
Vervet monkey .... c. pygerythrus
Mazambique monkey .... c. rufoviridis
Grivet monkey .... c. æthiops
Green monkey .... c. sabæus
Patas monkey.................... c. patas
Nisnas monkey.......... c. patas pyrrhonotus
Sykes's monkey .... c. albogularis
Black-bellied monkey .... c. leucampyx
Mona monkey .... c. mona
Bearded monkey .... c. pogonias
Erxleben's monkey................. c. grayi
Wolf's monkey .... c. wolfi
Campbell's guenon .... c. campbelli
Burnett's guenon .... c. burnetti
Dent's guenon.................. c. denti
Moustache monkey .... c. cephus
Red-eared monkey .... c. erythrotis
Hocheur monkey .... c. nictitans
Lesser white-nosed monkey ..... c. petaurista
Diana monkey .... c. diana
Rolloway monkey .... c. rolloway

GENUS 6
Mangabeys, or White-eyelid Monkeys—Cercocebus
SPECIES

Sooty mangabey ...... Cercocebus fuliginosus
White-collared mangabey .... c. collaris
White-crowned mangabey ....c. lunulatus
Crested mangabey .... c. albigena
Orange-bellied monkey ..... c. chrysogaster
Hagenbeck's monkey .... c. hagenbecki

GENUS 7
The Macaque Monkeys—Macacus
SPECIES

Bonnet monkey .... Macacus sinicus
Toque monkey .... m. pileatus
Crab-eating monkey .... m. cynomolgus
Lion-tailed monkey .... m. silenus
Bengal monkey .... m. rhesus
Chinese rhesus.................m. lasiotis
Himalayan macaque .... m. assamensis
Pig-tailed monkey...........m. nemestrinus
Burmese pig-tailed monkey ...... m. leoninus
Brown stump-tailed monkey......m. arctoides
Tibetan stump-tailed monkey .. m. tibetana
Moor macaque .... m. maurus
Japanese macaque .... m. fuscatus
Magot, or Barbary macaque .... m. inuus

GENUS 8
Black Ape—Cynopithecus
SPECIES

Black ape...............Cynopithecus niger

GENUS 9
Gelada Baboon—Theropithecus
SPECIES

Gelada .... Theropithecus gelada

GENUS 10
True Baboons—Papio
SPECIES

Mantled baboon...........Papio hamadryas
Chacma baboon .... p. porcarius
Anubis baboon................. p. anubis
Yellow baboon .... p. cynocephalus
Thoth baboon.....................p. thoth
Guinea baboon .... p. sphinx

GENUS 11
Mandrill and Drill—Maimon
SPECIES

Mandrill .... Maimon mormon
Drill......................... m. leucophæus

Fossil Old World Monkeys and Baboons
Pliocene macaque .... Macacus priscus
Grecian monkey .... Mesopithecus pentelici
Mountain ape .... Oreopithecus bambolii

FAMILY 4
American Monkeys—Cebidæ
GENUS 1
Sapajous, or Capuchin Monkeys—Cebus
SPECIES

White-cheeked sapajou .... Cebus lunatus
Brown sapajou .... c. fatuellus
Slender sapajou .... c. flavus
Tufted sapajou .... c. cirrifer
Weeper sapajou .... c. capucinus
White-fronted sapajou........... c. albifrons
White-throated sapajou .... c. hypoleucus
Smooth-headed sapajou.........c. monachus
Crested sapajou .... c. robustus

GENUS 2
Woolly Monkeys—Lagothrix
SPECIES

Humboldt's woolly monkey....Lagothrix humboldti

GENUS 3
Woolly Spider Monkeys—Brachyteles
SPECIES

Woolly spider monkey......Brachyteles arachnoides

GENUS 4
Spider Monkeys—Ateles
SPECIES

Red-faced spider monkey......Ateles paniscus
Chameck.............. a. subpentadactylus
Black-faced spider monkey ....a. ater
Grizzled spider monkey ....a. grisescens
Hooded spider monkey ....a. cucullatus
Chuva ....a. marginatus
Black-handed spider monkey .... a. geoffroyi
White-bellied spider monkey .... a. vellerosus
Variegated spider monkey ...... a. variegatus

GENUS 5
Douroucolis—Nyctipithecus
SPECIES

Three-banded douroucoli....Nyctipithecus trivirgatus
Feline douroucoli.................. n. felinus
Lemurine douroucoli ....n. lemurinus

GENUS 6
Squirrel Monkeys—Chrysothrix
SPECIES

Common squirrel monkey..Chrysothrix sciurea
Short-tailed squirrel monkey .... c. usta
Black-tailed squirrel monkey .... c. boliviensis

GENUS 7
Titi Monkeys—Callithrix
SPECIES

Red titi .... Callithrix cuprea
Collared titi .... c. torquata
White-chested titi .... c. amicta
Moloch titi .... c. moloch
Reed titi .... c. donacophila
Black-fronted titi .... c. nigrifrons
Brown titi .... c. brunnea
Black-handed titi .... c. melanochira

GENUS 8
Saki Monkeys—Pithecia
SPECIES

White-headed saki .... Pithecia leucocephala
Humboldt's or hairy saki ...... p. monachus
Red-backed saki .... p. chiropotes
Black saki, or cuxio .... p. satanas
White-nosed saki .... p. albinasa

GENUS 9
Uakari Monkeys—Uacaria
SPECIES

Bald uakari .... Uacaria calva
Red-faced uakari .... u. rubicunda
Black-headed uakari....... u. melanocephala

GENUS 10
Howling Monkeys—Alouatta, or Mycetes
SPECIES

Black howler .... Alouatta nigra
Yellow-handed howler............ a. belzebul
Vera Cruz howler.................. a. villosa
Red or golden howler .... a. seniculus
Brown howler .... a. ursina
Mantled howler .... a. palliata

Extinct American Monkeys

Miocene monkey .... Homunculus
Pleistocene monkey .... Protopithecus

FAMILY 5
Marmosets—Hapalidæ
GENUS 1
Short-tusked Marmosets—Hapale
SPECIES

Ouistiti marmoset .... Hapale jacchus
White-necked marmoset .... h. albicollis
Black-eared marmoset .... h. penicillata
White-eared marmoset .... h. aurita
White-shouldered marmoset.... h. humeralifer
Silver marmoset .... h. chrysoleucus
Black-tailed marmoset .... h. melanura
Pigmy marmoset .... h. pygmæa

GENUS 2
Long-tusked Marmosets, or Tamarins—Midas
SPECIES

Negro tamarin .... Midas ursulus
Red-handed tamarin .... m. rufimanus

Brown-headed tamarin .... m. fuscicollis
Black-and-red tamarin.......... m. rufoniger
Deville's tamarin................m. weddelli
Moustached tamarin .... m. labiatus
Red-bellied tamarin .... m. rufiventer
Pinché........................ m. œdipus
Geoffroy's marmoset ....m. geoffroyi
Silky marmoset................. m. rosalia
Lion marmoset .... m. leoninus
Golden-headed marmoset .... m. chrysomelas

SUBORDER 2
Lemuroids—Lemuroidea
FAMILY 1
Lemurs—Lemuridæ
GENUS 1
Indri Lemur—Indris
SPECIES

Indri .... Indris brevicaudata

GENUS 2
Sifakas—Propithecus
SPECIES

Diademed sifaka ...... Propithecus diadema
Verreaux's sifaka .... p. verreauxi
Crowned sifaka .... p. coronatus

GENUS 3
Avahi Lemur—Avahis
SPECIES

Avahi, or woolly lemur........Avahis laniger

GENUS 4
True Lemurs—Lemur
SPECIES

Ring-tailed lemur .... Lemur catta
Mungoose lemur .... l. mungos
Black lemur .... l. macaco
Ruffed lemur.................... l. varius
Red lemur .... l. ruber
Red-bellied lemur .... l. rubriventer
Sooty lemur .... l. nigerrimus

GENUS 5
Gentle Lemur—Hapalemur
SPECIES

Grey gentle lemur ...... Hapalemur griseus

GENUS 6
Weasel Lemurs—Lepidolemur
SPECIES

Weasel lemur ....Lepidolemur mustelinus
Red-tailed weasel lemur ...... l. ruficaudatus
Edwards's weasel lemur.......... l. edwardsi
Small-toothed weasel lemur ..... l. microdon

GENUS 7
Mouse Lemurs—Chirogaleus
SPECIES

Fork-headed mouse lemur..Chirogaleus furcifer
Lesser mouse lemur............... c. minor

Coquerel's mouse lemur .... c. coquereli
Dwarf mouse lemur .... c. smithi
Brown mouse lemur ....c. milii

FAMILY 2
Aye-Aye—Chiromyidæ
GENUS
Aye-aye—Chiromys
SPECIES

Aye-aye .... Chiromys madagascariensis

FAMILY 3
Galagos and Lorises—Lorisidæ
GENUS 1
Galagos—Galago
SPECIES

Great galago .... Galago crassicaudata
Monteiro's galago .... g. monteiri
Garnett's galago .... g. garnetti
Allen's galago .... g. alleni
Pale-coloured galago............. g. pallida
Senegal galago .... g. senegalensis
Senar galago .... g. senarensis
Demidoff's galago .... g. demidoffi

GENUS 2
Slow Loris—Nycticebus
SPECIES

Slow loris .... Nycticebus tardigradus

GENUS 3
Slender Loris—Loris
SPECIES

Slender loris .... Loris gracilis

GENUS 4
Pottos, or African Slow Lemurs—Perodicticus
SPECIES

Bosman's potto .... Perodicticus potto
Edwards's potto .... p. edwardsi
Bates's potto .... p. batesi
Uganda potto .... p. ibeanus

GENUS 5
Awantibos—Arctocebus
SPECIES

Awantibo .... Arctocebus calabarensis
Golden awantibo .... a. aureus

FAMILY 4
Tarsiers—Tarsiidæ
GENUS
Tarsiers—Tarsius
SPECIES

Malay tarsier .... Tarsius spectrum
Philippine tarsier .... t. fuscus

FOSSIL LEMURS

Microchœrus
Adapis
Archæolemur
Megaladapis

A GROUP OF HANUMAN MONKEYS
From a photograph by Gambier Bolton, reproduced by permission of the Autotype Company, 74, New Oxford Street London, W.

EXAMPLES OF THE MAN-LIKE APES

A YOUNG CHIMPANZEE

TWO YOUNG GORILLAS

A BABY ORANG-UTAN AND TWO BLACK BOYS

YOUNG ORANG-UTAN

SILVERY GIBBON

From photographs by permission of Carl Hagenbeck, Lewis Medland, and W. S. Berridge

ORDER I. APES, MONKEYS, LEMURS
GENERAL SKETCH OF THE MAN-LIKE APES
By R. LYDEKKER

EVERYBODY knows what an ape or a monkey is, and the proverb "mischievous as a monkey" reveals the estimation in which the latter animal is commonly held. The more or less human-like form, the frequent tendency to assume an upright position, coupled with their hand-like feet, would be amply sufficient to distinguish the group to which these animals belong from all others, were it not for the circumstance that there are the less familiar creatures termed lemurs, which, while evidently related to monkeys, yet differ from them in so many respects as to render it almost or quite impossible to give any characteristics which will absolutely distinguish the order to which they belong from all others. This is, however, a difficulty with which the zoologist has often to put up with, and to make the best of.

That the higher apes are closely related in their bodily structure to man is obvious to all, and it is a fact that the differences between some of these apes and man are, from a purely anatomical point of view, of far less importance than those by which the lower monkeys are separated from the higher apes. It has, indeed, been attempted to show that apes and monkeys are sharply distinguished from man by the circumstance that while man is two-handed, apes and monkeys are four-handed. The difference between the foot of one of the larger apes and that of man is, however, merely one of degree, and is much less than that between the apes and the lowest representatives of the order, as is well shown in the illustration on page 174, which exhibits the various forms assumed by the hand and foot of these animals.

Although the larger apes are those which come nearest to man in their general organisation, yet the strong ridges on the skulls of the adults, and the consequent overhanging and prominent eyebrows, give them an expression which, at the best, is but a gross caricature of the human countenance. It is, however, in the young of these animals, in which the ridges on the skull are much less developed, and the tusks, or canine teeth, of the males do not attain the dimensions reached in the adult state, that we find a much more human-like cast of expression. Moreover, some of the smaller apes, in which the great ridges on the skull are never developed, approach much more nearly in the shape of their skulls to the human type. The larger apes are, indeed, repulsive animals in the adult condition ; and it is usually only the smaller kinds of monkeys which are kept as pets.

Most of the Primates, with the exception of man, are adapted for living in warm climates, and are never found in regions which have not at least a hot summer. Some of them are, however, capable of withstanding a considerable amount of winter cold ; and it is no uncommon sight in the outer ranges of the Himalaya to see troops of monkeys leaping from bough to bough of the snow-laden pines.

Moreover, two species of monkeys inhabit the elevated regions of North-eastern China, where the winter is intensely cold. Excepting the apes found on the Rock of Gibraltar, which must either have reached their present habitation when Spain was united to Africa, or have been introduced by man at a later period, none of the Primates is found in Europe ; they occur, however, throughout the warmer regions of the rest of the globe, with the exception of the Australian region ; but whereas all the apes and monkeys of the Old World belong to two well-marked families, those of the New World represent two other families closely allied to one another, but markedly different from both those of the Old World. The lemurs, as we shall see later, are now wholly Old World forms, and are especially characteristic of Madagascar, although also represented in India, and on the continent of Africa, as well as in certain islands. In past times, however, lemurs were distributed over the greater part of the globe ; and monkeys even roamed over the ancient forest-lands of Essex, as is proved by the discovery of a few remains in the Brick-earth of Ilford ; and they were also abundant over the more southerly regions of Europe.

Nearly the whole of the Primates are adapted for a more or less completely arboreal life, most of them being inhabitants of forest regions. Aided by their hand-like feet, all of them are expert climbers, and many, like the Indo-Malay gibbons and the South American spider-monkeys, rarely leave the trees, leaping from bough to bough, and thus from tree to tree, far above the heads of the travellers below, to whom their presence is made known only by their continual howling or chattering. The climbing powers of many South American monkeys are largely aided by their prehensile tails, which serve the purpose of a fifth limb. Owing to the warmth of the regions in which most of them dwell, no monkeys ever hibernate. Contrary, however, to what is often supposed to be the case, several of the smaller species are expert swimmers, and will fearlessly cross comparatively wide rivers.

Summarising some of the more characteristic features which distinguish the order as a whole from other mammals, it may be mentioned in the first place that the hand and the foot are, as a rule, furnished respectively with five fingers and five toes, although in a few instances the thumb is wanting. Then, again, the hand is always adapted to act as a grasping organ, and, save in man, the same is the case with the foot, though it has been discovered that the foot of the newly-born human infant displays distinct traces of having been originally a grasping organ. In those cases where the hand attains its most perfect development, the thumb can be opposed to the fingers, but in some of the lower species this action is only possible in a limited degree. The great toe is, in a similar manner, opposable to the other toes, although in man, as is shown in the figure of his skeleton (page 148), this action has been more or less completely lost, and the bones of this toe lie parallel to those of the other toes. In this respect the foot of man is markedly different from that of the gorilla and the other man-like apes. With the exception of the orang-utan, in which the great toe is often devoid of any trace of such appendage, all the fingers are furnished with nails. In the higher species these nails are of a flattened shape in all the digits; and this flatness is always characteristic of the nail of the great toe, although the other toes of the lower species have curved nails. In order to form an efficient support for these nails, the bones of the terminal joints of the digits, with the exception of the index finger of the lemurs, are transversely flattened out, and are thus very different from those of rodents and carnivores. That the hand and foot should have perfect freedom of movement, it is, of course, necessary that the bones of the forearm and the lower part of the leg should remain completely separate from one another; and, as may be seen from the figure of the human limbs (p. 142), the radius and ulna in the forearm, and the tibia and fibula in the leg, are both equally well developed and capable of motion upon one another. Another important point as regards the free use of the arms is the presence of complete collarbones, which are always well developed in apes and monkeys, as they are in man.

On looking once more at the figure of the skeletons of man and the gorilla (p. 148), it will be seen that in the skull the sockets, or orbits, of the eyes are completely surrounded by a ring of bone, and that the sockets themselves look almost directly forwards. This complete bony ring round the eye-sockets at once serves to distinguish the skulls of all the Primates from those of most of the Carnivora.

In correlation with the herbivorous habits of the majority of the species, the teeth of the Primates are adapted for grinding; the cheek-teeth have broad, flattened crowns, which may be either, as in man, surmounted by tubercles or by transverse ridges. Except in one family of American monkeys, there are always three molar teeth in each side of both jaws, the last of which corresponds with the human "wisdom tooth"; and these molar teeth are invariably larger and more complicated than the premolars. Very generally, as in man, the number of the latter teeth is reduced to two on each side, and no living member of the order has more than three of these teeth. Frequently, and, indeed, invariably in the apes and

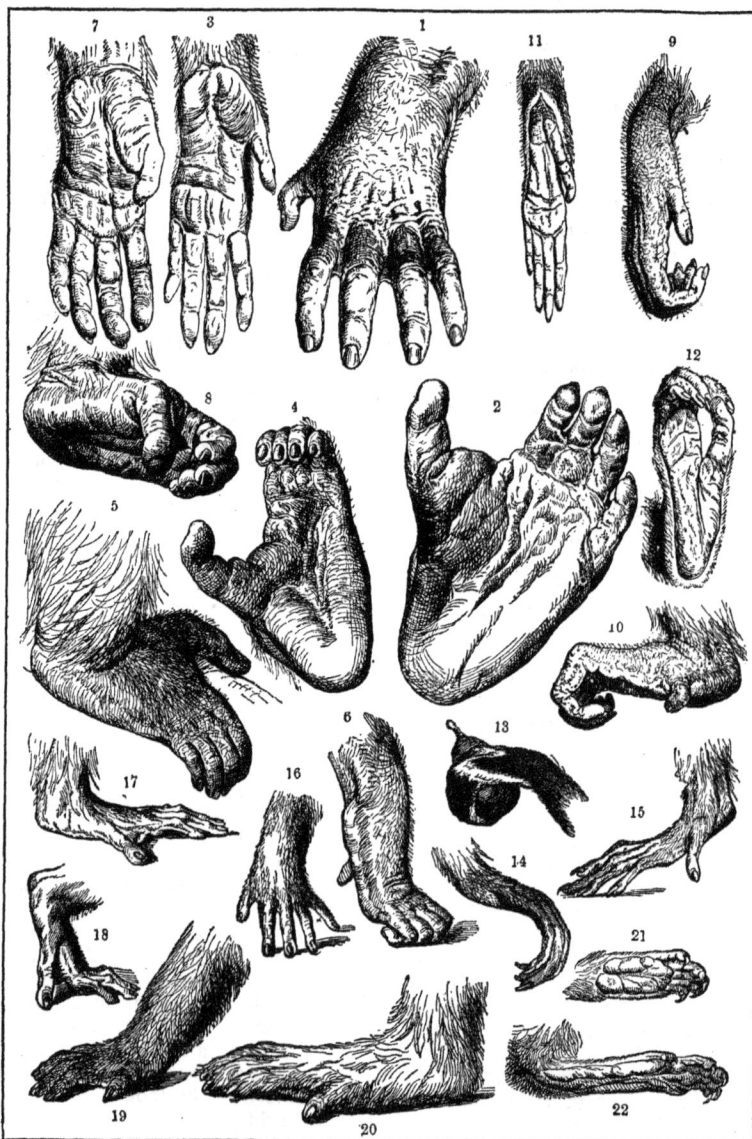

HANDS AND FEET OF APES AND MONKEYS

1, 2, Gorilla; 3-8, Chimpanzee; 9, 10, Orang-utan; 11-13, Gibbon; 14, 15, Guereza; 16-18, Macaque; 19, 20, Baboon; 21, 22, Marmoset

monkeys, there are but two incisor teeth on each side of both the upper and the lower jaws.

With the single exception of the curious aye-aye of Madagascar, there are at least two mammæ situated on the breast of the females of all members of the order.

These, then, are the chief common characters possessed by apes and monkeys, on the one hand, and lemurs, on the other; but, such as they are, they are considered of sufficient importance by many naturalists to justify the inclusion of both groups in a single order. The two groups constitute, however, separate suborders, of which the first is termed the Anthropoidea, and the second the Lemuroidea.

Apart from man himself, the man-like apes include the largest representatives of the Primates. They are found only in the dense forests of the warmest and dampest regions, and are all characterised by their strikingly human-like form, although none habitually walks solely on the hind limbs without obtaining additional support from the long arms.

In all the larger species the resemblance to man is more marked in the young than in the adult; while in the adult the human characteristics are more pronounced in the female than the male. Dr. R. Hartmann, who has devoted much attention to these apes, observes, for instance, that " in the gorilla, the chimpanzee, and the orang-utan, the external form is subject to essential modifications, according to the age and sex. The difference between the sexes is most strongly marked in the gorilla, and the differences are least apparent in the gibbons. When a young male gorilla is compared with an aged animal of the same species, we are almost tempted to believe that we have to do with two entirely different creatures. While the young male still displays an evident approximation to the human structure, and develops in its bodily habits the same qualities which generally characterise the short-tailed apes of the Old World, with the exception of the baboon, the aged male is otherwise formed. In the latter case the points of resemblance to the human type are far fewer; the aged animal has become a gigantic ape, retaining, indeed, in the structure of his hands and feet, the characteristics of the Primates, while the protruding head is something between the muzzle of the baboon, the bear, and the boar. Simultaneously with these remarkable alterations of the external structure there occurs a modification of the skeleton. The skull of an aged male gorilla becomes more projecting at the muzzle, and the canine teeth have almost attained the length of those of lions and tigers. On the upper part of the skull, which is rounded in youth, great bony crests are developed on the crown of the head and on the occiput. The arches above the eye-sockets are covered with wrinkled skin, and the already savage and indeed revolting appearance of the old gorilla is thereby increased."

In all the man-like apes the number of the teeth is the same as in man himself—that is to say, there are on each side of both the upper and lower jaws two incisors, one canine, two premolars, and three molars; the formula thus being : $i.\frac{2}{2}$, $c.\frac{1}{1}$, $p.m.\frac{2}{2}$, $m.\frac{3}{3}$, making a total of 32 teeth. Not only do the teeth agree in number with those of man; but, with the exception of the great size of the tusks, or canines, of the males, they likewise resemble them in structure. We are familiar with the form of our own molar teeth, which have wide crowns, with their angles rounded off, and surmounted by four main tubercles set somewhat obliquely to one another; and the molars of the man-like apes are of the same general type of structure (see figure on page 151). In the apes, however, the whole series of teeth does not present the horseshoe-like contour which is so characteristic of human teeth; but, on the contrary, the cheek teeth form nearly straight lines, having an angulated junction with the curved line of the front teeth.

None of these apes possesses the peculiar pouches in the cheeks occurring in many of the monkeys, and none of them has any trace of a tail. Moreover, the naked patches so often found on the buttocks of the other Primates are either absent, or, if present, are of very small size. All these animals agree, however, with monkeys, and thereby differ from man, in the great length of the arms as compared with that of the legs; this difference being clearly indicated in the figure of the skeletons of man and the gorilla (page 148). Another characteristic of the man-like apes shown in the same figure is the great breadth and flatness of the breastbone, or sternum, this being a feature in which they agree with man, and differ from baboons and monkeys. Then, again, some of the man-like apes differ from the latter and resemble man in the absence of a small bone occupying a central position in the wrist, and hence known as the centrale of the carpus.

In addition to the points already mentioned, man is distinguished from the man-like apes by the greater relative size of his brain and the portion of the skull in which it is contained, as compared with the face and muzzle. His canine teeth are, moreover, but little longer than the other teeth, and are thus quite unlike the huge tusks of the male gorilla and orang-utan. The great toe is also relatively longer, and is, at the most, only opposable in a very limited degree to the other toes. Indeed, the whole skeleton of man, as may be seen from the figure, is of a lighter and neater build, with certain peculiar curvatures of the lower part of the backbone, these permitting the assumption of the perfectly upright position without fatigue, and without need of any support from the arms, which do not reach below the middle of the thigh. Again, no ape has an ear modelled on the beautiful lines of that of the human species. The naked human body is not, however, a character which a naturalist would consider of any importance as distinguishing man from the apes.

The man-like apes are now restricted to the forest regions of the hottest parts of Asia and Africa; the chimpanzee and gorilla being exclusively African, the gibbons common to North-eastern India and Malaya, and the orang-utan solely Malayan. Chimpanzees and apparently orang-utans, however, formerly occurred in India, and gibbons and an extinct type at a somewhat earlier period in Europe, while the earliest-known member of the group is found in the Oligocene strata of Egypt. This distribution indicates the former existence of an extensive forest tract connecting India and Malaya, by way of Arabia, with Equatorial Africa ; the central part of this area being now either submerged or deforested. This explains the remarkable resemblances between the fauna of Southern India, Ceylon, and Malaya, on the one hand, and that of the forest region of Africa, on the other.

THE CHIMPANZEE

Of all the large man-like apes, the one which makes the nearest approach in bodily structure to man is the chimpanzee (*Anthropopithecus troglodytes*) of Western and Central Equatorial Africa, of which there are several geographical races. (See frontispiece.)

The chimpanzee has been long known in Europe ; and it has, indeed, been considered that the so-called " gorillas " met with by the Carthaginians of Hanno's voyage in 470 B.C. on the rocky coasts of Sherboro Island, off Sierra Leone, were chimpanzees. According, however, to Winwood Reade, who travelled in Western Africa for the purpose of obtaining authentic information about the chimpanzee and the gorilla, the creatures seen and captured by Hanno's party were neither gorillas nor chimpanzees, but dog-faced baboons. Be this as it may, that the chimpanzee was known in Europe so far back as 1598 is proved by an account brought back from the Congo by a Portuguese sailor named Eduardo Lopez, and published at Frankfort by Pigafetta in his account of the Congo district. In 1613 there appeared, in Samuel Purchas's " Pilgrimage," the history of the wanderings of an English sailor named Andrew Battel, who travelled in the lower part of Guinea in 1590, and appears to have heard of or seen, not only the chimpanzee, which he designates the enjocko (a corruption of n'djeko or n'schego), but likewise the gorilla, which he calls the pongo.

Battel's account may be quoted at length, as follows : " There are two kinds of monsters common to the woods of Angola ; the largest of them is called pongo in their language, and the other enjocko. The pongo is in all its proportions like a man (except the legs, which have no calves), but he is of gigantic height. The face, hands, and ears of these animals are without hair ; their bodies are covered, but not very thickly, with hair of a dunnish colour. When they walk on the ground, it is upright, with the hands on the nape of the neck. They sleep on trees, and make a covering over their heads to shelter them from the rain. They eat no flesh, but feed on nuts and other fruits ; nor have they any understanding beyond instinct. When the people of the country travel through the woods, they make fires in the night, and in the morning when they are gone the pongos will come and sit round it till it goes out, for they do not possess sagacity enough to lay on more wood. They go in bodies to kill many negroes who travel in the wood. When elephants happen to come and feed where they are, they will fall on them, and so beat them with their clubbed fists and sticks that they are forced to run away roaring. The grown pongos are never taken alive, owing to their strength, which is so great that ten men cannot hold one of them. The young hang upon their mother's belly with their hands clasped about her. Many of them are taken by shooting the mothers with poisoned arrows."

From that date our knowledge of these animals has been gradually added to, although there is still room for fuller authentic accounts of their habits in a state of nature. Young chimpanzees are frequently brought alive to Europe and exhibited in zoological gardens, but they require great care and attention, and even with these almost invariably die after a few years or months from the effects of the climate, which generally show themselves in various organic affections, although not, as has been supposed, in the form of tubercular disease of the lungs.

In addition to certain distinctive features in the teeth, such as the relatively small size of the tusks, or canines, of the males, and the circumstance that the upper " wisdom tooth " is smaller than either of the two molars in advance of it, the typical chimpanzee may be distinguished from the gorilla by the fact that the males are but slightly larger than the females. The skull of the male chimpanzee is also characterised by the absence of the enormous bony ridges which overhang the sockets of the eyes in that of the gorilla ; while in the lower jaw the length of the bony union between the two lateral branches is much less than in the latter. In both these respects the chimpanzee is decidedly nearer to man than is the gorilla ; and a further approximation to the human type is presented by the relatively shorter arms, which in the perfectly upright posture only reach a short distance below the knee. The hands and feet also are longer and more slender than those of the gorilla, as may be seen by comparing Figs. 3–8 with 1 and 2 of the illustration on page 174. Moreover, as is the case in man, the middle finger is longer than any of the others ; and although there is some degree of variation in the relative length of the thumb in different individuals, as a rule this digit reaches to the base of the first phalangeal joint of the index finger. The male chimpanzee does not usually exceed five feet

HEAD OF CHIMPANZEE

in height when full grown, and is thus considerably inferior in size to the male gorilla.

Dr. Hartmann remarks of the chimpanzee that, although the arched ridges above the eyes " are not so excessively prominent as in a gorilla of the same age, they are strongly developed, covered with wrinkled skin, and in this case also there is a species of eyebrow, stiff and bristly, with shorter hairs between. The large, wrinkled lids are furnished with thick eyelashes. A general physiognomical distinction between the gorilla and the chimpanzee consists in the fact that the bridge of the nose is shorter in the latter than in the former. In the chimpanzee this part of the organ is depressed, yet the depression is of a conical and convex form and is covered with a network of wrinkles of varying depth. In the chimpanzee the interval between the inner angle of the eye and the upper lateral contour of the cartilaginous end of the nose is shorter than in the gorilla. There is also some difference in the form of the nose ; it is on the whole flatter, the tip is less apparent, and the nostrils are not so widely opened nor so thickly padded. The external ear of the chimpanzee has, on the whole, less resemblance to the human ear, and its contour is larger than that of the gorilla. But this organ varies so much in individuals that it is difficult to lay down any rule for its average size. The skin of the chimpanzee is frequently of a light yet muddy flesh-colour, which sometimes verges upon brown. Spots, varying in size and depth of colour, sometimes isolated, sometimes in groups, and of a blackish-brown, sooty, or bluish-black tint, are found on different parts of the body of many individuals, especially on the face, neck, breast, belly, arms and hands, thighs and shanks, and more rarely on the back. The face, which, soon after birth, is of a flesh-colour, merging into yellowish brown, assumes a darker shade with the gradual development of the body. The hairy coat is sleek, or only in rare cases slightly curled, and the coarser and bristly hair is generally stiff and elastic. The parting on the

TWO CHIMPANZEES, " SUSAN " AND " JIMMY "
From a photograph by W. P. Dando

forehead is often so regular that it might have been arranged by the hairdresser's art. Close behind that part of the head at which the projecting ridges over the eyes of the gorilla generally meet there is in the chimpanzee [as is well shown in the figure on page 176] an altogether bald place, or often only a few scattered hairs. Round the face the growth of hair streams downwards like a beard. On the neck it is of considerable length, and it falls in the same long locks over the shoulders, back, and hips. The hair on the limbs is not so long and takes a downward direction on the upper arm, and an opposite direction on the forearm, while there is often a longitudinal parting on the centre of the inner surface of this part of the limb. On the back of the wrist the hair grows in a kind of whorl ; the upper hairs turn upwards and backwards, the middle ones turn backwards, the lower ones backwards and downwards. The back of the hands and the roots of the fingers are hairy. On the front of the thigh the hair takes a downward direction, while behind it grows backwards. On the shank it grows downwards in the region of the tibia, and turns back on the inside of the leg. The back of the foot and the roots of the toes are likewise hairy. There is a shorter growth of these hairs on the face, chin, and ears. In other cases the hair of the true chimpanzee is of a black colour. Short whitish hairs may be observed on the lower part of the face and chin, as well as round the posteriors ; and sometimes the colour of the hair is shot throughout with reddish or brownish black."

Hartmann's description applies to the typical chimpanzee (*Anthropithecus troglodytes*). As regards the local races of this species, the explorer Du Chaillu, in his " Equatorial Africa " (1861), described a chimpanzee known to the natives as nshiego mbouwe, and to naturalists as the bald chimpanzee (*A. troglodytes calvus*), and in the autumn of 1883 a young chimpanzee was purchased by the Zoological Society of London, which was recognised as being very different from the true or common chimpanzee, and was regarded as in all probability identical with Du Chaillu's bald chimpanzee.

YOUNG CHIMPANZEE AND ORANG-UTAN

D

P

THE CHIMPANZEE "MAFUKA"

Writing of this animal, Mr. A. D. Bartlett re-marked that, while "the colour of the face, hands, and feet in the true chimpanzee are white or pale flesh-colour, the same parts of the animal under consideration are black or brownish black. Another well-marked difference is to be found in the hair upon the head and face. In the true chimpanzee the hair on the top of the head and that passing down from the centre, where it divides, to the sides of the face or cheeks is tolerably long and full, forming what may be considered rather bushy whiskers; whereas in the new specimen the front, top, and sides of the head and face are nearly naked, having only a few short hairs on the head, quite destitute of any signs of the parting so very conspicuous in the chimpanzee. Another striking difference may be noticed in the size and form of the head and ears. Out of the number of chimpanzees I have seen and examined, both old and young, none has possessed the large flat ears that are so conspicuous in this individual. The form of the head, the expression of the face, the expanded nostrils, the thicker lips, especially the lower lip, together with the more elevated skull, cannot fail to distinguish this animal from the true chimpanzee. Again, the habits of this animal differ entirely from those of the common chimpanzee. She has always shown a disposition to live upon animal food. Soon after her arrival I found she would kill and eat small birds; seizing them by the neck, she would bite off the head and eat the bird, skin, feathers, and all; for some months she killed and ate a small pigeon every night. After a time we supplied her with cooked mutton and beef-tea; upon this food she has done well. I have never found any ordinary chimpanzee that would eat any kind of flesh.

"Another singular habit was the producing of pellets, or 'quids,' resembling the castings thrown up by birds of prey. They are composed of feathers and other indigestible substances that had been taken with her food. Moreover, she is an expert rat-catcher, and has caught and killed many rats that had entered her cage during the night. Her intelligence is far above that of the ordinary

chimpanzee. With but little trouble she can be taught to do many things that require the exercise of considerable thought and understanding; she recognises those who have made her acquaintance, and pays marked attention to men of colour, by uttering a loud cry of 'Bun, bun, bun.' She is never tired of romping and playing, and is generally in a good temper."

The bald chimpanzee is chiefly distinguished by the feature from which it takes its name. A second race, the kulu-kamba chimpanzee (*A. troglodytes kulu-kamba*), is characterised, on the other hand, by its marked approximation to the gorilla, both as regards its large size and the prominent ridges above the eyes. To this race belonged "Mafuka," a female formerly living in the Zoological Gardens at Dresden, which is figured on this page; and also the female named "Johanna," which was, for a time, one of the features of Messrs. Barnum and Bailey's show. This race has also been described under the name of *Troglodytes aubryi*, but its proper title is that given here.

As already mentioned, chimpanzees inhabit Western and Central Equatorial Africa, where they range over a considerable area of country. On the west coast their range appears to be limited to the north approximately by the River Gambia, while their southward range extends about to the River Coanza, which flows into the ocean at the boundary between Angola and Benguela. Their limits on each side of the Equator do not, therefore, exceed some twelve degrees, the northern range in latitude being greater than the southern. With regard to the extent of their range across the continent eastwards, chimpanzees occur to the north-west of the great lakes in the Niam Niam district, in 28° E., and they are also recorded from Monbottu, and range to about 32° E. in Uganda and Unyoro, the districts lying between Lakes Victoria and Albert.

Like all other man-like apes, chimpanzees are forest-dwelling animals, although on the coast of

SIDE VIEW OF THE HEAD OF "MAFUKA"

the Loango district they are found in the mountains. Their food is usually the various wild fruits which grow abundantly in these dense woods ; but, as already mentioned, at least the bald-headed species will take kindly to an animal diet in captivity.

Chimpanzees appear to be continually shifting their haunts in order to find fresh feeding-grounds, and will not unfrequently visit and pillage deserted native plantations. They utter loud cries, which may be heard resounding through the forests at all hours of day and night. Dr. E. Pechuel-Loesche, who accompanied the expedition sent to Western Equatorial Africa during the years from 1873 to 1875, observes that chimpanzees " are really accomplished in the art of bringing forth these unpleasant sounds, which may be heard at a great distance and are reproduced by the echoes. It is impossible to estimate the number of those who take part in the horrid noise, but we often seemed to hear more than a hundred. They generally remain upon the ground among the dense underwood and thickets of amomum [a member of the ginger family] and other scitamineous plants, and only climb trees for the sake of obtaining fruit. Their track may be plainly discerned on soft ground ; they stop short wherever grows the amomum, of which they are very fond, and the red husks of the fruit of which may be seen strewn around."

There seems to be no doubt but that chimpanzees build a kind of nest high up in the trees for their families ; and it is stated that the male of the family takes up his position for the night beneath the shelter afforded by the nest. It is probable that this habit has given rise to the idea that these animals construct pent-houses for themselves ; an illustration of the bower is given in Du Chaillu's " Equatorial Africa." According to Du Chaillu, the creature ascends the trunk by a hand-over-hand movement and with great rapidity, creeps carefully under the shelter, seats itself on the base of a projecting bough, on which both feet and haunches rest, and then puts one arm about the trunk for security.

It is said that chimpanzees will generally take to flight at the sight of man, but that when driven to bay, or when their retreat is cut off, they will attack him fiercely, and are then very awkward customers to deal with. Dr. Livingstone, in his " Last Journals," gave an account and sketch of a chimpanzee hunt by the Manyema tribe, describing these animals under their name of soko, but apparently confusing them with the gorilla. The doctor's graphic sketch shows four chimpanzees surrounded by natives, one of the former having received its death-wound, a second with a spear in its back, and a fourth making a vigorous onslaught on one of the hunters, whose hand it has seized in its mouth. Dr. Livingstone states that the chimpanzee " kills the leopard occasionally, by seizing both paws and biting them, so as to disable them ; he then goes up a tree, groans over his wounds, and sometimes recovers, while the leopard dies. The lion kills him at once, and sometimes tears his limbs off, but does not eat him. The soko eats no flesh ; small

THE CHIMPANZEE'S NEST, OR SHELTER
From the drawing by J. Wolf in Du Chaillu's " Equatorial Africa "

bananas are his dainties, but not maize. His food consists of wild fruits, and of these one is large—a large sweep sop, but indifferent in taste. The soko brings forth at times twins."

In captivity, chimpanzees, when in health, are gentle, intelligent, and affectionate, readily learning to feed themselves with a spoon, or to drink out of a glass or cup. One of the earliest accounts of the captive chimpanzee was given by W. J. Broderip, and related to a young male brought from the Gambia in the year 1835, which was deposited in the gardens of the London Zoological Society. Dr. Hartmann has also published an interesting description of the habits of another male, which was exhibited in the Berlin Aquarium in 1876, and was remarkable for its unusually lively and cheerful disposition.

More recent is the description by G. J. Romanes of the mental power of the bald chimpanzee " Sally," which lived for eight years in London. This account was written in 1889, after the ape had been nearly six years in the Zoological Gardens. The intelligence of " Sally " is compared by Romanes to that of a child a few months before emerging from the period of infancy, and is thus far higher than that of any other mammal, exclusive of man. In spite, however, of this relatively high degree of intelligence, the creature's power of making vocal replies to her keepers, or those with whom she was brought into contact, were of the

YOUNG CHIMPANZEE AT REST ON A BRANCH
Photograph by Lewis Medland

most limited kind. Such replies were, indeed, restricted to three peculiar grunting noises. One of these indicated assent or affirmation; another, of very similar intonation, denoted refusal or distrust; while the third, and totally different intonation, was used to express thanks or recognition of favours. In disposition "Sally" was, like many of her sex, apt to be capricious and uncertain; although, on the whole, she was good-humoured and fond of her keepers, with whom she was never tired of a kind of bantering play, which was kept up at intervals almost continuously. By singing in a peculiar kind of monotone, in imitation of her own utterance, her keepers were usually able to induce her to go through a series of remarkable actions, the meaning of which is not very apparent. First she would shoot out her lips into a tubular form, uttering at the same time a weird kind of howling note, interrupted at regular intervals. The pauses would, however, gradually become shorter and shorter, while the sing-song cry became louder and louder, until it finally culminated in a series of yells and screams, not unfrequently accompanied with a stamping of the feet and a violent shaking of the netting of her cage. After this climax the utterance of a few grunts ended the performance.

Some time before 1889 it occurred to Romanes that "Sally" would be a good subject to test the powers of the simian intelligence by a series of special experiments. It was found, however, that such experiments were seriously hampered by the effects on the creature of the visits of the numbers of people who were constantly passing in and out of the room in which she was kept; and there is consequently but little doubt that, in more favourable circumstances, the results obtained would have been more remarkable than they are.

Romanes, having secured the assistance of the keepers, caused them to ask "Sally" repeatedly for one, two, or three straws, which she was to pick up and hold out from among the litter strewing her cage. The number of straws asked for was constantly varied, and never followed any regular order; and when the correct number was presented the animal was rewarded by a piece of fruit, while if the number were incorrect her offer was refused. "In this way," observes the author, "the ape was taught to associate these three numbers with their names. Lastly, if two or three straws were demanded, she was taught to hold one or two in her mouth until she had picked up the remaining straw, and then to hand the two or three straws together. This prevented any error arising from her interpretation of vocal tones. As soon as the animal understood what was required, and had learnt to associate these three numbers with their names, she never failed to give the number of straws asked for. Her education was then extended in a similar manner from three to four, and from four to five straws.

"Here I allowed her education to terminate. But more recently one of the keepers has endeavoured to advance her instruction as far as ten. The result, however, is what might have been anticipated. Although she very rarely makes any mistake in handing out one, two, three, four, or five straws, according to the number asked for, and although she is usually accurate in handing out as many as six or seven, when the numbers eight, nine, or ten are named, the result becomes more and more uncertain, so as to be suggestive of guess-work. It is evident, however, that she understands

A CHIMPANZEE EVINCING INTEREST
Photograph by W. P. Dando

the words seven, eight, nine, and ten to betoken numbers higher than those below them; and if she is asked for any of these numbers, she gives some number that is above six and not more than ten; but there is no such constant accuracy displayed in handing out the exact number named as is the case below six. On the whole, then, while there is no doubt that this animal can compute any number of straws up to five, beyond five the accuracy of her computation becomes progressively diminished."

In addition to these attempts to determine "Sally's" capacity for numbers, a series of experiments was undertaken to test her powers of recognising and distinguishing between colours. "It appeared to me," wrote Romanes, "that if I could once succeed in getting her to know the names of black, white, red, green, or blue, a possible basis might have been laid for further experiments wherein these five colours could have been used as signs of artificially associated ideas. The result, however, of attempting to teach her the names of colours was so uniformly negative that I am disposed to think the animal must be colour-blind. The method adopted in these experiments was to obtain a number of brightly and uniformly coloured pieces of straw, each piece being either black, white, red, green, or blue. Offered the straws two by two of different colours on each occasion, the ape was invited to select the straw of the colour named from the one whose colour was not named, and, of course, on choosing correctly was rewarded with a piece of fruit. In this way she quickly learned to distinguish between the white straws and the straws of any other colour; but she never could be taught to go farther. Now, the distinction between the white straws and the straws of any other colour is a distinction which can be drawn by an eye that is colour-blind; and from the fact that the ape is always able to perceive this distinction, while she cannot be taught to distinguish any of the others, I conclude that her failure in this respect is not due to any want of intelligence, but to some deficiency in her powers of colour-perception."

A young chimpanzee, known as "Focke," lived for some time in the Royal Zoological Gardens at Dresden, where on several occasions he exhibited marked signs of more than average intelligence. On one occasion "Focke" was removed into the winter house, a large and spacious apartment with wooden benches fixed to the walls. Soon after his arrival he was seen to take up his position on one of these benches, where he remained for some time surveying his new abode. Next he examined all the utensils in the apartment, and more especially a jug of water. After some time he discovered a

THE CHIMPANZEE "SALLY"

knot-hole in a deal plank forming part of one of the benches, which seemed to puzzle him. He approached nearer to the hole, which he regarded attentively for some time. At first his object seemed to be to find to what depth it descended, and when he could not assure himself by other means of any limit in this direction he thrust his index finger as deep as it would go into the hole, and appeared to be astonished at finding no bottom.

He next seemed to be thinking the matter over for some time, after which he spat on the plank as near as possible to the hole, into which he afterwards guided the saliva with his finger. The fluid did not, however, suffice to fill the hole, the plank being laid a few inches above the uneven ground.

More puzzled than ever, the ape fetched the water-jug and poured its contents into the hole, when the water disappeared without filling the space between the plank and the ground. Thereupon "Focke," after returning the jug to its place, approached once more to the plank, near which he sat down in a thoughtful mood, from time to time casting glances at the hole. On the following day he paid several visits to the spot, and made other attempts to ascertain the depth of the hole, but without success. Ultimately he seemed to realise the fruitlessness of all his investigations, and gave the matter up as a bad job, taking no sort of notice of the hole.　R. LYDEKKER

STORIES OF THE CHIMPANZEE

"Sally," to whom reference has been made, was not the first, and will not be the last, tame chimpanzee to display intelligence of a high order. Among the seven chimpanzees which at the end of 1909 were housed at the London Zoological Gardens were one or two in whom uncannily human attributes appeared. "Mick," the largest of the chimpanzees in residence in the year of grace just named, had been a character of the gardens for the preceding twelve years, having been deposited there when two years old. Another chimpanzee had been in the gardens some half-dozen years. Under close observation and encouragement these two animals should have gone far towards showing to what extent the chimpanzee can be trained. Certainly the highest possibility has not yet been attained.

Two public performers, "Consul I." and "Consul II.,"

attracted attention in the early years of the twentieth century by tricks of a surprising character. They wore, with seeming satisfaction, clothes like those of a human being. They ate at table with knife, fork, and spoon, and observed the amenities with propriety (page 103). They showed sufficient knowledge of cause and effect to realise that by banging the keys of a pianoforte sound was produced, and apparently they enjoyed the result of their efforts. This was, however, to take them back to Nature, so to speak ; for in its wild state the chimpanzee thumps its hollow tree and howls with ecstasy. Possibly the same love of noise may account for the energy with which these animals tapped away at their typewriters. Their education stopped short, of course, of any intelligent attempt to order the sequence of the letters produced.

In common with the tricks of all other animals which perform in public, the feats of the highly educated chimpanzees which form part of a public entertainment are to be regarded with some suspicion. It is difficult to determine where obedience to protracted drilling ends and individual volition begins. For this reason the testimony of passengers who watched " Consul II." on board the liner by which it was carried to New York in November of 1909 is interesting. The animal roller-skated (page 183) and rode a bicycle on deck with delight, even when the ship was experiencing rough seas. To cycle and to skate are two of its public tricks, but here there was no compulsion, and the ape considered the exercise good fun. Moreover, it exhibited pleasure in helping to swab the decks after it had watched the sailors at their work. Nor were some of the occupations of the lady passengers less attractive to it, for in imitating them in the art of hemming handkerchiefs it is said to have displayed no mean ability. An animal which masters the feat of poising itself upon a bicycle and upon a pair of roller skates possesses intelligence that should not be lightly overlooked by the student of animal psychology.

The pity is that the subject has not received more systematic study. The chimpanzee in private custody stands a far better chance of making a good return for attention to its training. The life of an ape that performs in public is inevitably short. It is taught to smoke and to drink alcohol, and to indulge in food ruinous to the constitution of an animal whose natural diet consists of fruit and vegetables. A modern rival of the

YOUNG CHIMPANZEE RESTING

two " Consuls " was an American-trained chimpanzee, which rode a bicycle, carried a miniature hod of bricks up a ladder, and smoked pipe and cigar. The chimpanzee privately trained need not acquire the tobacco and, emphatically not, the drink habit, and will long outlive its fellow of the circus and the music-hall.

In its natural environment the chimpanzee undoubtedly displays a grim sort of humour. The love of fun is by no means limited to captive specimens. Dr. Livingstone has told us how one of his men, when hoeing, was stalked and caught from behind by a chimpanzee. The man howled with terror, whereupon the chimpanzee " giggled and grinned, and left him, as if it had been in play." No other interpretation than the latter is possible. The chimpanzee knows that it has the power grievously to injure a man, and it has sense enough to discriminate between the armed man and the unarmed. The latter need fear nothing ; but the man who carries a spear must expect attack. An adult chimpanzee will seize the spearman, and bite off his hand. Women, from whom experience has taught the chimpanzee no harm is to be feared, are in no danger from these animals, despite many sensational stories to the contrary. It is true that now and then a chimpanzee will snatch up a baby and carry it to the top of a tree ; but the motive seems to be the same spirit of innocent mischief which prompts a monkey to carry property up the rigging of a ship in order to enjoy the fun of seeing the sailor's perplexity and anger. The child abducted by the chimpanzee seems to suffer no harm. A bunch of bananas will tempt the offender to earth again, when it will release the child in order to take the fruit.

But it cannot be pretended that the nature of the chimpanzee is compounded solely of good humour; the beast has characteristics of an opposite description, as a striking experience narrated by Sir Harry Johnston suffices to demonstrate. " I remember travelling home from West Africa once," he wrote, " with a celebrated chimpanzee of many accomplishments, which, until we reached Madeira, travelled as, and was treated like, a first-class passenger, behaving in his freedom with absolute decorum. When the vessel reached Madeira, a number of passengers came on board, amongst others being a very young baby. This was sometimes placed by its mother in a cradle on

YOUNG CHIMPANZEE WAITING TO BE FED

the upper deck. One day while we were at lunch, the chimpanzee, which had become very jealous of the attention shown to this infant, disappeared. I followed it from the luncheon-table to see what it was going to do, and arrived just in time to prevent it from throwing the baby overboard. It had pulled the poor little child from its cradle, fortunately without hurt, and would certainly have hurled it into the sea if my arrival had not caused the guilty ape to drop the child on the deck and shamble away. After this it spent the rest of its journey caged, on deck." This well-authenticated story implies striking reasoning powers in the chimpanzee, but not more than those which characterise the dog which, jealous of a kitten, carries its tiny rival out of the house and drops it into a pond.

On the other hand, the chimpanzee is capable of actions which we should describe as heroic were they performed by men. An instance of the kind was that in which the chief figure was a chimpanzee which for nine years enjoyed a good deal of freedom in the French town of Grenoble, whither it had been brought by its captor, an explorer. The ape was allowed to roam the town with the liberty which is claimed by the monkeys of Benares. Thus, when a child fell down a well, the ape was at hand to see the accident, and to effect a rescue. Swinging itself over the top of the well, it lowered itself quickly, hand under hand, down the rope by which the buckets were wound up and down, drew the child from the water, and carried it up in safety. This was the act of an untrained ape, yet in it instinct may be traced. The chimpanzee is heroic in defence of its own young. To rescue one of its kind from peril would be the work of an instant. A like instinctive desire would be called forth by the sight of a child

"CONSUL II." ON ROLLER SKATES

that had fallen down a well. Is it too much to urge that we should accord to the chimpanzee as compassionate a nature as to the dog which plunges into a torrent to save the life of a drowning man ?

The chimpanzee, untrained, has a ready perception of human sympathy. Livingstone had a young chimpanzee which he found a most affectionate companion. Secured near his tent, it was content so long as he remained in sight. It would squat on a mat at his feet, happy. If he went for a walk, his pet must go, too. She would manifest the greatest anxiety to be taken, holding out her hands to him ; and, if disappointed by his going alone, she whimpered and wrung her hands with every evidence of grief. She was cleanly in her habits, and was careful to wipe her hands with leaves if they became soiled. If danger, real or imaginary, threatened, she retreated at once to Livingstone, and, standing upright against his legs, derived moral as well as physical support

to carry her through the trials and troubles that beset her.

Left to its own resources in captivity, the chimpanzee would probably develop but little intelligence. Its mental expansion appears to require external aid. One new idea will suggest others ; the difficulty lies in impressing the first idea upon the ape's mind. Taught, after some trouble, the use of a key, a chimpanzee at the Manchester Zoological Gardens turned the knowledge to account by securing its own liberty. Another, which was placed in a cage within a cage to keep it from actual contact with the public, was wont to receive surreptitious gifts of nuts. These often fell beyond the ape's reach, but it discovered a way of securing them by flinging its blanket over them and suddenly withdrawing it. The same ape was shown how to use a stick for the same purpose. From this it learned that where a nut was still too distant to be touched by a short stick, the latter could be employed to reach a longer stick by which the nuts could be gained. But although it became expert with these, it could never be taught to appreciate the use of the crooked handle for the task.

For another example of the curious limitations of the chimpanzee's intelligence we are indebted to Frank Buckland, who has recorded his experiences with one of the apes at the Zoological Gardens in the days when Mr. A. D. Bartlett was superintendent. The fine ape-house in which the man-like apes are now lodged was not in existence then, so, in order to give " Joe," the chimpanzee, exercise, it was the custom to let him have the run of the monkey-house before the public were admitted. Everything went well until the hour for the ape's return to his cage. Then he would hide on the top of the cages out of the reach of his keeper, and no amount of coaxing would entice him down.

Bartlett discovered that " Joe's " weak points were curiosity and cowardice, and worked upon them with excellent results. He went to the keeper, and, touching him gently on the shoulder, pointed in a mysterious manner to the dark passage underneath the gas-pipe which traversed the house, affecting to believe that some nameless horror lurked there. The two men would peer at the spot with looks of dire alarm, and the effect upon " Joe " was electrical. If his two best friends were frightened, then there must be good reason for dismay. He would slowly descend to the floor, and, uttering his alarm note, steal towards the point of danger. When he had reached the right position, the two onlookers raised cries of simulated terror, and fled as from some object which, they pretended, was emerging from the darkness towards the room occupied by the ape. The ape, not to be behindhand in flight, cried his loudest, and, his hair erect and bristling, bolted towards the open door of his cage, jumping, leap-frog fashion, over their heads in his anxiety to be first to reach it. The door was then promptly closed behind him, and he was safe.

This performance was repeated again and again. There was always the same difficulty about getting the ape back, but the trick never failed. His cowardice and curiosity were always successfully appealed to, for experience taught him as much as it taught the Bourbons. ERNEST A. BRYANT

THE GORILLA

We have seen (page 176) that in 1590 an English sailor, Andrew Battel, had heard of a gigantic ape haunting the forests of Guinea, and known to the natives as the Pongo, this ape being also called by such names as Jina, N'Jina, or Indjina, or N'Guyala, while by Europeans it is universally termed the Gorilla. Buffon appears to have lent credence to Battel's pongo (N'Pungu or M'Pungu, as it is variously spelt); but his account was summarily rejected by Cuvier as a mere traveller's tale. Still, however, vague rumours of the existence on the West Coast of Africa of an ape of larger size and fiercer habits than the chimpanzee from time to time reached Europe; and in 1819 T. E. Bowdich, in his account of the "Mission from Cape Coast Castle to Ashanti," definitely stated that among the many curious apes found in the Gabun district, the ingenu (gorilla) was by far the largest and strongest. But it was not until 1847 that precise evidence of the existence of this mysterious ape reached Europe. In that year Dr. Savage, an English missionary stationed at the Gabun, wrote to the illustrious anatomist Sir Richard Owen enclosing drawings of the skull of an ape from that district, which was described as being much larger than the chimpanzee, and feared by the negroes more than the lion or any other wild beast of the forest. These sketches showed the bold bony crests over the eye-sockets which mark the skull of the gorilla as distinct from that of the typical chimpanzee. "At a later date in the same year," writes Sir Richard Owen, "were transmitted to me from Bristol two skulls of the same large species of chimpanzee as that notified in Dr. Savage's letter; they were obtained from the same locality in Africa, and brought clearly to light evidence of the existence in Africa of a second larger and more powerful ape." In the following year these specimens were described by the English anatomist under the name of *Troglodytes savagei*.

It appears, however, that about the same time that Dr. Savage forwarded the sketches to Sir Richard Owen he also sent a skull of the unknown ape, together with a description of the animal itself, by the hand of a fellow-missionary, the Rev. Dr. J. Leighton Wilson, to Boston, in the United States, and in an American scientific journal for the year 1847 the new ape was described and named *Troglodytes gorilla*. Thus matters stood till the year 1851, when Captain Harris presented to the Royal College of Surgeons the first skeleton of a gorilla that had

ever been brought to England; while in the same year another skeleton was sent to Philadelphia by Mr. H. A. Ford, a medical missionary. This made a distinct advance in our acquaintance with the creature; and in 1852 a French naturalist came to the conclusion that the gorilla ought not to be included in the same genus as the chimpanzee, and he accordingly proposed for it the name of *Gorilla gena*. By the rules of nomenclature adopted among zoologists, however, he had no right to supersede the specific name proposed in America; and the gorilla may thus be known scientifically as *Anthropopithecus gorilla* (see page 157).

In 1856 Paul Belloni Du Chaillu arrived at the Gabun, preparatory to his expedition into the interior, and two years later the British Museum received from the Gabun an entire gorilla preserved in spirits, the skin of which was soon afterwards mounted and exhibited to the public.

Such is the history of the gradual acquisition of our knowledge of the largest of the apes. On his return from the Gabun to America, Du Chaillu set to work to publish an account of his travels and adventures; and in 1861 the world was startled by the appearance of his "Explorations and Adventures in Equatorial Africa," which gave a full and illustrated narrative of numerous personal encounters with gorillas. Somewhat later, an Englishman, Winwood Reade, made an expedition to the Gabun for the purpose of verifying these accounts; the results of his journey being given in a work entitled "Savage Africa," of which the first edition appeared in 1863. In this work it is asserted that neither Du Chaillu nor any other European had up to that date ever seen a living wild gorilla in its native haunts, though Reade possibly did not refer to those driven to the shore in 1851 (page 186); and his assertions were supported by the members of the German Loango Expedition in 1873-76.

The features distinguishing the typical race of the chimpanzee from the gorilla have been referred to under the account of the former species (pages 176-7); but as these characters will not serve to distinguish the kulu-kamba chimpanzee from the gorilla, recourse must be had to other characteristics, of which the following, apart from anatomical points, appear to be the most important. In addition to its sullen, untamable, and ferocious disposition, the gorilla is distinguished from the chimpanzee by the greater length of the nasal, or nose, bones, which descend below the level of the lower border of the sockets of the eyes. It has also characteristic deep folds running from the angles of the nose to the corners of the upper lips. The great size of the canine and cheek teeth, especially the last of the latter, is also distinctive, as are the broad, short, and thick webbed hands and feet, the long heel, and the great relative length of the upper half of the arm and the proportionately shorter forearm. There are also differences in the form of the neck.

Having distinguished between the chimpanzee and gorilla, mention may now be made of certain characteristics by which these two species are collectively distinguished from the lower man-like apes, and thereby agree with man. One of these is that the total number of joints in the backbone, or vertebræ, lying between the solid mass called the sacrum and the neck is seventeen, or the same as in man. It is true, indeed, that whereas in man only twelve of these vertebræ carry ribs, in the gorilla and chimpanzee thirteen are so provided ; but this is a matter of minor import, which is entirely over-balanced by the numerical similarity. The other point is the absence of the central bone in the wrist ; so that whereas in man, the chimpanzee, and the gorilla the total number of separate wristbones is nine, in all the other Primates it is ten. This is a very important characteristic in con-necting these two apes with man.

A full-grown male gorilla, if standing in a perfectly upright posi-tion, will generally measure rather more than six feet in height ; and since his body is much more bulky and his limbs are longer than those of a man, he is de-cidedly the largest repre-sentative of the Primates. As in the chimpanzee, there are distinct eye-brows on the forehead and lashes to the lids of the eyes. The nose has a relatively long bridge, and its extremity is high, conical, and widely expanded, the whole length being divided by a distinct furrow running down the middle line and becoming more marked as age advances. The upper lip is remarkable for its shortness, and the whole of the dark skin in the region of the nose, cheeks, and mouth is marked by a number of folds. The massive jaws are extremely projecting, and with their huge tusks, or canine teeth, complete the repulsive aspect imparted to the expression by the overhanging eyebrows. The lower jaw has scarcely any indication of the prominent chin which is such a characteristic feature in the human countenance, but slopes rapidly away from the middle line in front, so as to assume a somewhat triangular contour. The whole skin of the face is a deep black colour, of a glossy appearance, and sparsely sprinkled with coarse hairs. The ears are comparatively small, with their hind border sharply angulated in the middle,

and appear to be fastened above and behind to the sides of the face. Like the face and lips, the ears are of deep black hue. The head is joined to the trunk by a very short and thick neck, which gives the appearance of its being set into the shoulders ; and the term " bull-necked " is there-fore strictly applicable to the creature. This great thickness and power of the neck is largely due to the backward projection of the occipital region of the skull, and the tall spines surmounting the vertebræ of the neck. In correlation with the great development of this region, we find the muscles of the shoulders and chest equally powerful, as is essential for the move-ment of the mighty arms. On the latter the arrange-ment of the hair is the same as in the chim-panzee ; but there is a great difference in the form of the hands, as may be seen from the illustration on page 174. Thus, in marked contrast to that of the chim-panzee, the hand is re-markable for its great width and stoutness, coupled with the short-ness and generally clumsy make of the fingers, which are united together by a strong web reaching nearly to the end of their first joints. The thumb is short in proportion to the fingers, reaching only slightly beyond the middle of the metacarpal bone of the index finger, and is nearly conical in shape at its extremity. The fingers, on the con-trary, are somewhat flattened at their ex-tremities. There is but little difference, as seen in the figure, between the lengths of the index, middle, and ring fingers,

SKETCH OF A GORILLA AT REST IN A TREE

the first being in some specimens as long as, but in others shorter than, the middle finger. In all cases, however, the "little" finger is true to its name in being shorter than any of the others. The skin on the back of the wrist is thrown into a number of deep folds, with an oblique direction ; while a net-work of wrinkles covers the backs of the fingers, which have large wart-like callosities on the first and some-times also on the second joints, these callosities being produced by the animal walking, when on all-fours, with its fingers doubled on the palms of the hands. On the deep black and naked skin of the palms of the hands, which are hard and horny, there are generally numerous wart-like growths,

Apart from reference to its enormous bulk, especially in the lower part, no particular mention is necessary of the body of the gorilla, and attention may accordingly be directed to the hind limbs. One of the most important features in these is that the calves are more developed than in any of the other man-like apes. The foot, as contrasted with that of the chimpanzee in the illustration on page 174, is characterised by its great breadth and width, and also by the extreme shortness of the very thick toes. The big toe varies somewhat in length, as compared with that of the others, reaching in some individuals as far as the end of the first joint, and in others to the middle of the second. In contrast to the thumb, the great toe is expanded at the end, and, in opposition to the other toes, forms a grasping organ of remarkable power. None of the other toes is as thick as the big toe, the middle toe being slightly longer than either of the adjacent ones, while the " little " toe is considerably shorter. The sole of the foot is somewhat convex, but its upper surface is very flat, and there is no sort of resemblance to the human instep in the whole foot. The upper surface of the foot, as far as the beginning of the toes, is thickly covered with hair, but on the latter the hairs become thinly scattered ; while the sole is bare and covered with a thick, horny skin. Owing to its habit of sometimes walking with the toes bent under the sole of the foot, the gorilla has callosities on the upper surface of the toes.

With regard to the colour of the hair, of which, as already noted, the general hue is blackish, there is considerable individual variation, and likewise a change attendant upon age, very old gorillas becoming more or less completely grizzled. As a rule, a reddish brown tint is noticeable on the hair at the top of the head, although it may be dark brown or even black, the hairs generally being differently coloured in different portions of their length. On the sides of the face the hair is greyish at the roots and dark at the tips ; while on the neck and shoulders it tends to become lighter at the tips. A dark grey colour seems to characterise the tips of the hairs over the major portion of the body and the upper parts of the limbs ; but below the tips these hairs have a dark brown ring, beneath which they again become lighter. On the lower parts of the limbs and hands the hairs are darker at their tips, where they vary from brown to black ; but in some individuals these portions may be covered, like the trunk, with a mixture of grey and brown hairs.

The hair consists of an outer coat of long, stiff bristles, and of a shorter inner coat of fine, short, curly hairs, approximating to a woolly nature. The moderately long hair on the crown of the head is very stiff, and can be erected when the animal is enraged. Although the front and sides of the chin have but a short covering of hair, its upper portion has a distinct beard, or ruff. By far the longest hair on the upper part of the body is that growing on the shoulders, and hanging down thence on to the back and upper part of the arms. The length of this hair, however, is somewhat exceeded by that

growing on the thighs. On the chest and the rest of the under parts the hair is much shorter, that on the chest generally taking an upward and outward direction. The woolly under-hair is not very thick, and has no tendency to mat together. The long hair of the shoulders, back, and thighs communicates a shaggy appearance to the gorilla, although this is much less marked than in the orang-utan.

The female gorilla is much smaller than the male and, when standing upright, does not generally exceed about four and a half feet in height. The whole build is, moreover, relatively weaker, the tusks are but slightly developed, and the skull is proportionately smaller and more rounded, without the huge bony arches over the eyes. It appears, further, that in the adult female the bridge of the nose is shorter than in the male, while the cheeks are wider and the upper lip longer than is usually the case in the latter. The appearance of the female is therefore less ferocious and less repulsive than that of her lord and master.

The geographical range of the gorilla is much more restricted than is that of the chimpanzee, being limited to that district of Western Equatorial Africa lying between 2° N. and 5° S., and apparently not extending farther into the interior than 16° E. This hot and miasmatous region includes the mouths of the rivers Ogavai, Gabun, and Muni, and also the range of mountains running for about a hundred miles in a northerly direction between the Ogavai and the Cameroons, known as the Sierra do Cristal. According to Mr. H. A. Ford, gorillas are most common in the Sierra do Cristal, and have also been found a day's journey from the mouth of the Muni. During 1851 and 1852 numbers of gorillas, probably driven from the interior by want of food, were seen on the coast of the Gabun district, several of which were killed, the specimens sent by Captain Harris to London and by Mr. Ford to Philadelphia being probably some of these. Since 1852 they appear never to have been seen on the coast. According to the report of the German Loango Expedition, gorillas are very rare in the Loango district near the coast, but are met with in or near the mountainous region farther inland.

In regard to the actual mode of life of the gorilla there is a great dearth of authentic information. The old stories that these animals would seize with their foot natives passing beneath the trees on which they dwelt and drag them up, and likewise those to the effect that they gathered round the deserted camp fires of the natives, as well as the legends that they drove off the elephant with clubs, were disposed of once for all by Du Chaillu's account. The members of the German Loango Expedition frankly confess that they never saw a living wild gorilla, although they brought home a young one which had been captured by some native hunters ; neither did Winwood Reade himself ever come across these creatures in their native wilds. A later German traveller, Herr von Koppenfels, appears, however, to have been more fortunate, and states that he once observed a male and female with their two young quietly feeding.

From this account, and also from native report, it is known that gorillas habitually live in small families, having young ones of various ages with them; and that they frequent the most gloomy recesses of the forest, where the light of day is reduced to a twilight so dim that on cloudy days it might be supposed that the sun was eclipsed. The climate of these forests is hot and damp, suggestive of a Turkish bath or hothouse; and, as in most primeval forests, signs of animal life are extremely rare, although the stillness may be broken now and then by the voice of a bird. According to the account given by Herr von Koppenfels — although this does not appear to be supported by others— gorillas are in the habit of making a kind of nest in the trees by bending the boughs together and covering them with twigs and moss at a height of several yards above the ground. In this nest the female and young pass the night, while the male takes his station at the bottom of the tree, where he remains in a sitting posture during the night, ready to protect his family against the attacks of prowling leopards. The same writer likewise states that gorillas do not frequent the same sleeping-place for more than three or four nights consecutively; and this is but natural when we reflect that they must wander to considerable distances in search of fresh supplies of food.

Contrary to the custom of most wild animals, other than monkeys, gorillas roam the forest in search of food solely during the daytime, and are totally stationary during the night. As a rule, they walk on all-fours; and while in walking the fingers of the hand are usually doubled on to the palm, the whole sole of the foot is applied to the ground. They can, however, walk with the fingers extended, and likewise with the toes bent down on the sole of the foot.

Although to all appearance male gorillas are somewhat unwieldy creatures, yet, like all their kindred, they are active and indefatigable climbers, and are said to ascend to the very tops of the forest trees,

where they pass from tree to tree almost as readily as the far lighter spider-monkeys of Brazil. They also seem capable of taking leaps from great heights to the ground without damage to themselves, since Herr von Koppenfels asserts that he even saw an adult spring from a tree at a height of some thirty or forty feet, and on alighting rapidly vanish into the jungle.

Although when driven to close quarters the gorilla is doubtless one of the most terrible of foes, yet it appears certain that very exaggerated accounts have been given of its natural ferocity. Herr von Koppenfels, as quoted by Dr. Hartmann, mentions that so "long as the gorilla is unmolested he does not attack men; and, indeed, rather avoids the encounter." And when these creatures catch sight of men they generally rush off precipitately in the opposite direction through the underwood, giving vent at the same time to peculiar guttural cries.

Many gorillas are said to be killed by the natives with the aid of a weighted spear suspended in the creature's path by a cunningly devised system of cords. Others are, however, undoubtedly shot by the negroes, although it would seem that, at least in many instances, such animals have been accidentally met by the hunters as they travelled through the forest rather than deliberately sought out and tracked. As already mentioned, the members of the German Loango Expedition and Winwood Reade expressed their belief that up to the dates of their explorations of the West Coast no European had ever seen a gorilla, but according to letters from Herr von Koppenfels, referred to by Dr. Hartmann, that traveller claimed to have killed several. In April, 1900, a gigantic gorilla, measuring 6 ft. 9 in. from the crown of the head to the tip of the middle toe, was shot near Tonsu, in the hinterland of the Cameroons, by Herr H. Paschen. This ape was marked down by a party of natives in thick bush. When approached by Herr Paschen and his party it rushed up a tall tree with marvellous speed, where it was shot by the German. Its skin and skull are preserved in the museum at Tring.

A FINE EXAMPLE OF THE FULL-GROWN MALE GORILLA
From a stuffed specimen in the Zoological Museum, Tring, Herts.

As we have already stated (page 186), the members of the German Loango Expedition received in 1875 a young male gorilla captured by native hunters. From the account of this animal given in the report of the expedition by Herr Falkenstein, it appears that, when first received at the station of Chinxoxo, in Loango, the hardships which it had undergone in its transit down country had reduced the creature to a deplorable condition. By the aid, however, of a plentiful supply of wild and other fruits, and the milk of a goat, the young animal was restored to something approaching its normal state of health; and preparations were then made for its transport to Berlin. Having been thus gradually accustomed to eat fruits and other food which could be procured on board ship, as well as to the company of Europeans, this young gorilla was finally shipped for Germany. During the voyage it was never chained up, and was soon allowed to wander freely about the ship with very slight supervision. This animal seems to have been of a gentle disposition and, although self-willed, was never malicious. In taking its food it was remarkably well-behaved, helping itself from a plate with its thumb and two fingers, and even carrying small vessels of water to its mouth, and replacing them undamaged when empty. When larger vessels of liquid were put before it, it would lower its mouth to them and drink by suction. Its regard for personal cleanliness was also noteworthy; and, when foreign substances adhered to its skin, it either brushed them off or held out its arms in a manner clearly indicating that it wished to have them removed.

When not able to obtain any article it desired, or when otherwise thwarted in its wishes, the youngster had recourse to various clever devices by which its object might be attained. For instance, "when he felt a desire for the sugar or fruit, which was kept in a cupboard in the eating-room, he would suddenly leave off playing and go in an opposite direction to the room, only altering his course when he believed that he was no longer observed. He then went straight to the room and cupboard, opened it, and made a quick and dexterous snatch at the sugar-box or fruit-basket, sometimes closing the cupboard door behind him before beginning to enjoy his plunder, or, if discovered, he would escape with it; and his whole behaviour made it clear that he was conscious of transgressing into forbidden paths. He took a special and what might be called a childish pleasure in making a

noise by beating on hollow articles, and seldom missed an opportunity of drumming on casks, dishes, or tin trays, whenever he passed by them." Strange noises, more especially thunder, alarmed him much.

This gorilla arrived safely at Berlin, where it was for a considerable period an inmate of the aquarium. There it throve at first, and was docile, though inclined to be mischievous. Eventually, however, it succumbed to the malady which sooner or later carries off so many of the large man-like apes in a northern climate, dying of a rapid consumption in the autumn of 1877, after having lived for fifteen months in Berlin.

By the intervention of Pechuel-Loesche and Falkenstein, a second living gorilla was obtained from the Loango district, and safely transported to Berlin, where it arrived in the early part of 1883. The journey during the winter appears, however, to have left its mark on the constitution of this animal, and, after living for fourteen months in the Aquarium, it died of the same disease as its predecessor, in the spring of 1884. Dr. Hartmann states that there was a third live gorilla at Berlin in the autumn of 1881, which died soon after its arrival. There have also been of late years a few young gorillas in the London Zoological Gardens, but none survived long.

Curiously enough, so long ago as the year 1860, a travelling showman in England actually had a veritable living gorilla in his exhibition, which he considered to be a chimpanzee, no one suspecting till after the creature's death the nature of the treasure he had possessed. R. LYDEKKER

BABY GORILLA AT THE "ZOO"
Photograph by W. P. Dando

STORIES OF THE GORILLA

As we have seen, there has been not a little to unlearn concerning the gorilla. Travellers' tales invested the animal with terrors and wonders as fantastic as were attributed of old time to the basilisk. The supposed ferocity and the awe-inspiring hideousness of the creature caused the natives to worship it. The gorilla is worshipped to-day as it was worshipped when Savage secured the first gorilla skull. That skull had been mounted on a pole and was the local idol, as deeply reverenced as the bottles from which had been drawn the medicine with which a missionary had cured the natives whom he afterwards found prostrated before them. Du Chaillu was assured that the spirits of men inhabited gorillas. Millions of natives in India to-day believe the same thing of tigers; and millions of men in past ages have believed that their fellows were converted into wolves and other animals.

Probably not a native of the parts in which the gorilla has its home doubts to-day that gorillas make prisoners of men, and carry women away captive. The vitality of such legends is amazing.

But the credulity of the natives is not greater than the credulity of their betters. A European official reported that certain Tibetan apes had been discovered which were skilled potters and makers of wine. A picturesque account from a French explorer told of caravans held up by Titanic gorillas in Algeria, where one beast was said to have been captured measuring 7 ft. 6 in. in height, with a width of 4 ft. across the shoulders, and a weight of 720 lb., which is equal to the weight of five men of 10 st. each.

Nor is the store of travellers' tales exhausted. It is the character of the steaming climate in which it has its home that makes it so difficult to gain intimate knowledge of the habits of the gorilla in its wild state. For all the new facts that come to light, Africa might be as remote from us as before the days of steam. In one respect it is as well that this should be so, for the big-game hunter would speedily cause gorillas to be classed with many other extinct animals. The remoteness of this great ape from human observation is less surprising when we recollect that not until a few years ago had we ever heard of the okapi; that not until 1909 had a Tibetan takin been seen in Europe; that it took half a century to convince naturalists that the Australian duckbill platypus lays eggs.

Cleared of its reputation for gratuitous ferocity, the gorilla stands acquitted also of the charge of exceptional dislike of children. Herr Falkenstein's details of the life of a young gorilla in captivity suffice to show how quickly the young animal takes kindly to human beings. The following effective picture appeared in "The Times," describing an afternoon with a young gorilla in captivity: "I found the creature romping and rolling in full liberty about the private drawing-room, now looking out of the window with all becoming gravity and sedateness as though thoroughly interested in, but not disconcerted by, the busy multitude and novelty without; then bounding rapidly along on knuckles and feet to examine and poke fun at some new-comer, and playfully mumbling at his calves, pulling his beard—a special delight—clinging to his arms, examining his hat, not at all to its improvement, or curiously inquisitive as to his umbrella, and so on with visitor after visitor. If he become overexcited by the fun, a gentle box on the ear will bring him to order like a child—like a child, only to

BABY GORILLA AT PLAY
Photograph by W. P. Dando

be on the romp again immediately. He points with his index finger, claps his hands, puts out his tongue, feeds on a mixed diet, decidedly prefers raw, roast meat to boiled, eats strawberries with delicate appreciativeness, is exquisitely clean and mannerly."

Another writer describes the gorilla at play with a little boy and dog, his favourite companions, turning head over heels, chasing his friends, shaking hands, swinging on a trapeze side by side with his boy friend, giving him claret and water in a wineglass, and behaving with grave courtesy and childlike good nature and docility. This writer found him " as like a little negro boy in the face as a being not absolutely human can be."

Frank Buckland, however, who was worrying at the time over a misconstruction of the Darwinian theory, would not have the latter description. Seeing the gorilla at play with two children, he remarked: "When the two humans and the gorilla were sitting at play on the ground floor, I could not help seeing the amazing difference between the countenances of the gorilla and of the children; the one decidedly and purely monkey, the others decidedly human. I could not, in fact, help seeing what a vast line the Creator had drawn between a man and a monkey." Buckland bore testimony, though, to the gorilla's absence of ferocity while at play with the children. Finding the ape respectful, grave, and somewhat distant towards adults, he watched with the greater curiosity when the two young people entered to see the ape. "After a while they both began to play with 'Pongo' [the gorilla]. Gradually they fraternised, and began to play together after the manner of little children. Not being a child, I cannot enter into their funny sayings and doings about nothing at all. So these three, the little boy and girl and the gorilla, played together for half an hour after their own childish fashion, and I made the children experiment on him with ornaments, handkerchiefs, etc." So much for the legend that the gorilla, young or old, is specially fierce towards children.

The authorities of the London Zoological Gardens have never had an opportunity of experimenting to any extent with their gorillas. The first arrived at Regent's Park in 1887, and created a sensation in town. It was about three years old, and stood 2 ft. 6 in. in height. It arrived in miserable condition, and, refusing to leave its travelling cage, had to be forced into its permanent home. A young macaque monkey was assigned to it for company, but the ape was too indisposed to return its friendship. Skilful attention and a diet of bread, milk, and fruit kept it alive for two months, but then it died. In the meantime it had made friends with its keeper, and showed signs of considerable intelligence.

The second venture was made in March, 1896, when the largest gorilla ever imported alive into England was obtained. It was just acquiring its permanent teeth, but though for a time it thrived, it never showed much activity, and died within five months. Two young female gorillas were secured in 1904, but did not survive a month, dysentery being responsible for the death of both. Still persevering, the Society secured in the following year a female gorilla of tender age, which was known for a brief spell of fame as "Miss Crowther." She had to be forced from the case in which she had travelled, into the cage adjoining the chimpanzees. Her arrival caused the wildest excitement amongst the latter, who shrieked and shook their cage and danced with rare vivacity. The stranger gave much trouble to her keepers, seeming determined to starve, and while life remained in her she was transported to New York.

A sixth gorilla arrived in March, 1906. It was a young female, and was so weak and fragile that she was never exhibited in public, being kept in a cage in a secluded part of the Gardens, whence she would be carried in the arms of a solicitous keeper for daily airings in the sun. From the first her case was hopeless. Doctors skilled in infantile complaints attended her daily; her diet and medicine were the subject of anxious consultations among the learned of the medical profession. As a leading official of the Gardens remarked, the young gorilla could not have been more anxiously watched had she been a millionaire's daughter. Pneumonia caused her death, and, after all their experiences, the authorities reached the conclusion that it is almost an impossible task to attempt to rear a gorilla in their Gardens. It is now agreed that unless gorillas can be caught young and kept in captivity near their home for some time before being brought away, thus giving them a chance of becoming accustomed to hand-feeding and to confinement, we must take our gorillas on trust, and meet them only in works of natural history or in travellers' tales.

Seeing how protracted and angry was the discussion of Du Chaillu's gorilla data, it may be of interest to quote here an incident recorded by A. D. Bartlett, at that time superintendent of the Zoological Gardens. "At Du Chaillu's request," he writes, "I went to Mr. Murray's to see the skin [of the gorilla] unpacked, and conveyed it to my office for inspection. I called M. Du Chaillu's attention to the face of the animal which, I told him, was

HEAD OF A YOUNG ORANG-UTAN

not in a perfect condition, having lost part of the epidermis. In reply, Du Chaillu assured me that it was quite perfect, remarking at the same time that the epidermis on the face was quite black, and that the face of the skin being black was a proof of its perfectness.

"I, however, then and there convinced him that the blackness of the face was due to its having been painted black. Finding that I had detected what had been done, he at once admitted that he did paint it at the time he exhibited it in New York.

"The question that arose in my mind upon making this discovery was, did M. Du Chaillu kill the gorilla, and skin and preserve it? If so, he must recollect that the epidermis [outer skin] came off. Supposing he did forget this, he must have been afterwards reminded of the fact when he had to paint the face to represent its natural condition. These facts, to which I had a witness, led me to doubt the truthfulness of M. Du Chaillu's statement, and it occurred to me that he was not aware of the condition of the skin, and probably had not prepared it himself." The point is of no importance now, but it shows that Du Chaillu was himself to blame for much of the incredulity with which his statements were for a time received. E. A. B.

THE ORANG-UTAN

Partly from the reddish hue of its hair, and partly from the conformation of its face and skull, as well as the much greater proportionate length of its arms, the orang-utan (page 158) of Borneo and Sumatra (*Simia satyrus*) is a very different-looking creature from the chimpanzee and the gorilla.

The name orang-utan is a Malay word, signifying "man-of-the-woods"; and the ape so designated was known to Linnæus at least as far back as 1766. It was not, however, till a considerably later date that it became fully known in Europe. It is true, indeed, that in 1780 Baron Wurmb, then the governor of the Dutch settlement of Batavia, transmitted to Holland the entire skeleton of an orang-utan; but he did not recognise it as such, calling the animal to which it belonged the pongo—a name which belongs to the gorilla (page 184). In the year 1804, however, an orang-utan was living in the menagerie that belonged to the Prince of Orange; and this example was in that year described and depicted by a naturalist named Vosmaer. Subsequently to this the identity of Wurmb's pongo with the orang-utan was fully demonstrated; and from that period our knowledge of the structure and habits of this ape has gradually increased. Among those who have especially contributed to advance our knowledge of the orang-utan in its living condition we may mention Raja Sir James Brooke, of Sarawak, and Dr. Alfred Russel Wallace, the latter of whom has given an admirable account of the creature's habits, in his work, the "Malay Archipelago."

In the uncongenial climate of Europe, orang-utans are as difficult to keep for any lengthened period in confinement as are the large man-like apes of Western Africa. The case is, however, very different in the moist subtropical climate of Calcutta, where adults have thrived well in cages exposed to the open air, and have taught us many facts in relation to their habits. I once received from a friend in Singapore a photograph of an orang-utan remarkable for its coat of long hair; and the receipt of this set me wondering why a tropical animal like an orang-utan required such a profusion of hair. The only analogous instance that occurred

to me was the Indian sloth-bear, but the length and thickness of the hair were much less than in this orang-utan. Two suggestions arose with regard to the reason of the extraordinary capillary development in the latter species. The first was that it might serve as a protection against the deluges of rain in the wet season in the native home of the orang-utan, since these animals sleep in trees, where they must necessarily be much exposed, despite the protection afforded by the leaves and by the sleeping-place they construct for their abode. The second suggestion was that it must afford a protection against mosquitoes, which would presumably find some difficulty in reaching the skin through such a mass of shaggy hair.

The leading or, as naturalists say, generic characters distinguishing the orang-utan from the chimpanzee and gorilla are to be found in the proportionately greater length of the arms, which in the upright position reach to the ankles; and in the form of the skull, which is elevated almost into a point at the summit, as well as in a difference in the number of the joints in the backbone and of bones in the wrist. Thus there are sixteen (instead of seventeen) vertebræ in the backbone between the neck and the sacrum, twelve of these carrying ribs, as in man. In regard to the number of bones in the wrist, we find that the orang-utan possesses the central bone which is wanting in man, the chimpanzee, and the gorilla, and thus has nine, in place of eight, bones in the wrist. In this respect the Bornean ape agrees with the lower members of its order; but in the absence of callosities on the buttocks it shows its kinship with the gorilla and chimpanzee.

All these characteristic features clearly indicate that the orang-utan is decidedly lower in the scale than are the other two man-like apes.

An adult male orang-utan stands about 4 ft. 4 in. in height when in an upright position, in which posture it can almost touch the ground with its fingers. The legs are extremely short and thick, and twisted in such a remarkable manner that the knees are turned outwards, and the feet consequently set very obliquely to the line of the leg. From the peculiar structure of their legs and feet these apes walk entirely on the outer sides of their feet, of which the soles are turned inwards, so as almost to face one another. Although this arrangement is ill-adapted for walking rapidly on the ground, it is one admirably suited for climbing, in which these animals excel.

As shown in the illustration of the adult, the orang-utan has a tall,

ORANG-UTANS IN THEIR NATIVE FOREST

elevated forehead, very different from the retreating one of the chimpanzee ; and the whole aspect of the face is curiously flattened, and often laterally expanded in old males, although there appears to be considerable individual or racial difference in this respect, some old males having comparatively narrow and rounded heads like the young. Not unfrequently there is a well-marked prominence in the middle of the forehead. Although there are slight ridges over the eyes, these are much less developed than in the chimpanzee, and have, therefore, no sort of resemblance to the enormous ones of the gorilla. The extraordinary height of the crown is well exhibited in the figure (see page 190) of the head of a young specimen, the whole of this part of the head being curiously shortened and compressed from back to front. In the immature animal the jaws are not very prominent, but they become much more projecting in old males. The bridge of the nose is much depressed and flattened, but the whole nose is generally larger than in the chimpanzee and gorilla, and not so much expanded at its termination, the wings being arched and narrow, and the small oval nostrils separated from one another by a narrow partition. The mobile lips are comparatively smooth and thin, the upper one being characterised by its great length and breadth. In the adult of the orang-utan, as shown in the illustration on page 191, the neck is surrounded by a kind of collar formed of folds of skin containing an internal cavity communicating with the larynx, or upper expansion of the windpipe. In some very old males these pouches attain enormous dimensions, and by no means add to the beauty of their owner. The ear is small and well formed, being much more human-like than that of the gorilla. Frequently the sides of the cheeks of the males have a warty protuberance, or callosity.

The body is not nearly so powerfully built as that of the gorilla is ; and the sloping and stooping shoulders and extremely prominent abdomen make the whole shape of the animal ungainly in the extreme.

We have alluded to the great length of the very powerful arms, which vastly augment the animal's climbing powers. The hand (shown in figures 9 and 10 of the illustration on page 174) is even longer and more slender than that of the chimpanzee, and characterised by the remarkable shortness of the thumb, which scarcely reaches so far as the root of the first joint of the index finger. The fingers themselves are connected by a web, which extends for a third or nearly half of the length of their first joints. With regard to the relative lengths of the fingers, there is some amount of individual variation ; but the middle finger may exceed either of the others, while the ring finger is longer than the index, and the "little" finger relatively long. All the fingers are narrow and tapering, with well-formed arching nails.

The calves of the legs are less developed than in either the chimpanzee or gorilla, and the narrow,

flat heels are less projecting. The long and slender feet (shown in figure 10 of the illustration on page 174) are likewise of a lower type of structure, as is particularly seen in the very small size of the great toe, which is peculiar among the Primates in frequently having no trace of a nail in the adult. Curiously enough, old animals often lose the last joint of the big toe, apparently not through disease, but as a normal condition. Both the hands and feet on the backs, and the hands on their under surfaces, have wart-like callosities.

The general colour of the orang-utan's skin is bluish grey, although it may have a more or less decided tinge of brown. In marked contrast to the general slaty hue of the face, there often occur yellowish-brown rings round the eyes, nostrils, and upper lip. The full reddish-brown hair is long, shaggy, and bristly, with a small admixture of woolly under-hair. The hair of the head may either have a natural parting in the middle, as in the figure of the head (page 190), or may be tossed in wild confusion, in some individuals standing almost upright. Usually there is a well-developed beard on the cheeks and neck. On the whole of the under surface of the body the covering of hair is thin and scanty, and it is even less developed on the face and ears, and the backs of the hands and feet.

The canines, or tusks, of the male are of enormous size. In the female, however, they are much smaller ; and this sex is also characterised by the smaller development of the folds and pouches of skin around the neck.

As we have said, orang-utans appear to be confined to the islands of Borneo and Sumatra ; and there has been considerable discussion as to whether there is more than one species. It was once thought that the large orang-utan of Sumatra was specifically distinct from that of Borneo, and it accordingly received a separate scientific name. But later investigations indicate that this is not the case, and that the animal is common to both islands, although individuals vary considerably in their colour. Probably the Bornean orang-utan is racially distinct from the one inhabiting Sumatra, and there may be local races in both islands. The Dyaks of Borneo, by whom the orang-utan is generally designated the mias, distinguish two forms, calling the one provided with cheek-excrescences mias pappan, and the one without these appendages mias rambi. In addition to these two varieties of the true large orang-utan, the Dyaks recognise a third kind, which they distinguish as the mias kassir. These animals are much smaller than the typical orang-utan, and never have the excrescences on the cheeks. A young individual of this orang-utan was described by Sir Richard Owen as *Simia morio*.

Orang-utans are stated to be much more numerous in Borneo than in Sumatra ; and, since dense, low-lying forests are essential to their existence, they are not found in the neighbourhood of Sarawak, where the ground is hilly. The unbroken, large areas of primeval forests, occurring in many parts of Borneo, are the true home of the orang-utan ;

such forests, according to Dr. Wallace, being like open ground to these apes, since they can travel in every direction from tree to tree, as easily as the North American Indian traverses his native prairie. In all their movements these apes are slow and deliberate, this being especially noticeable in the perfectly healthy adults which have been exhibited in the Zoological Gardens at Calcutta, where they enjoyed a climate not unlike their own. This deliberation in their movements is noticeable in Wallace's description of the manner in which these apes travel through the forest when undisturbed and at ease. We are told that they proceed with great circumspection along the larger branches of the trees, in the half-upright position rendered necessary by the great length of their arms and the shortness of their legs. Almost invariably they select such trees as have their branches interlaced with the adjacent ones ; and, when such boughs are within reach, they catch hold of them with their arms as if to try their strength, after which they deliberately venture upon them. Although the orang-utan never leaps or jumps, and never seems to be in a hurry, yet he will make his way overhead in the forest as fast as a man can run on the ground below. In this progression the long, powerful arms are of the utmost service ; and it is by their aid that these apes pluck the choicest fruit from boughs too light to support their weight, and likewise gather the leaves and young shoots to form their sleeping-places.

Orang-utans, like gorillas, go in small family parties, consisting of the parents, accompanied frequently by from two to four young ones. Although they will devour leaves, buds, and young shoots— more especially those of the bamboo—their chief food consists of fruits of various kinds, the prime favourite being the luscious but ill-smelling durian, or jack-fruit. Of this fruit they waste a vast quantity, throwing the rejected rinds on the ground.

The nest, or sleeping-place, of the orang-utan is generally constructed in a comparatively small tree

ORANG-UTAN IN THE HAGENBECK COLLECTION

at a height of from 20 ft. to 50 ft. from the ground, at this elevation being protected from wind by the taller surrounding trees. The Dyaks believe that orang-utans construct a fresh nest every night ; but, if this were the case, deserted nests would be much more common than they are. These apes remain in their nests till the sun has risen sufficiently high to have dried the dew from the forest leaves. Their feeding-time is during the middle of the day ; but they seldom return for more than two consecutive days to the same tree for this purpose. It has been observed that the orang-utan must have a task of considerable difficulty in getting at the interior of the durian, since the fruit is protected by a thick and tough skin, covered with strong conical prickles. Probably, however, the ape first bites off a few of these prickles or spines, and then makes a small hole, into which it inserts its fingers, and thus manages to pull the fruit in pieces.

An instance of the ferocious nature of the orang-utan when angered or driven to bay is given by Wallace in the following words : " A few miles down the river there is a Dyak house, and the inhabitants saw a large orang feeding on the young shoots of a palm by the riverside. On being alarmed he retreated towards the jungle, which was close by, and a number of men, armed with spears and choppers, ran out to intercept him. The man who was in front tried to run his spear through the animal's body, but the mias seized it in his hands, and in an instant got hold of the man's arm, which he seized in his mouth, making his teeth meet in the flesh above the elbow, which he tore and lacerated in a dreadful manner. Had not the others been close behind, the man would have been more seriously injured, if not killed, as he was quite powerless ; but they soon destroyed the creature with their spears and choppers. The man remained ill for some time, and never fully recovered the use of his arm."

The same writer relates the history of a young orang-utan which he received in Borneo when it was only a foot high. When first carried home, this tiny creature took such a firm grasp of its owner's beard that it was with difficulty it could be made to loose its hold. At the time of its capture there were no signs of teeth in its mouth, but in the course of a few days two of the lower incisors were cut. Unfortunately, there was at the time no means of obtaining a supply of milk for the little ape; but this difficulty was overcome by feeding it with rice-water sucked from a bottle with a quill through the cork. The animal soon managed to suck comfortably from this contrivance; and when sugar and cocoanut milk were added to the mixture it thrived well enough on the diet. If its owner introduced his finger into the ape's mouth, it first sucked vigorously, but soon found out its mistake, and pushed the finger away, with angry screams like those of a disappointed child. When caressed, it was contented and happy, but when laid down, soon began to scream; and for the first two nights of its captivity was very noisy and restless. It was kept in a kind of cradle, made of a box, with a soft mat at the bottom. The little orang-utan seemed to appreciate a frequent bath; and, indeed, when it required one, announced the fact by loud screams. The process of drying and rubbing after each bath seems to have been a source of great enjoyment; and this was likewise the case when

YOUNG ORANG-UTAN CLIMBING
Photograph by W. S. Berridge

its hair was combed and brushed. At first it clutched vigorously by all four limbs at any object in its neighbourhood, so that its owner had continually to be on his guard to save his beard. When it could find nothing better to do, it would nurse its own foot. Little by little the strength of the tiny creature's grip decreased, probably owing to the want of sufficient exercise. In order to remedy this, a short ladder was constructed, from which the ape was suspended by his hands and feet for a quarter of an hour at a time. This exercise seemed at first to afford it pleasure, but afterwards it loosened its hold, first with one limb, and then with another, till it finally fell to the ground. These tumbles did not appear, however, to do it much harm.

Its owner endeavoured to construct a kind of artificial mother out of buffalo-hide, which the baby orang-utan might fondle. For a time this afforded satisfaction, but eventually was discarded, as the animal was nearly choked with the hair it had torn off the skin, and swallowed. After a week's captivity it was fed from a spoon, containing a mixture of soaked biscuit, egg, and sugar, or, at other times, sweet potatoes. This food was swallowed readily, and with apparent satisfaction, the creature making droll grimaces to express either pleasure or the reverse. When it had swallowed anything which appeared grateful, it drew in its cheeks, and screwed

up its eyes; while, when the food was distasteful from want of sufficient sugar or other cause, the ape, after turning it about in its mouth for a short time, finally ejected it. If this rejected food were again offered, it displayed marked displeasure by loudly screaming and throwing its arms about.

After three weeks a young macaque monkey was introduced to the orang-utan, and the two, although very different in demeanour, soon became fast friends. The helplessness of the young orang-utan, when compared with the macaque, was especially noticeable; and this character distinguishes the young of all the man-like apes from those of the lower monkeys. Even after the young orang-utan had been about a month in captivity, it was very unsteady when placed on its hands and feet, and would frequently overbalance and topple over. When it required attention it would cry loudly for a time, but if this met with no reply the young creature would remain quiet till a step was heard approaching, when its calls would be at once renewed. Although at the end of four weeks the two upper incisors had been cut, the little creature, doubtless owing to improper food, had not increased perceptibly in weight, and soon afterwards sickened and died of a kind of intermittent fever.

Young orang-utans love play quite as much as children do, and appear to have several games of which they are specially fond. Ill-health, however, is apt to make them dull in European menageries.

R. LYDEKKER

STORIES OF THE ORANG-UTAN

Attention has been directed to the difficulty experienced in keeping the orang-utan in captivity. Possibly an extension of the principle of greater liberty for these animals may produce better results. That, at any rate, is the opinion of Carl Hagenbeck, the chief importer of wild beasts in Europe. His unique experience leads him to say: "It is not so much warmth that animals and birds require as fresh air. I have photographs showing my zebras and flamingoes, my lions and antelopes, standing out in deep snow, preferring it to the stuffy atmosphere of an enclosed den." The apes are notoriously less hardy of constitution than any of the animals named by Mr. Hagenbeck; but when all is said, we have the fact that there flourished in the London Zoological Gardens at the end of 1909 an orang which had been there fully five years; while the illustrious "Jenny" survived four years of the trying climate of the English metropolis.

The power of endurance in the orang-utan when

wild seems to be exceptional, in so far as resistance to wounds is concerned. Dr. Wallace, in collecting specimens for museums, had to inflict severe injuries by means of gunshots before he could secure his prey. Some that had received several shots yet managed to escape. Even the young orang-utan which he succeeded in keeping alive for nearly three months survived a serious accident. When, after the death of the creature, he examined its skeleton, he found that it had suffered from a broken leg and a broken arm. These injuries were the result of a fall from a tree at the time of its capture. Yet the broken parts had united so rapidly that he had been unaware of the existence of the injury, and the orang-utan had regained comparative health upon an obviously inadequate diet.

The affectionate nature of this orang-utan is typical of the species. Fierce and vindictive if attacked, and fully conscious of the formidable power of its teeth, this ape in captivity is invariably gentle and docile, unless deliberately provoked to anger. Then it will unhesitatingly use its teeth. There are many instances of its curiously "human" characteristics. Mr. Frank Finn, who had considerable experience of the orang-utan in captivity at Calcutta, tells us that he found these apes less merry and less monkeyish than the chimpanzee, and more emotional and more exacting. "Disappointed, it will throw itself on the ground and roll about, crying like a spoilt child," as did Wallace's infant pet. Testimony to the orang-utan's preference for human company to that of the society of its kin is borne by an interesting experience which Mr. Finn had in the yard of a dealer. A cage in which a young orang-utan was confined was opened, giving the inmate freedom to choose its company. Disregarding all the monkeys in the yard, the ape sidled over to the dealer and quietly took its place in his lap. At the Calcutta "Zoo" was a female orang-utan which used to drop her food at the approach of Mr. Finn and affectionately put her arm round his neck. Another orang-utan of the same sex took a rooted dislike to a regular visitor because the latter had on one occasion given food to monkeys in the same house before visiting her.

"But," says Mr. Finn, in his interesting "Ornithological and Other Oddities," "it was, I think, on my very first introduction to the orang-utan that the hidden humanity of the creature first impressed me." This was when Mr. Abraham Bartlett, at the London Zoological Gardens, gave him a private interview with an orang-utan which had just arrived. "The first thing the little imp did was to climb on my knee, take off my hat, and put it on its own head,

TWO BABY ORANG-UTANS AT THE LONDON "ZOO"
Photograph by W. S. Berridge

after which it proceeded gravely to pinch one of the superintendent's eyelids. In short, it examined us with a scientific curiosity which, in a lower animal, was decidedly impressive. This little man of the woods could not have chosen a more striking way of claiming a kinship so often denied."

All who have seen the orang-utan in its native wilds agree that it is the most sluggish of the three great apes. Sir James Brooke, of Sarawak, had a poor opinion of the orang-utan's attempt to avoid human pursuers. "I never observed the slightest attempt at escape," he says, "and the wood which sometimes rattled about our ears was broken by their weight, and not thrown, as some people assert."

Against this, however, we have Dr. Wallace's evidence that on three occasions he saw female orang-utans, accompanied by their young, "breaking off branches and the great spiny fruit of the durian tree with every appearance of rage, causing such a shower of missiles as effectually kept us from approaching the tree." Sir James Brooke agrees with the common opinion as to the ferocity of the orang-utan when brought to bay: "One unfortunate man who, with a party, was trying to catch one alive lost two of his fingers, besides being severely bitten on the face, while the animal finally beat off his pursuers and escaped."

According to the same authority, when hunters wish to catch an adult they cut down a circle of trees round the one on which the orang-utan is seated, and then fell that also, and close in before he can recover himself, and endeavour to bind him. Sir James's experience goes some way towards justifying the stories which Wallace heard of many nests or arboreal huts made by the orang-utans. The former was surprised at the facility with which they formed these structures. "I had an opportunity of seeing a wounded female weave the branches together, and seat herself in a minute." There she received the fire of her pursuers without moving, and expired in her lofty abode. An adult male which Brooke killed reclined lazily on a tree, and, when approached, only took the trouble to interpose the trunks between himself and the hunters, peeping out and watching, and dodging as they dodged. Sad to say, he left an arm exposed and, being wounded in this limb, was incapacitated from flight, and was killed.

Happily, we are able now to study the orang-utan at close quarters without first seeing it suffer, for, despite the difficulties of catching and keeping it alive, specimens thrive in the Zoological Gardens, not only of London, but of Antwerp, Cologne, Frankfort-on-Main, Basle. and elsewhere. The specimen at Basle had its liberty, and was wont to pass the hottest hours of the summer days in the pleasant shelter of a tall tree, without abusing the confidence of its keeper. The sluggish disposition of the ape may have contributed to the sense of safety with which its custodian liberated it, but it must not be inferred that the orang-utan has not intelligence enough to make its escape. Darwin saw a young one put a stick into a crevice, slip its hand to the other end, and use the stick in the proper manner as a lever. He also records an orang-utan's manner of expressing fear of punishment. The little creature, expecting to be whipped for some fault, covered and protected itself with blanket or straw.

A newcomer to the London "Zoo" disarmed reproof by an even more diplomatic proceeding. Placed in the cage from which one of the hapless gorillas had been removed, it at once began to tear away some of the wire netting. A keeper had to go in to put an end to the destruction. The orang-utan, expecting to be punished for its misdoing, flung its arms around the man's neck, and hugged him with such a show of affection that the stoniest of hearts would have yielded.

Another tame orang-utan, when unable to reach any article it desired, used to drag a chair from one end of the room to the other, and then stand upon it. By the same means it was enabled to reach a latch which the ape's natural height would not permit it to unfasten.

Sluggish the orang-utan may be, but not so sluggish as not to be alarmed at unfamiliar forms of life. Thus one showed the greatest fear on being confronted for the first time with a live turtle. Nor is the animal too sluggish to betray definite evidence of a sense of the ludicrous. Monkeys dislike ridicule ; they hate to be laughed at ; but the orang-utan which Darwin studied afforded instances to the contrary, of which he mentioned one : " Her feeding-tin was of a somewhat peculiar shape, and when it was empty she used sometimes to invert it upon her head. The tin then presented a comical resemblance to a bonnet, and as its wearer generally favoured the spectators with a broad grin at the time of putting it on, she never failed to raise a laugh from them. Her success in this respect was evidently attended with no small gratification on her part."

One orang-utan at the Zoological Gardens in London became known to fame under the name of " Jenny." She was, indeed, one of the most distinguished of favourites, and numbered a multitude of celebrities among her visitors. " Jenny " would take her cup of tea with the nicest skill, and, invested with a smart cap by Hunt, her faithful keeper, spoon or sip the beverage "in the most lady-

like way." Queen Victoria and other members of the Royal House paid a special visit to the Zoological Gardens to see " Jenny," who, however, was denied the use of her bonnet that day, the keeper fearing that her illustrious guests might deem her vulgar.

" Jenny " had a warm affection for her keeper, and was seen at her best when her house was not open to the public. Such occasion was a certain Christmas when Sir Richard Owen and his wife paid her a visit. Lady Owen thus describes the incident : " ' Jenny ' certainly attempts speech as far as her powers admit. When she is fond of a person she puts her long, strong arms round his neck, and makes a curious noise, like an attempt to utter caressing words, opening the lips and moving them as though trying to make certain sounds. She produces a sort of murmur, which one might easily translate into kind expressions. To-day she took a fancy, when out of her cage, to look out of the window, and slyly crept along till she got there, under pretence of friendship. Hunt pretended to be offended at her not coming when he called, and she ran up to him, put her arms round his neck, whispering to him and kissing him, till he seemed to forgive her."
ERNEST A. BRYANT

AN ORANG-UTAN IN A TREE
Photograph by W. P. Dando

THE GIBBONS AND FOSSIL APES

With the gibbons we come to the last of the man-like apes, distinguished from those which we have already described not only by their smaller size, lighter build, and longer arms, but also by the presence of small naked callosities on the buttocks, resembling those of the lower monkeys. They are, moreover, the only apes accustomed to walk in an upright position, in which, as shown in the illustration on page 197, they are at times assisted by their long arms, although they can walk perfectly well with their hands clasped behind the neck. On account of these differences, naturalists regard the gibbons as representing a family by themselves—the *Hylobatidæ*.

The gibbons, or long-armed apes, comprise several species found in the warmer regions of South-eastern Asia, and more especially in and around the Malay Peninsula. The largest of all the species only slightly exceeds three feet in height, while the others are not more than about thirty inches. Their arms are so long that they reach to the ankle, so that these apes can actually walk upright and at the same time touch the ground with their fingers. The head is well-shaped, without the upward prolongation of the crown that is so characteristic of the orang-utan ; and the lower jaw is remarkable for the great development of the chin, which is more human-like than that of any other ape. Moreover, from the absence of prominent ridges and crests, and the nearly upright forehead, the whole skull approaches the human type far more nearly than do the skulls of the other apes. This must not, however, be considered as an indication that the gibbons are of a higher type than the other man-like apes, since the contrary is clearly demonstrated by their long arms and the callosities on the buttocks. The resemblance of their skulls to the human type is, indeed, merely a superficial one, due to the circumstance that small animals necessarily have proportionately larger brains than the bigger members of the same group ; and also to the absence of the strong ridges which are required for the powerful skulls of the larger forms, but would be quite useless in their smaller cousins. The superficial human-like characters of the skulls of the gibbons are, however, to a great extent masked by their long, slender tusks, or canines, which project far beyond the level of the other teeth. The long and narrow hands and feet of these animals (shown in figs. 11–13 of the illustration on page 174) are characterised by the extent to which the thumb and great toe are separated from the other fingers and toes, as well as by the flatness of all the nails. In colour, the gibbons vary from black to yellowish white, this variation occurring even in different individuals of the same species. The comparatively well-formed nose, as seen in the figure of the siamang (page 198), imparts to their physiognomy an expression far less repulsive and less forbidding than that which characterises the larger man-like apes.

Several of the gibbons are more or less black, among these being the Hainan species (*Hylobates hainanus*), which is often wholly so. Some years ago, however, a female example of this gibbon was deposited in the London Zoological Society's Gardens, obtained when believed to be about six months old. The ape was then dark smoky grey, which soon turned to black, and she remained black till after her arrival in London, when she was fully adult. A greyish tinge in the fur attracted attention in February, 1894, and by the following midsummer the lighter tint had so gained on the darker that no one would have recognised this ape as the one received in the previous January. In autumn she was silvery grey all over, except for a black band, fading on the sides and behind, down the middle line of the head ; but later became brownish or silvery grey, varying in tint according to the light, the black band being retained, but as a narrowed

patch ; while the hair on the chest turned pale and thin, showing the skin as a dark, triangular shield, that on the brows and the penultimate joints of the fingers and toes being black. This colour-change, however, does not appear to be characteristic of the species, for a jet-black gibbon in Hainan, said to be a male, had been in captivity long enough to justify the belief that it was about twelve years old ; and it has been suggested that the change in the Zoological Society's specimen was the result of a process of decoloration or bleaching, and probably exhibited only by mature females.

Another remarkable colour-variation in this group has been recorded in Annam and Tongking by a French traveller, M. Louis Boutan, who, writing in the " Comptes Rendus " of the Paris Academy of Sciences, notes it in the case of the white-cheeked gibbon (*Hylobates leucogenys*), a species originally described many years ago from Siam. These gibbons, according to M. Boutan, go about in small parties, comprising in most cases six or seven individuals, among which two types of very different appearance are frequently observable. In the first, or typical form, the otherwise black fur is relieved by well-marked white whiskers and a beard on the sides and lower part of the face. In the second form, on the other hand—which has been described as a distinct species, under the name of *H. henrici*, in honour of Prince Henri of Orléans—the general colour is golden yellow, and there are no white whiskers or beard. That the two are specifically identical is quite certain, and, according to our author, the light-coloured form does not pass into the black one as it grows older. The natives have, indeed, an idea that the pale gibbons are the females, but this M. Boutan denies ; and it seems to be disproved by the fact that individuals of this colour are much less common than the black ones. Assuming these observations to be correct, we have here a clear case of the existence of what is known as dimorphism, or rather dichroism, in the gibbons of this species.

In disposition gibbons are gentle and confiding, and, when captured young, can be readily tamed. Their constitution is, however, even more delicate than that of the other man-like apes ; and consumption soon terminates their existence in Europe, even when the greatest care and attention are bestowed upon them. In the Zoological Gardens at Calcutta they thrive excellently ; and one kept there years ago was accustomed to make its presence known to people living more than a mile away by the loudness of its morning and evening cries.

GIBBON, SHOWING LENGTH OF THE ARMS
Photograph by W. P. Dando

All the gibbons are thoroughly arboreal in habit, and, in the rapidity of their movements among the trees, offer a marked contrast to the more deliberate and somewhat sluggish motions of the orang-utan. So rapid and lightning-like are these movements that one species—the hulock—has been observed, when in captivity at Calcutta, to catch birds on the wing that had flown into its cage ; and there can be little doubt but that such habits are natural to these animals in their wild condition, when it is probable that birds thus captured constitute an appreciable portion of their food.

Although several of the species are found in the forests of the plains, the hulock appears to be almost, if not exclusively, restricted to those of the hilly districts. In marked contrast to the larger man-like apes, most gibbons go in large flocks or droves, which may comprise from fifty to a hundred, or even more individuals ; although, as with most gregarious animals, solitary males are occasionally observed. The long arms are the chief agents in their active movements among the trees, and by

A SIAMANG IN A SITTING POSTURE
Photograph by W. S. Berridge

their aid the distances they can swing from bough to bough, and thus from tree to tree, are surprising. When going downhill, they travel at an extremely rapid pace by swinging themselves in a downward direction from one bough to another on a lower level, from that to the next, and so on.

Although walking rapidly when on the ground, gibbons can easily be overtaken by men. When walking on the ground, the hulock rests on its hind feet alone, with the sole flat on the ground, and the big toe widely separated from the other digits. The arms are usually held upwards, sometimes horizontally, their great length (as shown in the illustration on page 197) giving the animal a peculiar aspect.

The fondness of the hulock for small birds has been mentioned, but, in addition to this kind of diet, gibbons subsist mainly on various fruits and leaves, as well as young and tender shoots ; they also feed on insects and spiders and the eggs and callow nestlings of birds.

The habit which makes the gibbons known to, as well as cordially hated by, all who dwell in the

districts which they frequent is their custom of uttering at morn and even cries of a peculiarly loud and somewhat unearthly nature. These cries consist in the repetition of two syllables in quick succession ; and the name hulock is given to the Indian representative of the group in imitation of its cry.

In all the gibbons the thumbs of both hands and feet are separated from the other digits to the base of the metacarpal and metatarsal bones, this character being found elsewhere among the Primates only in two genera of lemurs (*Indris* and *Propithecus*). The long-armed apes are true bipeds when on the ground, applying the sole flatly, with the big toe widely separated from the other digits ; the hands are held up to be out of the way, rather than for balancing, even when ascending a flight of steps, but they are always ready to seize hold of any object to assist themselves along, just as a man grasps a banister when ascending a staircase.

THE SIAMANG

The siamang (*Hylobates syndactylus*) is the largest of the gibbons, and since it also differs in certain structural peculiarities, it may be dealt with first. This fine species is apparently confined to Sumatra, and its habits were described years ago by the French naturalist Duvaucel. When fully grown, this gibbon stands a little over three feet in height when in the upright position. It is of a uniform glossy black colour, with the exception of a grey or whitish beard, the hair on the body and limbs being comparatively long. The hair on the forearm is directed upwards towards the elbow, as in the larger man-like apes, whereas the other members of the genus have it pointing towards the wrist. There are two other features in which this species differs from the other gibbons ; the first of these peculiarities, and the one from which the animal derives its scientific designation, being the circumstance that the second and third toes of the foot are joined together by a thin web of skin, reaching in the male as far as the last joint, but in the female only to the middle one. The second distinctive peculiarity of the siamang is the possession of a pouch formed by folds of skin round the neck and throat, resembling that mentioned as occurring in the orang-utan. The chin is also better developed than in the other gibbons.

Duvaucel says these apes assemble in troops in the forests, conducted by a chief, whom the Malays believe to be invulnerable, probably because he is more agile, more powerful, and more difficult to reach than the rest. Thus united, they salute the rising and setting sun with the most terrific cries, which may be heard several miles away and which, when near, deafen when they do not frighten. These apes are slow and heavy in gait ; they lack confidence when they climb and agility when they leap, and can be easily caught when surprised. Nature, however, in depriving them of the means of readily escaping danger, has endowed them with a vigilance which rarely fails. If they hear a strange noise, even though they be a mile off, fear seizes them, and they immediately take flight. When surprised on the ground, they may be captured without resistance, being either overwhelmed with fright, or conscious of their weakness and the impossibility of escaping. At first, indeed, they endeavour to avoid their pursuers by flight, and it is then that their awkwardness is most apparent. Their body, too tall and heavy for their short, slender thighs, inclines forwards, and, availing themselves of their long arms as crutches, they advance by jerks, which resemble the hobbling of a lame man whom fear compels to an extraordinary effort.

Their want of agility when surprised on the ground, however, is amply made up for when in the trees, where they take long flying leaps. According to a German writer, siamangs inhabit the forests in Sumatra at an elevation of 3,000 feet above the sea-level, rarely leaving the trees to descend to the earth. At any sudden fright they rush violently down the mountain-sides, by leaping from bough to bough and from tree to tree in the manner already described. Dr. Wallace maintains that the siamang is decidedly slower in its movements than the other gibbons, not taking such tremendously long leaps, and keeping at a lower elevation in the trees. The extraordinary relative length of its arms is indicated by the same writer, who observes that in one specimen about three feet in height, they measured 5 ft. 6 in. from hand to hand, when both arms were stretched out from the body. A young siamang that was brought to Wallace was at first somewhat savage, but soon became more amenable to discipline, feeding readily on rice and fruits. This individual did not, however, long survive in captivity; and it appears that the Malays, who are adepts in keeping and taming wild animals, find it exceedingly difficult to rear siamangs for any length of time. Siamangs have been exhibited alive in the Zoological Gardens at Calcutta. In disposition they are regarded by the Malays as stupid and dull. This species has been stated to occur in the Malay Peninsula, but this is disputed, and it appears to be but little known, even in Singapore, where the captive specimen mentioned above attracted a considerable amount of attention.

A white siamang is recorded from Sumatra.

THE WHITE-HANDED GIBBON

The first example of the more typical species of the group, all of which are very closely allied, is the white-handed gibbon (*Hylobates lar*). This species (page 159), like the other typical gibbons, is considerably smaller than the siamang, standing about thirty inches in height; and it is also of a lighter and more slender build. Although subject to great variation in colour, it may be recognised by the pale colour of the hands and feet, of which the upper surfaces are usually either white or yellowish white. Another distinctive characteristic is to be found in the presence round the black skin of the naked face of a complete ring of more or less nearly white hairs, which imparts a most peculiar aspect to the animal. Occasionally, however, this white ring is almost absent; different individuals showing a gradation in this respect, from those in which it is but very slightly developed to those in which it attains its full proportions. The general colour of

the body and limbs of this gibbon varies from a full black, through various fulvous shades, to a yellowish white. In opposition to what usually obtains in mammals, some individuals of this species have the back lighter than the under parts of the body; and it may occasionally be much variegated.

The white-handed gibbon is met with throughout the Malay Peninsula, as far north as the province of Tenasserim, and may possibly reach into Lower Pegu. It inhabits the forests skirting the mountains, at elevations varying from about three thousand to three thousand five hundred feet above the sea.

The white-handed gibbon is somewhat more heavily built and less agile than the hulock, and it walks on the ground less steadily. It is also said to differ from the hulock in its manner of drinking—scooping up water in its hands, and thus carrying it to its mouth, instead of applying its mouth directly to the surface of the water. There is a great difference in the voice of the two species. The white-handed gibbons are also stated to go about in smaller parties than the other species, the number in a drove being usually from six to twenty. They depend almost entirely on their hands in passing from bough to bough, and use their feet to carry food. A drove of these apes has been seen to escape in this manner with the plunder stolen from a garden made by a Karen tribe near the forests which they frequent. Like other species of the group, the white-handed gibbon almost

THE WHITE-HANDED GIBBON
Photograph by Lewis Medland

invariably has only a single young one at a time. The young are born at the beginning of the winter and cling to the body of the mother for nearly seven months, after which they shift for themselves.

THE HULOCK

One of the best known of all the gibbons is the hulock (*Hylobates hulock*), or white-browed gibbon, which takes its name from its characteristic two-syllabled cry. This is the only species which occurs in India, where it is confined to the north-eastern districts, being found in the hill ranges south of the Assam valley, as well as in the provinces of Sylhet, Cachar, and Manipur. Thence it ranges to the east and southwards into the hill forests of the Irawadi valley near Bhamo, in Upper Burma, and in the neighbourhood of Chittagong and Arakan. It may also occur near Martaban, in Upper Tenasserim ; but the extent of its range on the east side of the Irawadi river is not yet definitely determined.

The hulock may be readily distinguished from the white-handed gibbon by the presence of a white or grey band across the eyebrows, and also by the whole of the rest of the head, as well as the upper surfaces of the hands and feet, being of the same colour as the body. This general colour varies from black to a light yellowish grey, the females being usually paler than the males. Their build is rather lighter, while their habits are more active than those of the white-handed gibbon.

All who have written of the hulock agree as to its docile and engaging disposition, and the readiness with which, even when adult, it can be thoroughly tamed in a short space of time. Writing of a pet hulock formerly in his possession, Mr. R. A. Sterndale observes : " Nothing contented him so much as being allowed to sit by my side with his arm linked through mine, and he would resist any attempt I made to go away. He was extremely clean in his habits, which cannot be said of all the monkey tribe. Soon after he came to me I gave him a piece of blanket to sleep on in his box, but the next morning I found he had rolled it up and made a sort of pillow for his head, so a second piece was given him. He was destined for the Queen's Gardens at Delhi, but, unfortunately, on his way up he got a chill, and contracted a disease akin to consumption. During his illness he was most carefully attended by my brother, who had a bed made for him, and the doctor came daily to see the little patient, who gratefully acknowledged their attentions, but, to their disappointment, he died. The only objection to these monkeys as pets is the power they have of howling, or rather whooping, a piercing and somewhat hysterical ' whoop-poo ! whoop-poo ! whoop-poo ! ' for several minutes, till fairly exhausted."

Mention has been made of the wide distance over which the cries of a hulock kept in the Zoological Gardens at Calcutta could be heard. Dr. W. T. Blanford, writing of the cries of these animals, remarks that " at a distance the sound much resembles a human voice ; it is a peculiar wailing note, audible from afar, and in the countries inhabited by these animals is one of the most familiar forest sounds. The calls begin at daybreak, and are continued until 9 or 10 a.m., several of the flock joining in the cry, like hounds giving tongue. After 9 or 10 o'clock in the morning the animals feed or rest, and remain silent throughout the middle of the day, but recommence calling towards evening, though to a less extent than in the earlier part of the day."

Like the white-handed gibbon, hulocks have been exhibited, although less numerously, in the Gardens of the Zoological Society in Regent's Park, London.

The Hainan gibbon (*H. hainanus*), the coloration of which has already been discussed (page 197), is allied to the hulock, but differs from that and all other species, except the siamang, in the absence of a white band on the forehead, and is thus, when in the black dress. wholly sable.

THE AGILE GIBBON

The agile gibbon (*Hylobates agilis*) is subject to such an amount of individual variation that several so-called species, such as the Malay gibbon of Sumatra and the crowned gibbon of Siam, have been founded upon what appear to be nothing more than local races or varieties of one and the same species.

Inclusive of all these local varieties, the agile gibbon has a rather wide geographical distribution, ranging from Cochin China to Siam ; it is also found in Sumatra and Borneo, as well as in the small islands of the Sulu Archipelago lying between Borneo and the Philippines.

The activity of the agile gibbon, which is known to the natives of Sumatra as ungka, or ungka-puti, is sufficiently attested by its name. It was this species which was first observed to have the power of catching birds while on the wing (page 198) ; and the animals are capable of taking clear leaps of forty feet when passing from bough to bough. They are stated to live in pairs rather than in droves.

In the typical form of the agile gibbon from Sumatra the general colour is usually dark brown, the face being bluish black or brown, and surrounded by whitish hair, through which the ears are only partially visible, while the hands and feet are of the same general colour as the body. It may be distinguished by the prominent arches on the skull above the eyes, the comparatively flat nose, and the large nostrils. The colour of the back in the darker varieties is lighter than that of the under parts. The variety named after Sir Stamford

THE HULOCK GIBBON
Photograph by W. S. Berridge

Raffles (*H. a. rafflesi*) is nearly black, tending to brown on the sides and back. The Siamese variety, known as the crowned or tufted gibbon (*H. a. pileatus*), is likewise of a blackish colour, but differs in that the hands, feet, and a ring round the crown of the head are white. The white patch on the crown helps to distinguish this variety from the typical agile gibbon, although all these Malay gibbons are singularly alike, and often difficult to distinguish even by the practised zoologist. The so-called variegated gibbon (*H. a. variegatus*) appears to be merely another of the numerous phases of the agile gibbon.

THE SILVERY GIBBON

The grey or silvery gibbon (*Hylobates leuciscus*), or wou-wou—a name often incorrectly applied to the agile gibbon—comes from the island of Java, and most naturalists agree in regarding it as a distinct species. It is characterised by its general ashy or bluish grey colour ; the presence of a large square black patch on the top of the head ; and also by the white or grey fringe of hair surrounding the blackish face. The fur also appears to be longer, thicker, and of a more woolly nature than is the case in the other species ; and the colour is

THE HAINAN GIBBON
Photograph by Lewis Medland

stated to be usually lighter on the under parts than on the back. Specimens of this and the preceding species have been exhibited in the London Zoological Society's Gardens in Regent's Park.

FOSSIL APES

Allusion has already been incidentally made (page 156) to the fossil remains of the man-like creature known as *Pithecanthropus erectus*, which were discovered by Dr. Eugène Dubois in Java in 1891. These relics comprised a calvarium, or vaulted portion of the skull, two molar teeth, and a left femur, or thighbone. The first two fragments were found together, and the femur was brought to light about a year later at a distance of a few yards, and Dr. Dubois considered they all belonged to an animal that came between man and the anthropoid apes, and were of Pliocene or early Pleistocene age. Enthusiasts recognised in the remnants the long-looked-for "missing link," but more cautious observers doubted whether the cranium and femur belonged to the same skeleton, maintaining that the former was of a low human type, while the femur was human. The teeth are large and powerful.

It has always been a matter of surprise that no large man-like ape now inhabits the dense tropical forests of India or Burma, which would appear to be just as suitable for these animals as are those

of Borneo or Equatorial Africa. However, the discovery in the north-west of India, in rocks belonging to the Pliocene or later division of the Tertiary period, of a jaw of a large ape apparently belonging to the same genus as the chimpanzee shows that large man-like apes must have once roamed over the plains of Hindustan. Why chimpanzees, together with hippopotamuses and giraffes, which are likewise found fossil in India but are now confined to Africa, should have totally disappeared from the former country is one of those puzzling problems connected with the distribution of animals which there is little hope of answering satisfactorily. The fossil chimpanzee was found in the arid districts of the Punjab, and since we know that the living man-like apes dwell in the deepest gloom and solitude of primeval forests, where vegetation grows luxuriantly and offers a constant supply of fruits throughout the year, it may be inferred that the Indian chimpanzee inhabited a similar forest-clad country ; and that, consequently, the present area of the Punjab was, in parts at all events, clothed with forests in which dwelt this ape, instead of being, as now, a sun-scorched and somewhat desolate region. Evidence of the former existence of these forests is afforded by the occurrence of numbers of fossil tree-stems in various parts of the same series of rocks from which the remains of the fossil chimpanzee were obtained.

Curiously enough, the same strata have also yielded the broken tusk, or canine tooth, of another large ape, which there is every reason to believe was a species of orang-utan. If this be so, it is possible that India may have been the original home of the ancestors of all the large man-like apes of the present day, and that from this centre their descendants have gradually dispersed eastwards and south-westwards. At a still earlier date, however, Africa would appear to have been the home of the group, as the lower jaw of a small man-like ape from the Oligocene strata of the Fayum district in Egypt has been described as *Propliopithecus*, and regarded as representing the ancestor of both the man-like apes and man himself.

In addition to the fossil Indian apes, we have, moreover, sure evidence that at an earlier part of the Tertiary period, known as the Miocene age, one or more species of large man-like apes inhabited Western Europe. One of these extinct creatures has been named Dryopithecus, and its remains have been found in Miocene strata in France, in which beds have also been discovered relics of another genus, to which has been given the name of Pliopithecus ("Ancient Ape"), which is supposed to exhibit characters approximating more or less closely to the gibbons. It appears to have been about the same size as the chimpanzee, but differs from all the living man-like apes in the great length of the bony union between the two branches of the lower jaw. In this respect this ape approaches decidedly towards the lower monkeys.

In explorations conducted in the caves of Borneo, remains of gibbons, probably belonging to species still in the same regions, have been met with in a sub-fossil condition. This is only what we should naturally have expected to be the case. Very different, however, is the occurrence of fossil gibbons in freshwater strata belonging to the middle portion of the Tertiary period in France and Switzerland; for it is certain that these animals could not have lived in a climate at all approaching that which now distinguishes Europe. We shall therefore be safe in assuming that, at the epoch in question, portions of Southern Europe were clothed with dense forests, growing in a hot and moist climate closely resembling that of the Malay Archipelago of the present day. The evidence for the former prevalence of this tropical European climate does not, however, rest solely on the fossil gibbons, since many of the other animals found in the same strata are very similar to those now characteristic of the warmer regions of the East; while the presence of palms resembling those of tropical lands, as well as other plants, supplements the evidence of the animal remains in a convincing manner. After the Middle or Miocene division of the Tertiary period there is no evidence of the existence of gibbons in any part of Europe, although many kinds of monkeys were abundant until much later. R. LYDEKKER

STORIES OF THE GIBBONS

Admitting that the gibbon can bite at need, the weight of evidence in favour of its amiable and affectionate nature is overwhelming. Dr. H. O. Forbes, the well-known naturalist and explorer, in his "Handbook to the Primates," gives a very attractive picture of a young siamang which he kept in Sumatra: "The gentle and caressing way in which it clasps me round the neck with its long arms, laying its head on my chest, and watching my face with its dark brown eyes, uttering a satisfied crooning sound, is most engaging." This gibbon became as much a pet as was Livingstone's young chimpanzee, and demanded as full a return of friendship. Nor was it denied, for Dr. Forbes writes: "Every evening it makes with me a tour round the village square, with one of its hands on my arm. It is a very curious and ludicrous sight to see it in the erect attitude on its somewhat bandy legs, hurrying along in the most frantic haste, as if to keep its head from outrunning its feet, with its long, free arm see-sawing in the most odd way over its head to balance itself, and now and again touching the ground with its finger-tips or its knuckles."

In Calcutta the experiment was tried of allowing complete freedom to one of the hulocks, and the plan was for some time quite successful. Mr. Frank Finn, in the work to which reference was made on page 195, likens this gibbon, as he runs along the ground, to a little man in a fur coat. Food was daily put out for the gibbon, but he provided to a considerable extent for himself. He had sufficient confidence in mankind to pay frequent visits to the refreshment bars to be fed by sightseers. He got

THE AGILE GIBBON
Photograph by W. S. Berridge

his drink in quite a natural way, by sliding down a bough overhanging a pond and dipping up the water with his hand. His round was not confined to the grounds of the Zoological Gardens in Calcutta, but, by means of the telegraph wires, he would accomplish journeys to more or less distant parts of the city. In one of these rambles he turned up at the abode of the lieutenant-governor and hauled down the British flag. For this act of disloyalty and treason his liberty was for some time sacrificed, but eventually he was free once more to wander. In the end he turned savage over a love affair, like many a wiser and many a more foolish creature than himself. There were two female gibbons in one of the monkey-houses, and to these the hulock used to pay visits, and exchange confidences through the bars of the respective cages. "I do not know if the worry of this double ménage affected his temper," says Mr. Finn, "but ultimately he became vicious, and had to be shut up permanently."

We have not the opportunity of seeing the gibbons free in the London Zoological Gardens, but even the comparatively restricted quarters assigned to them in the new ape-house suffice to display their extraordinary agility and untiring activity. The briefest inspection of the gibbons here will reduce the most accomplished human gymnast to despair. Their flights are bewildering; it is so puzzling to realise how they get the impulse which carries them from point to point. The gymnast braces his muscles, holds his breath, and projects himself as a missile from a catapult when he springs for the high horizontal bar. A snapshot of a horse as he rises at a jump shows remarkable contraction of muscles. A bird bows its breast to the ground to get the necessary spring to launch itself into the air; and Herbert Spencer discovered that by making it impossible for a cockerel to raise himself to his full height upon his perch, the bird could be prevented from crowing untimely. But in the gibbon there is no evidence of exertion as it flies from tree to bar, from bar to chain. A Hainan gibbon which arrived at the Gardens in Regent's Park in 1904 had one arm useless through injury, yet its activity in climbing and swinging was phenomenal. How they acquire the power for their leaps with such little apparent exertion is as great a mystery to the layman as the light which the glow-worm kindles without any rise in temperature. The performance of one of the " Zoo " gibbons has been described in some detail by an observer :

" It is almost impossible to convey in words an idea of the quickness and graceful address of her movements; they may well be termed aerial, as

THE SILVERY GIBBON

she seems merely to touch in her progress the branches among which she exhibits her evolutions. In these feats her hands and arms are the sole organs of locomotion. Her body, hanging suspended as by a rope, sustained by one hand—the right, for example—she launches herself by an energetic movement to a distant branch, which she catches with the left hand; but her hold is less than momentary;

the impulse for the next launch is acquired; the branch then aimed at is attained by the right hand again, and quitted instantaneously, and so on, in alternate succession. In this manner spaces of twelve and eighteen feet are cleared with the greatest ease and uninterruptedly, for hours together without the slightest suggestion of fatigue being manifested; and it is evident that, if more space could be allowed, distances very greatly exceeding eighteen feet would be as easily cleared; so that Duvaucel's assertion that he has seen these animals launch themselves from one branch to another forty feet asunder, startling as it is, may well be credited. Sometimes in seizing a branch in her progress she will throw herself, by the power of one arm only, completely round it, making a revolution with such rapidity as almost to deceive the eye, and continue her progress with undiminished velocity.

" It is interesting to observe how suddenly this gibbon can stop, when the impetus given by the rapidity and distance of her swinging leaps would seem to require a gradual abatement of her movements. In the very midst of her flight a branch is seized, the body raised, and she is seen, as if by magic, quietly seated on it, grasping it with her feet. As suddenly she again throws herself into action. The following facts will convey some notion of her dexterity and quickness. A live bird was let loose in her apartment; she marked its flight, made a long swing to a distant branch, caught the bird with one hand in her passage, and attained the branch with the other hand, her aim, both at the bird and at the branch, being as successful as if one object only had engaged her attention. It may be added that she instantly bit off the head of the bird, picked its feathers, and then threw it down without attempting to eat it. On another occasion this animal swung itself from her perch across a passage at least twelve feet wide, against a window which it was thought would be immediately broken; but not so; to the surprise of all, she caught the narrow framework between the panes with her hand, in an instant regained the proper impulse, and sprang back again to the cage she had left—a feat requiring not only great strength, but the nicest precision."

This killing of the bird by the gibbon rather suggests a desire to kill for killing's sake, such as certain carnivora exhibit. But the gibbon, as we have already seen (page 198), has a taste for animal food, and, given a young bird, would readily eat it. An older bird would be seized with the same instinct; possibly lack of appetite would prevent its being eaten, or the age of the bird might make the gibbon reject it. An artist who was sketching the gibbons gave one of them a cockroach. The ape took it between its thumb and the lower part of its forefinger, and climbed with it up to its perch, and there greedily ate it, using the tips of the fingers of the hand which held the beetle to assist in climbing. The gibbon is a sure fly-catcher, and among the best of " mousers."

It is not surprising that gibbons should dwell together in unity when under restraint, since in their wild state they form large companies. They differ in this respect from the captive orangs, which, moody and moping when alone, do not get on well in company. Another distinction between the two animals is that, whereas the free orang will carry off its dead from the hunter, whenever possible, the gibbon takes no notice of its dead fellow. This is the more surprising, seeing that the gibbon is the most solicitous of nurses. That the mother should take its little one to the stream and there wash it does not surprise more than the acts of some of the monkeys, but the attention bestowed on a sick and ailing gibbon by its mates deserves notice. A writer who was so fortunate as to have a number of gibbons free in his garden witnessed a striking example of the kind.

One of the apes, a young agile gibbon, fell from a tree and dislocated its wrist. It received the greatest attention from the others, especially from an old female, which, however, was no relation.

Before eating her own plantains, she used every day to take up the first that was offered to her, and give it to the cripple, who was living in the eaves of a wooden house. The same authority noticed that a cry of fright, pain, or distress from one would bring all the others at once to the complainer, and they would then condole with him and hold him in their arms.

There may be in the nature of the gibbon something of the cat's love of sporting with its prey. A spider is a delicacy which no gibbon's palate can withstand, yet the ape will play for a long time with the spider, dropping it from its fingers and letting it spin a ladder of descent. But in the end the spider will go the way of other spiders—towards making a meal for its captor. Tastes in gibbons differ, as in other animals. One, while a passenger on board ship, behaved itself with the greatest politeness, excepting when a dish of carrots appeared at table. The gibbon's consuming passion for this vegetable mastered all sense of decorum. It would instantly leave its chair, mount the table, and, picking its way among the china, without overturning any article, pounce upon its favourite food, then remember its manners and retrace its steps to its chair, sated and penitent.

The tame gibbon is like a human child, in that it loves to do the things which, presumably, it knows it should not. One such is described in a familiar work. This ape had an irresistible fancy for disarranging the articles in the cabin in which it travelled with its master. Among these articles a piece of soap specially attracted its attention, and for the removal of this the ape had once or twice been scolded. One morning the author was writing, the ape being in the cabin. On glancing towards the ape, he saw it in the act of cautiously making for the soap. " I watched him without his perceiving that I did so ; and he would occasionally cast a furtive glance towards the place where I was sitting. I pretended to write ; he, seeing me busily occupied, took the soap and moved away with it in his paw. When he had walked half the length of the cabin, I spoke quietly, without frightening him. The instant he found I saw him, he walked back again and deposited the soap in the same place whence he had taken it." There was, however, something more than instinct in that line of conduct. Both by his first and last actions, the gibbon betrayed a consciousness of having done wrong.

F. A. B

THE CROWNED OR TUFTED GIBBON

OLD WORLD MONKEYS AND BABOONS

ALTHOUGH there is some degree of uncertainty as to the precise significance to be attached to the names apes, monkeys, and baboons, it is convenient to restrict the first term to the man-like apes, and the second and third to those other Old World Primates which do not belong to the group of lemurs. The name monkey, however, is also applicable to one family of the Primates of the New World. Using, then, the terms monkeys and baboons in this sense, it may be mentioned, in the first place, that naturalists include the whole of those inhabiting the Old World in a single family, for which they adopt the name *Cercopithecidæ* (which may be freely translated " tailed monkeys," from the Greek *kerkos*, " tail " ; *pithekos*, " monkey "), taken from a genus of African monkeys. The next point is to consider how the numerous species comprised in the various genera are to be distinguished as a whole from the man-like apes, on the one hand, and from American monkeys, on the other.

As regards the number of their teeth, all the Old World monkeys and baboons agree with the man-like apes ; the total number of teeth being thirty-two, among which there are two premolars and three molars on both sides of each jaw. This character, as we shall presently learn, will at once serve to distinguish any Old World monkey from an American monkey or marmoset. The Old World monkeys and baboons may, however, be distinguished from the manlike apes by the form of their cheek teeth. The premolar and molar teeth of the latter group, as we have already seen

HEAD OF THE BLACK APE
Photograph by W. P. Dando

(page 175), closely resemble those of man, the crowns of the molars being relatively broad and surmounted by four low main tubercles situated at the four corners of each tooth, but arranged somewhat obliquely to its iong axis. It may be added that the last molar in the lower jaw is of the same general form as the two teeth immediately in front of it.

If, however, the skull of any species of Old World monkey or baboon be carefully examined, it will be found that the molar teeth by no means accord with the foregoing description. We shall, indeed, recognise in these teeth the four tubercles at the corners ; but instead of these tubercles being low, and set obliquely to one another, without any connection between those forming the front and hind pairs, they are comparatively high, and placed in pairs opposite one another ; while each pair is connected by a low, imperfect transverse ridge. This two-ridged character of the molars, which is

more distinct in the lower than in the upper teeth, is therefore a readily available method of distinguishing between an Old World monkey or baboon and a man-like ape. Moreover, with the single exception of one African genus of monkeys, and one Oriental species of another, the last lower tooth of all the monkeys and baboons of the Old World may be distinguished from that of the man-like apes by having a kind of projection or heel behind the second transverse ridge.

There are, however, other characters distinctive of the present group which must now be mentioned. In the first place, in the nose of an Old World monkey the vertical partition dividing one nostril from the other is comparatively thin—this character affording a well-marked distinction from the monkeys and marmosets of the New World. It has been already stated that no man-like ape has a tail ; but there is great variation in this respect among the members of the present group, some having exceedingly long tails, others short tails, and a few no tails at all. In no instance, however, are the tails of this group endowed with the power of prehension, as they are in American monkeys. Here it may be observed that it has been often considered that the term monkey should be restricted to such species as have long tails, while those with short tails should be called baboons, and those with no tails at all apes. The arbitrary employment of terms, however, will not hold good when put in practice; since, if it were adopted, we should have to call certain of the different species of one single genus of monkeys by all the three names.

In all the monkeys and baboons of the Old World those peculiar patches of hard, naked skin on the buttocks, known as callosities, already mentioned as occurring in the gibbons, are invariably present. These callosities, which are not unfrequently bright-coloured, afford another character by which an Old World monkey may at once be distinguished from its American cousins. Their use is to afford a comfortable rest for the body in the upright sitting posture assumed by the monkeys and baboons of the Old World.

Another feature absolutely peculiar to the monkeys and baboons of the Old World, although by no means common to the whole of them, is the presence of those pouches in the cheeks with which all who have fed tame monkeys must be familiar. These cheek pouches are formed by folds in the skin, and when empty lie flat on each side of the face.

They can, however, be so distended as to contain a large quantity of food, and then they stick out prominently on either side, so as to communicate a peculiarly bloated appearance to the face. The possession of these pouches must obviously be a great advantage to the monkey, since by their means a large quantity of food can be hurriedly gathered, stowed away, and afterwards eaten at leisure in some place of security. It might, indeed, be urged that the monkeys which do not possess these convenient receptacles appear to get on in life quite as well as their relations which are thus provided, and that, consequently, these pouches are of no real benefit to the owners of them. To this it may be replied that such Old World monkeys as have no cheek pouches feed much more on leaves and shoots than on fruits; and that they are furnished with a peculiarly complex stomach in which this food can be rapidly stowed away before undergoing complete digestion.

With regard to the limbs of Old World monkeys and baboons, it may be observed that the arms never present that great excess in length over the legs which we have seen to characterise the man-like apes; and the legs may, sometimes, be the longer of the two. The thumb of the Old World monkeys and baboons can in all cases be fully opposed to the fingers, except, of course, in those African species in which it is either absent or rudimentary, and therein lies another marked point of difference from the American group.

Finally, the skeletons of all members of the present group may be readily distinguished from those of the man-like apes by the breastbone being narrow and flattened from side to side instead of broad and flattened from back to front. Moreover, all the species have a central bone in the wrist—a characteristic they have in common with the gibbons and the orang-utan among the man-like apes.

Such, then, are the leading features by which the monkeys and baboons of the Old World—forming a larger group than any other in the order—are distinguished from the groups immediately above and below them in the zoological scale; and the student ought to be able to tell at once whether any particular monkey should or should not be included in the present group. When we speak of the members of this group occupying a position immediately below that of the man-like apes, we must guard ourselves from conveying the idea that the latter can in any sense be regarded as the ancestors of the other. The difference in the structure of the molar teeth of the two groups is alone sufficient to prove that this cannot be the case, those of the man-like apes being of a more primitive type than are those of the monkeys and baboons. The common ancestor of the two groups should perhaps be sought in some long extinct type more nearly akin to lemurs.

Although the majority of Old World monkeys and baboons are inhabitants of the warmer regions of the Eastern hemisphere, yet the group is by no means so strictly confined to tropical and sub-tropical regions as is the case with the man-like apes, which, in a state of nature, are not met with outside of the warm regions of the earth. Indeed, some of the Asiatic species are capable of withstanding a very considerable degree of cold, and may be found among the snows of the Himalaya and North-Western China.

THE LANGUR MONKEYS

With this group of long-tailed Asiatic monkeys, which constitute the genus *Semnopithecus*, we come to the first of four nearly allied genera, characterised by their extremely slender and "lanky" build, by the excessive length of their tails, by the legs being longer than their arms, and by the absence of cheek pouches. All these distinctive features can be verified in the living animal, but there is one other for the examination of which we must turn to the anatomist. This internal character relates to the stomach, which, instead of having the simple bladder-like form which it assumes in all other members of the order, is divided into a number of pouches or sacs. When the peculiar pouched stomach was first described scarcely anything was known as to the habits and food of the monkeys in which it is found. Sir Richard Owen, however, suggested from the analogy presented by this peculiar type of stomach to the one characterising ruminants, as well as some other vegetable-feeding animals, that the food of these monkeys consisted in great part of leaves. This surmise was fully confirmed by subsequent observations; and although the habits of the langurs are still imperfectly known, yet it is stated that they are more purely herbivorous than those monkeys which are provided with cheek pouches, and that a very considerable portion of their food consists of leaves and the tender shoots and young twigs of trees. The presence of this remarkable kind of stomach is, indeed, a kind of compensation for the absence of cheek pouches, being more suited to the needs of these animals than pouches would be.

The langurs are so-called from the name applied by the natives of Northern India to those species of the group which inhabit the outer ranges of the Himalaya. Langurs, which are known in Germany as *Schlankaffen*, or slender monkeys, are found over a large portion of South-Eastern Asia, being especially abundant in India and Burma, and represented by one species in the highlands of North-Western China.

As their German name implies, the bodies and limbs of these monkeys are exceedingly slender; while the tail is so long that very generally, and invariably in all the species from India, Ceylon, and Burma, it is actually longer than the whole length of the head and body together. This is shown in the figure of the hanumán monkey on page 211. In all the species the thumb is well developed; this being a character of great importance, and the chief one by which the members of this group are distinguished from some closely allied African monkeys. The row of long, stiff black hairs, seen in the figure, projecting from above the eyebrows of the langurs is another feature by which these monkeys may be easily recognised. Further, the skulls of all the langurs may be readily distinguished from those of other monkeys, with the exception of the allied African group mentioned above, by the circumstance that the aperture for the nostrils, which is exceedingly narrow, extends upwards between the sockets for the eyes, instead of stopping at about the level of their lower border.

Almost the earliest account that we have of the langurs relates to those of Ceylon, and was given in the year 1681 by Robert Knox, an English seaman, who for nearly twenty years had been a prisoner in that island. Knox says that some of the Sinhalese monkeys " are as large as our English spaniel dogs, of a darkish-grey colour, and black faces, with great white beards round from ear to ear, which make them show just like old men. They do but little mischief, keeping in the woods, eating only leaves and buds of trees ; but when they are catched they will eat anything. This sort they call in their language wanderows." It was thought that Knox alluded to the lion-tailed monkey, formerly incorrectly called the wanderu ; but this is black, and there is no doubt that he described the langurs, which are the wanderus of the Sinhalese.

THE HANUMÁN

Perhaps the best-known of all the langurs, and the one which gives the scientific name (derived from the Greek *semnos*, " sacred ") to the genus, is the hanumán monkey, or true langur (*Semnopithecus entellus*). It is found throughout the northern part of India, from South-western Bengal and Orissa to Gujerat and Bombay, and also occurs in Kathiawar, and probably Cutch, although unknown in Sind and the Punjab. Southwards it ranges into the Bombay Deccan ; while its extreme northern limit extends to the outer ranges of the Himalaya, although there is still some doubt as to where the range of this species ends and that of the next begins.

The hanumán is one of several species of Indian langurs characterised by having the hair covering the crown of the head radiating in all directions from a central point situated on the forehead. It is distinguished from its allies by the absence of any crest of hair on the head, of which the colour is scarcely, if at all, paler than that of the back, and by the full black of the upper surfaces of the hands and feet. The hair of the cheeks does not cover the relatively large ears. The general colour is greyish brown, paler in some individuals than in

others ; but the face, ears, feet, and hands are coal black. In size a large male hanumán will measure some 30 in. in head and body ; but average specimens will be about 25 in., while their tail will measure as much as 38 in. The general appearance is so peculiar that no one who has once been told of a large, long-limbed, slender monkey, with a prodigious tail, black face, and overhanging brow of long, stiff, black hair projecting like a penthouse, can fail to recognise the hanumán (pages 171, 211).

Langurs are exceedingly common throughout a large part of India, and in most districts are held sacred by the Hindus, by whom they are allowed to plunder the grain-shops at will. The best times of the hanumán are, however, ended, as it is not allowed the free run of the bazaars so readily as it once was, while in some districts the aid of Europeans has been invoked to rid the natives of the devastations of these monkeys, which take their name from the god Hanumán, to whom they are sacred.

HANUMÁN MONKEYS

The protection accorded to the hanumán by the Hindus of Northern India has caused these monkeys to be so tame, and so utterly regardless of the presence of man, that there are but few mammals whose habits can be so well observed. The hanumán is usually found in smaller or larger communities, composed of individuals of both sexes and of all ages, the youngest clinging to their mothers, and being carried by them, especially when alarmed. As with so many other mammals, an old male is occasionally found solitary. The story that males and females live in separate troops appears to be fictitious, though females with very young offspring may keep together and temporarily apart from the remainder of the troop to which they belong.

In regard to the cry of these langurs, Dr. Blanford observes that " their voice is loud, and is often heard, especially in the morning and evening. The two commonest sounds emitted by them are a loud, joyous, rather musical call, a kind of whoop, generally uttered when they are bounding from tree to tree, and a harsh, guttural note, denoting alarm or anger. The latter is the cry familiar to the tiger-hunter, among whose best friends is the hanumán. Safely ensconced in a lofty tree, or jumping from one tree to another as the tiger moves, the monkey by gesture and cry points out the position of his deadly enemy in the bushes or grass beneath, and swears at him heartily. It is marvellous to observe how these monkeys, even in the wildest forests where human beings are rarely seen, appear to recognise men as friends, or at least as allies against the tiger. It is a common but erroneous notion of sportsmen that this guttural cry is a sure indication of a tiger or leopard having been seen, whereas the monkeys quite as often utter it merely as an expression of surprise ; I have heard it caused by the sight of deer running away, and I believe that it is frequently due to the monkeys catching sight of men."

The food of the hanumán consists largely of leaves and young shoots, and also grain of all kinds, especially in the towns. In disposition this monkey is gentle, and appears never to attack human beings. Its constitution is delicate when in captivity—probably from the want of suitable food,—but the species is generally well represented in the London Zoological Society's Gardens.

THE HIMALAYAN LANGUR

So closely related to the hanumán is the Himalayan langur (S. schistaceus) that it ought, perhaps, to be reckoned merely as a local race of that species. The Himalayan species is characterised by being somewhat larger—although there is probably no great difference between large individuals of both species ; by the head being much paler in colour than the back, and by the feet being little, if at all, darker than the limbs ; by the smaller ears, which are concealed by the long hair of the cheeks, and by the form of the skull.

This species is found throughout the greater part of the Himalaya proper, ranging from Bhutan in the south-east to the Kashmir valley and adjacent regions in the north-west. It appears not to be found below 5,000 feet, and in the interior of Sikkim it ranges as high as 12,000 feet. One of the first, if not actually the first, to record the occurrence of the Himalayan langur in the interior of Sikkim was Sir Joseph Dalton Hooker in his " Himalayan Journals." The author of that charming book of travel says, on arriving at a Tatar village, at an elevation of about 9,000 feet, he saw a troop of large monkeys gambolling in a wood of Abies brunoniana ; this surprised him, as he was not prepared to find so tropical an animal associated with a vegetation typical of a northern climate. Other writers have observed these langurs in the outer ranges of the Himalaya in the neighbourhood of the hill stations of Simla or Mussuri, leaping from bough to bough of the snow-clad pines and deodars. And I myself was once sufficiently fortunate to behold a similar sight when crossing a pass called the Rutten Pir, in the mountains to the south of the valley of Kashmir. On a sudden, when passing through a forest composed partly of pines and deodar cedars and partly of rhododendrons, a whole troop of these langurs dashed across the path, springing from tree to tree, and scattering in all directions the thick wreaths of snow with which the dark fir boughs were concealed, the season of the year being the middle of the spring.

In the autumn these langurs are to be found in large droves in the extensive forests of the higher valleys surrounding Kashmir. Here they are a decided nuisance to the hunter, as their cries will not unfrequently alarm the deer or bear which he may be pursuing. I was once tempted to shoot a male, but the cries and expression of the wounded brute were so human-like that I never again shot a monkey of any kind.

THE MADRAS LANGUR

In Madras and Ceylon the hanumán is represented by an allied species known as the Madras langur (S. priamus), distinguished by a distinct crest of hair on the crown of the head, and by the upper surfaces of the feet and hands not being black. The following account of the habits of this species is taken from Sir J. Emerson Tennent's " Natural History of Ceylon," in which, as we have said, the langurs are described as wanderus. The Madras langur " inhabits the northern and eastern districts, and the wooded hills which occur in these portions of the island. In appearance it differs both in size and colour from the common wanderu (S. cephalopterus), being larger and more often greyish ; and in habits it is much less reserved. At Jaffna, and in other parts of the island where the population is numerous, these monkeys become so familiarised with the presence of man as to exhibit the utmost daring and indifference. A flock of them will take possession of a palmyra palm ; and so effectually can they crouch and conceal themselves among the leaves that, on the slightest alarm, the whole party becomes invisible in an instant. The presence of a dog excites, however, such an irrepressible curiosity that, in order to watch his movements, they never fail to betray themselves. They may frequently be seen congregated on the roof of a native hut ; and some years ago the child of a European clergyman stationed near Jaffna, having been left on the ground by the nurse was so teased and bitten by them as to cause its death."

The Malabar langur (S. hypoleucus), which is common not only in the forests, but likewise on the cultivated lands fringing the Malabar coast, is another member of the group in which the hair of the crown of the head radiates from a single point on the forehead.

THE BANDED LANGUR

A rare monkey from Sumatra, Borneo, and the Malay Peninsula, extending as far north as Tenasserim, is the banded langur (S. femoralis). It differs from those already mentioned in that the hair of the crown of the head radiates from two distinct points on the forehead. The hair on the hind part of the head stands up so as to form a crest ; while that over the temples bends forwards to overhang the eyes. In colour this monkey is much darker than any of the foregoing species, varying from blackish brown to black over the greater part of the body, but white over a larger portion of the under surface of the body and inner sides of the thighs ; the white area always including the abdomen. The young are of a whitish hue throughout.

A closely allied, if not identical, kind of langur from the same regions is called S. chrysomelas, and differs merely in some details of colouring.

THE MANDRILL

The whole aspect of this creature, one of the largest of baboons, "is far more suggestive of the forms imagined during a nightmare than is the case with any other living mammal."

R

HANUMÁN MONKEYS

The hanumán monkey is very common in India, and in most districts is "held sacred by the Hindus, by whom it is allowed to plunder the grain-shops at will."

It is a curious circumstance that the skulls of both these species or races of langurs may be distinguished from those of all others by the form of the last molar, or " wisdom-tooth," in the lower jaw. In all the other langurs this tooth has five tubercles ; in the banded species it has but four, as in the undermentioned group of guenons.

THE NEGRO LANGUR

Far better known than the last species is the negro monkey (*S. maurus*), or buden, as it is called by the Javanese. This langur, which was originally obtained from Java, but is also found in Sumatra and the Malay Peninsula, takes its English name from the full black colour prevailing over all the body in the adult, except a portion of the under surface and the root of the tail, where it is replaced by grey. It agrees with the banded langur in the forward projection of the hairs on the front of the crown, as is well shown in the figure on page 214. The length of the head and body of this monkey is about 24 in., the tail being longer than the head and body, and frequently furnished with a small tuft at the extremity. The young are light-coloured, being of a yellowish or reddish tint, the dark colour of the adult appearing first on the hands, and then gradually spreading over the limbs and body. This light colour of the young shows that the dark tint of the adults is an acquired or specialised character.

Nearly allied to this species is another and much rarer monkey, found in Java, where it is called by the natives the luton. It is known scientifically as *S. pyrrhus ;* and it differs from the negro langur in being of a rusty red colour at all ages, and is therefore evidently a less specialised form. So like, indeed, are the two that the luton was long considered to be merely a light-coloured variety of the buden. It has, however, been shown that the skulls of the two present considerable structural differences, and indicate perfectly distinct species.

Although in Java these two monkeys have distinct names, the Malays call both by the name luton, distinguishing the negro langur as the luton itam, and the red species as the luton mora; the words " itam " and " mora " signifying respectively black and red.

The opinion that these two monkeys are distinct species is confirmed by a marked difference in their disposition, as was pointed out by Dr. Horsfield, from whose work on the " Zoology of Java " the following account, with some slight verbal alterations, is taken. After observing that the black buden is much more abundant than the red luton, the author observes that " the latter, both on account of its rarity and comparative beauty, is a

THE WHITE LANGUR
Photograph by Lewis Medland

favourite with the natives. Whenever an individual is obtained, care is taken to domesticate it, and it is treated with kindness and attention. The buden, on the contrary, is neglected and despised. It requires much patience in any degree to improve the natural sullenness of its temper. In confinement, it remains during many months grave and morose ; and, as it contributes nothing to the amusement of the natives, it is rarely found in their villages or about their dwellings. The buden is found in abundance in the forests of Java ; it forms its dwelling in trees, and associates in numerous societies. In meeting them in the forests, it is prudent to observe them at a distance. They emit loud screams on the approach of a man, and by the violent bustle and commotion excited by their movements, branches of decayed trees are not unfrequently detached, and thrown down on the spectators. They are often chased by the natives for their fur. In these pursuits, which are generally ordered and attended by the chiefs, the animals are attacked with cudgels and stones, and cruelly destroyed in great numbers. The skin is prepared by a simple process, which the natives have acquired from Europeans ; and they conduct it with great skill. It affords a fur of a jet black colour, covered with long silky hairs, which is usually employed, both by natives and Europeans, in preparing riding equipages and military decorations."

THE CRESTED LANGUR

The crested luton, or langur, of Sumatra and Borneo is closely allied to the negro langur, from which it appears to be chiefly distinguished externally by the blackish fur being usually grizzled, or washed with greyish white. A male from Sumatra has been described as of a brownish black colour, with a dusky tinge on the flanks, forearms, and crest ; the short crest on the vertex of the head being directed backwards, and the long, black hair on the temples coming forwards. The female is black, with the tips of the hairs on the head and body of a lustrous grey tint ; the hair of the limbs being yellowish grey, except on the hands and feet, where they are black. On the under parts the hair is paler, with yellowish grey tips ; while the tail is black, tipped with grey above but yellowing underneath, more especially near the root. The face is bluish black. The young is yellow ; and there is stated to be a race in which the colour of the adult is either light grey or whitish.

THE NILGIRI LANGUR

With the Nilgiri langur we come to the first of a large group of langurs in which the hair of the crown, instead of radiating from one or more points on the forehead, is uniformly directed backwards without any trace of parting. This species (*Semnopithecus johni*, which derives its Latin name from a former member of the Danish factory at Tranquebar in Madras, belongs to a subgroup characterised by the absence of a crest of hair on the crown of the head; the hair of the crown itself being not longer than that on the temples and the nape of the neck. The Nilgiri langur is a comparatively small species, the length of the head and body varying from about 21 to 23 in., and that of the tail from 32 to 35 in.; though larger individuals are occasionally met with. The hair of the body is long, fine, and glossy; and the general colour black to blackish brown, with the exception of the head and rump, of which the former is brownish yellow, and the latter ashy grey. The young of this monkey are black throughout, and this appears also to be the case in the purple-faced langur. This character serves, therefore, to distinguish these langurs very markedly from those of the preceding group, in which the young are light-coloured; and it may be taken as an indication that the present group is the most specialised of all the langurs, not only having acquired the black tint in the adult, but even in the earlier stages of existence.

As its name implies, this langur is found in the Nilgiri Mountains (or Hills, as they are commonly called by Anglo-Indians) of Southern India; and its range extends from the Wynaad southwards to Cape Comorin.

According to Dr. W. T. Blanford, this langur "is shy and wary, the result of human persecution. It inhabits the *sholas*, or dense but abruptly limited woods of the Nilgiris, and other high ranges of Southern India, and is also found in the forests on the slopes of the hills, usually in small troops of from five to ten individuals. It is very noisy, having a loud guttural alarm-cry, used also to express anger, and a long, loud call. Jerdon relates that when the *sholas* of the Nilgiri range were beaten for game, these monkeys made their way rapidly and with loud cries to the lowest portion, and thence to a neighbouring wood at a lower level." These monkeys are frequently shot for their beautiful skins, hence their shyness.

THE PURPLE-FACED LANGUR

The purple-faced langur (*S. cephalopterus*) is the representative of its group in Ceylon. It is known to be liable to considerable variations of colour, and

the Sinhalese langurs, known as the white monkey (*S. senex*), and the bear monkey (*S. ursinus*), may be regarded as nothing more than well-marked races or varieties of this species.

There is a ready means of distinguishing the purple-faced langur from the Nilgiri langur. In the latter the cheeks are of the same brown colour as the rest of the head; in the former they are always much paler than the crown. Typically, this species is of small size, the length of the head and body being only 20 in., and that of the tail 24½ in. The so-called bear monkey is, however, somewhat larger, the length of the head and body being 21 in., and that of the tail 26 in. In colour, the typical purple-faced langur varies from dusky to smoky brown and black, more or less tinged with grey on the back and upper parts, this grey being always present on the haunches. In the head the long whiskers on the cheeks stand out in striking contrast to the brown hue of the rest of the head. Some varieties are more decidedly brown; and in the bear monkey, dusky brown is the prevalent hue, with complete absence of grey on the haunches. The white monkey, which may be regarded merely as a variety of this species, is a curious-looking creature of a general yellowish white colour, with a faint brownish tinge on the head, and tending to a dusky hue on the shoulders and down the middle of the back. The face and ears retain the usual black hue, but the palms of the hands and the soles of the feet are flesh-coloured.

The typical form is found over the greater part of Ceylon at low or moderate elevations, and apparently not ascending above some thirteen thousand feet above the sea-level. The bear and white monkeys, however, are confined to the southern parts of Ceylon, and ascend to much higher elevations; the former variety being especially abundant in the lofty mountains in the neighbourhood of the town of Neuera Ellia. Sir Emerson Tennent, writing of the typical purple-faced langur, which he terms the wanderu of the low country, states that it is far the commonest of the Sinhalese langurs, and that "it is an active and intelligent creature, little larger than the common bonneted macaque, and far from being so mischievous as the other monkeys in the island. In captivity it is remarkable for the gravity of its demeanour, and for an air of melancholy in its expression and movements which are completely in character with its snowy beard and venerable aspect. In disposition it is gentle and confiding, sensible in the highest degree

THE NEGRO LANGUR

THE LANGUR MONKEYS

of kindness, and eager for endearing affection, uttering a low, plaintive cry when its sympathies are excited. It is particularly cleanly in its habits when domesticated, and spends much of its time in trimming its fur and carefully divesting its hair of particles of dust. Those which I kept at my house near Colombo were chiefly fed upon plantains and bananas, but for nothing did they exhibit a greater partiality than for the rose-coloured flowers of the red hibiscus. These they devoured with unequivocal gusto; they likewise relished the leaves of many other trees, and even the bark of a few more of the succulent ones."

Tennent goes on to mention that in the hills the typical black form of this monkey is replaced by the so-called bear monkey. "The natives, who designate the latter as the maha, or great, wanderu, to distinguish it from the kala, or black one (the typical purple-faced monkey), with which they are familiar, describe it as much wilder and more powerful than its congener of the lowland forests. It is rarely seen by Europeans, this portion of the country having till very recently been but partially opened; and even now it is difficult to observe its habits, as it seldom approaches the few roads which wind through these deep solitudes. At early morning, ere the day begins to dawn, its loud and peculiar howl, which consists of a quick repetition of the sounds 'how, how!' may be frequently heard in the mountain jungles, and forms one of the characteristic noises of these lofty situations." There is a record of one of these monkeys having attacked a native laden with a bag of rice.

THE CAPPED LANGUR

Of somewhat smaller dimensions than the hanumán is the capped langur (*S. pileatus*) of Assam and the neighbouring districts of North-Eastern India and Upper Burma. This species may be readily distinguished from the Nilgiri langur and the purple-faced monkey, with its varieties, by the hair of the crown of the head being longer than that on the occiput and temples, thus presenting somewhat the appearance of a cap, from which character the species derives its name.

In colour this monkey varies from dusky grey to a brownish ashy grey on the upper parts; the upper portion of the back, and sometimes also the crown of the head, being darker. The hands and feet are dark or black above, but occasionally some or all of the fingers may be yellowish. The tail is dark brown, but may be black at the tip. The face is

THE PURPLE-FACED LANGUR
Photograph by Lewis Medland

always black, but the sides and lower parts of the head, as well as the neck, vary from a golden brown or orange to a pale yellow or yellowish white tint. The light colour of the sides of the face extends backwards to a line just above the ears, so that, with the light-coloured nape of the neck, the dark cap is well defined, and gives to this monkey a peculiar and distinctive appearance.

The red-bellied langur (*S. chrysogaster*) is a little-known species, reputed to have been obtained from Tenasserim. In the adult, the upper parts, the limbs, and the tail are jet black, with the lower portions of the individual hairs ruddy, and their extreme bases white; the band on the forehead, as well as the cheeks to behind the ears, and the sides and front of the neck, together with the chin and the upper part of the breast, are pure white. The remainder of the under parts are of a deep bright ferruginous red, which also tinges the inner sides of the limbs, and gives the animal its distinctive appellation. The young are of a uniform reddish white colour. The head of the adult appears to have a small crest, and by this it is distinguished from the typical capped langur.

THE DUSKY LANGUR

The dusky langur (*S. obscurus*), found in Siam, the Malay Peninsula, and the Tenasserim provinces, while agreeing with the Nilgiri and the capped langur in the backward direction of the hair on the crown of the head, is characterised by a distinct crest of longer hairs on the occiput, arranged in a pointed form.

Closely allied to this species is Phayre's langur (*S. phayrei*), distinguished by the crest of hair being placed on the crown of the head instead of on the occiput, and by this same crest being compressed and longitudinal, instead of pointed, while the colour of the body is dark grey above, and whitish underneath. Phayre's langur inhabits Arakan, part of Pegu, and Northern Tenasserim.

HOSE'S LANGUR

This handsome and peculiarly coloured langur (*S. hosei*) from Borneo belongs to the group in which the hair of the crown extends evenly backwards. It is about the same size as the dusky langur. The crown has a longitudinal crest, starting about half an inch behind the centre of the forehead. The general colour of the body is a hoary grey, caused by the mixture of black and white hairs. The crest, as well as the centre of the crown of the head, the nape of the neck, and the eyebrows, are deep glossy black, and the hands and feet are of

215

THE ORANGE SNUB-NOSED MONKEY

the same jetty hue. In marked contrast to these sombre tints is the brilliant white of the forehead, temples, sides of the head and neck, and chin. This white is continued down the throat and chest to the under surface of the body, and the inner sides of the upper parts of the limbs. This handsome species differs from all the langurs yet mentioned in the marked contrast presented by its black crest to the brilliant white of the temples and cheeks.

Another monkey from Borneo has been named S. thomasi, but comes very close to Hose's langur.

THE DOUC

The douc, or variegated langur (S. nemæus), is an inhabitant of the forests of Cochin China, where it is found near the coast, as well as in the interior, and is remarkable for its brilliant colouring. The general form of the douc is so different from that of other langurs that it has been made the type of a distinct genus. The build is more robust, and the limbs are stouter, and of nearly equal length, whereas, in the typical langurs, the arms are considerably shorter than the legs.

The hair on the top of the head is directed backwards, without any crest ; and the brilliant white whiskers have likewise the same direction, and are closely pressed to the face. The general colour of the head is brown, but there is a narrow band of bright chestnut passing backwards under the ears, and the naked face is of a brilliant yellow, which makes a bold contrast to the pure white whiskers. Owing to the hairs of the body having alternate dark and light rings of colour, the general tint of the body is a mottled, grizzled grey, darker on the upper than on the under parts. The upper parts of the arms and legs, as well as the hands and feet, are deep black ; but the lower portions of the legs are chestnut, and the forearms white. A large patch on the rump near the root of the tail, as well as the tail itself, is likewise white.

As might naturally be supposed, fossil remains of langurs have been found in their native land of India. Some of these have been obtained from caverns in the Madras Presidency, and do not date back much, if at all, beyond the human period. Other remains occur, however, in the much older Siwalik sandstones forming the ranges on the flanks of the Himalaya, and belonging to the upper part of that division of the Tertiary period known to geologists as the Pliocene. This does not, however, by any means limit the range of extinct langurs, since their remains have been found in the Pliocene deposits of the Val d'Arno in Tuscany, and also in strata of equivalent age in the South of France. We have, therefore, evidence that these monkeys, which are now confined to the Indo-Malay and Chinese countries, were formerly widely spread over the Eastern hemisphere.

THE SNUB-NOSED MONKEYS

Perhaps the last place in which we should expect to find a living monkey would be the highlands of North-Western China. This monkey (Rhinopithecus roxellanæ) may be recognised among all its congeners by its " tip-tilted " nose. Although short and small, the nose is so much turned up that its tip reaches to the level of the lower border of the eyes. This monkey differs from the langurs in its stouter build, and relatively shorter limbs. The upper surface of the body, the crown of the head, the outer sides of the limbs, and the whole of the tail are an olive brown colour, flecked with golden yellow ; while the sides of the face, the lower part of the forehead, and all the under parts and the inner sides of the limbs, are brilliant yellow, tending to orange, the naked parts of the face being bluish grey. The males, which are nearly double the size of the females, develop very long winter coats, with beautiful tufts or tresses of long, silky, golden hair on the back. These monkeys inhabit the forests of the mountain region between Moupin, in Sze-chuen, and Lake Khoko, where snow lies for a large portion of the year. They are stated to live in numerous troops, always ascending the loftiest trees, and feeding on fruits, but, when pressed by hunger, eating also the leaves and shoots of the bamboo. From its colour, this species is called the orange snub-nosed monkey. A second member of the group, the slaty snub-nosed monkey (R. bieti), inhabits the Upper Mekon Valley, and has a coat of a uniformly slaty grey colour. A third species is R. brelichi, of Central China.

THE PROBOSCIS MONKEY

If the physiognomy of the snub-nosed monkey appear ludicrous, it is hard to say what epithet ought to be applied to the far more grotesque-looking creature represented in the figure on p. 217. The nose of the proboscis monkey (Nasalis larvatus) is, indeed, so enormous in proportion to the face that it presents the appearance of an absolute deformity, and it is very difficult to imagine of what possible advantage it can be to its owner.

THE PROBOSCIS MONKEY

The proboscis monkey is an inhabitant of Borneo, and its marked difference from other monkeys is one of the many proofs indicating the great antiquity of that island, and the long period during which it has been isolated from other lands. In general structure this monkey conforms so closely to the langurs that the peculiarity of its nasal organ would not alone justify its separation from that group as the representative of a distinct genus, although it was on this ground solely that the separation was originally made. Subsequent researches have, however, shown that the skull can be distinguished at a glance from that of any of the langurs, and also from those of the African genus *Colobus*, to be mentioned immediately, by the form of the aperture of the nasal cavity. Thus, whereas in the latter this aperture extends upwards between the sockets of the eyes, in the proboscis monkey the nose bones which roof over this aperture descend considerably below the lower margin of the eye sockets. In this respect the species under consideration resembles the macaques and their allies.

The proboscis monkey was first made known to European science in 1781 by Baron Wurmb, sometime Dutch governor of Batavia. Wurmb described it under the name kahau, a term apparently made up from a resemblance to its cry, but unknown to the native inhabitants of Borneo, by whom it is said to be called bantugan. Specimens were subsequently sent to Europe, which were considered to indicate two distinct species; but it was afterwards discovered that these supposed two species were founded upon the male and female of the one and only proboscis monkey, in which the two sexes differ considerably in point of size.

The proboscis monkey is a rather large animal, the combined length of the head and body of the male being about 30 in., while the tail measures some 27 in. The general colour is a kind of ochre yellow, the head and upper parts of the body being chestnut. The under parts are lighter; a large patch on the rump above the root of the tail, as well as the tail itself, together with the forearms and the lower portion of the legs, being greyish yellow. The forehead is very low, and the dark chestnut hair directed backwards from a nearly straight line immediately over the eyes,

while the hair of the temples is continued down the sides of the face as whiskers, which meet as a beard beneath the chin. The whole of the large naked face is, therefore, surrounded by a hairy frame. In stuffed or dried specimens the skin of the face fades to a dull leaden hue; but when the animal is alive the skin is of a reddish brown flesh colour.

The light-coloured area on the loins near the root of the tail usually takes the form of a number of large rectangular spots, producing a peculiar and characteristic kind of colouring, which is, however, absent in the female.

According to a photograph taken a few years ago from a living example the form usually given to the nose in mounted specimens and figures of the proboscis monkey is quite incorrect. Instead of being sharply pointed, compressed, and projecting straight forwards, this appendage is expanded and depressed at the extremity, which hangs down in front of the upper jaw so as to conceal the greater part of the mouth in a full-face view.

The enormous nose, from which the creature derives both its popular and scientific appellation, has the nostrils placed on its under surface, although separated by a much narrower septum than in man. This excessive development of the nose is, however, only reached in the adult male; it being much less throughout life in the female, while in the young of both sexes it is comparatively small, and upturned as in the snub-nosed monkey.

Accounts of the habits of the proboscis monkey in its wild condition are few. The following note is taken from a translation of the original account given by Baron Wurmb. After stating that these monkeys are found in large troops, the author says that " they assemble together morning and evening, at the rising and setting of the sun, and always on the banks of some stream or river; there they may be seen seated on the branches of some great tree, or leaping with astonishing force and rapidity from one tree or branch to another, at a distance of fifteen or twenty feet. It is a curious and interesting sight; but I have never remarked, as the accounts of the natives would have you believe, that they hold their long nose in the act of jumping; on the contrary, I have uniformly observed that on such

THE PROBOSCIS MONKEY

YOUNG PROBOSCIS MONKEY
Photograph by W. P. Dando

occasions they extend the legs and arms to as great a distance as possible, apparently for the purpose of presenting as large a surface as they can to the atmosphere. The nature of their food is unknown, which renders it impossible to keep them alive in a state of confinement."

THE GUEREZA MONKEYS

The langurs, which, as already mentioned, are widely distributed over South-Eastern Asia, and more especially that portion forming the Oriental region of zoologists, are replaced in Africa by a group of monkeys closely allied to them in all respects, but distinguished either by the total absence or rudimentary condition of the thumb. When present at all, this digit merely takes the form of a small tubercle, which may or may not be provided with a minute nail. Such a point of difference from the langurs is rightly regarded as worthy of generic distinction, and these African monkeys have accordingly been described under the name of Colobus, from a Greek word meaning " docked," in allusion to the feature in question. There is no popular name by which these monkeys are generally known, but it seems best to apply the native name guereza to the whole group, although this strictly belongs only to certain black-and-white species from Abyssinia and North-East Africa.

There are about a dozen species of this group known to science. Our acquaintance with their habits is, however, extremely imperfect, and few of them have been brought alive to Europe, since, like their cousins the langurs, they are delicate, and do not thrive well in confinement. The sacculated, or pouched, stomach indicates that their food, like that of the langurs, is in all probability largely composed of leaves and twigs. If, however, their habits at all resemble those of the group last mentioned, it is not easy to see why they should have lost their thumbs—unless, indeed, the small thumbs of their Indian cousins are practically useless.

Although several of the guerezas are met with in East Africa, these monkeys attain their maximum development on the west coast, where they were formerly very abundantly represented. Most of them are remarkable for the length and beauty of the silky hairs with which their bodies are clothed, their fur being largely imported into Europe for use as trimming for other furs and various kinds of apparel.

Before noticing some of the species of this group it may be mentioned that the hair of all the guereza monkeys is coloured uniformly, and by this character even a small piece of their fur may be distinguished from that of most other African monkeys, in which each individual hair is ringed with different hues.

THE BLACK-AND-WHITE GUEREZA

This handsome species (*Colobus guereza*) occurs typically in Central Abyssinia in the neighbourhood of Simien. In Southern Abyssinia it is comparatively common in the district of Gojam, and thence it extends southwards into the Galla country. From the Galla country and Somaliland this guereza appears to range to the south-west into the Niam Niam district, lying to the north-west of Albert Nyanza, and southwards as far as Kilimanjaro on the east coast. Westwards it ranges into the upper part of the Congo valley. The East African representative of the species is distinguished as *Colobus caudatus*, and the Congo and Unyoro form as *C. occidentalis*. Both, however, should perhaps be regarded as races rather than species.

The head, body, and limbs of the black-and-white guereza are covered with jet black hair of moderate length ; but on either side of the back there arises a line of long hair, hanging down below the flanks, and forming a kind of mantle of pure white. The dark face is also surrounded with a fringe of white hair, which forms long whiskers lying flat on the cheeks, and directed backwards. The long tail terminates in a white tuft. The contrast of the white of the mantle, of the cheeks, flanks, and tail against the velvety black of the rest of the body is most striking and without exact parallel among other mammals, although the colouring of the skunk is somewhat suggestive of it (see page 17).

Handsome as is the typical guereza in these

THE FLAG-TAILED GUEREZA

respects, it is exceeded by the aforesaid East African species or race, which occurs commonly at an elevation of about three thousand feet on the sides of Kilimanjaro. In the typical species the first twelve or sixteen inches of the tail are black and short-haired, the white tufted portion including only the last eight or ten inches; while the white mantle of hair depending from the back conceals only about one-third of the black portion of the tail. In the East African flag-tailed guereza (*C. caudatus*), only some three or four inches of the base of the tail are black and short-haired, while the remainder is covered with long white hair for a length of some twenty inches, each hair measuring from eight to nine inches. Moreover, the white hairs of the mantle entirely conceal the black of the root of the tail, so that the mantle and tail-brush practically become continuous. In another race (*C. sharpei*), inhabiting Northern Nyasaland, the white on the head and body is restricted to the forehead, and a long tuft of hair on each side of the nape. The guereza lives in small companies, and is constantly on the move, but is said to be completely silent. Its leaps from tree to tree are described as of tremendous length. It subsists mainly on various kinds of wild fruits, seeds, and insects, and retires to sleep high up in the trees. In Gojam it is hunted for the sake of its fur, which is used for covering the shields of Abyssinian soldiers.

These black-and-white guerezas illustrate the evolution of a remarkable type of colouring. Starting from the wholly black species a gradual transition can be traced to one in which the sides of the face, flanks, and hindquarters, together with nearly the whole of the tail, are furnished with long fringes of pure white hairs, apparently developed to accord with the pendent white lichens clotting the branches of the boughs among which these monkeys dwell.

THE BLACK GUEREZA

In marked contrast to the pied colouring of the preceding species is the sable hue of the black guereza (*C. satanas*), first described from specimens obtained at Fernando Po, on the West Coast of Africa, in 1838. The uniform black suffices to distinguish it at once from all its congeners. In addition to this black colouring, the crown of the head has a crest of long hair projecting over the temples and eyes, and the whiskers are long and expanded. The whole of the body is covered with long and rather coarse hair; but the tail is short-haired throughout the greater part of its length, and has no trace of a tuft at the end. The whole of the hair has a dull and shaggy appearance, recalling that of the sloth bear of India. The length of the head and body is 32 in., that of the tail 40 in.

THE KING GUEREZA

The king monkey (*C. polycomus*) of Sierra Leone is one of the few guerezas that have been exhibited alive in the Gardens of the London Zoological Society. It has no crest on the head, but a long mane on the throat and chest; the hair of the sides of the body

THE WHITE-THIGHED GUEREZA

being likewise long. The general colour is black, but the mane, the forehead, and the sides of the face, as well as the whole of the tail, are dazzling white. The tail has a well-marked tuft at the end; and the entire coat of hair is very glossy.

THE URSINE GUEREZA

Closely allied to the last species is the so-called ursine guereza (*C. ursinus*), from Fernando Po, in which the mane is greyish, and not longer than the hair on the sides of the body. Another nearly related West African monkey is the Angola guereza (*C. angolensis*), which differs from the king guereza in that the chest and two-thirds of the lower portion of the tail are black (page 220).

THE WHITE-THIGHED GUEREZA

More markedly distinct than the preceding from the king guereza is the white-thighed guereza (*C. vellerosus*) of Western Africa, distinguished by the absence of a mane on the head and throat, although it has a large fringe round the face. The general colour is glossy black; but that of the forehead, of the frill round the face, and on the chin is white. The tail is also white; but the most distinctive characteristic of the species is the silvery white of the thighs, from which it derives its name. The haunches are, moreover, generally grey. The white hair of the thighs is shorter than that on the body.

THE BAY GUEREZA

Very different in colouring from any of the species yet mentioned is the bay guereza (*C. ferrugineus*), known from the Gambia and the Gold Coast, from which a specimen was brought alive to England in 1890, but did not long survive its arrival. This handsome species has comparatively short hair, which, on the crown of the head and the back and upper part of the sides is blackish grey, while the cheeks and throat, as well as the under parts and the limbs, are of a rich ferruginous bay. The upper part of the root of the tail is blackish, and the remainder reddish brown. The ears and the greater part of the face are bluish, but the nose and lips are flesh-coloured. Altogether the bay guereza is a striking species, which, once seen, will always be easily recognised.

THE CRESTED GUEREZA

The last of the guereza monkeys to be noticed is the crested guereza (*C. cristatus*), which is likewise a West African species, distinguished by its short yellowish brown fur, which becomes greyer on the front of the body ;

HEAD OF THE ANGOLA GUEREZA
Photograph by W. P. Dando

the shoulders and outer sides of the arms, the throat, chest, under parts, and inner sides of the limbs being greyish white. It differs from all the other species in that the hair on the forehead radiates from two points on the temples, and in having a low erect crest of longer hairs running along the middle line of the head.

THE GUENON MONKEYS

In systematic zoology such of the Old World monkeys as have no cheek pouches, but possess sacculated stomachs, and in which the legs are longer than the arms, constitute the subfamily Colobinæ of the family Cercopithecidæ. There remain for consideration all the other Old World monkeys, together with the baboons, which, although belonging to the same great family, constitute the large separate subfamily of the Cercopithecinæ. This group is characterised by the circumstance that all its members are furnished with cheek pouches, but their stomachs are simple, and the arms and legs of nearly the same length.

Since there is no English name to distinguish this group of African monkeys from others of the same family, it will be found convenient to use the French name guenon, meaning " one who grimaces," which appears to have been especially applied to the monkeys of this group, as being those most familiar in menageries and shows.

As already mentioned, these monkeys are strictly confined to Africa, where they are represented by a great number of species, of which a large proportion is found on the western side of the continent. None is of large size, and they all present the following features by which they are characterised as a genus (*Cercopithecus*).

In build they are comparatively slender, and their muzzle is short, or, at least, not very long. The tail is invariably long and slender, and the naked callosities on the buttocks are of small size. For another important point of distinction recourse must be had to the dried skulls, an examination of which will show that the last molar, or wisdom tooth, on each side of the lower jaw consists of four tubercles only, and of these the front and hind pairs are connected by a pair of transverse ridges. In this respect the guenons differ, not only from the langurs and guerezas, but from all those to be subsequently noticed, in which the last lower molar has a fifth tubercle forming a kind of heel projecting from behind the second transverse ridge.

In general appearance, more especially as regards their slender build and long tails, the guenons are the members of the present sub-family which make the nearest approach to the langurs and their allies. Like most of the African monkeys to be noticed later, they are characterised by having each hair marked by a series of different-coloured rings, which imparts to the fur a peculiar mottled appearance.

In disposition these monkeys are docile and easily taught, and so well do they thrive in captivity that it is not uncommon for them to breed in menageries. In consequence of this docile disposition, and their comparatively hardy constitution, as well as from the facility with which they learn tricks and obey the word of command, it is these monkeys, or the members of the next genus (the mangabeys), which are generally chosen as companions by organ-grinders. " Mischievous as a monkey " is truer of the guenons than of any other members of the order, and it is largely to them that the monkey-house at the Zoological Society's Gardens in London owes its popularity.

In saying that the guenons are docile, it should be added that they are teachable for monkeys, since in the strict sense of the word all monkeys are far less docile and less susceptible of education than many other mammals. This, however, by no means implies that monkeys have not a very high degree of intelligence. In regard to this point the following paragraph from the writings of Dr. W. T. Blanford may be quoted : " It is the commonest mistake among superficial observers, and even among naturalists, to confound docility and intelligence among animals, and to measure their intellectual powers by the facility with which they can be taught. Hence the very common, but, as it appears to me, very incorrect notion, that monkeys are of inferior intelligence to such animals as dogs and elephants. In reality they are less docile, less willing to learn, and less adapted to captivity ; moreover, being of but little use to man, far less trouble has been taken in studying their habits. Thus, while dog and elephant breaking engage all the time and mental resources of particular classes of men, the instruction of monkeys is left to the unaided efforts of amateurs and organ-grinders. The negro race among men appears to be far better adapted for slavery than most savage races, being more docile in a state of captivity ; but it is scarcely proved to be more intelligent on that account. The same reasoning will, doubtless, apply to animals. I have often seen dogs and monkeys kept together, and in every instance it has appeared to me that the monkey ruled the dog, and that the dog, although the more powerful animal, feared the monkey ; and I can only account for this by the superior intelligence of the monkey."

In their native condition guenons associate in separate families or droves, each under the leadership of an old male. And it appears that each drove has its own particular limits of territory beyond which it cannot go without intruding on the domains of another drove, an invasion which is treated at once as a *casus belli*. Indeed, this principle of territorial rights appears to be so deeply implanted in guenon nature that it persists even in captivity, when it is not uncommon to see two or more of these creatures zealously guarding one portion of the cage from all intruders.

As they were the common monkeys of Africa, the guenons were well known to the ancient Egyptians, and it is probable that most of the long-tailed monkeys we see on old sculptures are either guenons or mangabeys. Not only were the guenons familiar to the Egyptians, but they appear to have been imported into classic Greece and Rome ; and it is believed that the Greek and Latin term *Cebus* was used to designate them, although this name is now applied to a South American group, the sapajous, or capuchins.

THE MALBROUCK MONKEY

The malbrouck monkey (*Cercopithecus cynosurus*) of Western Africa may be taken as the first representative of a group of guenons characterised by their oval heads and somewhat long muzzles, as well as by their stiff and backwardly-directed whiskers. The fur is grizzled, each hair being marked with greenish or reddish rings.

The malbrouck is distinguished from the other members of this group by the large and broad face being flesh-coloured. The general hue of the

fur is yellow, grizzled with black ; a distinct band on the forehead, as well as the whiskers, throat, the under parts, and the inner surface of the limbs being whitish. The first specimen exhibited alive in England had an unusually mild and gentle expression of countenance, and was calm, circumspect, and inactive in its habits. It did not, however, appear anxious to encourage familiarity on the part either of its keepers or of strangers, and was always ready to resent any interference with its liberty

THE VERVET MONKEY

Still better known than the malbrouck is the South African vervet monkey, or black-chinned vervet (*C. pygerythrus*), as it has been called, in which, as in all the other members of this group, the rather small and narrow face is entirely black or blackish.

The fur of the vervet is of a greyish green colour, finely speckled with black on the greater part of the body. The face, hands, and feet, and the terminal third of the tail, are of a deep black ; while the cheek, throat, and under parts of the body are reddish white, and the root of the tail and adjacent regions are red. The white band on the forehead is distinct. The red root and black tip of the tail, which has no basal tuft, and chin are absolutely distinctive of this species.

In size the vervet is somewhat smaller than the mona. It is common in forest districts throughout Cape Colony and adjacent regions, more particularly along the tract of coast extending from Cape Town to Algoa Bay, and thence through Kaffraria and Natal. It is said to feed chiefly on gum from the acacias known as camel-doorn and rhinaster-bosh ; and its habits are like those of the green monkey.

The Mozambique monkey (*C. pygerythrus rufoviridis*), from the Mozambique and Zambesi districts, is a local race of the vervet, with which it agrees in having the root of the tail and adjacent regions of a ferruginous red, but differs in the more yellowish grey tint of the fur of the upper parts, which tends to a blacker hue on the crown of the head, the tail, and the outer sides of the limbs ; while the under parts and the inner sides of the limbs are pure white, instead of reddish white.

THE GRIVET MONKEY

The grivet (*C. æthiops*) is a member of the group inhabiting North-Eastern Africa, inclusive of the Upper Nile districts, Abyssinia, Senar, and Kordofan.

THE VERVET MONKEY

The fur is olive green, speckled with yellow and black, while the chin, whiskers, and under parts of the body are white, and the root of the tail and adjacent regions grey. The forehead has a broad whitish band, more or less strongly marked; and the face is black. The white chin and grey root to the tail serve to distinguish this monkey from the vervet, with which it agrees in size. Writing of grivets in Abyssinia, Blanford says they are rarely seen, and then only in forests. "On the highlands I only once saw a flock—this was near Dildi, south of Lake Ashangi. I met with large numbers on the Anseba, where they inhabited the high trees on the banks of the stream. The flocks seen were small, not exceeding twenty to thirty individuals. I had but few opportunities of observing their habits, but they appeared to differ but little from those of macaques, except that the former is a quieter animal and less mischievous. In captivity they are well known as excessively docile and good-tempered, and fairly intelligent."

BLACK-BROWED GREEN MONKEY

One of the commonest of the guenons in menageries is the West African green monkey (C. sabæus), in which the colour of the fur is a mixture of black and yellow, giving a general dark green hue to the upper parts ; the crown of the head, the hands, feet, and the upper part of the root of the tail being blacker. There is no distinct light band on the forehead, and if this be represented at all, it is very narrow. The whiskers, throat, and the under side and end of the tail are yellowish, sometimes tending to orange ; and as in the grivet, the base of the root of the tail is grey (see page 160).

This green monkey is closely allied to the grivet, from which it may be distinguished by the more yellow-green hue of its upper parts, the yellowish whiskers, and the absence of a distinct white band on the forehead. It is about the size of a large cat, the length of the head and body being 16 or 18 inches, and that of the tail rather more. It is one of the hardiest of the guenons, on which account it is frequently seen in confinement, as it bears the English climate well. Although gay and gentle during youth, it usually becomes morose and vicious when old, and is therefore not one of the species usually selected for exhibition by organ-grinders. Like the other members of the group, the green monkey never utters a sound when in captivity, and from an early account of the species it appears to be similarly silent in its wild condition.

Senegambia, Sierra Leone, and Liberia, are the home of this species. In Nigeria it is represented by the white-browed green monkey (C. tantalus), which has a similar black face, but a conspicuous white brow-band.

THE PATAS MONKEY

The patas, or red monkey (C. patas), of Senegambia, is a large, long-legged monkey, differing from all the other members of this group yet noticed in the red colour of the fur of the greater part of the body, the whitish lower portions of the limbs, and the pink-coloured ears and face ; the nose, an arched band on the forehead, and the outer surfaces of the arms being blackish. In addition to these leading features, it may be mentioned that beneath the large ears there are thick bushy tufts of light grey hair, which extend forwards on to the cheeks and lower jaw, so as to cause the naked part of the face

to be limited to a narrow space between the eyes and the upper lip. From these tufts the greyish-coloured hair is continued on the whole of the under surface of the body, as well as on the inner sides of the limbs. The hands are of a dusky brown colour, with very short fingers, and the thumb is reduced almost to the condition of a tubercle. The black of the nose continuing upwards to the arched band of the same colour above the eyes communicates a very peculiar and characteristic physiognomy to the patas, which led Buffon to describe it as the monkey à bandeau noir.

On the opposite side of Africa, in Nubia and Somaliland, the place of the patas is taken by a closely allied monkey, known as the nisnas, which may best be regarded as a local race of the patas, under the name of C. patas pyrrhonotus. The nisnas is distinguished from the typical patas by the nose of the adult being white instead of black, and by the red colour of the body being continued on to the shoulders and the outer sides of the arms, instead of these parts assuming a blackish tinge. The nisnas is the monkey so frequently represented on the ancient Egyptian monuments ; and it appears to be undoubtedly the cebus of the ancients, which, on the authority of Pythagoras, was described by Ælian as inhabiting the Red Sea littoral, and was said to be of a bright flame colour, with whitish whiskers and under parts.

Other races of the patas inhabit North Central Africa, and thus connect the habitat of the typical patas with that of the nisnas. These monkeys are believed to be inhabitants of open country instead of forests, which accounts for their peculiar and distinctive type of colouring.

SYKES'S MONKEY

This East African monkey (C. albogularis) is the type of a group of some eight species of large guenons allied to the black-bellied monkey, but distinguished by the smaller amount of black in the colouring. The head, nape, shoulders, and cheeks are very uniformly coloured, generally speckled yellowish grey and black ; and the back is usually reddish or yellowish. The present species is distinguished by the pure white colour of the under parts and chest, which extends on to the throat, and thus suggests the scientific name. This monkey was originally brought to England by Colonel William Henry Sykes, afterwards M.P. for Aberdeen, by whom it was described in 1831. Its describer stated that its manners were grave and sedate ; its disposition was gentle, but not affectionate ; free from that capricious petulance and mischievous irascibility which characterise so many of the African species, but yet resenting irritating treatment, and evincing its resentment by very sharp blows with its hands. Other members of this group are C. kolbi, from the Keelong Escarpment ; C. molonyi, of Nyasaland ; and C. stairsi, of Zambezia.

THE BLACK-BELLIED MONKEY

The black-bellied Pluto monkey (C. leucampyx) is a West African species, representing a group characterised by the prevalence of black on the head,

CHARACTERISTIC SPECIES OF THE GUENON MONKEYS

Malbrouck Monkey

Grivet Monkey

Mozambique Monkey

Young Nisnas Monkey

Patas Monkey

Green Monkey

Mona Monkey

Hocheur Monkey

Sykes's Monkey

Schmidt's and Red-eared Monkeys

Moustache Monkey

Burnett's Monkey

Photographs by L. Medland, W. P. Dando, W. S. Berridge, H. Irving, and others

nape, shoulder, belly, and thighs. When the head is strongly speckled, the belly and thighs are black; and when the two latter are speckled, the head, except for the light brow-band, is black. The black-bellied species has no beard, but large bushy whiskers. The colour of the fur is black, finely grizzled with grey; the forehead has a whitish band, and the sides of the forehead, as well as the shoulders, chest, tail, and limbs, are entirely black, with no white on the haunches. The range of this monkey extends from West to East Africa, the eastern races including *C. leucampyx stuhlmanni*, of the district of the great lakes, and *C. l. carruthersi* from Mount Ruwenzori. The other members of the group are *C. opisthostictus*, of the Lake Mweru district, and *C. kandti*, from the neighbourhood of Lake Kivu.

THE MONA MONKEY

One of the most familiar of all the guenons is the mona monkey (*C. mona*), a beautiful little creature which may be recognised by the presence of a large and distinct white spot of an oval shape situated on each hip immediately in front of the root of the tail, this feature being peculiar to the species.

The mona, which is a West African monkey, has no real title to its name, which is merely the Moorish word for monkeys in general. The general hue of the fur is blackish olive, finely grizzled with yellow; this gradually darkens towards the hind part of the body, so that the tail and the outer surfaces of the limbs are nearly black. The under surface of the body is nearly pure white, these white parts being separated from the darker regions by an abrupt division; and there is also the distinctive white spot on each side near the root of the tail. The naked portions of the face are purplish, with the exception of the lips and chin, which are flesh-coloured. The bushy whiskers, which come forward so as to conceal a large part of the cheeks, are straw-coloured, with a mixture of a few black hairs. A black transverse band, surmounted by a thin streak of grey, extends from above the eyebrows to the base of the ears; the latter, together with the hands and feet, being of a livid flesh-tint.

The mona is the typical representative of a group of guenons collectively characterised by the following features. The pale colour of the inner surface of the arms forms a marked contrast with the black or dark grey tint of the outer side. A black band extends from ear to ear across the temple, sharply sundering the colour of the cheeks from that of the crown of the head. There is also a well-defined pale brow-band, which sometimes extends over a considerable portion of the summit of the head. In colour the face is slaty blue, but the middle of the lips is flesh-coloured. The shoulder and the fore portion of the back are redder or greyer than the posterior area, which tends to blackness.

There are rather more than half a dozen species in this group, of which the collective range extends from the West Coast through the Equatorial forest-zone to the Ituri forest of East Central Africa.

Among the members of the group is the bearded guenon (*C. pogonias*), which has the fur of the upper parts either greyish or olive brown, finely grizzled with grey or yellow. From above the eyes to the ears there is a black streak, while there is also a stripe of the same hue down the back; and the hands and feet, as well as the tail, are likewise black. The forehead, the whiskers, the small moustache, and the under parts of the body and the inner sides of the limbs are yellowish. This species has been obtained from Fernando Po and the Gabun.

This bearded monkey and the nearly allied Erxleben's monkey (*C. grayi*) represent a subgroup characterised by their yellowish or rufous ear-tufts and the presence of three longitudinal black stripes on the forehead.

Another member of the mona group is the West African Wolf's monkey (*C. wolfi*), which has light under parts, and differs from the other species in the ferruginous colour of the legs, as well as in the yellowish white bar across the forehead, and in the red on the inner sides of the arms, thighs, and belly being pronounced only when the white passes into the coloured outer or lateral surface.

All the above-mentioned members of the present group, with the exception of the mona itself, have a rusty yellow tinge on at least the borders of the under surface of the body, and also on the hairs of the ears. In the mona and three other closely related species the whole under surface and the under sides of the limbs are white, while the hair on the ears is generally speckled.

Campbell's guenon (*C. campbelli*), which is one of the commonest West African monkeys, is one of these three immediate relatives of the mona, all of which lack the white spot on the thigh near the root of the tail. Campbell's guenon, which is found in troops and is stated to take readily to water, is characterised by having the basal two-thirds of the tail yellowish grey, and a white brow-patch. Burnett's guenon (*C. burnetti*), of the African West Coast, differs by the yellow brow-patch, which is nearly obsolete in Dent's guenon (*C. denti*), of the Ituri forest.

THE MOUSTACHE MONKEY

In all the round-headed and short-muzzled guenons mentioned above the nose and body are of the same colour. There are, however, other groups of guenons in which the nose usually has a spot or patch of white, red, or blue. Most of the members of these groups are West African, but they are also to be found in Eastern Equatorial Africa. The leading characters of the moustache monkey (*C. cephus*) are a triangular blue mark on the nose, and yellow whiskers. Its colour is olive green, speckled with yellow, the throat and under parts being grey, the face and temples black, and the feet and hands blackish. The members of the moustache group resemble the lesser white-nosed monkey in the almost uniformly speckled colouring of the head, back, and sides of the body, and usually in the presence of a dark brow-band extending backwards to the ears, and of a second black stripe separated from the former by a patch of yellowish hair running from the neighbourhood of the upper lip for a variab e distance towards the lower rim of the ear, but having the throat and under parts, and at least the upper portion of the inner surface of the limbs, dark, ashy grey, instead of white or greyish white. All the members of the *cephus* group are West African. In addition to the type species, they include *C. erythrotis*, of Fernando Po and the Cameroons, and *C. sclateri*, of Benin.

THE HOCHEUR MONKEY

The hocheur monkey (*C. nictitans*) is one of two species belonging to the group of guenons known as white-nosed monkeys, from the circumstance that the nose in all of them is covered with white hairs. This particular species is distinguished by the blackness of the fur on the upper and lower surfaces of the body, that of the back being finely speckled with yellow. The naked part of the face is of a bluish black colour ; the upper eyelids are flesh-coloured, and the hands and feet jet black. It inhabits the Cameroons and neighbouring districts, and may be distinguished from its relative, *C. martini*, ranging from Liberia to the Cameroons, by the under parts and inside of the arms being blackish instead of whitish. The hocheur, sometimes known as the larger white-nosed monkey, may be distinguished from the lesser white-nosed monkey not only by its superior size, but also by its more prominent nose. It has been described as lively and good-natured, but not so gentle and familiar as the next species, and more resembling in its temper and general character the mona monkey.

THE LESSER WHITE-NOSED MONKEY

This elegant monkey (*C. petaurista*), which comes from the West Coast of Africa, is one of the smallest of the guenons. It was described in the works of Buffon under the well-chosen name of blanc-nez, and is distinguished from the hocheur by its smaller size and the flatness of its nose, as well as by the lighter tint of the under parts. In colour, the fur of the back is olive green, speckled with yellow ;

THE HOCHEUR MONKEY
Photograph by W. S. Berridge

the face black ; the white spot on the nose small and nearly triangular ; and the cheeks, chin, under parts of the body, the inner sides of the limbs, and the under side of the tail are white. It represents a small group, which includes *C. ascanius*, of the Congo, with a local race, *C. a. schmidti*, in Uganda, and the West African *C. signatus*.

The following account of the habits of the lesser white-nosed monkey in confinement is taken from an anonymous writer, who states that its manners " are playful and engaging beyond any other species we have ever observed, and it has an amiability and innocence in its conduct and expression which, united to its lively and familiar disposition, never fail to make it a prime favourite with its visitors. An individual of this species, which formerly lived in the gardens of the London Zoological Society, was confined in the same cage with a young hanumán, whose gravity was sorely disturbed by the unwearied activity and playfulness of its mercurial companion. Whilst the white-nose was frolicking round the cage or playing with the spectators, the hanumán would sit upon the perch, the very picture of melancholy and apathy, with his long tail hanging down to the bottom ; but his attention was roused and his security endangered every moment by the tricks of the restless little creature, which in its sports and gambols continually caught the hanumán's tail, either to swing itself out of the reach of the spectators, or, like a boy at his gymnastic exercises, to assist it in climbing up to the perch. All this, however, was done with great good-nature on both sides, and it was highly diverting to see the playful innocence of the one and the gravity with which the other regarded it, like a parent enjoying the follies of a child."

THE DIANA MONKEY

The well-known Diana monkey (*C. diana*), which derives its name from the white crescent on the forehead above the eyebrows, typifies a group of two West African species characterised by the longer or shorter white beard, the white streak on the haunches near the root of the tail being also distinctive.

THE ROLLOWAY MONKEY
Photograph by W. S. Berridge

The general colour of the fur is black, finely speckled with white, thus producing a greyish grizzle. In addition to the white beard and the crescent on the forehead, the cheeks, chin, throat, chest, front of the shoulders, as well as the inside of the thighs and the streak across the haunches, are likewise white. On the other hand, there is a broad streak of a bay colour down the back, and the same tint also prevails on the rump. The face, tail, and the outer sides of the wrists and legs are black, as well as the hands and feet. All the colours are sharply defined from one another; and the long, narrow, black face, terminating below in the white beard, and surmounted by the crescent above the eyebrows, gives it a peculiar and characteristic expression. The whole length of the head and body is about 18 in., while the length of the tail reaches to some 24 in.

In the Diana monkey the beard is relatively short, with its basal portion covered to a considerable extent by black hairs; but in the rolloway (C. rolloway) it is much larger, and wholly white (page 225). The former species is confined to Liberia, while the latter ranges from the Gold Coast to Guinea.

THE MANGABEYS

The typical mangabeys, or, as they are often called, white-eyelid monkeys, comprise a small group of West African species, which, while agreeing in general characters with the guenons, are distinguished by the presence of a projecting heel at the hind end of the last molar tooth on each side of the lower jaw, so that the crown of this tooth carries five, in place of four, tubercles. In this respect the mangabeys agree with the great group of macaques, and on this ground these monkeys have been separated from the guenons to form a distinct genus under the name of Cercocebus.

The name mangabey, it may be observed, is taken from the district Mangabe, or Manongabe, in Madagascar, and was applied by Buffon to these monkeys from the mistaken idea that they came from that island, which in his time appears to have been a kind of refuge for the destitute in regard to animals whose habitat was unknown. In spite, however, of this totally erroneous origin, the name is a convenient one, and has been almost universally adopted for this group.

In respect of general characteristics it may be said that the typical mangabeys have an oval-shaped head, with a somewhat long muzzle; and that they may be readily recognised in the living condition by the flesh-coloured eyelids, and by the circumstance that their hairs differ from those of the guenons in not being ringed with different colours.

THE SOOTY MANGABEY

This monkey belongs to a small group characterised by the hair of the crown of the head being directed backwards, without any prolongation into a crest. As its name implies, the fur of the sooty mangabey (Cercocebus fuliginosus) is of a deep and dull black hue; the chin and under parts being ashy. The face is livid, marked with dark brown blotches about the eyes, nose, muzzle, and cheeks; the ears, as well as the palms of the hands and the soles of the feet, being of a blacker brown.

At least in captivity, this species is characterised by the unusual habit of keeping its long tail turned forwards over the body. In confinement this mangabey is docile and good-tempered, and more amenable to instruction than is the case with the majority of the larger guenons. A specimen which lived many years ago in the Zoological Society's Gardens in London was a most importunate beggar; but instead of snatching in the customary simian fashion the contributions of his visitors with violence or anger, he solicited them, and, of course, obtained them the more readily, by tumbling, dancing, and a hundred other entertaining tricks. He was very fond of being caressed, and used to examine the hands of his patrons with remarkable gentleness and gravity, trying to pick out the tiny hairs, expressing satisfaction during the process by smacking his lips and uttering a low, surprised grunt.

The white-collared mangabey (C. collaris) may be easily distinguished from the sooty mangabey by its blackish grey colour, the white round the neck, and the bay on the crown of the head; the white of the collar extending on to the cheeks, throat, and chest (page 227).

A third representative of this group is the white-crowned mangabey (C. lunulatus), which takes its name from a characteristic white spot on the crown, and is also distinguished by a white streak running down the middle of the back.

THE CRESTED MANGABEY

The fact that the hair of the crown of the head is lengthened so as to form a crest affords a ready means of distinguishing the crested mangabey (C. albigena) from the true mangabeys; a further distinction being the black upper eyelids (page 227). The colour is blackish, but the creature's name is founded on the greyish hairs on the sides of the throat and cheeks. It was first made known to science in 1850 from specimens sent home to England from the West Coast of Africa by Du Chaillu before his celebrated expedition of 1856. Since that date several local races have been distinguished, one of which (S. albigena johnstoni) is from the Tanganyika district; while

THE SOOTY MANGABEY
Photograph by W. P. Dando

THE WHITE-COLLARED MANGABEY
Photograph by W. P. Dando

C. albigena rothschildi is probably from the Congo. Some naturalists consider that these monkeys should be separated from the typical mangabeys under the name of *Semnocebus*.

THE ORANGE-BELLIED MONKEY

This remarkable West African monkey (*C. chrysogaster*), probably from the Congo, appears to connect in some degree the guenons with the mangabeys. Having brilliant orange under parts, this monkey resembles the guenons in the differently-coloured rings on the hairs of the crown of the head and back, but agrees with the mangabeys in the flesh-coloured upper eyelids.

HAGENBECK'S MONKEY

Another interesting West African species, likewise probably from the Congo, is Hagenbeck's monkey (*C. hagenbecki*), in which the hair of the crown of the head and neck is ringed with black and yellow, like that of the guenons, but that on the back uniformly yellowish grey, as in the mangabeys. In the male the upper eyelids are dark, although in the presumed female they are of the flesh-coloured type characteristic of the typical mangabeys.

THE MACAQUE MONKEYS

Reference has already been made to the curious origin of the term mangabey, and it appears from the sequel that there is a kind of fatality in regard to the misapplication of names among these animals. So far as can be learnt, the name macac, or macaque, seems to be a barbarous word which, in Marcgravs' "Natural History of Brazil," published in the year 1648, is given as the native name of a monkey from the Congo and Guinea. Buffon, however, with the facility for misappropriation for which he was

notorious, transferred this name to the Indian group forming this part of our subject, and to them it has ever afterwards clung, having been Latinised into *Macacus*. In spite of its origin, the name is good enough, and so must remain.

In the account of the mangabeys it was shown that these monkeys differ from the guenons in having a heel, and thus five cusps, to each of their last lower molar teeth, and frequently also in the uniform colouring of the individual hairs (page 226). As this is also the case in the macaques, it is obvious that in this respect the mangabeys form a transition to them from the guenons ; and it is now desirable to point out how the macaques and mangabeys are distinguished from one another.

In the first place, macaques are always of stouter build than mangabeys ; and secondly, they are characterised by the considerably greater prolongation of the muzzle and the larger size of the naked callosities on the buttocks. In some macaques the tails are as long as those of the guenons and mangabeys ; in others these appendages are very short, while in few they are actually wanting ; thus showing that the presence or absence of a tail is of no import either as a generic character, or as indicative of a higher or lower degree of organisation. In common with those of all the monkeys hitherto considered, the nostrils of the macaques do not reach as far forwards as the extremity of the muzzle.

From these characters it will be apparent that while the macaques are sufficiently distinguished from the mangabeys to be entitled to rank as a separate genus, yet both groups are closely allied. In the opposite direction the macaques are intimately connected through one singular intermediate form with the baboons of Africa, so that an almost complete transition exists from the guenons through the mangabeys to the macaques, and thus to the baboons.

THE CRESTED, OR GREY-CHEEKED MANGABEY
Photograph by W. S. Berridge

In speaking of the macaques as Asiatic monkeys, it should be mentioned that one solitary outlying species is found in the mountains of North-west Africa, and also on the opposite coast of Gibraltar. The great majority of the species, however, are confined to India, Burma, the Malay Peninsula, and the islands of Borneo, Sumatra, etc. Some range as far east as China, and one is found even in Japan. To the north, macaques extend into the outer ranges of the Himalaya Mountains, while a single species inhabits the secluded highlands of North-western China.

The whole of the monkeys reckoned as macaques seem to have much the same general habits, being always found gathered together in troops, which may include considerable numbers, and always comprise individuals of both sexes, and of all sizes and ages. They are forest-dwelling animals, and, while active and rapid in their movements, are less markedly so than their compatriots the langurs. As regards food, macaques have a varied appetite, most, if not all, eagerly eating insects as well as seeds and fruits. Moreover, they have occasionally been observed to devour lizards, and it is reported that frogs also form part of their food on rare occasions ; while one species is known to subsist partly on crustaceans. Their cheek pouches are of very large size, and it is the general habit of these monkeys to stuff these receptacles as full as they will hold on every available opportunity.

The voice and gestures of all the macaques are similar, and differ markedly from those of the langurs. According to Colonel Tickell—an excellent observer of the habits of Indian animals— " Anger is generally silent, or, at most, expressed by a low, hoarse ' *heu*,' not so gular or guttural as a growl. Ennui and a desire for company are rendered by a whining ' *homi*' ; invitation, deprecation, entreaty, by a smacking of the lips and a display of the incisors into a regular broad grin, accompanied with a subdued grunting chuckle, highly expressive, but not to be rendered on paper. Fear and alarm are indicated by a loud, harsh shriek, ' *kra*,' or ' *kraouk*,' which serves also as a warning to the others who may be heedless of danger. Unlike the langurs and gibbons, they have no voice of call to one another."

Most of the species are docile if caught young, but old males that have been captured when fully grown are sometimes exceedingly spiteful ; and I have a very vivid recollection of a pig-tailed macaque formerly in the Zoological Gardens at Calcutta

THE BONNET
MONKEY

that was very ferocious, and would fly, with open mouth and the most menacing gestures, at every visitor who approached his cage. In their wild state it also appears that they will occasionally show fight. Mr. R. A. Sterndale states, for instance, that on one occasion during the Indian Mutiny he came across a party of rhesus macaques, among which there were several females with young ones. He endeavoured, without success, to run them down in order to capture the latter, when he was deliberately charged by the old males of the party, the leader of which he had to despatch with a pistol bullet. Several of the species will breed in captivity. As a rule, their manners under restraint are the reverse of pleasant.

THE BONNET MONKEY

One of the best known of the longer - tailed macaques is the South' Indian bonnet monkey (*Macacus sinicus*), which is one of two closely allied species characterised by the fact that the hair of the crown of the head is lengthened and arranged in a radiating manner from the middle line, and by the large ears.

This species takes its name from the crest of hair on the crown, which, instead of coming over the forehead, as a rule stops short of that part of the head, and thus assumes a toque-like form (page 212). On the forehead the short hair is usually parted down the middle line. The fur, which is of moderate length, straight and smooth, is brown or greyish brown above, and pale brown or whitish on the under parts. The face and ears are flesh-coloured, and in some examples the ends of the hairs are ringed. The tail is generally nearly or quite as long as the head and body, the length of the two latter being about 20 in.

This macaque, which occurs all over Southern India and extends westwards to Bombay, is the common monkey of those regions, being found not only in the forests, but likewise in the towns, where it pillages the shops of the *bhanias*, or native grain-sellers. It is exceedingly mischievous, and a ready mimic, although the rhesus monkey is its equal in these respects.

In Ceylon this monkey is replaced by the closely allied toque monkey (*M. pileatus*), which appears only to differ in colour, although the long hair of the crest of the head seems to be more generally continued on to the forehead.

Among the Sinhalese this monkey is known as the rilawa. Sir Emerson Tennent speaks of it as the

universal pet and favourite of natives and Europeans alike, the Tamil conjurers teaching it to dance and carrying it from village to village, clad in a grotesque dress, to exhibit its lively performances. After all, the mimicry and amusing tricks of a monkey in captivity are a mere shadow of what they are in its native forests, so that persons who have only seen these animals in confinement have but a faint idea of their faculty for fun.

THE CRAB-EATING MACAQUE

This species derives its name (*M. cynomolgus*) from its habit of feeding on crabs from the brackish waters of the lagoons and swamps on the coast. It is the true macaque of Buffon, and is known to the Malays, apparently from its cry, as the kra.

It may be at once distinguished from the bonnet monkey by the circumstance that the hair on the crown of the head is neither longer than the rest nor distinctly radiated from the middle. In some individuals there is, however, a trace of a crest, with slight radiation of the hair from one or more points on the forehead. As a rule, the general colour of the fur of the upper parts varies from a dusky or greyish brown to rufous or golden brown; the under parts being either light greyish brown or nearly white. The hairs of adult individuals vary in colour in different parts of their length, and are ringed at their tips. The naked parts of the face and the callosities on the buttocks are flesh-coloured or dusky. The eyelids are either white or bluish white. The tail is nearly as long as the head and body, the combined length of the two reaching 22 in.

In the dark and smaller variety of this common monkey the fur is dusky; while in the lighter or golden rufous variety the face is flesh-colour.

The range of the crab-eating macaque is a very wide one, extending from Siam in the east, through the Malay Peninsula into Lower Burma and the Arakan coast. It is also found in the Nicobar Islands in the Bay of Bengal, although it has probably been introduced there by human agency. Local races inhabit many of the Malay islands.

What induced the ancestors of this monkey to forsake the usual simian food and take to a diet of crabs and insects is difficult to conceive; unless, indeed, they may have been driven to it during a season of scarcity, and found it so much to their

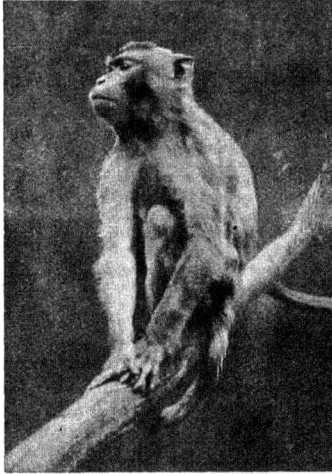

THE RHESUS MONKEY

YOUNG CRAB-EATING MACAQUE
Photograph by W. P. Dando

liking that they have continued it ever since. Be this as it may, however, there is no doubt whatever as to the crustacean-devouring proclivities of this species. Besides stating that they frequent the banks of salt-water creeks and devour shellfish, Sir Arthur Phayre says that in the cheek-pouches of a female were found the claws and body of a crab. There is not much on record regarding the habits of this creature in a wild state beyond what is stated concerning its partiality for crabs, a *penchant* which Sir Arthur believes to be shared by the rhesus in the Bengal Sundarbans. Crab-eating macaques are common on the tidal creeks and rivers of Burma and Tenasserim, especially in the delta of the Irawadi. They go usually in small family parties of from five to fifteen, including an old male and four or five females with their offspring. Their home is among the roots and boughs of the mangrove trees, and they spend a large portion of their time in searching for insects and crabs. From the constant presence of human beings on the waterways near which they dwell, these monkeys become very tame, and can be easily approached. They will even pick up rice or fruit thrown down to them. Still more remarkable is the facility with which they can swim and dive. On one occasion a male of this species that had been wounded and placed for security in a boat jumped overboard and dived several times over to a distance of some fifty yards in order to prevent recapture. Like most macaques, this species is gentle if captured at a sufficiently early age, but the old males always become fierce and morose. On account of its white eyelids, care must be taken not to confound this monkey with the mangabeys, noticed previously (page 226).

THE LION-TAILED MONKEY

The lion-tailed monkey (*M. silenus*) of Western India is the first of the macaques in which the length of the tail is less than three-quarters of that of the head and body taken together. It is often called the wanderu—a term which, however, should be restricted to the langurs of Ceylon—and may be distinguished from all the other species by its general black colour, and the enormous grey beard and ruff which surround the black face (page 212), with the exception of the middle of the forehead, where the ruff stops short. The fur is long, and the slender tail, which is tufted at its

T

THE LION-TAILED MONKEY

extremity, measures from half to three-quarters the combined length of the head and body. The thin and tufted tail, like that of a lion, is one of the characteristic features of this species, and that from which it derives its name. The enormous ruff, totally concealing the ears, is, however, that which especially attracts attention, and gives the owner somewhat the appearance of a black-faced old man with shaggy whiskers and beard.

These monkeys inhabit the Malabar, or western coast of India, from Cape Comorin to about 14° N., being especially abundant in the districts of Travancore and Cochin. They restrict themselves to the forest lands on the range of trappean mountains known as the Western Ghats, and are always found at a considerable elevation above the level of the sea. They associate in troops of from twelve to twenty or more in number, and are excessivly shy and wary, and when caught are sulky and savage in captivity, so that it is only with great difficulty that they can be taught feats of agility or mimicry.

THE BENGAL MONKEY

Perhaps the best known of the macaques is the Bengal or rhesus monkey (*M. rhesus*), the bandar of the Hindus, found all over Northern India (pages 229, 236). It presents little resemblance to the last species, having no trace of a beard or ruff, and its colour being brown, with a tinge of grey. As a species, it is characterised by the straightness of its moderately long hair, and also by the buttocks being naked for some distance round the callosities. The tail is about one-half the length of the head and body, and tapers regularly from base to tip, without any trace of a terminal tuft. The face, as well as the callosities on the buttocks, is flesh-coloured, except in the adults, in which both are bright red.

In India the Bengal monkey is found continuously northwards from the valley of the Godaveri to the Himalaya, extending to the west coast of Bombay. It inhabits the valley of Kashmir and surrounding regions, at elevations of and above four thousand feet. In the neighbourhood of the hill station of Simla these monkeys are found at an elevation of between eight and nine thousand feet above the sea; and it is one of the regular excursions from Simla to ride or walk to see the monkeys on their own hill, which bears the appropriate name of Jako. Here they are regularly fed by a fakir, who has taken up his abode on the same mountain, and they come down in troops at his call. Indeed, these monkeys are almost invariably found in large droves, usually in the forests or more cultivated lands, but occasionally near and in the towns. Although not regarded as sacred, it appears that the rhesus monkey is frequently protected by the Hindus, and in Kashmir the writer has seen them forming part and parcel of the appanages of the temples. In several parts of India the Hindus have, indeed, a strong objection to the slaughter of these monkeys.

The rhesus is an intelligent creature, and, if captured young, is easily taught. It is the common monkey carried about by itinerant jugglers in Northern India, by whom it is taught many amusing tricks. Old animals, more especially the males, become vicious and spiteful.

In their wild state these monkeys make a hideous noise with their incessant chattering, and are always mischievous. In addition to the consumption of large quantities of fruit and seeds, they also subsist on insects and spiders, and parties of them may frequently be seen carefully searching the ground for these delicacies. The rhesus, like the crab-eating macaque, swims well, and takes readily to the water.

It is probable that the Bengal monkey ranges to the north-east into Assam and Upper Burma, and thence into the province of Yunnan, in Western China. In Sze-chuen, and eastwards into the interior of China, it is replaced by the allied but tailless hairy-eared macaque (*M. lasiotis*, page 231).

Another nearly related species is the Himalayan macaque (*M. assamensis*), found at considerable elevations in the Eastern Himalaya, Assam, the Mishmi Hills, and parts of Upper Burma. This species is distinguished from the Bengal monkey by the wavy nature of the hair, which in the Himalayan species assumes a decidedly woolly texture. This Himalayan macaque is also larger than the Bengal species, and more powerfully and more compactly built, and thus approaches the pig-tailed monkey. It is likewise, whether wild or tame, more sluggish in its movements than the Bengal monkey; and there is a slight difference between the voice of the two species.

THE PIG-TAILED MONKEY

The next species of macaque for notice is the pig-tailed monkey (*M. nemestrinus*). It is distinguished from most of those mentioned above by the shorter tail, which is thin and whip-like, and only about one-third the length of the head and body. It is a comparatively stout and long-limbed monkey, recognisable by the hair radiating from the centre of the head, the slender pig-like tail, and the very projecting muzzle, which approximates to that of the baboons. An adult full-grown male may be compared to a good-sized mastiff, both as regards size and strength. This monkey has been long known to science, and was described by Buffon as the maimon. It inhabits the province of Tenasserim, and thence extends southwards into the Malay Peninsula, and is also found in the islands of Borneo and Sumatra. The voice and manners of this monkey are described as being very similar to those of the Bengal monkey. Its habits were long ago described by Sir Stamford Raffles from specimens observed in Sumatra ; and it is stated that the inhabitants of that island train these monkeys to ascend the cocoa-palms, and select and then throw down the ripest fruit. If the story be true, it would seem probable that it must be only young or female individuals that are thus taught to serve their masters, since the old males are exceedingly fierce and vicious, and from their size and powerful build are formidable antagonists.

THE BURMESE PIG-TAILED MONKEY

In Arakan and Upper Burma the place of the pig-tailed monkey is taken by a nearly allied species, the Burmese pig-tailed monkey (*M. leoninus*). This macaque may be distinguished from its relations by its shorter limbs and muzzle, and longer hair, as well as by the black horseshoe-like crest on the temples above the eyes, a feature that stands out in marked contrast to the brown colour of the rest of the fur. Moreover, the short tail, which is generally carried over the back, is more hairy and more or less distinctly tufted at the end. The males are dark brown above, but the females somewhat lighter ; the face in both sexes being of a dusky flesh-colour, while the combined length of the head and body is about 23 in. ; the tail only measures some 8 in., exclusive of the hair at its extremity, which adds another 2 in. to its length. Mr. E. Blyth, who alluded to this species as the long-haired pig-tailed monkey, in contradistinction to the short-

THE PIG-TAILED MONKEY

haired pig-tailed monkey (*M. nemestrinus*), states that it does not appear to be at all common, and that it chiefly inhabits the range of limestone mountains from the north of Arakan to an undetermined distance southwards. The Burmese pig-tailed monkey serves to connect the other species with the Bengal monkey.

THE BROWN STUMP-TAILED MONKEY

The brown stump-tailed monkey (*M. arctoides*) may be taken as an example of another group of macaques inhabiting Burma and the Malayan region, and thence ranging into China and Japan, and characterised by the reduction of the tail to a mere stump. The species is distinguished by the length of its dark brown or blackish brown hair, which may measure more than 4 in. ; and also by the bright red hue of the naked portions of the face and buttocks. As in the last-named species, the terminal portions of the hairs of old individuals are decorated with rings of different colours. The length of the head and body is about 24 in., while that of the tail does not exceed 1 or 2 in. This monkey appears to range from the southern parts of Assam into Upper Burma, and is also found in Cochin China. We have not, however, full information on the subject of its geographical range, and absolutely none as to its habits, but it is said to be an inhabitant of hilly districts. In old individuals the tail is sparsely clad with hair, or naked. In the forests of Sze-chuen the stump-tailed macaques are represented by a species (*M. tibetanus*) characterised by its larger size and the thickly-haired tail. The Moor macaque (*M. maurus*), which has received several names—*M. ochreatus*, for instance—represents the stump-tailed monkeys in Borneo, Celebes, and probably some of the other Malayan islands, and is a dark and black-faced species. In Japan the group is represented by the Japanese

THE HAIRY-EARED MACAQUE
From the "Proceedings" of the Zoological Society

231

macaque (*M. fuscatus*), which is one of those in which the tail is thickly haired (page 237). There is, however, still much to learn as to the number of species of these stump-tailed macaques, and their exact geographical distribution ; while information as to their mode of life is likewise desirable.

THE MAGOT, OR BARBARY MACAQUE

In the preceding pages it has been pointed out that a gradual shortening of the tail can be traced as we pass from the bonnet macaque, through the rhesus monkey and its allies, to the pig-tailed, and thence to the stump-tailed group. From the latter it is but a step to the total loss of the tail ; and the magot, or Barbary macaque (the Barbary ape of many authors), is the culminating member of the series. This absence of a tail was regarded as a reason for separating the magot (*M. inuus*) as a distinct genus from the other macaques, but there is no justification for this view, especially as the Chinese rhesus is tailless.

In addition to being the one of two tailless macaques, the magot is the sole existing species which is not Asiatic. The magot inhabits the north-western corner of Africa, in the districts of Morocco and Algeria, being especially common in the latter country in the neighbourhood of the city of Constantine. It is also found, across the strait, in Gibraltar and some of the neighbouring parts of Spain, but whether indigenous there, or introduced from the opposite continent by human agency, has not been clearly made out. The wide separation of the macaque from its Asiatic congeners suggests that it is the direct descendant of those extinct species which are found in the later geological deposits of various parts of Europe, at a date when we know that the genus was already in existence in India.

That the magot is the *Pithecus* of the ancients there is no doubt, as the description given by Aristotle is enough to identify it. This species was, indeed, in all probability, the only tailless member of the order with which the ancients were acquainted. It was, moreover, the animal from which the ancient Greeks obtained such knowledge as they possessed of human anatomy ; and an account of its anatomy, given by Galen, has been handed down to our own times. The name magot is of French origin, and was applied by Buffon.

This monkey is as large as a good-sized dog ; and the upper parts of its body, and the outer sides of the limbs, are of a light yellowish brown, becoming somewhat deeper on the head, and also along a line bordering the cheeks. The under parts are a dull yellowish white, while the naked portions of the face, hands, and feet, as well as the callosities on the buttocks, are flesh-coloured. The rudiment of the tail consists merely of a little fold of skin, having no sort of connection with the end of the backbone (pages 35, 233).

One of the best early original accounts of the magot is given by the French naturalist René-Luiche Desfontaines, who resided for some time in Algeria, during the closing decades of the eighteenth century. This writer observes that the magots " live in troops in the forests of the Atlas Mountains nearest to the seaboard, and are so common at Stora that the surrounding trees are sometimes covered with them. They live upon the cones of the pine, sweet chestnuts, and the figs, melons, pistachio nuts, and vegetables which they steal from the gardens of the Arabs, in spite of all the pains taken to exclude these mischievous animals. Whilst in the act of committing these thefts, two or three detach themselves from the general body, and keep watch from the tops of the surrounding trees or rocks ; and as soon as these sentinels perceive the approach of danger, they give warning to their companions, who presently scamper off with whatever they have been able to lay hands on."

An officer, writing in 1880, in " The Field," gave the following excellent account of the magots at Gibraltar. " These apes were formerly very numerous on the Rock, and there were several gangs of them, but they were so predaceous in their habits, coming down to the gardens in the upper part of the town, and stealing fruit, especially figs, that they were killed by trap or poison, so as nearly to bring about their extinction. In November, 1856, a garrison order was published for the guidance of the signalmaster," which forbade the destruction of the monkeys, and gave directions as to their being counted at regular intervals. " From that time," continued the officer, " the register has been very regularly kept by the signalmaster. There were only four or five at this time, and but three in 1863, when General Sir W. Codrington, who was then governor, saved them from destruction by a fresh importation from Africa. The following note occurs in the Journal of May 26, 1865 : 'Turned out four apes, wild from Barbary, two males and two females, all young.' After some time the new-comers made friends with the apes of the old stock ; and the band increased, but very slowly, however, owing to the great preponderance of females, until the present time, and it may be expected, as the signalmaster observes, now that there are two adult and rival males, that it will divide. Those who wish to see them will do well to remember that their haunts on the Rock are determined by the direction of the wind. They prefer the ledges of the [to man] inaccessible, abrupt escarpment of the Mediterranean face ; but cannot stand the cold, damp Levanter wind, which, as its name indicates, blows from the east, and compels them to resort to the western slopes on the town side of the Rock. At the bottom of Charles V.'s wall, overhanging the Alameda Gardens, is a favourite spot. On the western side, the Monkeys' Alameda, a small bushy plateau half-way down the precipice, is another choice resort, as is also Monkeys' Cave, close to the sea. Of late years they have become sufficiently confident in their friend and protector, the signalmaster, frequently to enter the enclosure of the station, especially in the summer drought, when they come for water. In a letter to me, of May 3rd,

Sergeant Brown, the signalmaster, says : ' The monkeys are sitting on the wall of the station as I write this—the first time this season that they have come up for water.'

" Their food consists of grass, the young blades of which I have seen them eating with avidity, and of a variety of roots and bulbs ; those of the yellow Cape oxalis being much sought after. The fruits of the palmetto—' monkey dates,' as the Gibraltar urchins, who also much appreciate the little brown viscous clusters, call them—are greedily devoured when ripe. The signalmaster has never observed them take any food left in their way but a few grapes, of which they seemed very fond. . . .

" Sometimes a fight occurs among the monkeys, when it is surprising to witness the rapidity with which they will follow an offender down the stupendous precipice of the eastern face ; tumbling one after another, and catching at bits of bush or protecting ledges on their way, they descend hundreds of feet in a moment or two. Sometimes the sergeant dresses wounds on them, probably from this cause, but they soon heal up."

According to recent reports, these apes are now nearly, if not completely, exterminated in Gibraltar. In captivity the magot, at least during youth, is lively, active, intelligent, and good-tempered ; but with advancing years it becomes sullen and capricious, and finally spiteful and vindictive. The French naturalist Frédéric Cuvier observes that the natural instinct which causes these monkeys when in a wild condition to associate in troops leads solitary individuals in confinement to make friends of such animals as they are thrown in contact with. Such animals, if sufficiently small, are carried about by the magots, which hug and caress their burdens, and become furious when any attempts are made to remove them.

The magot is perhaps brought oftener to Europe than any other monkey ; its native climate being such as to permit of its existing with tolerable comfort in more northerly regions.

THE BARBARY APE, OR MAGOT

EXTINCT MACAQUES

In describing the magot, incidental reference was made to the occurrence of fossil species of macaques ; but as this is a subject of considerable interest to the present geographical distribution of these monkeys, we must say a few words more. Asia being the headquarters of the group, it would only be natural to find these monkeys represented in a fossil state on that continent. As a matter of fact, with the exception of India, comparatively little is known of the geology of Asia. In India, however, remains of macaques are found in the caverns of Madras, and in certain Pliocene deposits in the Punjab.

In Europe fossil macaques occur in freshwater deposits belonging to the same Pliocene period, in the south of France, Switzerland, and the north of Italy, in the valley of the Arno. The occurrence of these extinct monkeys need not imply any very great change of climate in those regions. The case is, however, very different with the single fragment of the jaw of a macaque which has been found fossil near the village of Grays, in Essex, in strata which belong to the latest or Pleistocene epoch of geological history. This monkey must have lived in England during the time when man had already made his appearance ; and there is no doubt but that the climate must then have been milder than it is at the present day. As we have said, these extinct European macaques may be the progenitors of the magot.

In addition to remains of macaques, there occur in the Pliocene rocks of Attica and the south of France those of other monkeys which appear to indicate a transition from the macaques to the langurs. These monkeys, known as the mesopitheque and the dolichopitheque, have indeed short and stout limbs like those of the macaques, but skulls resembling those of the langur. As we shall never know the structure of their soft parts, their exact relationships cannot be determined.

THE BLACK APE

The island of Celebes is remarkable for several peculiar types of mammals, among which is the black ape (*Cynopithecus niger*), the sole representative of a genus in some respects connecting the preceding group of the macaques with the following one of the baboons. It was represented many years ago by a living example in the old menagerie at the Tower of London, and by another in that of Exeter 'Change, in the Strand. At that time, however, the true habitat of this animal was quite unknown, Cuvier suggesting that it came from the Philippines ; but its home was subsequently found to be Celebes. This monkey (page 235) is a striking animal, the whole of the fur, as well as the naked parts of the face, hands, and feet, being of an intense black, the only exception to this colouring being the large callosities on the buttocks which are flesh-coloured.

The hair of the body is long and woolly, but that on the limbs shorter. The tail is represented by a mere tubercle, not more than an inch in length. The face is characterised by the marked production of the muzzle, which terminates bluntly, with the nostrils opening obliquely, and placed some distance behind the extremity of the muzzle. It is this position of the nostrils which connects this monkey with the macaques, and distinguishes it from the true baboons, in which the nostrils are situated at the very end of the still more produced muzzle. The sides of the face have the peculiar longitudinal swellings characteristic of the latter, and the cheek pouches are very capacious. On the top of the head the black ape has a broad tuft of long hairs, curling backwards, and forming a very characteristic crest.

The earlier specimens of this monkey brought to England are described as being rather violent in temper, and tyrannising over the other monkeys with which they were placed in company. Others, however, are stated to have been more gentle in disposition, and thus very different from the fierce baboons. But few specimens of this monkey have been exhibited of late years in the Zoological Society's Gardens in London.

The black ape is very common in the forest near Wallace Bay, in Celebes, where these animals may be seen swinging from bough to bough in small flocks. This monkey is also found in the small island of Batchian, lying to the east of Celebes, and forming a part of the Molucca group. On account of the circumstance that none of the other mammals of Celebes extends to Batchian, it has been suggested that the black ape has been accidentally introduced by Malays, who often carry about with them monkeys and other animals. This is rendered more probable by the fact that the animal is not found in Gilolo, which is only separated from Batchian by a very narrow strait. The introduction may have been very recent, as in a fertile and unoccupied island such an animal would multiply rapidly. In its arboreal habits and predilection for fruit, the black ape is essentially a macaque, and not a baboon.

R. LYDEKKER

STORIES OF THE MONKEYS

It is always difficult to decide whether the performances of monkeys are the outcome of cruelty on the part of their trainers, or whether the animals themselves derive pleasure from the show. They have skill enough to achieve wonders, but the net result in such environment is not equal to that accomplished by, say, a dog of ordinary intelligence. This suggests that the monkey may not be a willing pupil. As a mimic he is, of his own initiative, almost incomparable. He seems, however, to have something of that pride of spirit which makes an elephant kill itself rather than resign its task to another.

In the Ceylon Village at the White City, in 1908, were many natives with performing monkeys. The latter did nothing more remarkable than was to be seen at an ordinary monkey show, but on one occasion an interesting exhibition took place that was not in the programme. A youngish Sinhalese squatted down before the little school, and set a small monkey to work. The latter appeared to have friends among the children, for instead of going through its round of tricks in the usual way, it watched the bairns at their lessons and seemed anxious to get to them. Its master, after one or two remonstrances, pulled it up to him by its chain, gave it a gentle cuff, and set it down, retaining its chain upon his wrist. Then he produced a second and larger monkey, and prepared to put it through its facings.

The lesser monkey, which had borne its punishment with resignation, became furious at the appearance of a rival performer. Like a flash it darted at the seated man, sprang on his shoulders, and tried to bite him on the back of the neck. He good-humouredly pulled it down and threw it on its feet, a yard away from him, and turned his attentions to the other monkey. The insulted performer, however, would not be put off in this way. With screams of rage it attacked its master again and again, biting at his hands and arms as he endeavoured to keep it at bay. It was in vain that the man scolded and cuffed; every time he tried to make the other monkey perform, the little one, with cries of fury, dashed at him. The larger monkey looked on with hair on end and jaws agape, the picture of suppressed rage and perplexity. It did not know whether it ought to attack its rival, or make common cause against its master.

For two or three minutes this curious contest lasted, and ended in the victory of the small monkey. For an elderly member of the Sinhalese party, drawn to the spot by the uproar, and seeing what had happened, ordered his countryman to rise, while he himself took charge of the two monkeys. At the first sound of the dice-box drum which the newcomer rattled, the rebel sidled up to him with a look upon its strange little face which could only be interpreted as a plea for permission to perform. Leave being given, the monkey went through its round of tricks like the most obedient child. Its master stood by, looking on and laughing. Every time the monkey saw him it bared its teeth and chattered with rage, and, but for its anxiety to acquit itself with credit before its new master, would have flown at him.

There are depths of intelligence in the monkey's mind which few, if any, of us can plumb. How much they know, or of how much knowledge they are able to make themselves master, it is impossible to say. There is something uncannily keen in their perceptive faculties, something like intuition, as was said of Frank Buckland's baboon (page 249). This faculty discovers itself in divers ways. We are indebted to Darwin for one admirable example. A keeper at the London Zoological Gardens was attacked by a savage baboon, which fixed its terrible teeth in the back of his neck as he knelt on the floor of its cage. In the same cage was a little American monkey, with which this keeper was a special favourite. Here was a position of the like of which the creature could not previously have had experience; there was nothing to guide its course of action. It saw its friend being slowly killed by the one animal of which it went in deadly terror, the despotic lord of the cage—this fierce baboon. In ordinary circumstances the monkey would have avoided the baboon as it would have avoided fire, but its love for its human friend gave it courage. It leapt upon the shoulders of the baboon, and, by screaming and biting, so distracted its attention that the man was able to escape from the cage. But for the intervention of the ordinarily timid little monkey the man would in all likelihood have been slain.

Another way in which the monkey displays this curious intuitive faculty is illustrated by Mr. Frank Finn in his " Ornithological and Other Oddities." In one of the cages of the Calcutta Zoological Gardens were three monkeys—two plump females and a gaunt and feeble-looking male. Noticing the miserable condition of this animal, Mr. Finn remarked one day to an assistant that it had better be put out of its misery. " Scarcely, however, were the words out of my mouth," he says, " before both female monkeys sprang straight at my face, and nothing but the intervening wire saved me from getting bitten. Of course, they could not have understood what I said, but I feel sure their sudden attack was not an accident, and that in some strange way they had divined that I was meditating something against their companion."

A further example of the monkey's alertness may be furnished from the writer's own experience of an intelligent rhesus, whose temper had deteriorated as his wits had been sharpened by a long course of misery on tramp with an Italian. This monkey paid regular visits to a certain garden, and knew exactly the spot to which he must go to be fed ; where he might climb and where not. He would be regaled with all sorts of dainties, but his chief joy was a piece of raw potato. After his meal, he would stretch himself on his side on the grass, sated and not very grateful. He quickly knew that tricks were not expected of him in that garden, and a tug at his tether, ordinarily the signal for the performance of some duty, provoked him to rebellion.

One day, when in a specially perverse mood, the rhesus insisted, after his rest, upon climbing the house by way of a creeper, the upper reaches of which were still young and tender. " Better not let him go any higher," remarked the writer to the monkey's

THE BLACK APE

owner. A slight pull at the chain was given. The monkey looked sharply round, took a couple of steps down, then leapt ten feet to the ground, bounded across the border with incredible swiftness, and did his best to make his teeth meet through the calf of the well-intentioned host. The latter was several yards from the owner of the monkey, and the owner held the chain which had been tugged. But the host had spoken the word that had caused the owner to pull the chain, and punishment must be meted out to him. The victim forbade reprisals, but during the remainder of his stay the monkey avoided him, not now in anger, but with a demeanour which suggested a feeling of shame, if the application of such a term to a monkey may be pardoned.

But we must not expect to find all monkeys imbued with the same wonderful sense of affection and gratitude which saved the keeper's life. The spirit of mischief rules their small minds. Mr. Gambier Bolton, when in India, went to great trouble to conciliate a number of the sacred hanumán monkeys, which he found wild. After the expenditure of vast patience and sugar and meal, several came about him, and when they had eaten their fill, one reached down from a ruin to which it had climbed and, with a hearty thump, expressed its gratitude by sending his pith helmet flying.

Sir Samuel Baker's pet monkey, a red one from Abyssinia, which he called " Wallady," was as big a rascal, and put his owner rather in a fix upon a certain fatiguing march. The traveller had caused a large quantity of black pancakes to be prepared for food for himself and escort on the march. The store was carried in a hamper on the back of the camel on which " Wallady " rode. When the food was sought for at a point when the whole party were famished, and beyond the reach of other supplies, it was found that the pancakes had all disappeared. The monkey had eaten as many as he wanted, and thrown the rest away. " Wallady " atoned, however, in a thousand ways. He was better than any watch-dog in preventing natives from becoming too pressing in their attentions upon the party. With white folk " Wallady " was affectionate and tolerant, but he had no respect for black men, and attacked the calves of dark strangers with such good will that they were only too glad to escape beyond his reach. He was useful, too, as evidence of the peaceful mission of his master, for no man in his senses, it was argued, would travel with hostile intent taking a monkey with him.

The life of the monkey in many points touches the human. In several respects the average monkey is like a growing child, and in nothing more than its insatiable passion for mischief. Frank Buckland's monkey, which so scandalised the dons at Christ Church College, Oxford, rewarded the kindness of its master by rending to shreds an important treatise which he had written for his examination. One of equally mischievous disposition was kept by an enthusiastic entomologist who had devoted years to collecting the rarest of beetles and butterflies. One luckless day he left open the door of his study, and, in the master's absence, the monkey invaded the sanctum. The naturalist returned some time afterwards, to find that the little imp, giving rein to its appetite for beetles and love of mischief, had not only eaten the whole entomological collection, pins and all, but had pinched out and swallowed all the coloured pictures in a valuable work on insects which had been left open. Needless to say, the pins were too much for the digestion of the poor monkey, which shortly afterwards died. Its destruction of the pictures is interesting, as showing that it had mistaken them for the genuine insects. It has been regarded as wonderful that an intelligent, well-trained dog should recognise the portrait of its absent master, but this monkey was at least as complimentary to the artists who had drawn the beetles and butterflies.

A certain "Jeremiah," an Indian specimen, has received the dignity of a biographical notice in "The Spectator." It was called "Jeremiah" because of its intensely dismal expression, but a world of mischief lurked at the back of its little head, and in its heart an abiding affection for its master. But no sense of affection or of gratitude could prevail against its love for the thing which should not be done. Hence, after nestling with pathetic joy in its owner's arms, it would immediately steal some treasured vase or other precious nick-nack, climb with it to the top of some tall tree in the garden, and hurl it to the ground. The scene was India, and its master, having to move from one station to another, magnanimously permitted "Jeremiah" to ride in the new English dogcart which he had just bought. The cavalcade had not gone far when a couple of buttons were twisted off one of the cushions, a hole was bitten in the cloth, and the monkey's paw was drawing out the stuffing

in handfuls. "Jeremiah" was therefore condemned to walk, and a very weary, travel-stained little monkey arrived, hungry and thirsty, at the end of the long day's journey.

"Jeremiah" had evidently pondered this indignity, and decided upon a plan of campaign. When it was given its supper, the men noticed that it sought out a biggish black dog from among the crowd of outcast dogs which hung about the camp, and offered it part of its meal. The next morning "Jeremiah" fed the dog again. Thus it succeeded in beguiling it into following the march, and, by and by, jumped on to the dog's back and rode in triumph. Every day the proceeding was repeated. The monkey rode the whole of the remaining eighty miles on the dog's back. Every night it shared its supper with its courser, and slept curled up close to it.

"Jeremiah's" parting from civilisation was in keeping with its association with it. Its master, having again to move, decided to give the monkey away. As he mounted his horse when quitting the bungalow, he saw the deserted "Jeremiah" come from one of the rooms where packing was going on, and speed across the carriage-drive. Under its arms were two cherished Venetian glass vases; a metal ornament

THE RHESUS MONKEY

was in one hand; a small pair of bellows in the other. Its cheek pouches were distended with stolen dainties. When it saw that it was observed, it darted up a tree, flung down the vases with an angry scream, bit a hole in the bellows, hurled them down after the vases, then, with a triumphant chatter, climbed to the upper branches and was seen no more.

A more amiable monkey was committed to the charge of Mr. Arthur Macleane, who was at that time principal of Brighton College. He was returning from an Indian trip, and a friend who was moving up-country begged him to convey the monkey home. Consent was readily given when the owner said: "I have so high an opinion of her qualities that I have appointed her guardian of four little orphan puppies." The new owner had the monkey and her foster children taken into his room, and, as he was busy, gave her some nuts. Nut-receiving time is always a moment of embarrassment for a monkey, for, after filling its cheeks and its four hands, its powers of receptivity, but not its desire for more, are exhausted. The position was especially difficult now, for she was chained to her pole, and four rampageous pups might escape from her while she ate her nuts. She puckered up her little face in study for a moment, then inspiration came to her. She took pup number one, and laid it on the floor; number two she laid with its head in the opposite direction; numbers three and four were placed in positions to complete the square. Their tails all pointed inwards, making together a cross. Then she sat down upon the tips of all four tails. She was at liberty then to eat her nuts in peace. If one of her charges managed to wriggle away from her, she had time to collect it again and restore it to its original position, not, however, without first giving it a gentle smack by way of rebuke.

It would have enhanced the interest in the story of the aforesaid "Jeremiah" if the author could have rounded it off with some glimpse of the hero in a state of freedom. To turn into the wilds a monkey which has been domesticated is not always a kindness to the animal. Dr. H. O. Forbes kept in captivity for some time a wow-wow gibbon (*Hylobates leuciscus*), a most engaging creature. When the calling of its free mates reached its prison, the ape used most pathetically to place its ear close to the bars of its cage and listen with such intent and eager wistfulness that it was impossible any longer to retain it in durance. It was accordingly set free on the margin of its old forest home. Strange to say, its former companions, perceiving, perhaps, the odour of captivity about it, seemed to distrust its respectability, and refused to allow it to mingle with them. Possibly "Jeremiah" may have found his reception as chilly.

There are always a certain number of outcast males to be found where the hanumáns make their home. The explanation is curious. The lord of the herd tolerates no rivals in his seraglio, and when a young male is born he kills it. When this has happened once or twice, the females appear to realise the situation, and, seizing the infant male monkey, disappear into the jungle, or high up the mountain-side, there to hide until the young male is able to look after himself. The youthful outcast bides his time. He dwells in solitude apart, and waxes strong. When he reaches maturity, he sets out to conquer the leader of a herd. There is a terrible battle, and, if the young one be fully grown and the old champion past his prime, victory lies as a rule with the challenger, and the seraglio thus gets a new lord.

The lethal instinct will out even among captive monkeys. Colonel Rice, whose circus included a number of American monkeys, was puzzled by a series of deaths in this interesting family. The deaths were not due to illness, for a monkey which was quite well overnight would be found dead in the morning, hanging by its tail from its perch. The wiseacres of the show declared that the victims had committed suicide; but Rice set a man to watch, and discovered that not suicide, but murder, was being committed. The victims had all been males; their murderer was their own sire. The culprit was removed, and there were no more deaths. Rice declares that the murderer, after committing his crimes, deliberately hanged his victims upon their perches to divert suspicion from himself.

The red-faced spider monkey, or coaita, of which Bates gives some extremely interesting details is among the best-tempered of monkeys, but a shocking thief of anything which it can convert to the purpose of bedclothes. The coaita does not monopolise this unenviable distinction. There was another of different breed to whose account was laid, in 1905, the destruction of scrip representing a considerable amount in the Bank of France. At an earlier date an unfortunate member of the simian family was accused of ravages among the vouchers of the London Zoological Society, so making it impossible satisfactorily to account for a deficiency. However, the reputation of that monkey was cleared.

Monkeys are blessed with good memories as well as ready, perceptive faculties. No matter how ample the supply of food from visitors, the monkeys at the London Zoo are all attention the moment their keeper puts his head through the doorway of their chamber. Monkeys temporarily lodged in the Gardens recognise their owners when the latter visit them. A striking example of the monkey's memory was given during a performance in a variety theatre at Copenhagen. In the middle of its "turn" a monkey suddenly sprang off the stage and threw itself into the arms of a man in the audience. The man, it proved, had owned this very monkey four years before.

A short acquaintance with a brown capuchin served to make it the faithful friend and admirer of Romanes. This was the monkey which he borrowed from the Zoo in order that he and his sister might keep it under observation. The monkey took a violent liking to him. The sound of his voice two flights of stairs away would set it yelling with delight, and when he entered the room the capuchin screamed with such power that it was impossible to converse even by shouting, until the monkey was taken up in his lap, when it settled down placidly with every sign of intense affection. The monkey was detained in the house under surveillance for about ten weeks, and it lived about two years longer. Every time Romanes entered the Zoo monkey-house, no matter how crowded it might be, the monkey would instantly espy him and dash across its cage towards him, and thrust its hands between the bars with every expression of joy. When he moved away it would follow him to the end of its cage, and remain watching as long as he was in sight.

JAPANESE MACAQUES AT THE ZOO

No more interesting chapter on monkey intelligence has been written than that which Miss Romanes furnished for her brother's book on her experiences with this animal. Untaught, it learned the secret of the screw, puzzled for two hours over the problem of lock and key, threw things when angry, and, when good-tempered, insisted on sharing its food with its mistress, popping a pawful of milk and bread into her hand when she was not looking. Frank Buckland complained that, though he had lived with monkeys about him for years, he had never seen one attempt to put anything on the fire. This one, however, delighted to thrust sticks into the fire, and to smell their smoking ends; to pass hot cinders over its head and chest to feel the warmth. A piece of paper was given to it, and as the chain was too short to allow it to reach the fire with the paper in its existing shape, it rolled it up in the form of a spill, and then thrust the end into the fire, withdrawing it when it was alight. Miss Romanes next gave him a newspaper. He tore it in pieces, rolled up each piece to make it long enough to reach the fire, and so burnt all.

Curiously enough, what most terrified this monkey was a toy model of himself. With a live monkey doubtless he would have fought, but this imitation one scared him extraordinarily. His behaviour before a mirror was very different from that which we generally expect of monkeys. Lieut.-General F. H. Tyrrell has described the antics of his pet monkey before a mirror placed on the ground. Seeing his own reflection in the mirror, the monkey imagined it to be another monkey, so sprang round the glass, to find nothing there. Returning, he looked intently at the reflection, and again rushed round to disappointment. Once more he returned, but this time he fixed his gaze upon the monkey in the mirror, and, keeping him in sight, stretched his paw round to the back of the glass to feel for the elusive stranger. Romanes' monkey behaved differently. At first he was a little afraid of the reflection in the mirror, but in a short time he gained courage enough to approach and try to touch it. Finding that he could not do so, he went round to the back and then to the front, repeating this a number of times. He appeared to mistake the sex of the image, and began in the most ludicrous manner to pay it the addresses of courtship. First placing his lips against the glass, he rose to his full height on his hind legs, retired slowly, and, looking over his shoulder at the image, with a preposterous "pinch" in his back, strutted up and down before the glass with the most absurd conceit.

We have seen that whereas the gibbon's interest in its kind is limited to the life of the latter, the dead being ignored, the orang-utan will, whenever possible, carry off its fellow which the hunter has shot. This affectionate instinct is discovered in certain monkeys also. Some species will carry their dead young about with them for weeks, until the corpses become mummies. As the mothers fondle and endeavour to feed them, it is likely they do not realise that death has taken place. Wallace's young orang-utan was as content with a stuffed dummy of its dam as was the cow which, refusing her milk unless her calf were before her, was satisfied with one made of hay stuffed into a calf-skin. When the latter got damaged and the hay protruded, the cow ate the hay, in blissful ignorance that such a substance should not issue from the hide of her little one.

In the affection of some animals there is a blind instinct, and the devotion of female monkeys to their dead young seems to be of that order. But we touch a higher note in a case cited by Mr. James Forbes in his "Oriental Memoirs." One of a party had shot a female monkey, and carried it to his tent, which was soon surrounded by forty or fifty monkeys. These made a great noise and were disposed to attack the aggressor.

"They retreated when he presented his fowling-piece, the dreadful effect of which they had witnessed, and appeared perfectly to understand. The head of the troop, however, stood his ground, chattering furiously. The sportsman, who perhaps felt some compunction for having killed one of the family, did not like to fire at the creature; and nothing short of firing would suffice to drive him off. At length the monkey leader came to the door of the tent, and, finding threats of no avail, began a lamentable moaning, and by the most expressive gesture seemed to beg for the dead body. It was given him. He took it sorrowfully in his arms, and bore it away to his expectant companions. They who were witnesses of this extraordinary scene resolved never again to fire at one of the monkey race."

That resolve has been expressed by all but the most hardened of hunters. The look of agony which comes to a stricken monkey's face, the human gesture with which it displays its wound to its murderer, are more than any man of feeling can endure. All hunters present the same report—the death of one of the Primates, be it gorilla, orang-utan, or common monkey, vividly recalls the death of a human being. It is the repugnance of hunters to take the life of creatures so nearly like man that accounts for the paucity of monkeys in museums.

In the foregoing incident there is evidence of a determination to rescue a companion, alive or dead. It agrees with the action of baboons bent on reclaiming their captured companions; but reveals no higher example of solicitude than is revealed by a bear that will follow a vehicle at startling speed to recover its captured cub. In Captain Hugh Crow's "Narratives of My Life," however, in an example furnished by Sir James Malcolm, we see discrimination and initiative as well as anxiety. On board the steamer by which Sir James travelled were two common monkeys from India; the one full grown, the other younger. The smaller monkey one day fell overboard. The larger one became frantically excited and, running to the side, held on by one hand, leaned over and, with the other hand, extended to the monkey in the water the cord by which the would-be rescuer was tied by the waist. The attempt was unsuccessful, but the little monkey was able to keep swimming until a sailor threw a rope, enabling it to climb on board.

That a monkey should swim so well in the sea need not surprise us. Man is the only animal that does not swim naturally upon compulsion. In August, 1907, pleasure-seekers on the Clewer racecourse at Windsor were surprised to see a monkey, dripping with water, make its way to the grand stand. The animal, which had escaped from captivity on an island up the Thames, had ravaged the neighbouring gardens, and then swam the river, as if curious to see what was attracting so many excited people on the racecourse. Having had several days of liberty—during which it had given full play to its fancy for destroying garden flowers—it now submitted to be captured with the best of grace, and was carried back to its home none the worse for its long swim.

The monkey-house at the Zoological Gardens in Regent's Park affords an endless entertainment to the observer. The sociable monkeys are kept in the cages which occupy the centre of the hall; the less amiable ones fill the compartments flanking the house. The inmates enjoy the excitement of a full house, and provide a wonderfully interesting study of the ways of our poor relations. In one particular the monkey is a much misunderstood animal. When we see him sedulously rummaging the fur of his neighbour, we are not to suppose that he is in quest of insects, or that the searcher carries them upon him. As a fact, the monkey keenly relishes insects as part of his diet, even as certain races of men do; but the object of those everlasting searches which the public witness is a salty secretion from the skin of the monkey which the searcher enjoys to eat.

Curiously childlike in his whims, in his likes and dislikes, in his impulses and persistent if futile industry, the monkey is not to be commended as fit to be the inmate of every home. But he has so many good and interesting qualities that at least we must raise our voices against the abominable cruelty to which he is submitted in his life of slavery and misery as the adjunct of the organ-grinder's paraphernalia.   E. A. B.

THE GELADA BABOON

The baboons, or dog-faced monkeys, are so called from the great prolongation of their muzzles, which far exceeds that obtaining in the black ape, and gives to them an expression quite different from that of any other members of the order. Their other peculiarities are mentioned in the description of the true baboons, and it will accordingly suffice to indicate here the chief characteristics of the extraordinary-looking gelada.

The gelada (*Theropithecus gelada*) is an inhabitant of the southern parts of Abyssinia, and is distinguished from the true baboons by the fact that the nostrils are placed some distance behind the extremity of the snout. In this respect, therefore, the gelada forms a connecting link between the black ape of Celebes and the true baboons.

The gelada is of comparatively large size, and dark in colour; the shoulders, back, rump, and forearms, and most of the naked parts, being deep black, whereas the head, whiskers, neck, and sides are a sooty grey, sometimes tinged with brown. The most peculiar feature about the creature is the great mantle of long black hair growing from the neck and flowing over the shoulders. The flesh-coloured chest is naked, while the moderately long tail is cylindrical, and furnished with a long black tuft at the end.

Taken altogether, the aspect of the gelada forcibly suggests a big black poodle dog, with an unusually abundant mane. The mode of life of this species is very similar to that of the true baboons;

THE GELADA BABOON

geladas living in large troops, and being especially addicted to rocky regions, whence they descend to plunder the cultivated grounds of the natives, occasionally entering into conflict with troops of the mantled baboon. Examples of the gelada have been exhibited in the gardens of the Zoological Society in Regent's Park.

THE TRUE BABOONS

With the true baboons and their allies forming the genus *Papio* we come to some of the most hideous members of the primates; their repulsive appearance being only equalled by the fierce and untamable disposition of several species of the group.

All the baboons are confined to Africa and the countries on the north-east of the Red Sea, so that they are totally absent from the Oriental region. They are found over the whole of Africa, but it is only on the west coast that the short-tailed baboons, which form a genus by themselves, are met with. Next to the man-like apes, the baboons include the largest members of the order, some of the species being as large as a pointer dog.

While agreeing with the gelada baboon in the great length of their snouts, the true baboons are readily distinguished from that species by the nostrils being placed at the very extremity of the snout; indeed, in the mantled baboon they actually project slightly beyond the upper lip, as is the case in most dogs. This canine form of countenance led the ancient Greeks and Romans to apply the name *Cynocephalus* ("dog-headed") to this animal, and this name was for many years adopted in scientific phraseology as the distinctive appellation of the group. This great prolongation of the snout shows that the baboons are the lowest of the Old World monkeys, and they bear the most marked signs of relationship with the inferior orders of mammals.

In addition to their long snouts, baboons are distinguished by the large proportionate size of their skulls, this being most obvious in some of the West African kinds. Moreover, the bones forming the upper jaw are greatly inflated, so as to give a swollen look to this part of the face in some of the species. They may also carry prominent oblique ridges, which form the support for the peculiar fleshy tumour-like structures occurring in two West African species which are assigned to a genus by themselves.

In all the baboons the callosities on the buttocks are unusually large, and may be very brightly coloured. In the true baboons the tail is of moderate length. The arms and legs, or, as they may be better termed, fore and hind legs, are nearly equal in length, and thus far better adapted for progress on the ground than for climbing. Indeed, none of the baboons is an adept at climbing, and many pass almost all their time on the ground. Several species show an especial predilection for rocky quarters, and are accustomed to go in large troops—this association being necessary for defence against the attacks of leopards and other carnivores, to which their terrestrial habits expose them (see page 210).

Their defence does not, however, rest solely on the strength of numbers; for the male baboons, which are superior in size and strength to their consorts, are armed with tusks of the most formidable dimensions. Indeed, a bite from one of these animals must be almost, if not quite, as severe and dangerous as a leopard's; and there are instances where leopards have been attacked and mastered by a few old male baboons.

The great size of the head, coupled with their general bodily conformation, renders all the baboons much less capable of assuming and maintaining the erect posture than any of the other Old World monkeys. They are, indeed, accustomed to go almost invariably on all-fours; and when on tolerably flat ground can gallop at a pace that requires a horse to overtake them. When brought to bay, a baboon will, however, sit up on its hind quarters to defend itself more readily.

In the wild state scarcely any kind of food comes amiss to baboons; and although the bulk of their nutriment may take the form of seeds, fruits, roots, and the gum which exudes from the stems of many of the African acacias, they also search for and eat insects, lizards, and birds' eggs. In regions of cultivated land much harm is done by the nocturnal excursions of baboons. During such raids many travellers state that certain members of the troop are selected to act as sentinels to give timely warning of the approach of an enemy. How much credence is to be lent to the assertions that on these occasions the marauders range themselves in long lines leading from the cultivated ground to their homes, and pass the stolen plunder from hand to hand, it is difficult to decide.

In disposition the baboons are the reverse of amiable, and fly into paroxysms of fury at any object which enrages or excites them; but some of the species are capable of being more or less completely tamed, and even learning a certain number of tricks; and it appears that members of one species were habitually tamed by the ancient Egyptians.

To show that baboons were known in Europe at least two centuries ago, the following account may, because of its interesting and quaint character, be quoted. It is extracted from a work by Job Ludolph, relating to ancient Ethiopia (the modern Abyssinia); the English translation being published in the year 1684. "Of apes," writes Ludolph, "there are infinite flocks up and down in the mountains, a thousand and more together; there they leave no stone unturned. If they meet with one that two or three cannot lift, they call for more aid, and all for the sake of the wormes that lye under; a sort of diet which they relish exceedingly. They are very greedy after emmets [ants]. So that having found an emmet hill, they presently surround it, and, laying their forepaws with the hollow downwards upon the ant-heap, as fast as the emmets creep into their treacherous palms, they lick 'em off with great comfort to their stomachs; and there they will lie till there is not an emmet left.

"They are also pernicious to fruit and apples, and will destroy whole fields and gardens unless they be carefully look'd after. For they are very cunning, and will never venture in till the return of their spies, which they send always before ; who giving information that all things are safe, in they rush with their whole body, and make a quick dispatch. Therefore, they go very quiet and silent to their prey ; and if their young chance to make a noise, they chastise them with their fists ; but if they find the coast clear, then every one hath a different noise to express his joy. Nor could there be any to hinder them from further multiplying, but that they fall sometimes into the ruder hands of the wild beasts, which they have no way to avoid but by a timely flight, or by creeping into the clefts of the rocks. If they find no safety in flight, they make a virtue of necessity, stand their ground, and, filling their paws full of dust or sand, fling it into the eyes of their assailant, and then to their heels again."

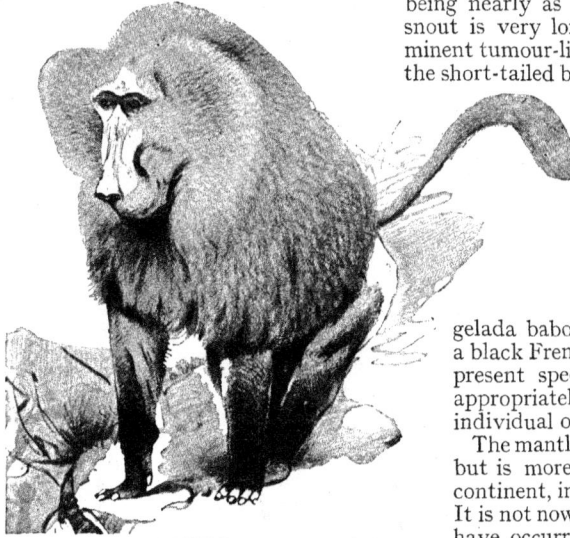

THE MANTLED BABOON

Although, having regard to the date at which he wrote, it is possible that Ludolph may have mixed up some other monkeys with them, there can be little doubt but that in the main this account refers to the mantled baboon, still so common in Abyssinia. This identification is supported by his mention of the large number of individuals in a troop, by the reference to rocks, by the search after insects, and also by the allusion to encounters with leopards. It must, however, be confessed that the figures of monkeys with which Ludolph's narrative is illustrated bear but little resemblance to baboons, although this may well be explained by the degree of licence which the engravers of his epoch allowed themselves in such matters.

THE MANTLED BABOON

The mantled, or sacred, baboon (*Papio hamadryas*) is the species often represented on the ancient monuments of Egypt, and may be recognised by its ashy grey colour, and the large mantle or mane with which the neck and shoulders of the males are covered. The males of this species are about as large as a good-sized pointer dog. The tail is of considerable length, and terminates in a tuft of long hair. The face has long whiskers of a slaty colour, and is itself, like the ears, flesh-coloured. The hands are black, and the large naked callosities on the buttocks bright red. The shaggy mane on the neck and shoulders of the males extends backwards over a considerable portion of the body, and all the hairs are ringed with different colours, so as to produce that speckled appearance common to so many African monkeys. The females and young are quite devoid of this mane ; the former being nearly as large as the males. The snout is very long, and has not the prominent tumour-like swellings characterising the short-tailed baboons. The nostrils project somewhat in front of the plane of the upper lip, like those of a dog, and are similarly divided by a vertical furrow. The eyes are surrounded by light-coloured rings, and the whiskers are brushed back so as to cover the ears. If the gelada baboon be rightly compared to a black French poodle, the males of the present species might still be more appropriately likened to a grey individual of the same breed.

The mantled baboon inhabits Arabia, but is more common on the African continent, in Abyssinia and the Sudan, It is not now found in Egypt, but may have occurred there in ancient times ; although, on the other hand, it is quite probable that it may have been imported by the ancient Egyptians from the Sudan. The typical Abyssinian representative of the species is much larger than the Arabian race, which is distinguished as *P. hamadryas arabicus*.

Among the ancient Egyptians this baboon occupied a prominent place in the long series of sacred animals, and was consecrated to the god Thoth. When scupltured by itself, it is the male that is represented, and it is always placed in a seated position, with the hands resting on the knees, the mane investing the body like a huge cloak. Hermopolis, the city of Thoth, was especially devoted to the cult of these animals ; while in Thebes a special necropolis was arranged for the preservation of their mummified bodies. In spite, however, of its sacred character, the ancient Egyptians, if we may trust their sculptures, were not averse from making use of the sacred baboon in the ordinary affairs of life. For instance, a bas-relief represents a fruit-bearing sycamore, in the branches of which are three monkeys, which from their long snouts, well-developed tails, and thickly-haired shoulders and necks, may be at once recognised as mantled baboons. On either side of the tree are two slaves, with baskets laden with sycamore figs, others of which they are receiving from the hands of the baboons. It thus appears that the ancient Egyptians had succeeded in training these intractable animals to gather fruit and hand it to their masters, precisely after the fashion that the modern Malays are said to have trained langurs in Sumatra to perform a similar kind of service ; the fruit in the one case being sycamore figs, and in the other coco-nuts.

The food of this baboon consists mainly of small fruits, berries, and seeds ; although young shoots and buds of trees form a portion of its diet. Like the rest of its kind, it avoids forests and trees, and keeps mainly to the open country, preferring rocky spots. When it climbs, and it does so in a heavy ungainly manner, very unlike the active movements of the generality of monkeys. Its move- ments, when on the ground and in a hurry, partake more of the nature of a steady gallop than of the bounding motion of other monkeys. The association of these baboons in large troops is doubtless for the purpose of mutual protection. The old males are sometimes said to threaten men, and there is one well authenticated instance of a troop combining to attack a leopard which had carried off one of their number.

Writing of his experiences in the Egyptian Sudan, Sir Samuel Baker observes : " Troops of baboons are now exceedingly numerous, as, the country being entirely dried up, they are forced to the river for water, and the shady banks covered with berry- bearing shrubs induce them to remain. It is very amusing to watch these great male baboons stalking majestically along, followed by a large herd of all ages, the mothers carrying the little ones upon their backs,

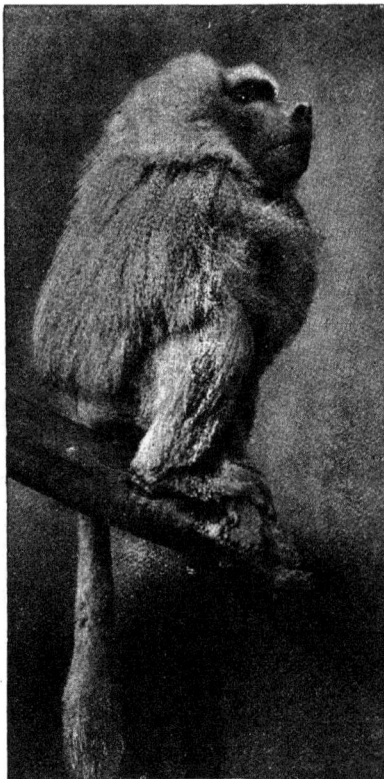

THE MANTLED BABOON

the latter with a regular jockey-seat riding most comfortably, while at other times they relieve the monotony of the position by sprawling at full length and holding on by their mother's back hair. Suddenly a sharp-eyed young ape dis- covers a bush well covered with berries, and, his greedy munching being quickly observed, a general rush of youngsters takes place, and much squabbling for the best place ensues among the boys ; this ends in great uproar, when down comes a great male, who cuffs one, pulls another by the hair, bites another on the hind quarters just as he thinks he has escaped, drags back a would-be deserter by his tail and shakes him thoroughly ; and thus he shortly restores order, preventing all further disputes by sitting under the bush and quietly enjoying the berries by himself."

THE CHACMA BABOON

The mantled baboon is an inhabitant of the countries bordering on the Red Sea littoral and the Upper Nile valley, but to reach the habitat of the chacma, or pig-tailed baboon (P, porcarius), it is

necessary to travel to the southern extremity of the African continent. The name chacma, it may be observed, is a somewhat euphonised rendering of the word t'chackamma, by which the Hottentots of South Africa designate this hideous baboon.

Like all the remaining representatives of the long- tailed baboons, the chacma differs from the mantled baboon in the absence of the mane on the neck and shoulders of the males. There exists, indeed, in this respect a gradual descending series from the gelada baboon, in which both sexes are maned, through the mantled baboon, in which only the males are so orna- mented, to the chacma, in which both males and females are mane- less. In size the chacma is one of the largest of the group, and it has been compared in this respect, as well as in its bodily strength, with an English mastiff.

The general colour of this animal is greyish black ; but there is often a kind of greenish reflection in the fur when seen in certain lights. The head, as well as the hands and feet, is deep black, but the small whiskers on the sides of the face, which do not conceal the ears, are greyish. All the hair of the body is com- paratively long and shaggy, while that on the nape of the neck, more especially in old males, forms a slender crest. The roots of the hairs are dun-coloured, but their extremities are ringed. The tail differs from that of the mantled baboon in the absence of any dis- tinct tuft at the end. The muzzle is perhaps even more prolonged than in the last- named species ; but the nose is similarly extended beyond the upper lip. The naked callosities on the buttocks are smaller than is generally the case among the baboons. The bare part of the face is of a purplish hue, with the exception of a white ring round each eye, and the whole of the upper eyelids, which are also white. In the latter point, curiously enough, this species resembles the African mangabey monkeys (page 226). Like the other members of this group, the chacma carries its tail at first curved somewhat upwards, and then hanging straight down. The chacma is essentially a dweller in mountainous tracts, and is found in all the ranges of the Cape district, such as the Sneeuwberg and the Drakenberg. How far it extends northwards it is difficult to ascertain, since travellers and sportsmen are, as a rule, reticent on the subject of monkeys and their kindred. In Abyssinia the species is represented by the closely allied doguera baboon (P. doguera), often called the Abyssinian chacma, which is much lighter-coloured than the Cape species.

THE TRUE BABOONS

A female baboon allied to the last has been named *P. lydekkeri*; its hair, like that of *P. doguera*, being not ringed with different colour. The Abyssinian chacma collects in large droves to raid the cornfields.

Mrs. Annie Martin, in " Home Life on an Ostrich Farm," gives a most excellent description of the habits of the chacma in the Cape district. " On mountain excursions," writes this lady, " you frequently hear his surly bark, and sometimes see him looking out defiantly at you from behind a rock or bush, where possibly you have disturbed him in the midst of an exciting lizard-hunt, or careful investigation of loose stones in search of the centipedes, scorpions, and beetles hidden beneath. These creatures, uninviting though they appear to us, are among his favourite dainties, and he catches them with wonderful dexterity. In the silence of night his voice is so distinctly audible from the homestead that you would imagine him to be close by, though in reality he is far off in one of the kloofs of the mountains. One night, as we strolled up and down near the house, enjoying the bright moonlight, a loud chorus of distant baboons to which we were listening was suddenly interrupted, evidently by the spring of a hungry leopard, the moment's silence being followed by the agonised and prolonged yells of the victim. . . . No vegetable poison has the slightest effect on the baboon's iron constitution; and, indeed, if there exists any poison at all capable of killing him, it is quite certain that, with his superior intelligence, he would be far too artful to take it; and when the fiat for his destruction has gone forth, a well-organised attack has to be made on him with dogs and guns. He can show fight, too, and the dogs must be well trained and have the safety of numbers to enable them to face him; for in fighting he has the immense advantage of hands, with which he seizes a dog and holds him fast while he inflicts a fatal bite through the loins. Indeed, for either dog or man, coming to close quarters with Adonis, as the chacma is ironically called by the Boers, is no trifling matter. One of our friends, travelling on horseback, came upon a number of baboons sitting in solemn parliament on some rocks. He cantered towards them, anticipating seeing the ungainly beasts take to their heels in grotesque panic; but was somewhat taken aback on finding that, far from being intimidated by his approach, they refused to move, and sat waiting for him, regarding him the while with ominous calmness. The canter subsided into a trot, and the trot into a sedate

walk, and still they sat there; and so defiant was the expression on each ugly face that at last the intruder thought it wisest to turn back and ride ignominiously away."

The most general food of the chacma is afforded by the bulbous roots of an iris-like plant, known as ixia, of which there are several South African varieties, one of which is specially known as the baboon's ixia. These bulbs the chacmas dig up with their strong hands, and carefully peel before eating. Other kinds of bulbous and tuberous roots are also eaten by these animals; while buds and young twigs form a less important part of their food. In addition to this vegetable diet, the chacmas also search for and devour various kinds of insects, especially locusts, and also allied animals, such as scorpions; the latter being carefully deprived of their stings before being consumed. Lizards and frogs are dainties less commonly eaten; while birds' eggs, together with various worms and grubs, practically complete the chacma's bill of fare.

These baboons are well represented in menageries, where they thrive well. When young they are fairly tractable, but their temper steadily deteriorates with advancing age.

THE ANUBIS BABOON

The anubis baboon (*P. anubis*), together with the two following species, may be distinguished from the chacma by the fact that the hairy parts of the hands and feet are much darker than the legs and back, instead of both feet and legs being of the same dark hue.

The general colour of the present species is olive green, whence it is sometimes known as the olive baboon. There is a small crest on the nape of the neck, and the hairs are grey near the roots, and ringed with black and yellow at the tips, while the face is dusky in colour. The typical locality of this baboon is Guinea, but it appears to extend across the continent to East Central Africa, where it is represented by a local race known as *P. anubis neumanni*, distinguished by the great development of the ridges on the upper jaw.

In habits these baboons appear to be very similar to the other species of the genus. They associate in large troops, and inhabit rocky mountainous regions, being especially common at the Black Rocks, some two hundred miles in the interior of Angola. Away from the river valleys the country is arid in the extreme, and these thirsty districts are the chosen abode of the baboons. Here they subsist largely

YOUNG CHACMA BABOON AND YELLOW BABOON
Photograph by W. P. Dando

THE ANUBIS BABOON

on that remarkable West African plant known as the welwitschia. Indeed, so extraordinary is this plant that it is worth a brief description. The welwitschia, in its earlier stages, consists of two ordinary seed-leaves. These grow considerably and extend horizontally outwards in opposite directions, raised but little above the surface of the sand; while the intervening stock thickens and hardens, assuming a somewhat conical shape, flattened at the top, and rapidly tapering below into the roots. In time the original pair of seed-leaves, having attained their full size and acquired a hard and fibrous structure, instead of dying, gradually split up into shreds ; at the same time the woody mass upon which they are borne, although rising but little in height, increases in width both below and above the insertion of the leaves, so as to clasp their bases in a deep slit on the margin. Every year several short flowering stalks are developed from the upper side of the base of the leaves. Each of these stalks forms an erect jointed stem, dividing in a fork-like manner, varying in height from six to twelve inches, and carrying at the end of each branch a cone, with the flowers and seeds beneath its scales. The result is that the country is studded with these tabular or anvil-like masses of wood, whose flat tops, pitted with the scars of old flowering-stems, never rise to more than a foot above the ground, but vary, according to age, from a few inches to upwards of five or six feet in diameter. Even those which are not more than 18 in. in diameter are supposed to be fully a century old, although still retaining their original seed-leaves, which, albeit torn and tattered by the wear and tear of time, are,

when entire, fully six feet in length. It is upon the stems and exposed portions of these extraordinary plants that the anubis baboons feed, tearing and ripping the woody tissue with their powerful tusks.

THE YELLOW BABOON

The yellow baboon (*P. cynocephalus*) may be distinguished from the anubis by the absence of a crest of hair on the nape of the neck and by its colouring. It takes its popular name from the pale brownish yellow hue of the fur, which is rather darker on the sides of the back than elsewhere, while it tends to a whitish tint on the cheeks. The hair on the crown of the head is somewhat elongated. Unlike the anubis baboon, the hairy parts of the hands and feet are of the same yellowish colour as the legs and body, while the face and other bare parts are flesh-coloured.

It was long thought that the yellow baboon came from Nubia and the Sudan, but it is now known to occur on the west coast. Writing of a baboon from the Kilimanjaro district, provisionally, but perhaps incorrectly, identified with the yellow species, Sir H. H. Johnston observes that these baboons generally frequent the outlying parts of the plantations of the natives, subsisting largely on the maize and other products stolen therefrom. In certain localities they are extremely numerous, going about in troops composed of from about fourteen individuals of both sexes and all ages. They have little fear of man, and instead of running away will turn round and face an intruder, with threatening gestures, at a distance of only a few yards. The natives are in the habit of driving them away from the crops, when the baboons retreat in a leisurely manner, with their cheek pouches crammed full, and often dragging off some of the plunder in their hands. In one instance it is related that a troop of these animals pursued a native lad for some time, until he had placed a river between himself and his pursuers.

The Thoth baboon (*Papio thoth*), of Abyssinia, agrees with the yellow species in the livid flesh-colour of the face, as well as in the absence of a long crest of hair on the nape, and may indeed be only a local form of that species.

THE GUINEA BABOON

There are few species of mammals that have given rise to more confusion in natural history literature than this, of which examples have been described under at least two distinct names and regarded as different species, though it is a well-ascertained fact that the common baboon, or papio, belongs to one and the same species as the sphinx, or Guinea baboon (*P. sphinx*).

The Guinea baboon is characterised by the uniformly reddish brown colour of its fur, which is washed with a yellowish tinge, more especially on the head, shoulders, back, and limbs ; the cheeks and throat being paler, and the whiskers fawn-coloured. The naked parts of the face are bluish black, and the hairy parts of the hands and feet are much darker than the legs and back. The nose projects rather beyond the upper lip, but is somewhat less elongated than in the chacma, and has small swellings corresponding with those so enormously developed in the mandrill. It is the smallest of all the baboons and is frequently carried about by showmen. As its name implies, it is a native of Guinea, and large numbers are imported into Europe.

THE MANDRILL

The mandrill (*Maimon mormon*) brings us to the first of two West African species of baboons distinguished from those hitherto considered by the reduction of the tail to a short stump, and also by the long tuberculous swellings on each side of the muzzle, which communicate the peculiarly ugly expression to the face. Moreover, the whole head is larger in proportion to the body than in other baboons, and as the fore quarters also appear to be relatively higher in proportion to the hind parts, the general appearance is ungainly in the extreme. In fact, the whole aspect is far more suggestive of the forms imagined during a nightmare than is the case with any other living mammal (page 209).

Many years ago it was suggested that these two species ought to be separated from all the other baboons, and placed in a genus by themselves; and although this view did not at first meet with general acceptance, there is little doubt but that it should be adopted. It is true that, as compared with ordinary baboons, the small size of the tail in the drills is merely analogous to the condition which obtains in certain members of the macaque monkeys, while the huge swellings on the face are only exaggerated developments of the smaller ones found in the Guinea baboon, but the general form of the short-tailed baboons is sufficient to justify the proposed separation.

GUINEA BABOONS
Photograph by W. S. Berridge

The mandrill is one of the largest of baboons, and is, in truth, a brute of tremendous power and ferocity. Its leading characteristics are to be found in the facts that the short and tuberculous tail has its under surface naked, and that the swellings on the face are ornamented with a brilliant colouring in the adult male, and are of enormous dimensions.

From the great development of these swellings on the sides of the muzzle, the English naturalist Thomas Pennant (1726-1798) gave to the mandrill the name of rib-faced baboon, but this has generally been discarded by modern writers in favour of the former term. Here it may be mentioned that the word "mandrill" signifies "man-like baboon," the term "drill" being an old English word of which one meaning denotes "baboon," or "ape."

The limbs of the mandrill are characterised by their relative shortness and powerful build, and in correlation with these the form of the body is likewise powerful and robust. The ugly and massive head has scarcely any distinct forehead, the profile sloping almost uninterruptedly upwards from the muzzle to the occiput.

The nose, instead of projecting in front of the upper lip, as in the mantled baboon, is somewhat truncated; while the projecting eyebrows and deeply sunk eyes communicate a forbidding expression to the whole countenance. The tubercular swellings on each side of the muzzle are supported on ridges arising from the swollen bones of this part of the skull, and are themselves almost the size of a man's fist. As a whole, they are somewhat sausage-shaped, and are marked with a series of prominent transversely disposed ribs of light blue, with deep purple in the

THE YELLOW BABOON

U

grooves, while the middle line and the tip of the nose are brilliant vermilion. The contrast between such brilliant colours and the general hue of the fur and the hazel eyes is most marked. The stump of a tail, which is naked on the under side, is carried erect and bent over the back somewhat after the manner of that of a pug dog. The general colour of the fur is a blackish olive, darker on the crown of the head, the middle line of the back, the nape of the neck, and the flanks, and lighter on the cheeks. The summit of the head is crowned with a crest of dark hair directed backwards in a pointed and peaked form, while the chin is ornamented with a small pointed beard of an orange-yellow colour. The long hair of the under parts in old males is white. To add to the strange effect of all these varied tints, the large naked callosities on the buttocks are of a bright blood-red colour. The pointed crest on the crown gives to the whole head a somewhat triangular form, and, in harmony with this peculiar contour, the naked bluish black ears are angulated at their fore-and-aft borders, suggesting the appearance of having been cropped. The truncated muzzle is surrounded by a raised border like that of the swine, from which circumstance it has been considered by some writers that the mandrill is the problematical animal alluded to by Aristotle as *Chœropithecus* (" hog-ape "), but this identification is by no means certain.

Such are the colours of the adult male mandrill, but the brilliant scarlet of the middle and end of the muzzle is not assumed until the first, or milk, teeth have been replaced by the permanent series, while, at a still younger age, the whole of the face is black. Moreover, it is only in the adult of the male sex that the swellings on the sides of the snout assume the enormous dimensions above noticed. Both in the young males and in the females of all ages these swellings are only of moderate dimensions, and in the female they are coloured plain blue. In correlation with the smaller size of the fleshy swellings, the skulls of females and young males are characterised by the much slighter development of the bony ridges underlying these structures, which form such prominent features in the skulls of old males.

In the wild state on the western coast of Africa mandrills have habits very like those of other baboons, living in large troops, and, on this account, as well as from their size and strength, being exceedingly formidable antagonists. The accounts given by the earlier travellers of their attacking men without being provoked require to be confirmed; and we are in want of full information as to their habits in general.

In confinement, the chief characteristic appears to be that the ferocity and moroseness common to the old males of all baboons is intensified. There is also a marked liking for spirituous liquors, which is likewise a trait exhibited by other species of the genus. One of the earliest examples of an adult male mandrill exhibited in London was the famous " Jerry," immortalised by Mr. W. J. Broderip, which

was kept first in the menagerie at Exeter 'Change, in the Strand, and then transferred to the Surrey Zoological Gardens, on the southern side of the Thames. This animal had learnt to drink daily a pint of porter, which he seemed thoroughly to appreciate, and he had also been taught to smoke tobacco in a short clay pipe, although this accomplishment did not appear to be so much to his taste.

Of late years the mandrill has often been represented in the Zoological Society's Gardens in Regent's Park. An extraordinary animal was born in the Society's menagerie in the autumn of 1878, being a female hybrid produced by a cross between a female mandrill and a male of the crab-eating macaque (*Macacus cynomolgus*).

The mandrill is strictly confined to the tropical parts of West Africa ; the Gabun district being, perhaps, its headquarters.

THE DRILL

Although described by Frédéric Cuvier so far back as 1807 as a distinct species, the baboon known as the drill (*Maimon leucophæus*) had for many years before, in spite of a figure given by Pennant, been considered to be merely the young of the mandrill which had not acquired the characteristic colouring of the face. The acquisition of adult specimens of the drill by museums and menageries proved, however, the correctness of the English and French naturalists' determination. It is exclusively West African, but its range in latitude appears to be somewhat more extensive than that of the mandrill.

The drill may be distinguished from its larger cousin the mandrill by the absence of bright colours on the naked parts of the face, which are entirely black. The short tail is covered with hairs over the whole of its surface, while the general build, and especially that of the limbs, is of a much more slender type. Again, although the face has the long sausage-like swellings of the mandrill, these are considerably smaller and less inflated. The drill is ugly enough, but it is one degree less repulsive than the male mandrill.

The general colour of the fur is brown, tending to a whitish tint on the forehead and the crown of the head, and darker on the shoulders and limbs. The under parts are also lighter, being either of a pale brown or a silvery grey tint. The hair of the upper parts is very long and fine, and is of a light brown colour at the root, but ringed with black and yellow at the tips ; these rings of two colours giving a greenish tinge to the fur when seen under certain lights. The whiskers are thin, and directed backwards like those of the mandrill ; and the drill also resembles that species in the peaked crest on the crown of the head, as well as in the small yellow beard beneath the chin. The apology for a tail terminates in a small tuft of hair. The naked face and ears present a jet black appearance, and the swellings on the snout are not marked by the oblique transverse furrows and grooves which characterise those of the mandrill. The naked portions of the hands and feet are copper-coloured, while the bare callosities on the buttocks are bright red. The colour of such portions of the skin as are covered with hair is of a uniform dark blue. The female drill is distinguished from her lord by her smaller size and the relatively shorter head and paler colouration, in which the young males resemble her.

Nothing is known of the drill in its wild state. In confinement, however, its habits are very similar to those of the mandrill, and doubtless there is the same similarity in the wild condition. Of late years it has been found that representatives of most of the mammals of the West Coast extend through the forest-zone to East Africa; but the mandrill and drill, like the gorilla, appear restricted to the former area.

With the drill is concluded the notice of the living monkeys of the Old World; but, before passing to those of the New World, a short space may be devoted to a few extinct baboons.

EXTINCT BABOONS

The survey of the long series of Old World monkeys has shown that as we pass from the man-like apes through the true monkeys to the baboons, we have been gradually receding farther and farther from a marked approximation to the human type, until we reach forms that show a decided resemblance in their projecting muzzles and general contour to the lower orders of mammals. These lowest forms being the baboons, it is but natural to assume that they are likewise old in the history of the animal kingdom, so that we should expect to find them in a fossil state. In Europe, however, no traces of fossil baboons have yet been discovered; while in Africa we only know of them as occurring in the superficial deposits of Algeria. The latter circumstance must not, however, be taken as an indication that other species of fossil baboons will never be found in Africa, since our knowledge of the geology of the greater part of that continent is of the most limited nature. At present it is only in the extreme north of India that evidence of the existence of fossil baboons belonging to a period antecedent to that during which man has existed on the globe has been obtained. And it is in the sandstones forming the outer flanks of the mighty Himalaya, which contain the remains of the extinct Indian chimpanzee and orang-utan, that those of the fossil baboons occur. These rocks, as already stated, belong to the Pliocene division of the Tertiary

period. The remains of the Pliocene Indian baboons are, like those of all the primates, extremely few, yet they are sufficient to prove the existence in that country of two distinct species. Both of these appear to have been closely allied to some of the long tailed African species; and it may therefore be concluded that the Indian species were allied to the mantled baboon or the chacma. There is, more-over, evidence that baboons continued to exist in India until either the early human or Pleistocene period, since a single tooth has been obtained from deposits in a cavern in Madras which has likewise yielded remains of man.

These and the aforesaid remains of man-like apes afford decisive proof that at a former epoch of the earth's history such an assemblage of primates was gathered together on the plains of India, at a time when the Himalaya did not exist, as has been seen nowhere else beyond the walls of a menagerie. Side by side with langurs and macaques closely resembling those now found in that region were chimpanzees and baboons as nearly related to those of modern Africa; while the extinct Indian orang-utan recalls the existing species of Borneo and Sumatra. India, therefore, in the Pliocene period, may per-haps have been the central point whence the main groups of Old World primates dispersed to their far-distant homes.

The generalised character and the large size of the baboons have suggested that we should look to them as the original ances-tral stock from which the man-like apes took their rise. There is, however, found in strata of the Lower Pliocene period of Europe a baboon-like ape known as the mountain ape (*Oreopithecus*), which combines to a certain extent the features now characteristic of the man-like apes and the baboons. It is this monkey, therefore, which has more claims to be regarded as near akin to the ancestral stock of the man-like apes; the baboons being survivors from a still older stock, from which the mountain ape was itself derived. On the other hand, as mentioned in an earlier page (201), remains of gibbons and man-like apes occur in Pliocene strata; a fact pointing to a still earlier separation of the two groups, which may have been independently derived from lemur-like creatures. In this connection it may be stated that Dr. A. Keith has expressed the opinion that the gorilla and chimpanzee are co-descendants of an early Miocene ape, for which the name Protroglodytes is suggested. It is estimated that more than five million years have elapsed since the separation of the human stock

THE MANDRILL

as a distinct type ; the body having probably been adapted to the upright posture before the close of the Miocene period, while development during the Pliocene and Pleistocene epochs has, in this naturalist's opinion, been mainly restricted to brain-expansion.   R. LYDEKKER

STORIES OF THE BABOONS

Although the baboon has so black a reputation as thief and savage, he has some redeeming features. His amiability does not always vanish with his youth. Doubtless, temperament varies in baboons as it varies in men and other animals. Hence, while it would be unwise to expect good nature in all baboons, it would be equally unjust to condemn the whole family as implacably ferocious. True, one nearly killed a keeper at the Zoological Gardens in London ; but another, an anubis baboon, which still flourished in 1909 in the same gardens, was so docile and trustworthy that it was successfully subjected to a ticklish operation. In a fight it had sustained a badly lacerated arm. Possibly to-day, now that the hospital at the Zoo includes a chamber for administering anæsthetics, the creature would be chloroformed before the surgeon operated. As it was, the keeper and the baboon were on such excellent terms that the man was able to hold the injured limb while the surgeon cleaned, stitched, and dressed the wound.

Herr Schillings had a tame baboon, a big and powerful creature, when he was out in Africa, and it was this animal which always sighted him as he returned after some long journey. " His sight is infinitely keener than man's," says the traveller, and though the natives are blessed with powers of vision to which Europeans are strangers, the baboon's eyesight was as much superior to theirs as theirs to ours. When his master was still but a speck upon the horizon the baboon would see him, and, " almost mad with joy," would be the first to give the signal.

The keen-sighted baboons which haunted the defences of beleaguered Ladysmith, it will be remembered, were always the first to apprise the British garrison of an advance of the Boers. But Herr Schillings knew another baboon which also displayed a lively sense of the dangers of war-time. He was chained up at the entrance to the fort at Moshi, where his chief human friend was a little negro child, with whom he played every day in the drollest fashion. The baboon was at his post one night when the garrison was expecting an attack. Suddenly the inhabitants of the station poured into the fort. The baboon, sharing their fears, managed to break loose, and was one of the party which rushed for protection into the stronghold.

The memory of the baboon for its friends is not short-lived, for we have it from Sir Andrew Smith that one recognised him with joy after an absence of nine months. The same exact observer gave Darwin another instance of the baboon's memory, this of a very different character. The animal had been plagued by a certain officer, and one day,

seeing him approach, poured water into a hole and hastily made some thick mud, which it skilfully dashed over the officer as he passed by, to the infinite amusement of the onlookers. For long afterwards the baboon rejoiced and triumphed whenever he saw his victim.

A human weakness discovers itself in the baboon's liking for strong drink ; but usually, after the miseries consequent on one debauch, the baboon rises superior to the temptation. Brehm saw captive baboons which were under the influence of liquor. On the following morning they were dismal and morose ; they held their aching heads with their hands, and wore a most pitiable expression. When beer or wine was offered them, they turned away with disgust, but relished the juice of lemons.

The baboon, when not captive, does not easily succumb to the influence of liquor, even when poison is added. An experience of Herr Windhorn in 1906 goes to confirm the statement of Mrs. Martin (page 243) as to this animal's immunity from vegetable poison. Herr Windhorn was bringing over a number of animals in the Comrie Castle, when a powerful baboon escaped from its cage into the hold of the vessel. Though shut off from the deck the baboon defied all attempts at capture for two or three days. Then Herr Windhorn and a keeper attempted to catch it with nets. These the ape eluded, clearing the obstacles by tremendous leaps, fifteen feet at a time. Herr Windhorn ventured at last into the hold, but tripped over one of the nets and fell. Instantly the baboon sprang upon him, and made its terrible teeth meet through one of his legs. He struggled to beat off the animal, but it seized him by the hand, and badly mutilated that limb. The keeper, assisted by the boatswain, came to the prostrate man's assistance, and both were severely bitten before they succeeded in making their escape with him from the hold. After this, a bottle, containing half a pint of whisky and opium, " enough to poison ten men," was lowered into the hold. The baboon swallowed the mixture without showing the least ill result. Eventually it was captured by a trick. Lured to a grating by a display of fruit, it put its hands between the bars to reach the food, and they were lassoed and bound, and the baboon was carried to its cage. After four days of renewed captivity, however, the animal died, " broken-hearted at being mastered," said its owner.

Herr Schillings pays his tribute to the skilful organisation of the baboons for the detection of enemies, and mentions that, fearing the appearance at their drinking-places of their dreaded enemy, the crocodile, they never drink without having the water watched and guarded by some veteran of the troop. At the first suggestion of danger the whole troop rush to safety, and do not descend until from their watch-towers the old baboons see that the course is clear. Very often they will all proceed to some shallower spot where it is more difficult for a crocodile to reach them. Herr Schillings says, in his " Flashlight and Rifle," that to him " it seems a fact that baboons have a language of their own, and that in danger the old animals give their commands by means of some simple method of speech."

Herr Hagenbeck, in his " Beasts and Men," seems to incline to the like belief. Describing the capture of baboons in a hut furnished with a spring door, he says that the herd outside surround the prisoners and urge them to escape. When the captives are taken away, their comrades do not desert them, but climb the palm trees and howl out unintelligible words, which are answered with mournful cries by the prisoners. Herr Hagenbeck adds a startling story of a successful raid by three thousand powerful Arabian baboons upon a party of hunters who had captured a number of these animals. Sheer weight of numbers routed the captors, armed though they were with firearms, and the victorious baboons soon made short work of the cage, and released their imprisoned friends.

If experience does not teach, intuition, if the word may be used, serves the baboon very well. Frank Buckland had an experience with one which bears upon the point. The baboon which he kept while living with his father at the Deanery of Westminster escaped one day from her pole, and darted off, carrying her chain with her. Away she went, over the roofs, with Buckland courageously following. Although her attention was distracted by the crowds of people gathered in the streets below, she managed to elude her pursuer every time he drew near. Her flight carried her to the roof of a house from which her chain hung down in front of a window at which a woman was standing. Buckland made no sign, but said quietly to the woman : " Please put out your hand and catch hold of that chain." Before the woman could make the attempt the animal anticipated her, and, drawing up her chain, hand over hand, as a sailor hauls a rope, was off again. " This is curious," remarks Buckland, " because I was careful to make no motion or sign of my intentions, and merely spoke the words. The animal seemed to understand what I said, for she hauled up her chain even before the woman put her hand out of the open window to catch hold of it." Eventually, when the truant had run to the end of the houses in Dean's Yard, she made a terrific leap on to some leads below, and, dashing off for home, made straight to her pole again, as if now satisfied to return after defeating all attempts to capture her.

It is a much disputed point whether any animal has sense enough to set a bait to catch its prey. Romanes had no doubt that a dog-headed baboon, which he watched, set a snare for an anubis baboon in an adjoining cage. In the one cage were the Arabian baboon and an anubis baboon ; in the next, a solitary dog-headed baboon. The anubis baboon passed its hand through the wire partition to reach a nut which the watcher believed to have been deliberately left as a bait by the big dog-headed baboon. " The anubis baboon knew perfectly well the danger he ran, for he waited until his bulky neighbour had turned his back upon the nut, with the appearance of having forgotten all about it.

The dog-headed baboon, however, was all the time slowly looking round with the corner of his eye, and no sooner was the arm of his victim inside the cage than he sprang with astounding rapidity and caught the retreating hand in his mouth. The cries of the anubis baboon immediately brought the keeper to his rescue, when, by dint of a good deal of physical persuasion, the dog-headed baboon was induced to leave go his hold."

There seems in that a strong case for those who maintain that certain animals learn to lay snares. The after proceedings were particularly interesting. The injured baboon retired to the centre of its cage, moaning piteously, and holding the damaged member against his chest, while he rubbed it with the other. Its companion, the Arabian baboon, now approached him from the top of the cage, and, while making a soothing sound very expressive of sympathy, folded the sufferer in its arms, exactly as a mother would her child in similar circumstances. Romanes adds that this expression of sympathy had a decidedly quieting effect upon the sufferer. His moans became less piteous as soon as he was enfolded in the arms of his comforter ; and the manner in which he laid his cheek upon the bosom of his friend was as eloquent as could be of sympathy appreciated.

There was at one time a chimpanzee at the Dresden Zoological Gardens which found great satisfaction in setting a bait for a little mona monkey which occupied the adjoining cage. Here, however, there was no sinister motive behind the trick ; the intention was only to tease. The ape would plant a biscuit near the netting where the monkey, by putting its arm through, could reach it. But every time that the monkey approached to seize the lure, the ape would snap it up in triumph. Once when the monkey, tired of the teasing, had retired to the other side of the cage, the chimpanzee put the biscuit through, and the monkey got it. Thereupon the ape seized the wires and shook them, and pouted with vexation like a child.

The baboon is a good hater, as we have seen. He will when young make friends with a host of people, but if one offend him he does not forget or forgive. He takes offence at trivial things, and broods over them like a sulky child. A little thing gave mortal offence to a great baboon which some years ago flourished in the Dublin Zoological Gardens. Its special favourite was Dr. Ball, the superintendent, who always made a point of paying it a visit, and petting it whenever he was near. One day, however, when showing the lord-lieutenant round the gardens, he was too much occupied when in the monkey-house to pay its customary visit to the baboon. When next Dr. Ball went to the cage the baboon would not take any notice of him, and refused to resume the old friendly relations until illness came upon it. Consumption seized it, and in spite of every care the animal grew worse and worse. The day before it died, seeing Dr. Ball, it crawled to the front of its cage, and held out its paw to him in token of reconciliation.   E. A. B.

AMERICAN MONKEYS

THE monkeys of Tropical America differ so remarkably from those of the Old World that they cannot be included in either of the families already treated of. The ordinary monkeys of the New World form, indeed, a perfectly distinct family by themselves, known as the Cebidæ. In addition to these, there is another group of American primates called marmosets, which, although nearly related to the Cebidæ, constitutes a second family.

The better-known representatives of the ordinary monkeys (Cebidæ) of the New World are popularly designated howlers, spider monkeys, sapajous, and titis, to which alone the term "American monkeys" should be restricted. In regard to the characters by which these monkeys are distinguished from their cousins of the Old World, it may be mentioned in the first place that no New World monkey has naked callosities on the buttocks. This character will at once serve to distinguish any American monkey from all those of the Old World except the larger man-like apes, with which there is not the slightest fear of its being confounded. Then, again, all the monkeys of the New World are characterised by the absence of cheek pouches ; so that whenever a monkey is seen cramming nuts into his cheeks we may be perfectly sure that he does not come from America. It is true, indeed, that this absence of cheek pouches will not help us to distinguish an American monkey from an Indian langur or an African guereza monkey, but then both the latter have naked callosities on the buttocks. Moreover, if we were to dissect an American monkey it would be found that it had a simple stomach, quite different from the sacculated one which characterises the langurs and guereza monkeys.

BROWN SAPAJOU
Photograph by Lewis Medland

Another peculiarity of some, although by no means all, of the American monkeys is that the tail is prehensile, and capable of being coiled round a bough so as to form a most efficient aid in climbing. These prehensile tails are characteristic only of the howlers and the spider monkeys, and their kin ; the tails of the titis and their allies being non-prehensile, like those of Old World monkeys. The reader may note, however, that whenever he sees a monkey swinging suspended by its tail, he may at once put that animal down as an American.

A more important feature of the American monkeys, as being common to the whole of them, is the great width of the vertical partition between the two nostrils, of which mention has already been made (page 35). This partition causes the end of the nose to be expanded ; and a comparison of full-faced figures of New World monkeys with those of the Old World will show the marked difference in this respect between the two groups.

Another important character is that in those of the American monkeys which are furnished with a thumb, this digit cannot be opposed to the other digits of the hand. The American monkeys agree, however, with their cousins of the Old World in having all their digits provided with well-developed nails.

The most important and perfectly constant distinction between the monkeys of the Old and New World can only be observed in the skulls. In the monkeys of the Old World the total number of teeth is thirty-two ; but if we examine the skull of any American monkey (excluding the marmosets) and count the teeth, we shall find that their total number is thirty-six. A closer examination will show that the additional tooth on each side of both jaws belongs to the premolar series—the so-called bicuspids of human dentistry. Thus, whereas all Old World monkeys have but two bicuspids on each side of both the upper and lower jaw, the American monkeys have three of these teeth ; and the number of teeth in the latter may accordingly be expressed by the formula $i.\frac{2}{2}, c.\frac{1}{1}, p.m.\frac{3}{3}, m.\frac{3}{3}$; total, 36.

If we care to carry our examination a little further, it will not fail to be noticed that the upper molar teeth of the American monkeys differ very decidedly in the form of their crowns from those of the monkeys of the Old World, so that a single detached specimen of one of these teeth is amply sufficient to decide to which of the two groups its owner belonged. Thus, whereas in the Old World monkeys, exclusive of the man-like apes, the crowns of these teeth are tall and narrow, with the four tubercles arranged in pairs nearly at right angles to the long axis, and each tubercle nearly conical, in the monkeys of the New World the crowns of these teeth are much shorter and broader, with their pairs of tubercles arranged obliquely to the long axis; the outer tubercles being much flattened, and the inner crescent-shaped. Those acquainted with the details of anatomy will also find characters by which the skulls of Old and New World monkeys can be mutually distinguished.

A German naturalist, Dr. Hermann Klaatsch, has made the interesting discovery that, with the exception of the man-like apes, none of the Old World monkeys possesses a muscle corresponding with the one known in human anatomy as the short head of the *biceps flexor cruris*. And since the representative of the same muscle, under two different modifications, occurs in American monkeys, the question whether these are more nearly related to man and the man-like apes than are the ordinary monkeys of the Old World is opened up.

The headquarters of the American monkeys are the great forest regions of the lower Amazon valley, known as the selvas ; although monkeys are also abundant in many other parts of Brazil, and likewise in the Orinoco valley in Venezuela. All these animals are truly tropical and subtropical, although they extend to a longer distance south of the equator than they do to the north. Northwards, indeed, monkeys do not extend beyond the Tropic of Cancer in the southern half of Mexico ; whereas in South America they are known to range as far as the Rio Grande do Sul, in 30° S.

American monkeys are divided into ten distinct groups, the first of which includes :

THE SAPAJOUS

The long and prehensile-tailed monkeys commonly seen in menageries, and known respectively as sapajous or capuchin monkeys and spider monkeys, may be regarded as the typical representatives of the family Cebidæ ; and, together with two other genera, constitute a group which can be easily recognised and distinguished from all their cousins. With the exception of the howlers, this group is indeed the only one furnished with prehensile tails ; and, altogether apart from the question of voice and the presence of certain structures connected therewith, all its members differ from the howlers by their rounded heads and the nearly vertical plane of the face.

The sapajous (genus *Cebus*) may be distinguished from the other genera in this group by the fact that their tails, which are comparatively stout and of only moderate length, have no naked part on the lower surface of the extremity. In this respect they are not so perfectly adapted for the purpose of prehension as are those of the other genera. Another feature is that the hair does not partake of a woolly nature. Moreover, the general build of the body is rather stout, the arms and legs according in this

THE WEEPER SAPAJOU

respect with the body, and not being excessively long or excessively slender.

The native name of these monkeys on the Amazon is caiarara, or " macaw-headed," the word " arara " meaning " macaw "—caiarara being abbreviated frequently into cai ; the name sajou or sapajou has been evolved by a curious modification, originally due to Buffon using the word sai (the equivalent of cai) for the weeper capuchin, and sajou for another species of the genus. The term capuchin takes its origin from the cowl-like appearance of the hair on the forehead of the species of that group.

The sapajous are represented by a large number of species, ranging from Central America to the south of Brazil. Our knowledge as to the real number of species is, however, still incomplete, as there is a great amount of individual and racial variation.

Together with the spider monkeys, the sapajous are the most docile and the most readily taught of all American monkeys, and, since they bear confinement and the European climate well, they are the commonest monkeys carried about by organ-grinders.

THE WHITE-CHEEKED SAPAJOU

The white-cheeked sapajou (*Cebus lunatus*) is an inhabitant of Brazil, characterised by the length of the hair on the head, which is directed backwards, while that round the face is longer, and curved so as to form a kind of crest on each eyebrow. On the cheeks the hair is short and flattened down. The fur of the body and head is long, soft, and silky, its general colour being blackish, but that on the cheeks and temples is yellowish white. It is this light hair on the cheeks that gives its distinctive name to the species. The head is relatively large.

THE BROWN SAPAJOU

In Guiana the sapajous are represented by a species known as the brown sapajou (*C. fatuellus*, page 250), which presents a variation, due either to differences of age or to individual peculiarity, in regard to the form of the hair on the head, which has led to the supposition that there were two distinct species. In one of these forms the hair on the crown of the head is nearly flat, and directed backwards, this form having been described as *C. apella*. In the other

SMOOTH HEADED SAPAJOU

type the hair on the sides of the crown of the head is lengthened, so as to form a pair of more or less distinct longitudinal crests, this type being hence known as the horned sapajou. It is in winter, when the fur is longest, that the crests of the horned variety become most prominent, these never making their appearance until the animal has cut its permanent canine teeth. Like that of its congeners, the disposition of this species in captivity is mild and affectionate.

Although subject to great variation in this respect, the general colour of the thick and rather harsh fur is reddish brown, becoming darker on the middle of the back, as well as on the legs and tail. The fore arms, together with a broad spot on the crown of the head and the whiskers, are nearly or quite black, while the front of the shoulders is yellowish. It is on either side of the dark spot on the crown of the head that the crests are situated in the horned variety. The face and other naked parts have a violet tinge.

THE SLENDER SAPAJOU

The slender or white sapajou (*C. flavus*) is a rare local species allied to the preceding, and inhabiting Bolivia in the neighbourhood of Santa Cruz. It is distinguished by the smaller size and lighter colour of the dark spot on the crown of the head, which is generally brown, and often has a small crest on each side. The general colour of the fur is fulvous or greyish fulvous, the limbs and tail being of a darker brown and the beard a golden yellow. There is also a nearly white variety. Bates, who alludes to the slender sapajou as the caiarara branca, heard of its reported existence in the forests of the Tapajos River, which flows into the Amazon from the Cordillera Goral, on the Bolivian frontier of Brazil. His search was, however, in vain ; and he was afterwards informed that the species only occurred across the watershed in Bolivia.

Another nearly related monkey, more widely spread in South America, is the tufted sapajou (*C. cirrifer*), in which the general colour of the short fur is black, but yellowish white on the cheeks, chin, sides of the forehead, and a narrow band over the eyebrows. Two long, recurved tufts of hair, which often occur on the side of the head, give the distinctive name to this monkey.

A monkey known to the natives of the Lower Amazon valley as the macaca prego was provisionally identified with this species by Bates, who described it as frequenting the cultivated lands, where it commits wholesale depredations with the most unblushing effrontery. The worst of these thefts is that, from the hasty and random manner in which the fruit is broken and plucked, the creature wastes far more than it can eat. When about to return to its native forest, it carries away as much plunder as it can hold in its hands and under its arms.

THE WEEPER SAPAJOU

One of the most common species is the weeper sapajou, or capuchin (*C. capucinus*), of Brazil (page 251). It is characterised by the hairs on the crown of the head being short and directed evenly

backwards, without any tendency to form crests on the sides. The colour of the fur is brown, with a golden tinge, the side of the forehead, cheeks, throat, and chest, as well as the front of the shoulders, being pale yellow ; while a black or dark brown line extends from the base of the nose to the back of the neck, gradually expanding as it goes backwards.

These sapajous have a wide range, extending across Brazil, from Bahia in the east to Colombia in the north-west. With the exception of the occasions when they descend to drink, their whole life is spent in the trees of those regions of the forest where there is no underwood. They generally live in small parties, numbering from about six to ten or twelve individuals, of which the majority are females. From their shy and timid habits they are very difficult to observe. Their cry appears to be limited to a kind of low whistle, which serves to attract attention to them.

When captured young, the capuchin is always easily tamed, but older animals refuse all food, become mopish, and do not survive more than a few weeks. The younger ones soon take to their masters, and exhibit remarkable fidelity. They are, however, more readily attached to coloured than to white people, and are generally very fond of other animals, so that in Paraguay it is a common custom to bring them up with a young dog, upon which they ride. To some persons they at once conceive a rooted dislike, which cannot be eradicated. Their intelligence is shown by the manner in which they learn to open an egg, most of the contents being lost at the first trial, but carefully secured at the second attempt. Although they flourish in captivity if well attended to, they must be protected from cold and damp. Their average term of life is about fifteen years.

Like most other monkeys, captive capuchins are the very spirit of mischief, and are also prone to theft, more especially of eatables. When detected in the act of stealing, they cry out before being touched ; but if not caught they affect perfect innocence, going about as if nothing had happened. When disturbed, small substances are hidden in their mouths and afterwards consumed at leisure. They are extremely covetous, and this greed is taken advantage of to capture them. The negroes are in the habit of removing the pulp of some gourds through a small aperture, and then putting sugar inside ; such prepared gourds are placed near the haunts of the capuchins, who come down, inspect the gourds, and insert their hands to extract the sugar. During the operation the blacks suddenly appear, and the monkeys, in their alarm, open their hands inside the gourds, and, thus hampered, fall an easy prey to their captors.

THE WHITE-FRONTED SAPAJOU

The white-fronted sapajou (*C. albifrons*), a common monkey in many parts of South America, is allied to the last species, but distinguished by its pale reddish brown colour, which becomes redder on the back and the outer surfaces of the limbs. The most characteristic colouring is the white which occupies the face, forehead, throat, shoulders, and chest.

Mr. Bates, who described this species as being of a light brown colour, states that it is pretty generally distributed over the forest lands of the level parts of Brazil ; and he saw it in large flocks on the banks of the Upper Amazon. The members of such a flock are described as affording a wonderful sight when leaping from tree to tree ; for these monkeys, like their fellows of the same genus, are the best performers in this gymnastic exercise. "The troops," observes Bates, " consist of thirty or more individuals, which travel in single file. When the foremost of the flock reaches the outermost branch of an unusually tall tree, he springs forth into the air without a moment's hesitation, and alights on the dome of yielding foliage belonging to the neighbouring tree, maybe fifty feet beneath ; all the rest following his example. They grasp, on falling, with hands and tail, right themselves in a moment, and then away they go along branch and bough to the next tree." Bates had one of these monkeys as a pet, which kept the house in a perpetual uproar, screaming in a piteous manner when alarmed, excited, or hungry. It was always making a noise of some kind ; often screwing up its mouth and uttering a succession of loud whistling notes. Frequently this young sapajou, when following its master, would walk upon its hind legs alone, although it had never been taught to do so. One day, however, in endeavouring to wrest some fruit from an owl-faced night monkey, it attacked the latter so fiercely that it cracked its skull with its teeth, upon which its owner considered that he had had enough of pet sapajous.

THE WHITE-THROATED SAPAJOU

This monkey (C. hypoleucus) is an inhabitant of Central America. It belongs to the same group as the preceding species, from which it is distinguished by its colouring. The general colour of the fur is black, but the forehead and part of the crown of the head, as well as its sides, together with the throat and neck, are white, while the naked portion of the face is pale flesh-colour.

THE SMOOTH-HEADED SAPAJOU

The smooth-headed, monk, or yellow-headed sapajou (C. monachus), is a species from Rio Janeiro and other places in South-Eastern Brazil, deriving all its three names from the extremely close and short yellow hair with which the front of the head is covered (page 251).

The fur of this species is very short and stiff.

In colour the crown of the head, the whiskers, and chin, together with the shoulders, haunches, limbs, and tail, are pure black ; the sides and back, more especially in the hind half of the body, are yellow, more or less mixed with black ; while the sides of the neck, the chest, and the front of the shoulders are yellow, the forehead and temples being whitish yellow. Such are the striking colours of the typical form of this species, but there are several variations.

THE CRESTED SAPAJOU

The last member of this genus noticed here is the crested sapajou of Brazil (C. robustus), a species distinguished from those previously mentioned by

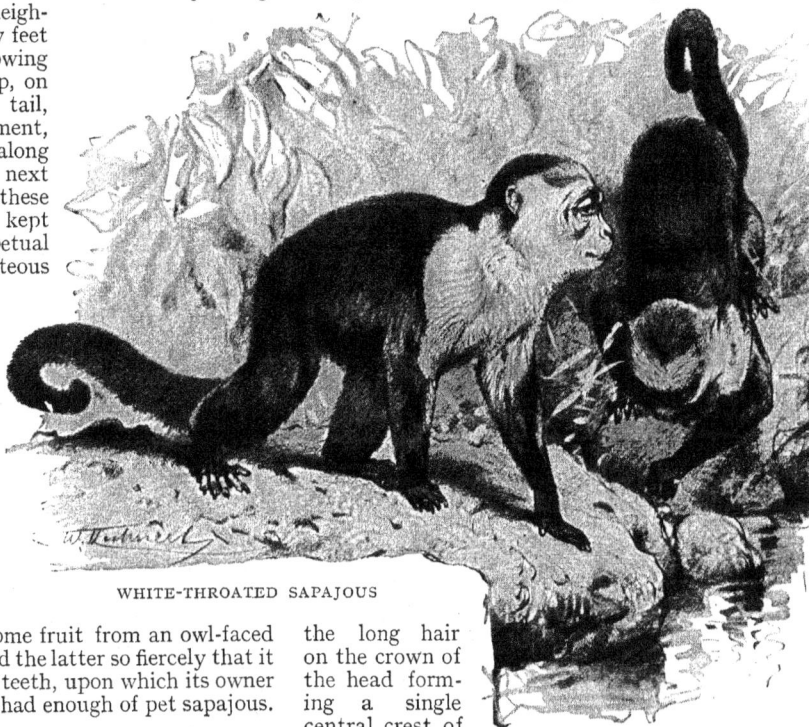

WHITE-THROATED SAPAJOUS

the long hair on the crown of the head forming a single central crest of a more or less conical shape. The general colour of the fur is bright red, with a black spot on the top of the crown, while the limbs and tail are blackish.

THE WOOLLY MONKEYS

The members of this group (genus Lagothrix) are best known by Humboldt's woolly monkey, first discovered on the Orinoco by the traveller whose name it bears, but also common in the upper part of the valley of the Amazon (page 254). This species is the only one that need be mentioned, although three or four others have been recognised.

The woolly monkeys take their name from the thick coat of woolly fur which is found beneath the longer hairs. This is one of the points by which they are distinguished from the sapajous. A more important point of difference is to be found in the

naked skin on the under part of the end of the tail—
a character in which these monkeys resemble those
of the next two genera. The woolly monkeys, however, have the same robust build as the sapajous,
and thereby differ from the other members of this
group of genera. They have well-developed thumbs,
and the great length of the tail, along with its
naked tip, renders it a prehensile organ of the most
perfect type.

To the Portuguese colonists of Brazil these animals
are known as macaco barrigudo, frequently abbreviated into barrigudo, the full name signifying
" big-bellied monkey," and being applied to them
in allusion to their bulky build, as contrasted with
the slender form of their cousins, the spider monkeys.
The ordinary form of Humboldt's woolly monkey
has a general blackish grey colour, with the head,
chest, under parts, and tail
black. The hairs are dark
grey with very short black
tips on those parts of the
body which are not black.
Younger animals are more
grey. Another phase, which
has been regarded as a
distinct species by some
naturalists, differs from the
typical form in having grey
fur on the head. Bates states
that both these monkeys
live together in the same
places, and are probably only
differently coloured indi-
viduals of one and the same species. In one of the
largest examples obtained by that traveller, the
length of the head and body was 27 in., and that of
the tail 26 in. ; these dimensions only being exceeded
among American monkeys by the black howler,
whose head and body may measure 30 in. in length.

THE WOOLLY SPIDER MONKEYS

The woolly spider monkeys, of the genus *Brachyteles* or *Eriodes*, form a kind of connecting link
between the woolly monkeys, on the one hand, and
the true spider monkeys, on the other ; having the
woolly under-fur of the former but the slender
build of the latter, while their thumbs are rudimentary. They differ, however, from both in that
their nails are extremely compressed from side to
side, and sharply pointed at the ends ; while the
partition between the nostrils is narrower.

Not much appears to be known of these monkeys,
which are confined to South-Eastern Brazil, and have
been divided into three species, mainly according to
the degree of development of the thumb. Naturalists
are, however, of opinion that these are merely
varieties of a single species (*Brachyteles arachnoides*),
since some individuals have a rudiment of a thumb on
one hand and not a trace of one on the other. In the
typical variety the general colour is ashy brown,
often tending to ferruginous at the base of the root
of the tail, with the naked parts of the face flesh-
coloured ; the females being of a paler hue. The
thumb is totally wanting (page 255). There is

another variety, or race, which has a distinct
rudiment of a thumb, and is of a dark brown
colour, with white on the sides of the face.

THE SPIDER MONKEYS

The spider monkeys (genus *Ateles*) may be regarded as those members of the group most specially
adapted to a purely arboreal life, as is shown by
their slight bodies, the long, prehensile tail, naked
below at the end, and the long, spider-like limbs,
from which they derive their popular title. In the
rudimentary condition, or total absence of the thumb,
the spider monkeys may be regarded as holding the
same relationship to the sapajous as is presented
by the guereza monkeys (*Colobus*) of Africa to the
langurs (*Semnopithecus*) of India ; and it is probable
that in both instances the
abortion of the thumb is due
to the uselessness of this digit
in a hand fitted to act merely
as a kind of hook in swinging from branch to branch.

In general characters the
true spider monkeys agree
with the woolly spider monkeys, but are readily distinguished by the total absence
of the woolly under-fur, the
comparatively slight degree
of compression in the nails,
and the greater width of the
partition between the nos-

HUMBOLDT'S WOOLLY MONKEY
Photograph by W. P. Dando

trils ; the thumb being generally absent. They are,
moreover, of far more active habits ; and in this
respect are only equalled by the langurs and gibbons
of the Old World, over which they have the advantage of the prehensile tail. They not unfrequently
use their tails to convey fruit and other articles of
food to their mouths. Those who have seen spider
monkeys swinging from rope to rope and leaping from
side to side of their cages in menageries can, when
the cage is sufficiently large, gain some idea of their
marvellous activity, although in such confined spaces
their movements bear no comparison to what the
creatures are capable of when they enjoy the
boundless freedom of their native forests.

It is noteworthy that the stomachs of the spider
monkeys have a trace of the sacculated condition
which distinguishes those of the long-limbed and
long-tailed langurs and guerezas of the Old World.
Their fur is generally smooth and stiff ; and, as a
rule, the hair on the crown of the head is directed
forwards.

The number of species is very large, and mention
need be made only of some of the better-known
types. The genus has a wide geographical range,
extending from Uruguay to Southern Mexico.

THE RED-FACED SPIDER MONKEY

The red-faced spider monkey (*Ateles paniscus*),
named by the natives of Brazil coaita, is one of
the best-known representatives of the genus. It is
found over a large area of Brazil and Guiana ; and,

although exceeded in bulk by the woolly monkeys, is in point of length of body the tallest of all the monkeys of these regions. It has been long familiar to zoology, for its scientific name was given by Linnæus, while its native designation, coaita, was in use in Europe in Buffon's time. The coarse fur is black in colour, and short on the crown of the head, although long and projecting on the forehead. The distinctive feature of the species is the tawny flesh-coloured hue of the prominent naked portions of the face, from which it derives its name (page 256).

In Brazil this species is found all over the lowlands of the valley of the Amazon, but it does not range southwards beyond the limits of the river plains, where it is replaced by the white-whiskered spider monkey. Like the other species, it lives in small parties, and is comparatively silent. Its flesh is much esteemed by the natives of Brazil, who capture it alive by shooting it with arrows tipped with weak urari poison, and restoring it, when fallen, with salt.

"Coaitas," writes Bates, "are more frequently kept in a tame state than any other kind of monkey. The Indians are very fond of them as pets, and the women often suckle them, when young, at their breasts. They become attached to their masters, and will sometimes follow them on the ground to considerable distances. I once saw a most ridiculously tame coaita. It was an old female, who accompanied its owner, a trader on the river, in all his voyages. By way of giving me a specimen of its intelligence and feeling, its master set to and rated it smartly, calling it scamp, heathen, thief, and so forth, all through the copious Portuguese vocabulary of vituperation. The poor monkey, quietly seated on the ground, seemed to be in sore trouble at this display of anger. It began by looking

earnestly at him, then it whined, and lastly rocked its body to and fro with emotion, crying piteously, and passing its long, gaunt arms continually over its forehead; for this was its habit when excited, and the front of the head was worn quite bald in consequence. At length its master altered his tone, 'It's all a lie, my old woman; you're an angel. a flower, a good, affectionate old creature,' and so forth. Immediately the poor monkey ceased its wailing, and soon afterwards came over to where the old man sat." The disposition of these monkeys is mild in the extreme, as they have none of the painful, restless vivacity of their cousins, the sapajous, and none of the surly and untamable temper of their more distant relatives, the howlers.

In the typical form of the red-faced spider monkey the thumb is absent. There is, however, a monkey similar to it in all respects, with the exception that it has a rudiment of the thumb on one or both hands. This monkey, which is known as the chameck, has been regarded as a distinct species, under the name of *A. subpentadactylus*, but it seems preferable to consider it merely as a variety of the red-faced spider monkey.

OTHER SPIDER MONKEYS

In Eastern Peru the place of the red-faced spider monkey is taken by a closely allied species (*A. ater*), in which the face is of the same black tinge as the fur. From this feature the species derives its distinctive title of the black-faced spider monkey; but it is further distinguished from the red-faced species by the shorter hair on the forehead.

Passing by one or two kinds, such as the grizzled spider monkey (*A. grisescens*), we come to a very well-marked species, known as the hooded or black-capped spider monkey (*A. cucullatus*, page 257), first described from an example in the Zoological Society's Gardens in London, of which the exact

THE WOOLLY SPIDER MONKEY

habitat was unknown, although believed to be Colombia. This species is distinguished by the length of the flaccid hair, which is a mixture of black and silvery grey in colour, and on the crown of the head is elongated, so as to form a large hood or penthouse over the eyebrows. The fur of the hands, feet, crown of the head, and nape of the neck is deep black, while the naked parts of the face are flesh-coloured.

The chuva, or white-whiskered spider monkey (*A. marginatus*), has already been incidentally mentioned as found in the Lower Amazon valley, to the south of the river plains which are inhabited by the red-faced species. It is of nearly the same size as the latter, but has moderately long hair, of a uniform black colour, with the exception of that on the forehead and the whiskers, which is white ; the face being flesh-coloured. As in the last species, all trace of the thumbs has disappeared. It does not appear to be common, since Bates seems only to have come across one pair, both of which were shot, while specially searching for it in the valley of the Tapajos. Being at the time unable to obtain other animal food, this traveller was fain to try the flesh, and describes it as being the best-flavoured meat he had ever tasted, although it was with difficulty that he persuaded himself to make the attempt.

In Central America, northwards of the Isthmus of Darien, this group of monkeys is represented by *A. geoffroyi*, noticeable as being found at very high elevations. This species, also known as *A. melanochirus*, is the black-handed spider monkey, readily recognised by the hands, feet, and the crown of the head being of a

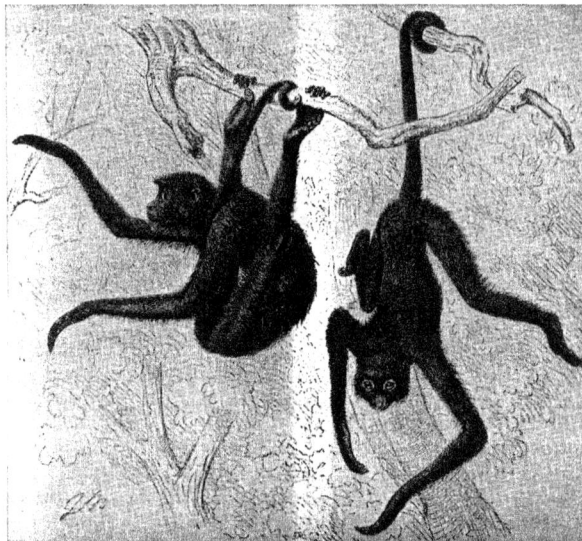

RED-FACED SPIDER MONKEYS

full black, while the fur of the body is generally some shade of pale or reddish brown, although more rarely yellowish, or even yellowish white, and, indeed, extremely variable in coloration. There appears to be a complete gradation from the typical form, with flesh-coloured rings round the eyes, to one in which the face is wholly black, while the whiskers are pale reddish yellow, the patch of erect black hair on the forehead is yellowish at the base, and most of the other parts are brownish red.

With the long-haired white-bellied spider monkey (*A. vellerosus*) of Central America and Mexico we come to a species closely allied to the last, but distinguished by the under surface of the body and the inner sides of the lower legs and forearms being white or greyish white. The Mexican species is very variable in colour, shading from black to reddish, with the loins paler, and the under parts and inside of the forearms, together with the front of the thighs,

the inner surface of the legs, and the under side of the tail, whitish ; the hair being rather long and somewhat limp. It is commonly found on the volcanic mountain known as Orizaba, near Vera Cruz, in the south of Mexico, where it ranges up to an elevation of some two thousand feet, living in troops in the forests of the deep valleys. In the neighbouring district of Oajaca, however, it reaches to a height of four thousand feet.

This is the only member of the group certainly known to range as far north as Mexico, the most northern locality for monkeys being apparently San Luis, Potosi, about 25° N. on the Tampico. These monkeys have been long known, and were described by Cuvier under the name of coaita à ventre blanc. Besides inhabiting Brazil, they were met with by Humboldt on the Orinoco. They are stated to assemble in considerable troops.

THE VARIEGATED SPIDER MONKEY

The variegated spider monkey (*A. variegatus*) is remarkable for its brilliant colouring and its wide geographical distribution ; the latter extending from the upper reaches of the Amazon in Peru to the banks of the Rio Negro, flowing from Venezuela into the lower portion of the Amazon, and northwards into the Andes of Ecuador and Colombia. It was originally described by the German naturalist Wagner, from specimens obtained on the Rio Negro. Later a monkey from the river Tigre, which flows from the north into the Amazon soon after it takes its great easterly bend on leaving Peru, was described as a new species, under the name of Bartlett's spider monkey (*A. bartletti*), but it proved to be identical with the variegated spider monkey of the Rio Negro.

The variegated spider monkey is characterised by its thick, long, and soft hair, of which, although the general colour is black, the cheeks are white. There is a band of bright reddish yellow passing across the forehead a little distance above the eyes ; while the under surface of the tail, the under parts of the body, and the inner surfaces of the limbs are yellow in the male and greyish white in the female. With the above variations in colour it will readily be imagined that the male of this species is a striking animal. In addition to more sombre tints, the female has the lines of separation between the colours less marked (page 257).

On the Tigre these monkeys live in small parties and travel rapidly through the forests in search of their favourite food, which is a berry resembling a gooseberry in size, but with a hard stone inside. Mr. A. D. Bartlett states that he had to ascend to the very summits of the ranges bordering the Tigre valley before his search was rewarded. "Here," he wrote, "we came across a number of them—about eight or nine. I shot the male that is now in the British Museum, and my Indians brought down another with a poison dart. Having obtained two of them, I was satisfied that I had found a new species. While, however, I was busy preparing the first specimen, my Indians had quietly placed the other on the fire, and, to my great horror and disgust, they had singed the hair off, and thus spoilt the specimen. Of course, I was obliged to keep the peace, for they had not tasted meat for some days, and the monkey proved a very dainty dish."

The first example of this monkey brought alive to England came from the Upper Caura River, in Venezuela, and arrived at the Gardens of the London Zoological Society on July 14, 1870; but its stay was of the briefest, as it died on August 18 following.

THE VARIEGATED SPIDER MONKEY
Photograph by W. S. Berridge

THE DOUROUCOLIS

The douroucolis, or nocturnal owl-faced monkeys (genus *Nyctipithecus*), belong to a group of three genera distinguished from the American monkeys hitherto noticed by their long tails not being prehensile. All have well-developed thumbs, and their form is massive. They agree with the members of the preceding group in the upright profile of the face, and also in the fact that the front or incisor teeth of the lower jaw are placed vertically.

The douroucolis are distinguished from the other genera of the group by their rounded heads and enormous eyes. The latter are, indeed, so large that in the dried skull the sockets occupy almost its entire width, being separated from one another by a mere line of bone, considerably narrower than the opening of the nostrils. Another distinctive feature is the narrowness of the partition between the two nostrils, which is more like that of the monkeys of the Old World than that of the other New World monkeys.

The ears are short, and the hair round the eyes is

THE HOODED SPIDER MONKEY

arranged in a radiating manner, after the fashion of the discs of feathers round the eyes of an owl. All the species are of relatively small size, and of purely nocturnal habits.

Apparently from their small ears the traveller Humboldt gave them the title of earless monkeys (*Aotus*), but the name nocturnal monkey (*Nyctipithecus*) had been applied at an earlier date. I am unacquainted with the origin of the name douroucoli. To the inhabitants of Ega, on the Upper Amazon, these monkeys are known by the name of ei-a.

At Ega, wrote Bates, "I found two species closely related to each other, but nevertheless quite distinct, as both inhabit the same forests, namely, those of the higher and drier lands, without mingling with each other or intercrossing. They sleep all day in hollow trees, and come forth to prey on insects and eat fruits only in the night. They are of small size, the body being about a foot long and the tail 14 in., and are thickly clothed with soft grey and brown fur, similar to that of a rabbit. Their physiognomy reminds one of the owl or tiger-cat. The face is round and encircled by a ruff of whitish hair; the muzzle is not at all prominent; the mouth and chin are small; the ears are very short, scarcely appearing above the hair of the head; and the eyes are large and yellowish in colour, imparting the staring expression of nocturnal animals of prey. The forehead is whitish and decorated with three white stripes."

THE THREE-BANDED DOUROUCOLI

This species (*Nyctipithecus trivirgatus*) was first discovered by Humboldt on the banks of the Cassiquiare, near the headwaters of the Rio Negro, in Venezuela, but it also occurs in Guiana and Brazil. Its chief distinctive character lies in the fact that the three bands on the forehead continue distinct from one another on to the crown of the head. The fur is relatively short and the tail cylindrical. The general colour is greyish brown, with a darker stripe down the middle of the back; the chest and under parts being rusty colour, and the tail blackish brown, except on the under part of its root, where it becomes yellowish (page 258).

The account given of this monkey by Alexander von Humboldt accords very closely with the descriptions of later naturalists. Humboldt refers to the difficulty of taming it, and states that one kept in his possession for nearly five months could not be reconciled to captivity. It slept during the day, concealing itself in the darkest corner it could find, and when awake could but seldom be induced to play with its master. Its agility in capturing flies was very remarkable. If irritated it hissed and struck out with its paws after the manner of a cat, at the same time inflating its throat. Its voice, for so small an animal, was very powerful, and Humboldt compares its cry on some occasions to the roar of the jaguar, while at others it is described as a kind of mewing, accompanied by a deep guttural sound.

Bates, who kept a specimen for many months when on the Amazon, observes that "these monkeys, although sleeping by day, are aroused by the least noise ; so that when a person passes by a tree in which a number of them are concealed, he is startled by the apparition of a number of little striped faces crowding a hole in the trunk. It was in this way that my companion discovered the colony from which the one given to me was taken. I was obliged to keep my pet chained up ; it never became thoroughly familiar." After referring to an individual of the next species, the same naturalist states that his douroucoli "was kept in a box, in which was placed a broad-mouthed glass jar ; into this it would dive head foremost when anyone entered the room, turning round inside, and thrusting forth its inquisitive face an instant afterwards to stare at the intruder. It was very active at night, venting at frequent intervals a hoarse cry, like the suppressed barking of a dog, and scampering about the room, to the length of its tether, after spiders and cockroaches. In climbing between the box and the wall, it straddled the space, resting its hands on the palms and tips of the outstretched fingers, with the knuckles bent at an acute angle, and thus mounted to the top with the greatest facility. Although seeming to prefer insects, it ate all kinds of fruit, but would not touch raw or cooked meat, and was very seldom thirsty. I was told by persons who had kept these monkeys loose about the house that they cleared the chambers of bats as well as insect vermin. When approached gently, my ei-a allowed itself to be caressed ; but when handled roughly, it always took alarm, biting severely, striking out its little hands, and making a hissing noise like a cat."

The feline douroucoli (*N. felinus*) is closely allied

THE THREE-BANDED DOUROUCOLI

to the last species, from which it is distinguished by the fact that the three dark bands on the forehead meet on the top of the forehead instead of continuing separately to the crown (page 259). It is an inhabitant of Brazil, dwelling in the same forests as the three-banded douroucoli, but always remaining separate, and being apparently much more rare. Bates described a perfectly tame specimen as being as lively and nimble as any sapajou, but far less mischievous and more confiding in its disposition, delighting in being caressed by visitors to the house of its owner, the judge of Ega, with whom it was a favourite for its pretty appearance and gentle ways.

The broad-tailed or lemurine douroucoli (*N. lemurinus*) derives its name from its broad and bushy tail, in which the hairs spread out on each side like those in the tail of a squirrel. It is further characterised by the greater length of the hair on the head and body, and also by the presence of a round pale-coloured spot over each eye, separated by a broad dark median line ; the three frontal bands of the first two species being wanting. This douroucoli is an inhabitant of Colombia, but we have not met with an account of its habits, which are, however, doubtless, much the same as those of the other species.

THE SQUIRREL MONKEYS

The elegant saimiris, or squirrel monkeys, of the genus *Chrysothrix*, form a small group of species closely allied to the titis— under which name those of this group are often included—but distinguished by several important features. In the first place, the eyes are very large, approaching in this respect those of the douroucolis, from which these monkeys are distinguished by the wide partition between the nostril and the peculiar form of the head. The peculiarity in the shape of the head consists in its great elongation from front to back ; the aperture by which the spinal cord passes out from the brain to the backbone being situated far in advance of the hinder occipital region of the skull, which projects backwards behind the neck in a manner unknown in any other monkeys. Other characteristic points are to be found in the

relatively large size of the tusks, or canine teeth, and also the comparatively short hair clothing the tail. The squirrel monkeys, or saimiris, as they were called by Buffon, also differ from the douroucolis in their diurnal habits.

The common squirrel monkey (*C. sciurea*), far the best-known representative of the genus, is an inhabitant of Brazil and the valley of the Orinoco. It is a small animal, not much larger than a squirrel, with the head grizzled grey, tending to blackish, and the fur of the body also grey, with a black mottling, but more or less tinged with gold in the region of the back. The outer sides of the forearms are yellowish, the paws whitish, and the long and slender tail is tipped at the end with black.

Writing of this species, Humboldt observes that no other monkey has so much the physiognomy of a child; it exhibits a similar expression of innocence, a similar playful smile, and a similar sudden change from joy to sorrow, or the reverse. When seized with fear, its eyes are suddenly suffused with

THE FELINE DOUROUCOLI
Photograph by Lewis Medland

tears. The one in possession of Humboldt was extremely fond of spiders and insects, and when shown uncoloured figures of wasps, etc., in a book of natural history, darted forwards as if to seize the insect. It remained, however, perfectly indifferent to figures of heads and skeletons of mammals.

Another species is the short-tailed squirrel monkey (*C. usta*), distinguished from the last mainly by its shorter tail and naked ears, and inhabiting the same regions. In some specimens the outer side of the forearm is of the same colour as the body, but in others it is shot with gold, as in the typical squirrel monkey.

In Bolivia the squirrel monkeys are represented by a well-marked species (*C. boliviensis*) differing from both the above by its black head, and by the hairs of the body being yellow with long black tips. These parti-coloured hairs cause the general hue of the fur to be golden brown. The upper part of the body, however, is of the same black hue as the head; and this colour likewise prevails on the tail, which is of moderate

THE COMMON SQUIRREL MONKEY
Photograph by W. S. Berridge

length. The face, throat, and the inner surfaces of the thighs are, on the contrary, yellowish grey.

THE TITI MONKEYS

The titis of the genus *Callithrix*, which form the last members of the present group of American monkeys, are distinguished from the squirrel monkeys by their round and well-formed heads, which are not elongated posteriorly, by their smaller eyes, less developed canine teeth, and the much longer hair clothing the tail. They are chiefly inhabitants of Brazil and other parts of the Amazon valley, and are represented by several species, some of which have been exhibited alive in England. Only a few of the best-known species call for notice.

The red titi of Brazil (*C. cuprea*), belonging to a group in which the fur is soft but intermingled with a number of long, stiff hairs, takes its name from the reddish bay colour of its hands, which forms a ready means of distinguishing it from the next species. The colour of the upper parts is blackish mixed with grey, but the cheeks, throat, under parts, feet, and legs are of the same reddish bay hue as the hands; the tail being generally rather darker than the back, although instances are known in which it has a white tip.

The second Brazilian species, *C. torquata*, known to the natives as the whaipu-sai, is readily distinguished from the preceding by the white hair of the hands. In general colour it is reddish brown tending to black, the hairs being red at the root and black at the tips. The face is surrounded by a narrow band of pure white hairs, and there is a narrow reddish white collar round the neck, from which the species takes its name. The forehead, feet, and tail are quite black. The remarkable colouring of this species has obtained for it among the Creoles of Brazil the name of the widow monkey; the white rim round the face, the whitish collar, and the white hands being likened to the veil, handkerchief, and gloves worn by widows in its native country.

In Guiana this species is replaced by the closely allied white-chested titi (*C. amicta*), which is distinguished by the presence of a pure white spot on the chest, the general colour being black tinged with red.

Another Brazilian species is the moloch titi (*C. moloch*), which, while agreeing with those just noticed in the nature of the fur, differs in that the colour of the hands and feet is of nearly the same grey hue as that of the back. The general colour of the upper parts is dark grey, with a grizzle of black and red; the cheeks, chest, and under parts being reddish, and both the hands and feet dark grey.

The reed titi (*C. donacophila*) is a paler species, closely allied to the moloch. Bates states that while on the Lower Amazon, when going ashore early one morning, he found the forest resounding with the

259

yelpings of a flock of whaipu-sai monkeys, which he thought probably belonged to this species. Although unsuccessful in obtaining a specimen, he was enabled to see them for a moment, and describes them as of small size, and clothed with long fur of a uniform grey colour.

The black-fronted titi (*C. nigrifrons*) differs from any of those yet noticed in its rigid and bristly fur, and also in both hands and feet being black. Its general colour is grey, tinged with black; but it takes its name from the black forehead; the ears, a spot on each side of the neck, as well as the hands and feet, and the inner surfaces of the forearms and legs being of the same sombre tint. The fur of the tail has a reddish tinge; while the back of the crown of the head and the nape of the neck are whitish grey.

The nearly related brown titi (*C. brunnea*), which is also known by the name of the masked titi, is subject to a great amount of individual variation in colour.

The last representative of these monkeys to be noticed is the black-handed titi (*C. melanochira*), one of two species which, while agreeing with the one last mentioned in its black hands and feet, is readily distinguished from the whole of those here noticed by the fur being soft and woolly, without any intermixture of long, stiff hairs. The general colour is reddish, but the crown of the head, the throat, and the inner surfaces of the limbs are a mixture of black and grey. There is a variety known in which the fur is bright red. This species has been obtained from Bahia, on the eastern side of Equatorial Brazil, but little or nothing seems to have been recorded with regard to its habits.

THE SAKI MONKEYS

The saki monkeys (genus *Pithecia*) typify a group of two genera, which, while resembling the dourou-colis and their allies in the non-prehensile character of the tail, are distinguished from them, and likewise from all other American monkeys, by the fact that all the front or incisor teeth of the lower jaw, instead of being vertical, are inclined forwards. In this respect these monkeys resemble lemurs. Like the titis, they approximate to the howling monkeys by having the sides of the hind part of the lower jaw considerably expanded.

Most of the sakis are characterised by having long hair on the crown of the head, which may either be divided in the middle line, or may radiate from the centre; and they all have whiskers and a beard, the latter being either broad and single, or separated by a division in the middle, and inclining back on each side. While in some species, like Humboldt's saki, the long hair covers the head, body, and tail, in others this long hair is confined to the head, where it may be present on the crown, cheeks, and chin, or only on the last two parts.

The headquarters of the sakis are Guiana and the valley of the Amazon, although they are also found in other districts. They are delicate animals, difficult to keep in captivity, and gentle and inoffensive in disposition when in confinement. Very little is known of their habits in the wild state, although it appears that they are normally silent.

THE WHITE-HEADED SAKI

The white-headed saki (*Pithecia leucocephala*) is an inhabitant of Guiana, and may be regarded as the typical representative of the group. It is characterised by its white or yellowish forehead, marked by a central streak of black, the rest of the long fur being black, and the hairs of the same colour throughout. In common with the two following species, the hair on the crown of the head is arranged in a radiating manner, the beard is broad and single, and the tail clothed with long hair, which, like that on the body, is stiff and coarse.

THE WHITE-HEADED SAKI
Photograph by W. S. Berridge

HUMBOLDT'S SAKI

In the Amazon valley, as far west as Ecuador, the white-headed saki is replaced by a species distinguished by having no black streak down the middle of the white or yellowish forehead, and also by the greater length of the hair covering the head and body. This species is Humboldt's saki (*P. monachus*), also known as the hairy saki (*P. hirsuta*), and—by the inhabitants of the Upper Amazon—as the parauacu (page 265). The general colour is black with a grey grizzle, and the tip of each hair white. There is, however, a paler variety (*P. albicans*), in which the general colour is greyish white, with only a large patch on the back and the tail black, each hair being tipped with pure white, as in the ordinary variety.

During his sojourn at Ega—or Teffé, as it is more frequently called nowadays—on the right bank of the Amazon, Bates saw several specimens of this monkey. He describes it as being " a timid, inoffensive creature, with a long, bear-like coat of speckled grey hair. The long fur hangs over the head, half concealing the pleasing, diminutive face, and clothes also the tail to the tip, which member is well developed, being 18 in. in length, or longer than the body. The parauacu is found on the *terra firma* lands of the north shore of the Solimoens, from Tunantins to Peru. It exists also on the south side of the river, namely, on the banks of the Teffé, but there under a changed form, which differs a little from its type in colours." The variety here alluded to is the whitish race of which mention was made a few sentences back.

RING-TAILED LEMURS AT PLAY

Noisy creatures, their whereabouts "are discoverable by the loud cries they are continually uttering"

X

AN INDIAN TIGER AND ITS PREY

"The food of tigers varies greatly, but the typical jungle tiger lives chiefly upon deer, wild pigs, and antelopes

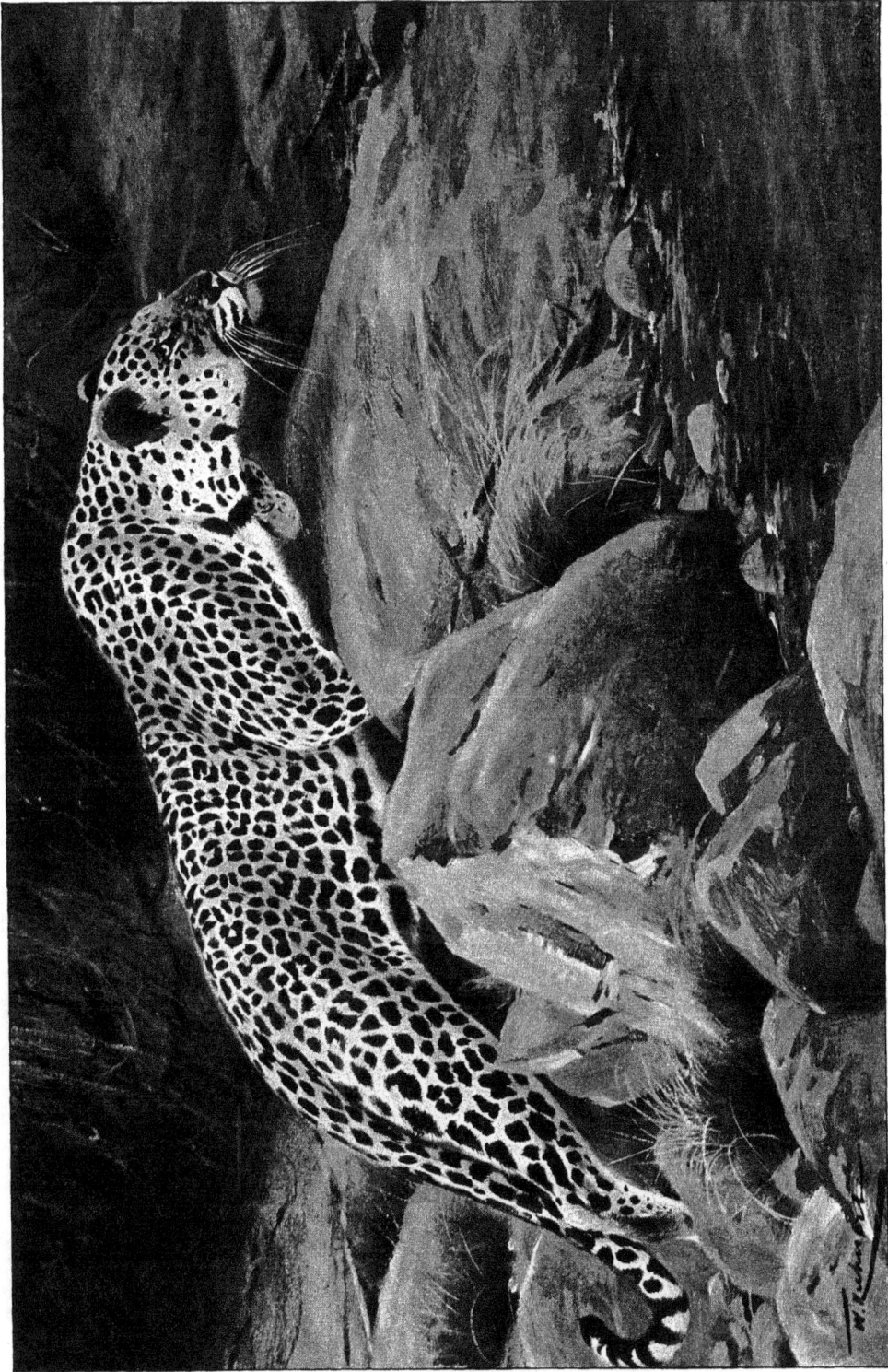

LEOPARD STALKING A VICTIM

From its stronghold the great spotted cat "descends with remarkable celerity and stealth, under cover of the rocks to cut off any straggling animal"

LYNX MAKING A RECONNAISSANCE

"In disposition the lynx is extremely savage, and will often kill more animals than it can devour"

Bates has several interesting notes about this saki. It is, he says, "a very delicate species, rarely living many weeks in captivity; but anyone who succeeds in keeping it alive for a month or two gains by it a most affectionate pet." He then notices another specimen of the pale variety which belonged to a French inhabitant of Ega. This animal "became so tame in the course of a few weeks that it followed him about the streets like a dog. My friend was a tailor, and the little pet used to spend the greater part of the day seated on his shoulder, while he was at work on his board. It showed, nevertheless, a great dislike to strangers, and was not on good terms with any other member of my friend's household but himself. I saw no monkey that showed so strong a personal attachment as this gentle, timid, silent little creature. The eager and passionate cebi [sapajous] seem to take the lead of all the South American monkeys in intelligence and docility, and the coaita [spider-monkey] has perhaps the most gentle and impressible disposition; but the parauacu, although a dull, cheerless animal, excels all in this quality of capability of attachment to individuals of our own species. It is not wanting, however, in intelligence as well as moral goodness, proof of which was furnished one day by an act of our little pet. My neighbour had quitted his house one morning without taking the parauacu with him, and the little creature having missed its friend, and concluded, as it seemed, that he would be sure to come to me, both being in the habit of paying me a daily visit together, came straight to my dwelling, taking a short cut over gardens, trees, and thickets, instead of going the roundabout way of the street. It had never done this before, and we knew the route it had taken only from a neighbour having watched its movements. On arriving at my house, and not finding its master, it climbed to the top of my table, and sat with an air of quiet resignation waiting for him. Shortly afterwards my friend entered, and the gladdened pet then jumped to its usual perch on his shoulder."

This little creature did not long survive; and if the life of these sakis is thus short when in comparatively free captivity in their native land, it must be doubly so in a cold northern climate. This is confirmed by the record of two specimens exhibited in the Zoological Gardens in London, one of which was received on May 15, 1866, and died on the 26th of the following June.

It is curious that there appears to be a marked difference in the habits of some of the sakis in regard to drinking. The black saki, for instance, is reported to drink freely, always bending down on its hands and putting its mouth to the water, heedless of wetting its head and regardless of the presence of spectators; whereas the next species has a totally different mode of drinking.

THE RED-BACKED SAKI

The red-backed saki (*P. chiropotes*), first obtained by Humboldt on the banks of the Orinoco, but also occurring in Guiana, is the first member of another group of the genus, distinguished in several respects from the species already described.

In the first place, the hair of the head, although radiating from a central point in the young, in the adult is divided by a median parting, and falls down on each side. The long beard is also divided by a gap in the middle of the chin into two lateral moieties, while the fur on the body, instead of being long and harsh, is short and soft. The tail has shorter hair than in the last group, and is thick and club-shaped. Finally, the hind part of the lower jawbone is more expanded than in the typical group. The peculiar form of the neatly divided and flattened hair of the head gives these animals the appearance of wearing a wig.

The general colour of the fur is blackish brown, but there is a considerable area on the back and shoulders of a yellowish red tint, from which the species derives its name. The tail is very thick and bushy, and the beard greatly developed.

Humboldt describes the red-backed saki as "a robust, active, fierce, and untamable animal; when irritated it raises itself on the hinder extremities, grinds its teeth, rubs the end of the beard violently, and darts upon the person who has excited its displeasure. In confinement it is habitually melancholy, and is never excited to gaiety, except at the moment of receiving its favourite food. It seldom drinks, but when it does so the operation is performed in a peculiar manner. Thus, instead of putting its lips, after the manner of other monkeys to the water or the vessel containing it, this species conveys it to its mouth in the hollow of the hand, at the same time bending forward its head. It is not, however, easy to witness this singular trait of character, since the animal is unwilling to satisfy its thirst when watched or likely to be observed." In their wild state the same traveller relates that these animals live only in pairs. Their voice, which is but seldom heard, is described as a kind of disagreeable grunt.

HUMBOLDT'S SAKI
Photograph by W. S. Berridge

Closely allied to the above is the black saki (*P. satanas*), which is an inhabitant of Brazil. It is distinguished by the absence of the yellowish red on the back and shoulders; the whole of the fur being of a uniform blackish brown colour, generally tending to a more decided black in the males, and browner in the females. In a male example in the Paris Museum the back is brown and the wig black, while in a female both the back and the wig are more fulvous. Unusually black' individuals have been described as a distinct species, under the name of *Chiropotes ater*.

The black saki, or cuxio, as this species is termed in Brazil, appears to be restricted to the lower parts of the Amazon valley. It has been observed at Cameta, on the southern side of the Amazon delta, and is stated to dwell in the most retired parts of the forests, in regions where the ground is not subject to inundations.

The last representative of these monkeys is the white-nosed saki (*P. albinasa*), which is likewise an inhabitant of the dense forests of the valley of the Amazon, and is deep black in colour, with a paler tinge on the tips of the hair, except on the nose, which is pure white, thereby rendering the species easily recognisable.

THE UAKARI MONKEYS

Just as among the monkeys of the Old World a great variation is found in regard to the relative length of the tail in closely allied forms, so in the New World there is a group of monkeys closely allied to the sakis, but distinguished by the extreme shortness of this appendage, and thereby differing from all other American monkeys. From their peculiar colouring two of the uakaris, as these monkeys are called, are among the most remarkable mammals in the world.

All the three species of uakari have long and silky hair, directed forwards on the forehead, but there is scarcely any distinct beard. The tail is very short, never being more than about one-third of the length of the body, and sometimes being reduced to a stump. From this feature they were first called *Brachyurus*, but since this term had been already applied to another group of animals it had to be changed, and the uncouth name *Uacaria* was adopted. The shelving forward of the lower incisor teeth, noticed as characteristic of the sakis, is still more marked in the uakaris.

THE BALD UAKARI

The bald uakari (*Uacaria calva*) is one of two closely allied monkeys found in the valley of the Upper Amazon, distinguished by their brilliant scarlet faces and the light colour of the long hair of their bodies. The length of the head and body of

THE BLACK SAKI
From the "Proceedings" of the Zoological Society

this species is about 18 inches, the whole of the body, from the neck to the tail, being clothed with long, straight, shining, whitish hair. The head is nearly bald, having only a very thin crop of short grey hairs (page 35). Beneath the chin and on the sides of the face are bushy, sandy whiskers; while the tint of the eyes is reddish yellow. The contrast between these colours and the vivid scarlet of the naked part of the face must be very striking when the animal is alive, but, owing to the fugitive nature of the face-pigment, is lost in museum specimens.

This monkey has an extremely limited distribution, being found only on the left bank of the Amazon, in the latitude of about 3° S., its small area being limited to the east by the Japura and to the west by the Putumayo, or Ica, as it is often called. Bates states that in this area the uakari " lives in small troops among the crowns of the lofty trees, subsisting on fruits of various kinds." Hunters say it is pretty nimble in its movements, but is not much given to leaping, preferring to run up and down the larger boughs in travelling from tree to tree. The mother, as in the other species of the monkey order, carries her young on her back. Individuals are obtained alive by shooting them with the blowpipe and arrows tipped with diluted urari poison.

They run a considerable distance after being pierced, and it requires an experienced hunter to track them. He is considered the most expert who can keep pace with a wounded one, and catch it in his arms when it falls exhausted. A pinch of salt, the antidote to the poison, is then put into its mouth, and the creature revives. The species is rare, even in the limited district which it inhabits.

Uakaris are never known to descend of their own accord to the ground, the forests inhabited by them being inundated during the greater part of the year. Hence the shortness of their tails is no indication of their habits being more terrestrial than those of the long-tailed sakis.

THE RED-FACED UAKARI

On the western side of the Putumayo the bald uakari is replaced by an allied species, the red-faced uakari (*U. rubicunda*), which appears to have an equally confined area, although the exact western limits of its range are unknown (see page 267). This uakari differs from the preceding in the hair of the body and the limbs being of an almost uniform rich deep chestnut hue, only becoming rather paler on the neck. This is in marked contrast to the pale sandy white, tending slightly to rufous, on the under parts and the inner surfaces of the limbs, characteristic of the bald uakari. Both species agree, however, in their brilliant scarlet faces, and in having hair of a rich chestnut tint beneath the throat ; and there can be no doubt but that they are extremely closely related, and have acquired their slight differences of coloration by being now completely separated from one another, although descended, as may reasonably be conjectured, at no very distinct epoch from a common ancestor.

THE BLACK-HEADED UAKARI

The most northerly representative of these monkeys is the black-headed uakari (*U. melanocephala*), which is found in the forests to the north of the Rio Negro, especially on the Cassiquiare and the Rio Branco. It thus becomes a zoological feature of the basins of both the Amazon and the Orinoco, so that it has a larger distributional area than either of the other species, from both of which it is widely different in coloration. The general colour is blackish, but the back and sides of the body are yellowish, while the loins, the outer surface of the thighs, and the tip of the

RED-FACED UAKARI

Photograph by Lewis Medland

tail are reddish chestnut, the face, hands, and feet being completely black.

THE HOWLING MONKEYS

The howling monkeys, or howlers (genus *Alouatta*, or *Mycetes*), derive their name from their vociferous cries, which are sufficient to distinguish them from all other American monkeys. To produce this extraordinary noise, there is a peculiar hollow shell of bone joining on to the upper part of the windpipe, corresponding to the so-called hyoid bone of man, which is a very small and solid structure. The resonance of the voice within this cavity communicates to the cry its peculiar intensity. In order to provide space for this bony shell the sides of the lower jawbone are extremely deep, and by this character, as well as by the extreme flatness of the part containing the brain, the peculiar skull may always be recognised. Bates surmises that their "harrowing roar," as he calls it, may serve to intimidate enemies. The howlers differ from the two preceding groups, and agree with the spider-monkeys and their allies, in having prehensile tails, in which the under surface of the extremity is naked. In addition to the presence of the large bony swelling at the top of the windpipe, they may be at once distinguished from all other prehensile-tailed monkeys by the extreme obliquity of the plane of the face and the projecting muzzle.

This obliquity is connected with the flattening of the hind part of the skull, already referred to, and is so marked that the profile inclines backwards almost in a straight line from the muzzle to the crown. Like all American monkeys, except the spider-monkeys and some of their allies, the howlers have well-developed thumbs. Their face is naked, with the muzzle very projecting, the naked parts being surrounded by a fringe of long hair on the forehead, cheeks, and chin. On the forehead this long hair may be directed either backwards or forwards, but that of the whiskers and beard always hangs down. The hair of the body, although shorter than that surrounding the face, is relatively long.

The howlers are especially abundant in Brazil, but they also range into Central America. They are represented by a considerable number of species, but since these are chiefly distinguished from one another by the colour of their hair, and there is considerable individual and sexual variation in this respect, it is in some cases difficult to decide which variations ought to be regarded as indicating distinct species, and which merely local races. The food of these monkeys is stated to consist entirely of leaves. Humboldt says that, when travelling in the neighbourhood of the Orinoco, the rising of the sun was always heralded by the cries of the howlers. Frequently this traveller and his companion, Bonpland, observed troops of these monkeys moving slowly in procession from tree to tree. A male was always followed by a number of females, several of the latter bearing their young on their shoulders. The uniformity with which they perform their movements is very remarkable. Whenever the branches of neighbouring trees do not touch one another, the male, who leads the party, suspends himself by the naked prehensile part of his tail, and, letting fall the rest of his body, swings himself till in one of his oscillations he reaches the neighbouring branch. The whole file performs the same movements at the same spot. The Indians informed Humboldt that when these monkeys filled the forests with their howling there was always one that chanted, as leader of the chorus. During a long interval one solitary and strong voice was generally distinguished, till its place was taken by another of a different pitch.

THE BLACK HOWLER

We select as our first example of that group of howlers in which the hair of the forehead is directed forwards so as to overhang the eyes, and the crown of the head is smooth, with radiating hairs, the black howler (*A. nigra*).

It is a native of Brazil. The adult males have their fur mainly uniform black, interspersed with red hairs on the flanks and loins. The females and young males are of a dingy white, and were described as belonging to a distinct species under the name of *A. straminea*. At one time there were young males of the black howler in the Jardin des Plantes, at Paris, which actually changed from the white into the black state.

The yellow-handed howler (*A. belzebul*) is another Brazilian representative of this group, which has been known since the time of Linnæus. It appears to vary considerably in colour, so that one variety was described as a distinct species (*A. flavimanus*). The general colour of the fur may be either uniform black or reddish, with some brown hairs on the shoulders ; but the hands and feet, as well as a line running down the middle of the upper surface of the tail, the tip of the same, together with a spot in front of each ear, and another on the knee, are invariably reddish yellow. The variety which is reddish-coloured all over is the prevalent type of howler in Para, on the southern side of the delta of the Amazon ; while in the island of Marajo, or Macajo, in the middle of the delta, this form is replaced by the darker one with yellowish hands and feet.

The red-and-yellow howler is a variety of *A. nigra* from Brazil in which the general colour is dark chestnut, with the back and sides golden yellow, and the beard somewhat darker.

Speaking generally, Bates says that howlers are the only monkeys the natives of the Lower Amazon forests have not succeeded in taming. Even in captivity they do not survive long, as a rule.

VERA CRUZ HOWLER

As the black howler is the most southerly representative of the genus, so the Vera Cruz howler (*A. villosa*) is its most northerly example. This species differs from the black howler in its long, soft hairs, which near their bases show a rusty tinge, in the hair of the face being inclined forwards instead of reversed, and also in the colour of the female and young being black, like that of the male.

THE RED HOWLER

The red or golden howler (*A. seniculus*) is perhaps the best-known representative of the group in which the hair is bent back so as to form a ridge across the centre of the crown of the head. The colour is reddish chestnut, but golden yellow in the middle of the back. In young individuals the hairs are short and stiff, without under-fur, and uniformly coloured throughout their length. In older individuals, however, they become long, soft, and silky, and are brown at the roots, and golden or chestnut-coloured at their tips ; while at the same time a thick under-fur is developed. Old individuals with this long, silky kind of hair have been described as a distinct species, under the name of the silky howler (see page 165).

This howler appears to be mainly a northern form, occurring in Colombia on the west, and in Guiana on the east side of South America ; while, according to Bates, who describes its fur as being of a shining yellowish red colour, it is the sole representative of the howlers in the Upper Amazon valley. It also occurs in Ecuador, and is represented by a pale variety in Bolivia.

The red howler is one of the two species of this genus that have been exhibited in the gardens of the Zoological Society in London. It is, however, difficult to keep alive for any length of time, and of two specimens received from the Dekka River, near Cartagena, on August 28, 1863, the one died on September 25, and the other on October 7 of the same year. Writing of these howlers, which he states are known to the natives as ouarines, and on the Demerara, in Guiana, are commonly called red monkeys, Charles Waterton states " that nothing can sound more dreadful than the nocturnal howlings of

BLACK HOWLERS

this red monkey. Whilst lying in your hammock amid these gloomy and immeasurable wilds, you hear him howling at intervals from eleven o'clock at night till daybreak. You would suppose that half the wild beasts of the forest were collecting for the work of carnage. Now it is the tremendous roar of the jaguar, as he springs upon his prey ; now it changes to his terrible and deep-toned growlings, as he is pressed on all sides by superior force ; and now you hear his last dying groan beneath a mortal wound. Some naturalists have supposed that these awful sounds, which you would fancy are those of enraged and dying wild beasts, proceed from a number of red monkeys howling in concert. One of them alone is capable of producing all these sounds ; and the anatomists, on an inspection of his trachea [windpipe], will be fully satisfied that this is the case. When you look at him, as he is sitting on the branch of a tree, you will see a lump in his throat the size of a hen's egg. In dark and cloudy weather, and just before a shower of rain, this monkey will often howl in the day-time ; and if you advance cautiously, and get under the high and tufted trees where he is sitting, you may have a capital opportunity of witnessing his wonderful powers of producing these dreadful and discordant sounds. Thus one single solitary monkey, in lieu of having others to sit down and listen to him, according to the report of travellers, has not even one attendant. Once I was fortunate enough to smuggle myself under the very tree on the higher branches of which was perched a full-grown red monkey. I saw his huge mouth open ; I saw the protuberance on his inflated throat ; and I listened with extreme astonishment to sounds which might have had their origin in the infernal regions."

THE BROWN HOWLER

The brown howler (*A. ursina*) is a Brazilian species, apparently found only or chiefly south of the Amazon. Its usual colour is a blackish brown, more or less washed with yellow ; but some varieties are almost entirely yellow, this being most marked on the limbs and tail. The howler described as *A. fusca*, of which specimens have been exhibited in the Zoological Society's Gardens in Regent's Park, is merely a variety of this species. It has been observed that the specimens of this monkey from the more northerly regions of Brazil are rufous or ferruginous in colour, while the females and those from the more southern regions of the country are brown or blackish brown. This species is very closely allied to the red howler.

THE MANTLED HOWLER

In Costa Rica, and probably also in other districts of Central America, the howling monkeys are represented by a well-marked species, known as the mantled howler (*A. palliata*). This monkey is characterised by the presence of a fringe of long brownish yellow hair running along the lower part of the flanks, so as to form a kind of mantle on each side of the body. The general colour of the

MANTLED HOWLER
From the "Proceedings" of the Zoological Society

fur is blackish brown, the hairs on the middle of the back, as well as on the upper parts of the sides, being yellowish brown, with black tips.

EXTINCT AMERICAN MONKEYS

The extinct monkeys which have left their remains in the great caverns of Brazil, or in the freshwater superficial deposits which cover such large areas of country in Argentina and other parts of South America, all belong either to existing genera, and in some cases even species, of American monkeys, or to extinct types of the same great family.

At the time when the huge ground-sloths known as megatheres and mylodons roamed over the pampas of South America, the forests of Brazil re-echoed as now with the cries of howling monkeys, apparently identical with the species still living ; while titis and sapajous are known to have existed at the same epoch, and remains of other living genera will doubtless also be found in the same deposits, which belong to the Pleistocene period. At the same time, with these existing genera there also lived an extinct genus of monkeys, known as *Protopithecus*. These monkeys appear to have been nearly related to the modern howlers, but were considerably larger than any living American monkey. In Argentina and Patagonia remains of monkeys apparently related to the *Cebidæ* occur in much older strata, which are probably equivalent to the Miocene of Europe ; these have been described under a number of generic names, such as *Homunculus* and *Anthropops*, and have been assigned to a separate family. Marmosets are likewise represented in the superficial South American deposits.

MARMOSETS

THE last, and at the same time the smallest, of all the true Primates are the tiny and beautiful creatures popularly known as marmosets and tamarins. These elegant little animals, many of which are much smaller than a squirrel, are confined to South and Central America, and, although agreeing in many points with the American monkeys (*Cebidæ*), yet differ in so many others as to render it necessary to refer them to a distinct family, the *Hapalidæ*.

The most important point by which marmosets are distinguished from the American monkeys relates to their teeth. It will be remembered that the American monkeys are distinguished from all their Old World cousins by having thirty-six in place of thirty-two teeth, the increase being due to the presence of an additional bicuspid or premolar on each side of each jaw. Now, if we take the skull of a marmoset and count its teeth, we shall find that their number is the same as in the Old World monkeys, namely, thirty-two. If, however, we compare the cheek teeth with those of an Old World monkey, we shall find that there is a very important point of difference. Thus, whereas in an Old World monkey there are on each side of both the upper and the lower jaw two bicuspids or premolars (teeth which are preceded by milk teeth) and three molars, in a marmoset there are three premolars and only two molars. That is to say, in place of there being two cheek teeth with one pair of cusps on the crown, which are preceded by milk teeth, and three teeth with four cusps which are not so preceded, there are three of the former type and only two of the latter. Although, then, a marmoset agrees with an Old World monkey in the total number of its teeth, yet in the much more important character of the number of premolars it resembles an American monkey, from which it differs in the comparatively unimportant feature of the loss of the last molar in each jaw. A marmoset may, indeed, be defined as a small American monkey which has lost its wisdom teeth; and the dentition of these animals may be expressed by the formula $i.\frac{2}{2}, c.\frac{1}{1}, p.m.\frac{3}{3}, m.\frac{2}{2}$; total 32.

The next most important feature in which the marmosets differ from the true American monkeys is that, with the exception of the great toe, all their fingers and toes are furnished with pointed claws, instead of more or less flattened nails; this character, like the presence of the additional premolar tooth in each jaw, clearly allying them to the lower types of mammals. It is in this group, moreover, that we for the first time find the tail ringed with alternate dark and light bands; a feature occurring also in the lemurs, and in some of the lower mammals. As in the American monkeys, the thumb of the hand cannot be opposed to the fingers, neither are there naked callosities on the buttocks, nor pouches in the cheeks. None of the marmosets has a prehensile tail. The hind limbs are always considerably larger and more robust than the front ones, and the great toe is invariably of such small dimensions that, in a literal sense of the term, it has no sort of right to its name.

Many marmosets have the ears fringed with long pencils of hairs, which give them a very peculiar and unmistakable appearance. Both in size and habits they are more like squirrels than monkeys, and climb in the same way. They are, indeed, essentially arboreal animals, subsisting not only on fruits, but likewise to a large extent on insects. As mentioned later, marmosets usually live in small parties, and all of them appear to be gentle in disposition, although frequently requiring a considerable amount of trouble and patience before they can be tamed. Whereas other monkeys usually give birth to a single young one at a time, marmosets normally have litters of two or three, and in this respect, therefore, show decided signs of affinity with animals of inferior rank in the zoological scheme. They retain, however the expressive and mobile faces characteristic of the higher monkeys.

There is a large number of kinds of marmosets, although there is still some uncertainty as to how many are entitled to rank as valid species. All of them are very similar in appearance, but they may be conveniently divided into two genera, according as the lower tusks, or canine teeth, are or are not longer than the front teeth, or incisors.

SHORT-TUSKED MARMOSETS

The marmosets of this group are characterised by the tusks not being longer than the incisors in the lower jaw, so that all the teeth present an even series. It is only in this genus (*Hapale*) that we meet with species in which the hair of the tail is marked by darker and lighter rings.

THE OUISTITI MARMOSET

The ouistiti, or common, marmoset (*Hapale jacchus*), is one of the best known of the family, having been described by Linnæus (page 271). It is an inhabitant of Brazil, more especially of the south-eastern regions, and belongs to a group in which the ears are large and bald over the greater part of their expanse, but furnished with a pencil of long hairs, which forms an expanded tuft on the front edge of their aperture; the hair on the sides of the crown of the head being likewise elongated. The tail is alternately ringed with bands of black and white, and the back has likewise darker and lighter cross-bands.

The ouistiti marmoset is of a generally blackish colour, but the back and outer surfaces of the thighs are marked with transverse bands of grey, and the head has a white spot on the upper part of the nose. The especial point of distinction, however, is that while the head is black and white, the tufts of hair on the ears are pure white.

The contrast between the black face and the white ear-tufts gives a very peculiar expression to this species, reminding us somewhat of a white-haired negro. It is frequently brought to England as a pet.

Bates, who compares the ouistiti to a kitten, banded with black and grey all over the body and tail, and having a fringe of long white hairs around the ears, only observed this marmoset in the neighbourhood of Para. On a certain occasion he observed one comfortably seated on the shoulders of a mulatto girl, whom he met walking in Para, and, on inquiry, learnt that it had been captured in the island of Marajo, at the mouth of the Amazon.

Another closely allied form from Brazil has been named the white-necked marmoset (*H. albicollis*), and is distinguished from the common kind merely by the circumstance that the hind part of the head and the back of the neck are grey instead of black.

In South-Eastern Brazil there is a third nearly related kind known as the black-eared marmoset (*H. penicillata*). The distinctive feature of this creature is to be found in the fact that not only the whole of the head and neck, but likewise the tufts of long hairs on the ears, are completely black. It is, however, doubtful whether either of these is more than a variety of the common species.

There are still other varieties or species, differing somewhat from any of the above in the colouring of the head and ears.

THE WHITE-EARED MARMOSET

The white-eared marmoset (*H. aurita*), which is also a Brazilian species, is the representative of a second group, in which the pencil of hairs on the ears is much more slender than in the common marmoset, while the hair on the back is generally somewhat speckled, although faint traces of banding are occasionally observable. The tail is ringed like that of the common marmoset.

The general colour is blackish, minutely speckled with yellow or a reddish tint on the back, the sides

THE OUISTITI MARMOSET
Photograph by W. S. Berridge

THE BLACK-EARED MARMOSET
Photograph by W. S. Berridge

of the head, the limbs, and the hind part of the body being pure black; while the crown of the head is brown, and a spot on the forehead, as well as the tufts on the ears, grey. In some instances, where the back is more decidedly red than usual, there are faint, paler cross-bands in this region, and more especially on the loins.

The white-shouldered marmoset (*H. humeralifer*) is a closely allied Brazilian kind, distinguished by the face, shoulders, chest, and arms, as well as the tufts on the ears, being white; the thighs being a mixture of brown and white in colour.

THE SILVER MARMOSET

With the silver marmoset (*H. chrysoleucus*) of Brazil we come to the first of a small group of species distinguished from those yet noticed either by the absence of rings of colour on the tail, or by the arrangement or absence of the longer hairs on the ears. They are all tiny, not much larger than a rat, and have no bands of colour on the back.

The silver marmoset has large and nearly naked ears, covered on both sides near the margin with long hairs, forming a double fringe instead of a pencil. The fur of this elegant creature is soft and silky, and either pure white or yellowish white in colour. In the white variety the limbs and tail, however, are invariably yellowish; while in the variety in which the fur of the body is yellowish, that covering the limbs, tail, and under parts may be chestnut brown. These two varieties have been regarded as distinct species.

THE BLACK-TAILED MARMOSET

The black-tailed marmoset (*H. melanura*) is distinguished from the preceding by the absence of the fringe of hairs on the large and flesh-coloured ears, and likewise by the black tail. Usually the general colour of the fur is ashy brown, paler on the front of the body, and whitish on the front of the thighs and loins, while the head and limbs are dark brown (page 272). There is, however, a variety which is entirely white, with the exception of the characteristic black tail.

Of this species, which he mentions under the name of *Midas argentatus*, Bates observes that it is one of the rarest of the American marmosets. " Indeed,

I have not heard of its being found anywhere except near Cameta, where I once saw three individuals, looking like so many white kittens, running along a branch in a cacao grove; in their movements they precisely resembled *M. ursulus*," of which a description is given later. " I saw afterwards a pet animal of this species, and heard that there were many so kept, and that they were esteemed as great treasures. The one mentioned was full-grown, although it measured only 7 in in length of body. It was covered with long, white, silky hairs; the tail being blackish and the face flesh-coloured. It was a most timid and sensitive little thing. The woman who owned it carried it constantly in her bosom, and no money would induce her to part with her pet. She called it Mico (the native name of these animals). It fed from her mouth, and allowed her to fondle it freely, but the nervous little creature would not permit strangers to touch it. If anyone attempted to do so, it shrank back, the whole body trembling with fear, and its teeth chattered whilst it uttered its tremulous frightened tones. The expression of its features was like that of its more robust brother, *M. ursulus*; the eyes, which were black, were full of curiosity and mistrust, and were always kept fixed on the person who attempted to advance towards it."

THE PIGMY MARMOSET

A third representative of the diminutive species constituting this group is the pigmy marmoset (*H. pygmæa*), found in the primeval forest regions of Brazil. This species is distinguished by the smallness of its short ears, which, although slightly hairy on their outer surface, have no tuft or fringe of long hairs, and are entirely concealed beneath the backwardly directed and elongated fur of the crown of the head. A further distinction is to be found in the presence of darker and lighter rings on the tail. The general colour of the body is a tawny or ferruginous brown, more or less varied with black and red on the back; the neck, under parts, and inner surfaces of the limbs being yellowish, and the hands and feet yellowish brown.

The pigmy marmoset, which is not uncommon on the Upper Amazon, at San Paulo, near Ega, measures only 7 in. in length, exclusive of the tail. The tiny face is furnished with long brown whiskers, brushed back over the ears; the general colour of the body being brownish tawny, but the tail elegantly ringed with black. This marmoset ranges as far north as Mexico, and is one of the few South Amazonian

THE BLACK-TAILED MARMOSET
From the " Proceedings " of the Zoological Society

members of the Primates that wanders far from the great river plain; a second being the Silky Marmoset, which has also been recorded from Mexico.

LONG-TUSKED MARMOSETS

The marmosets of this group (genus *Midas*) are distinguished from those of the preceding genus by the fact that the tusks, or canine teeth, of the lower jaw are considerably longer than the front, or incisor teeth; so that the whole series of lower teeth does not present the even and regular height characteristic of the short-tusked marmosets. None of the tamarins, as Buffon named these creatures, has pencilled ears; neither have they ringed tails, although some of the species have the back marked with dark and light cross-bands.

THE NEGRO TAMARIN

One of the best-known species is the common or negro tamarin (*Midas ursulus*), found in Guiana and the lower part of the Amazon valley. It belongs to a group in which both the forehead and face are hairy, and the hair of the head is not longer than that of the body, the ears being large and naked. The colour is nearly uniform black, especially on the nose, lips, and hands; but the hind part of the body has the fur more or less mottled with greyish white. Although not known as a distinct species at the time of Linnæus, this marmoset was described by the early French naturalists and distinguished by Buffon as the *tamarin nègre*.

Bates mentioned that the negro tamarin " is never seen in large flocks, three or four being the greatest number observed together. It seems to be less afraid of the neighbourhood of man than any other monkey. I sometimes saw it in the woods which border the suburban streets of Para, and once I espied two individuals in a thicket behind the English consul's house at Nazareth. Its mode of progression along the main boughs of the lofty trees is like that of the squirrel; it does not ascend to the slender branches, or take the wonderful flying leaps which the cebidæ do, whose prehensile tails and flexible hands fit them for such headlong travelling. It confines itself to the larger boughs and trunks of trees, the long nails being of great assistance to the creature, enabling it to cling securely to the bark; and it is often seen passing rapidly round the perpendicular cylindrical trunks. It is a quick, timid, restless little creature, and has a great share of curiosity, for when a person passes by under the trees along which a flock is running, they also stop for a few moments to have a stare at the intruder. In Para, *M. ursulus* is often seen in a tame state in the houses of the inhabitants. When full grown, it is about 9 in. long, independently of the tail, which measures 15 in. The fur is thick, and black in colour, with the exception of a reddish brown streak down the middle of the back. When first taken, or when kept tied up, it is very timid and irritable. It will not allow itself to be approached, but keeps retreating backwards when anyone attempts to coax it. It is always in a querulous humour, uttering a twittering, complaining noise, its dark, watchful eyes, expressive of distrust, observant of every movement which takes place near it. When treated kindly, however, as it generally is in the houses of the natives, it becomes very tame and familiar. I once saw one as playful as a kitten, running about the house after the negro children, who fondled it to their hearts' content. It acted somewhat differently towards strangers, and seemed not to see them seated in the hammock which was slung in the room, leaping up, trying to bite, and otherwise annoying them. It is generally fed on sweet fruits, such as the banana; but it is also fond of insects, especially soft-bodied spiders and grass-hoppers, which it will snap up with eagerness when within reach. The expression of countenance in these small monkeys is intelli-gent and pleasing. This is partly due to the small facial angle, which is given as 60°; but the quick movements of the head, and the way they have of inclining it to one side when their curiosity is excited, contribute very much to give them a knowing expression."

THE RED-HANDED TAMARIN

The red-handed tamarin (*M. rufimanus*), the true tamarin of Buffon, is an inhabitant of Dutch Guiana, or Surinam, and differs from the preceding in its yellowish or orange red hands, its habits being, doubtless, precisely similar. Like the negro tamarin, it has been exhibited alive in England.

THE BROWN-HEADED TAMARIN

The brown-headed tamarin (*M. fuscicollis*) is the Brazilian representative of several species or races distinguished from the two preceding species by the face being brownish, with a few grey hairs, though the nose still remains black. The colour is black, with a white mottling on the hind part of the back, the head being pale brown, with black

DEVILLE'S TAMARIN
Photograph by W. P. Dando

markings. In the male the outer surface of the limbs generally has a bright rufous tinge, while the under parts and the inner surfaces of the limbs are reddish brown. The so-called black-and-red tamarin (*M. rufoniger*) appears to be only a brighter-coloured variety of this species, in which the back, loins, thighs, and legs are of a bright chestnut red. It occurs in Brazil, and has been met with on the Upper Amazon in the neighbourhood of Ega. In referring to the marmoset provisionally identified with this form, Bates wrote: "One day, whilst walking along a forest pathway, I saw one of these lively little fellows miss his grasp as he was passing from one tree to another along with his troop. He fell head-foremost from a height of at least fifty feet, but managed cleverly to alight on his feet in the pathway; quickly turning round, he gave me a good stare for a few moments, and then bounded off gaily to climb another tree." The habits of this species are precisely similar to those of the negro tamarin.

Deville's tamarin (*M. weddelli*), from Peru, is another nearly related species, with the head, neck, front of the back, fore limbs, and tail black; the hind part of the back marked with grey and black transverse bars, and the loins and legs bright chestnut red.

MOUSTACHED TAMARIN

The little moustached tamarin (*M. labiatus*) belongs to a well-marked small group of species recognised by having the tip of the nose and the lips covered with white hairs, giving a very peculiar look to the face. It is found in the Upper Amazon valley, both in Brazil and Peru, and is black, with a brownish tinge on the back and thighs, the white hairs on the nose and lips being long and forming a broad tuft. When seen from a short distance, it looks as though it were holding a ball of snow-white cotton in its teeth.

The red-bellied tamarin (*M. rufiventer*) is an allied Upper Amazonian species, with a very smooth and glossy coat of a deep blackish brown colour on the back, while the under parts are a mixture of rich black and reddish hues. The white hairs on the nose and lips are much shorter and less conspicuous than in the moustached tamarin, those on their lips merely forming a thin line on the margins in-stead of a distinct tuft.

THE PINCHÉ

In Colombia (New Granada) and Panama the tusked marmosets are represented by two closely allied species differing in certain points from all

THE PINCHÉ MARMOSET
Photograph by W. S. Berridge

those found in the more southerly or easterly regions. Both have the face and sides of the head sparsely haired, while there is a distinctly marked patch of hair, different from the rest, on the crown of the head, and the hair on the neck is elongated.

The present species is restricted to Colombia, and has been long known in Europe ; it received its name of pinché—on what grounds I know not—from Buffon. It is of a greyish brown colour on the back ; the outer surface of the limbs and the root of the tail being tinged with red, while the long tuft of hair which forms a crest on the top of the head, as well as the throat, under parts, arms, and the front of the legs, are white ; the tip of the tail being black.

Geoffroy's marmoset (*M. geoffroyi*), which is the representative of the pinché in Panama, is distinguished from that species by the hair on the crown of the head not being elongated into a crest, but being short, and forming a narrow patch of an oblong shape (page 275). Its coloration is nearly the same as that of the pinché, excepting that the hair on the nape of the neck is chestnut-coloured.

THE SILKY MARMOSET

The last group of the marmosets is represented by the silky marmoset (*M. rosalia*) and the golden-headed marmoset, both of which inhabit the forests of South-Eastern Brazil, and are commonly exhibited in the menageries of Europe. They are distinguished by having the head and part of the neck covered with long hair, forming a kind of mane, the hair round the face being directed backwards. The face itself is only sparsely haired, and the naked ears are partly concealed by the mane. The colour is a golden yellow, more or less tinged with red ; but there is a variety in which the head, hands, feet, and the end of the tail are blackish.

The silky marmoset (page 167) was known to Buffon under the name of the marikina, and has also been described as the lion marmoset (*M. leoninus*), a name due to the long mane of brown hair

hanging from the neck and giving it somewhat the appearance of a miniature lion. Bates mentions that he saw a tame individual of this species on the Upper Amazon. After commenting on its playful and intelligent disposition, he observes that it was familiar with every person in the house where it was kept, and seemed to take particular pleasure in climbing about the bodies of the various visitors who entered. "The first time I went in, it ran across the room straightway to the chair on which I had sat down and climbed up to my shoulder ; arrived there, it turned round and looked in my face, showing its little teeth, and chattering, as though it would say, 'Well, and how do you do ?' It showed more affection towards its master than towards strangers, and would climb up to his head a dozen times in the course of an hour." These marmosets are described as keeping to the very top of their cages—a habit probably retained from the native one of living in the tree-tops. When descending they always come down backwards, with the tail pendent. Their ways are very similar to those of the common marmoset, but they are reported to have a habit of bounding from tree to tree with incredible rapidity, which is scarcely consonant with the account generally given of the movements of marmosets. They are stated to utter sharp but weak cries of alarm when frightened. The total length of this marmoset is rather less than two feet, of which one is occupied by the long tail.

The golden-headed marmoset (*M. chrysomelas*) may be regarded as a black representative of the silky species, its general colour being black, with the face, forearms, hands, feet, and the base of the tail tawny   R. LYDEKKER

STORIES OF THE MARMOSETS

Marmosets are not uncommon pets in England. With warmth and abundant variety of diet they are not difficult to keep. One lady frequenter of the

THE SILKY, OR LION, MARMOSET
Photograph by W. S. Berridge

London Zoological Gardens carries in her muff a tiny marmoset, which, in warmer weather, makes itself snug in a cunningly contrived pocket in the lady's belt, where it is hidden by a feather boa or lace ruff. In these days of painful realism in lady's furs and other articles of apparel it is hard to tell whether this living marmoset may not be merely some particularly extravagant fad of adornment.

The marmoset is not so intelligent a pet as many other members of the order to which it belongs. Its relatively big brain has not been cultivated to the same extent as have been the brains of other primates. Even so, it is apt to incur censure which it should escape. The female of a pair in captivity gave birth to three young ones. She bit off the head of one of them. The others managed to make their way close to her, where she at once fed them. Thereafter she proved the best of mothers, while the male, when the feeding was over, showed his paternal devotion by taking the little ones on his back or under his body, where they remained until they whined for another meal.

The observant owner of these marmosets attributed the destruction of the marmoset to the inexperience of the dam, to whom the cares of maternity were new. That explanation, however, will not do. The probability is that the animal killed her offspring because she was under observation. Long as the rabbit has been domesticated, the mother will destroy her young if she be interfered with during the first few days following their birth. A sow will destroy her litter for the same reason. Not inexperience drove the marmoset to her deed, but discontent with her surroundings, and the consciousness of her accessibility. Later litters met a similar fate, because, it is to be presumed, the dam remembered how she had used the first. There is always a danger, when an animal has once destroyed a litter, that a similar course will be pursued with regard to later offspring.

Frank Buckland managed to keep a marmoset in happiness. She occupied a squirrel's cage at one end of the mantelpiece of his study, while a lusty rat was the tenant of another at the opposite end. Being nocturnal, they remained inactive during the day, but, as night fell, they emerged from their cages. The two animals would meet on neutral territory midway between the two cages, where the marmoset would make some pretence at recognition, while the rat, with a supercilious toss of his nose, would merely acknowledge the salutation, and pass on his way. The programme never varied. The rat would go straight to the vacant cage of the marmoset and steal the food which the tenant had left behind, consisting of grapes, sugar-plums, meal-worms and what not. In the meantime, while her own premises were being ravaged, the marmoset was quietly robbing the home of the rat. Once the rat, despatching his booty hurriedly, returned unexpectedly to his cage and found the marmoset there, and attacked her so fiercely that Buckland had to interfere to save her from serious injury.

Experience failed to make this marmoset wise. She would hang for hours sloth-like from the gas-bracket. Every now and then she would whisk her tail into the flame, and as quickly whisk it out again. She never got over the habit, and the consequence was that her tail at last became quite bare. If anybody could have tamed a marmoset

GEOFFROY'S MARMOSET
From the " Proceedings " of the Zoological Society

that man surely was Buckland, but, whenever he attempted to pick her up, or even to touch her, she glared at him and screwed up her face into the nearest resemblance to ferocity to which she could attain, to the intent that terror might seize him and she escape his hand. Her voice on these occasions he compared to a watch whose mainspring had suddenly snapped. Probably his monkeys would have given the marmoset a very uncomfortable time could they have had her in their cage, but when they heard her cries of protest they always did their best to go to her rescue, shaking the bars and making the most fearsome grimaces at their master, who never harmed them, but whom they now judged guilty of assault and battery upon their cousin.

The South American natives make pets of the marmosets, as we have seen. One extraordinary means of keeping them has not been mentioned. A tiny tamarin (*Midas illigeri*) made its home on the head of the woman to whom it belonged, concealed within her luxuriant tresses. Having become tame, it was wont to hop out to feed, or, having captured a spider or two, it would scamper back again and hide under the abundant locks of its owner. Imagination suggests that this is an unpleasant method of housing a pet.   E. A. B.

CHARACTERS OF THE LEMUROIDS

All the animals treated of in the preceding sections as possessing many characters in common are regarded by naturalists as collectively constituting one great group of the Primates. And since this group is also taken in zoological classification to include man himself, it is spoken of as the Anthropoidea, or man-like group, the individual members thereof being referred to as anthropoids.

We now come to another and lower group of animals, which, while sufficiently nearly allied to the above to be included in the order, are yet so different as to be entitled to stand as a group of equivalent rank. These animals are primarily represented by the lemurs and lorises; but the group also includes two other creatures which form the types of two families by themselves. As it is desirable to have a common name for all the members of this group, and as it would be incorrect to allude to the whole of them as lemurs, the term lemur-like creatures, or, shortly, lemuroids, will be found convenient.

Although these lemuroids may always be distinguished at a glance from the apes and monkeys by their foxy, expressionless faces, it is difficult to point out the important structural features by which they differ from the former without entering into anatomical details unsuited to a popular work. In spite, however, of these differences, there are such resemblances between the two groups as to suggest that the lemurs and their allies are not far removed from the group from which we may presume apes and monkeys to have originated.

Considerable diversity of opinion has obtained among naturalists as to the affinity of lemuroids. As the result of the study of a large series of remains of extinct lemuroids from Madagascar, Mr. H. F. Standing and Dr. Elliot Smith consider that there is no justification for regarding the Anthropoidea and the Lemuroidea as distinct groups, since some of the extinct Malagasy forms are intermediate between the two. On the other hand, Professor A. A. W. Hubrecht, as the result of the study of their embryological development, is of opinion that lemuroids should not be included in the same ordinal group as anthropoids.

That the lemurs are much lower in the zoological scale than the apes and monkeys is shown by the simpler structure of their brains, which have far fewer foldings on their surface than is the case with those of the latter; the amount of such foldings, as giving a larger extent of superficial surface, being indicative of the mental powers of the owners of the brains.

A peculiar feature of all the lemurs and their allies is to be found in the fact that the second toe of the foot (corresponding to the index finger of the hand) is always furnished with a sharp claw. All lemurs have a well-developed thumb and great toe; but, curiously enough, in some of them the index finger of the hand is rudimentary. They may or may not have tails, but these are never prehensile, although, as in some of the marmosets, they may be marked by alternate dark and light rings.

A point of resemblance to the monkeys and apes is shown in the number of incisor, or front, teeth being very frequently two on each side of both jaws, in place of the three which are so commonly present in other mammals. In the apes and monkeys, however, the central pair of incisors in the upper jaw are in contact with one another, while in the lemurs they are almost invariably separated by a gap in the middle line. This affords a ready means of distinguishing the skull of a lemur at a single glance from the skulls of almost all other mammals, except bats and some of the Insectivora. The lower front, or incisor, teeth of the lemurs shelve forwards, after the manner already mentioned as characteristic of one group of the American monkeys (page 260).

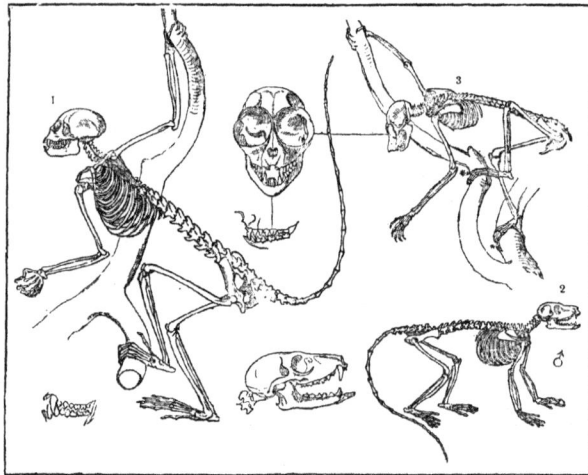

SKELETONS OF SQUIRREL MONKEY (1), MUNGOOSE LEMUR (2), AND SLENDER LORIS (3)

Many lemurs are purely nocturnal animals, and it was probably from this circumstance, coupled with their silent habits and stealthy movements, that Linnæus was induced to give them the name by which they are now universally known. The name "lemur" is taken from the Latin *lemures*, which, together with that of *larvæ*, was applied by the ancient Romans to such "shades of the dead" as were supposed to be of malignant propensities. It is curious that both terms should have been introduced into zoological nomenclature—the former to denote the animals of the present group, the latter the grub stage of insects.

Altogether, there are about fifty living species of lemur-like animals, all restricted to the southern regions of the great land masses of the Old World, none being found to the north of the Tropic of Cancer, while the Tropic of Capricorn very nearly limits their southward range. Within this area a few species are found respectively throughout the warmer regions of Africa and in Southern India and Ceylon, while their eastern limits are marked by the island of Celebes and the Philippines. In all these regions the number of species is comparatively few, and they form but an unimportant element in the general fauna. The case is, however, very different in Madagascar, the headquarters of the whole group. Here they constitute no less than one-half of the entire mammalian fauna of the island, being represented by about six genera, which include more than thirty

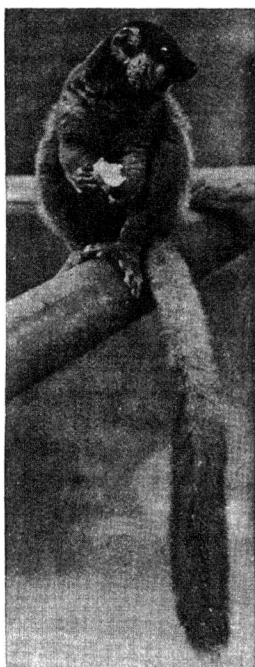

THE YELLOW-WHISKERED VARIETY OF FULVOUS LEMUR

species, most of the other mammals of the island being comparatively small forms, unknown either on the continent of Africa or in Asia. The true lemurs occur only in Madagascar, and it is very remarkable that all the species of the group found in that island scarcely show any closer relationship to those of the African mainland than they exhibit to those of Asia. So abundant, indeed, are lemurs in Madagascar that at least one individual is almost sure to be found in every little copse throughout the island.

Almost all of the lemurs are essentially of arboreal habits. Indeed, except when compelled to descend to the ground to obtain water, or for the purpose of crossing from one plantation or coppice to another, they but rarely leave the trees. Their diet is extremely mixed, scarcely anything coming amiss to them, leaves, fruits, insects, reptiles birds' eggs, and birds being eagerly consumed by most of these animals.

By the natives of Madagascar lemurs are looked upon with superstitious awe, and are consequently seldom molested. This is doubtless due to their nocturnal habits and ghost-like movements, while the large eyes essential to these and all other nocturnal creatures have perhaps contributed to this feeling. In Ceylon and India, as we shall learn, the large glaring eyes of one of the prettiest of the lemuroids used to lead to the unfortunate creatures being put to a cruel death. None of the existing lemuroids attains any very large size, and

THE BROWN VARIETY OF THE BLACK LEMUR

all, when unmolested, are harmless and inoffensive, except to the birds, reptiles, and insects upon which they prey. The nostrils of a lemur, which are always situated at the extremity of the muzzle, differ markedly in form from those of a monkey. In all the latter, whether they be thin-nosed, like the Old World kinds, or broad-nosed, like those of America, the nostrils are always more or less rounded in form, and thus approach to the human type. In lemurs, on the other hand, the nostrils are always in the form of curved slits, widest above, and with the convexity directed outwardly. The nostrils are, in fact, almost precisely similar to those of a dog or a cat, and we have in this another proof of a relatively low zoological position.

THE LEMURS

IN the true, or typical, lemurs (family *Lemuridæ*), which are exclusively confined to Madagascar, the first point to be noticed is that the upper front, or incisor, teeth are always two in number on each side of the jaw, and that the middle pair are separated from one another by a gap. The upper premolar teeth may be either two, as in the Old World monkeys, or three, as in their cousins of the New World, the molars being invariably three in number. The front teeth in the lower jaw, together with the one corresponding to the tusk, or canine, always shelve forwards, and are small. This small size and shelving direction of the lower tusks render it necessary that some other tooth should be enlarged so as

to bite against the upper tusk. Accordingly, the first premolar in the lower jaw takes on the form and size of a tusk, and bites against the true tusk, or canine tooth of the upper jaw. It has been mentioned in an earlier page (141) that whereas true tusks, or canines, have usually only a single root, premolar teeth nearly always have two roots, except when there are four of these teeth, in which case the first generally has but one root. Now, the tusk-like lower premolar of the lemurs has the usual two roots, and hence we have a ready means of distinguishing a lemur's skull from that of most other mammals ; that is to say, by the lower tusk having two distinct roots.

Another distinctive feature of the true lemurs is that, with the exception·of the second toe of the foot, all the fingers and all the toes have well-formed flattened nails like those of the majority of monkeys.

It is also important to notice that all the Malagasy lemurs, inclusive of the aye-aye, agree with one another in regard to the structure of the drum, or tympanum, of the ear, and thereby differ markedly from the lorises and galagos.

THE INDRI

The peculiar-looking indri (*Indris brevicaudata*) is one of the numerous lemuroids from Madagascar, and occupies the position of being the largest member of the entire group. It is likewise the sole representative of its genus. The name indri, or indris, is a corruption of the native name endrina, applied in certain districts to this animal. In other districts, however, it is designated baba-koto, or " little old man."

The indri is the first of a group of three genera which present certain characteristics in common not found in other lemurs ; the members of these three genera are accordingly grouped together in a separate subfamily—the *Propithecinæ*. Among these characters the most obvious is the large proportionate size of the legs as compared with the arms. Another is that, with the exception of the great toe, which is capable of being fully opposed to the others, the toes of the foot are joined together by webs as far as the end of the first joints. The total number of teeth in the adult condition is limited to thirty, the series being represented by the formula $i.\frac{2}{2}, c.\frac{1}{1}, p.m.\frac{2}{2}, m.\frac{3}{3}$. All the members of this group differ from the other lemuroids in that they do not give birth to more than a single young one at a time. From this circumstance, together with certain features in their structure, these indris are regarded as the most highly organised of all the lemurs, and are accordingly placed at the head of the list. They subsist exclusively on a vegetable diet.

The indri is sufficiently distinguished from the other two genera included in the group by its mere stump of a tail, although there are also certain other features which support its right to stand as the representative of a distinct genus.

As already mentioned, the indri is the largest of the lemurs, and in a fully adult animal the length of the head and body is about two feet. Although there is great individual variation in this respect, the indri is very strikingly coloured. Very frequently the forehead is blackish, but, like the cheeks and throat, it may be grey. The head, shoulders, back, and arms are of a full velvety black, and the black ears are large and prominent and covered with longer hairs than those on the head. From the loins to the tail there is a large triangular patch of either pure white or of a yellowish tinge ; this patch terminates in front in a sharp point, and is bordered on all sides with black. The flanks are also light-coloured, and the dark bands which usually separate the light area of the loins from that of the flanks are continued down the front of the legs ;

but the sides of the legs are in general whitish, and their hind surface is grey, the heel being reddish. The hands and feet are black, and, as a rule, almost denuded of hair.

Such are the common colours in the larger number of specimens of the indri. In almost every drove, however, individuals are found in which the light-coloured areas intrude more or less extensively upon those which are usually black ; and from these intermediate forms a complete transition can be traced to others in which the whole of the fur is white. The intermediately coloured individuals very generally retain the broad black streak down the front of the leg, and the black ears.

Instead of being distributed over the whole of Madagascar, the indri is confined to the forests on the east coast, this restricted distribution being due to the great range of mountains running longitudinally through the island, which cuts off these animals from the plains on the western side.

In contradistinction to most of the lemurs, the indris are purely diurnal in their habits, and are commonly found in small parties of four or five, although during the day single individuals, more or less widely separated from their companions, may frequently be seen. Their general habits appear to be similar to those of their near relatives the sifakas. Unless injured so badly as to be unable to make its escape, an indri does not give utterance to the least sound when wounded ; if, however, it is so severely hit as to fall to the ground, which it will only do when its extraordinary powers of holding on to the branches of the trees are exhausted, it gives vent to piercing shrieks.

THE SIFAKAS

The sifakas, as they are called by the native inhabitants of Madagascar, constitute the only genus (*Propithecus*) of the present group of lemurs which is represented by more than a single species. Although closely allied to the indri, they are distinguished by their long tails; the muzzle is also rather shorter, and the ears are considerably smaller, and partly concealed by the fur. Their skin is of a deep black ; but the general colour of the fur is usually white, more or less tinged with yellow, and, in some individuals, passing into red or even black. The fur on the breast is always much thinner than that of other parts of the body.

Three species of the genus are recognised, which are restricted to different parts of the island ; but of these species there are several more or less distinct races, which are likewise confined to particular localities. It has been observed that while those individuals of the several species which tend to assume a black colouring are found in the damper parts of the island, those which are most completely white frequent the drier regions at the northern extremity of Madagascar.

Sifakas live in bands of from six to eight. They are completely diurnal in their habits, and may be observed at morning and evening, when the heat is not too great, leaping in the forests from tree to tree in search of food. At sunrise they may often be seen sitting on the horizontal bough of a tree, close to where it branches off from the main stem, with their long legs bent so as to touch their chins, and their hands resting on their knees. At other times they will be noticed sitting in the same position, but with their arms extended, so as to receive the genial warmth of the rising sun on their bodies. During the heat of the day they conceal themselves in the depths of the foliage. When sleeping, they incline the head forwards on the chest, and cover it with their arms ; at the same time the tail is either curled up spirally between the legs or allowed to hang straight down.

Their shelving lower front teeth are admirably adapted for removing part of the rind of the fruits on which they so largely subsist, and thus making an aperture through which the pulp is removed piecemeal. The skins of the fruits are always rejected ; and it is stated that sifakas exhibit a marked preference for green rather than ripe fruit.

In all ways they are admirably adapted for a purely arboreal life. So strong, indeed, are their hind limbs that they can readily take leaps of from ten to eleven yards in passing from bough to bough ; and so rapid are their motions that they appear to fly rather than leap. On the rare occasions when they descend from their favourite trees, they advance by means of long leaps, as, owing to the shortness of their arms, it is not easy for them to walk on the ground on all-fours like the majority of monkeys. To see them resting on their hind feet, and at each leap throwing up their arms in the air, the spectator might be led to think for a moment that he was looking at children at play. Indeed, a troop of these creatures advancing across the plains in the manner described is said to be a truly ludicrous sight. Not only are the hands of the sifakas of no use to their owners in walking, but they are almost equally useless as organs of prehension ; and when a sifaka has occasion to pick up a fruit from the ground, it will usually stoop down and seize the prize in its mouth. When conveyed to the hand, such an object is grasped between the bent fingers and the palm, and not between the fingers and thumb. As grasping organs, adapted to afford a firm hold to the branches of trees, both hands and feet, however, are perfect.

In disposition, sifakas are gentle and seldom attempt to bite, while if they do so the wound they inflict is not serious. At certain seasons, however, the males are wont to engage in contests among themselves, the results of which are frequently visible in their torn and tattered ears. Unlike many other lemurs, they are, as a rule, silent ; but when frightened or angry they give vent to a low cry somewhat resembling the clucking of a fowl. In

THE RING-TAILED LEMUR

short, so far as character goes, these animals may be described as being not very active, not very restless, and not very intelligent.

THE DIADEMED SIFAKA

The diademed sifaka (*Propithecus diadema*), known to the natives of Madagascar as the simpona, is the largest of the three species, and the one first brought to the notice of science, having been described by Mr. E. T. Bennett in 1832. It takes its name from the band of white hairs running across the forehead, which, with the grey fringe of hair on the cheeks and chin, surrounds the black face, and thus gives to the animal a peculiar and striking physiognomy. The crown and back of the head, together with the outer surface of the ears and the nape of the neck, are dark brown, and the same tint extends over the shoulders, so as to give somewhat the appearance of a mantle, and ends in a point on the back, this point in some individuals being only just below the neck, while in others it reaches as far back as the loins. Occasionally this dark mantle-like area, instead of being dark brown, is of a grey tint. The loins and flanks are generally grey, varying considerably in different individuals ; the grey passing gradually into the brown of the back and the orange round the tail, and extending on to the upper part of the arms, or even enveloping the whole of the upper arm. The fore-arms, together with the region round the tail and the legs, are generally of a bright orange yellow, although occasionally yellowish white with some intermixed black hairs. The hands are mainly black, but the feet have a good deal of yellow in them ; the basal half of the tail is yellowish, while the rest of it is grey.

Such are the colours of the typical form of this species. In the moist regions of the south of Madagascar, however, there is a nearly or quite white race of this lemur, while in the dry regions of the north there is a black race, in each case immediate forms occurring which connect these varieties with the ordinary type.

The diademed sifaka inhabits the narrow strip of forest land extending along the whole length of the eastern coast, and bordering the chain of granite and slaty mountains which dips down towards the sea on the east, and is the cause of almost daily rain. It is where this chain almost dies out at the northern end of the island that the black race occurs.

VERREAUX'S SIFAKA

This and the next species, which are smaller than the diademed sifaka, and are those which are known to the natives as sifakas, are restricted to the western and southern coasts of Madagascar. Here they are only found in the thick forests which occur at intervals among the desolate solitudes of the western and southern sides of the island—regions of sandy plain where fertilising rains seldom occur.

The fur of Verreaux's sifaka (*P. verreauxi*) is woolly and soft to the touch, its colour being typically white with a faint tinge of yellow. The summit and hind part of the head, however, are often of a maroon colour, and more rarely reddish, while some individuals show more or less marked grey tints in various regions of the body. In no case, does the brown area of the head extend on to the neck and back, as it does in the diademed species. There are two well-marked varieties of this species, one being pure white, with the exception of patches of bright red on the arms and thighs.

THE CROWNED SIFAKA

The crowned sifaka (*P. coronatus*) agrees in size with the preceding, to which it is closely allied. It has, indeed, a crest of long blackish hairs on the forehead, from which it derives its name ; but, since a similar crest is found in some individuals of Verreaux's sifaka, it is evident that this cannot be taken as the ground for specific distinction. Neither can its colouring, peculiar though it be, form the distinction, since the difference in this respect from the typical race of the latter species is scarcely, if at all, greater than that occurring between the various races included under that heading.

In point of fact the crowned sifaka is ranked as a distinct species, mainly from the characters of the skull, which is altogether larger than that of the preceding species, in addition to which it has a proportionately larger muzzle, besides other distinctive features.

In colour, the forehead, the crown of the head, and the cheeks are blackish brown, in bold contrast to which stands out the white fur with which the ears are covered. The neck and upper parts of the body, as well as the limbs, are of the same white hue, having a more or less distinct rosy tinge on the limbs and at the root of the tail, this rosy tint being most distinct in the more southerly race of this species, in which it may extend on to the back. There is a patch of grey or brown, varying in size, on the nape of the neck. The tail and hands are invariably pure white.

This species is restricted to a small area on the north-west coast of Madagascar, situated to the north-east of Cape St. André, and bounded to the east by the River Betsiboka and by the Manzaray to the west.

Messrs. Milne-Edwards and Grandidier remark how curious it is that the various races and species are so sharply separated from one another that it is sufficient to cross a river—it may be of no great width—in order to find that, while on one bank all the sifakas belong to one race, on the opposite bank they will be of another race, if not of a distinct species. No satisfactory explanation of these peculiar features in distribution suggested itself to the authors named.

THE AVAHI LEMUR

The third and last genus of the group of lemurs is represented only by the avahi, or woolly, lemur (*Avahis laniger*), a species discovered in 1780, at the same time as the indri, by the French traveller Pierre Sonnerat. The avahi, although furnished with a long tail like the sifakas, is distinguished by the still shorter muzzle, and also by the ears being completely concealed by the fur, which is of a woolly instead of a silky nature. Although these differences are ample to distinguish the avahi from the sifakas when they are seen together, it is not on these alone that naturalists rely when referring them to distinct genera. There are, indeed, well-marked differences in their

THE WHITE-FRONTED VARIETY OF THE FULVOUS LEMUR

teeth, but it will be sufficient for our purpose merely to record the existence of these points of distinction. The avahi differs, moreover, from all the other members of the group to which it belongs in being nocturnal instead of diurnal in its habits.

The avahi is the smallest member of all this group of lemurs, its dimensions being rather less than two-thirds of those of the diademed sifaka. In colour, the long hairs on the forehead immediately above the eyes are grey at the base and pinkish at the tips, while there is in some individuals a small white or yellowish band, more or less irregular, across the crown of the head. The rest of the head, the neck, the back, and the arms are covered with woolly fur, of which the hairs are grey at the roots, reddish in the middle, and black at the tips, an arrangement which communicates a peculiar appearance to the whole fur. The concealed ears are reddish and the cheeks grey. The loins and flanks are of a much lighter colour than the back, especially in the region of the tail, where there is a large triangular patch of pinkish white running forwards into the dark area of the body. The hind limbs are still lighter in colour, and, as the hairs here tend to grow into bunches, or tufts, they reveal their grey bases and pinkish tips, thus giving to the fur a mottled appearance. The bushy tail is of a decidedly pink tint, more especially for the first third of its length. The hands and feet are reddish.

There are, however, great variations of colour among different individuals of the avahi inhabiting even the same district, some having the fur almost uniformly reddish, while in others all the parts above the thighs are nearly pure white.

Avahis, instead of living in small troops like indris and sifakas, are found either solitary or in pairs. They are completely nocturnal, sleeping during the day curled up in the fork of a branch, and issuing forth in search of food with the falling shades of night. Like their allies, they are sluggish in their movements, and seldom descend to the ground, and, when they do so, they walk in the peculiar manner of the sifakas.

The avahis are found in two parallel bands of forest on the east side of Madagascar, and also in the woods of a small area on the north-west, and are totally unknown on the west and south coasts, where the vegetation and climate are quite different. The members of the colony on the north-west coast are of smaller size and somewhat different colouring from those on the east side of the island. From their smaller size and nocturnal habits avahis are less noticed by the natives than are the other members of this group, and do not figure conspicuously either in their legends or their superstitions. The name avahi is the one by which they are known to the Antanala tribe. By other tribes, however, they are termed ampongi, fotsi-fe, or fotsi-afaka, the last two respectively meaning " white legs " or " white fork," in allusion to the peculiar colouring of the hind parts of these animals.

THE TRUE LEMURS

The true lemurs (genus *Lemur*) form the first of a group differing in several respects from those already noticed. The first and most easily recognised feature by which they and their allies may be distinguished from the group containing the indri and the avahi is that the toes of the foot are not connected at their bases by a web. In none of the true lemurs are the legs so long in proportion to the arms as is the case in the members of the preceding group, while all of them have long tails. Then, again, it may be mentioned that the members of this group are distinguished by the presence of an additional front tooth on each side of the lower jaw, and likewise by having one more premolar tooth on each side of both jaws, thus bringing up the total number of teeth from thirty to thirty-six. The formula is $i.\frac{2}{2}, c.\frac{1}{1}, p.m.\frac{3}{3}, m.\frac{3}{3}$; which may be compared with that given above as distinctive of that of members of the indri group of lemurs. They form the subfamily *Lemurinæ*.

The true lemurs are confined to Madagascar and the Comoro Islands, situated half-way between the former and Zanzibar. Although some are nocturnal and others diurnal in their habits, all differ from the indri group in subsisting on a mixed diet, insects, small reptiles, birds' eggs, and the callow young of birds forming at least as important a part of their food as fruits. It is probably owing to this diet that they are of a much hardier disposition than are the members of the indri group, so that they flourish in confinement in Europe so well as not infrequently to breed ; the number of young produced at a birth being either one or two.

In consequence of their arms being longer in proportion to their legs than in the indri group, the true lemurs and their allies, when on the ground, are in the habit of going on all-fours, although also capable of taking leaps of great length. The true lemurs may be distinguished from the other members of the group to which they belong by the length of their snouts, and the large size of their tufted ears, as well as by their diurnal habits.

THE RING-TAILED LEMUR

One of the best known of all the true lemurs is the ring-tailed lemur (*Lemur catta*). This animal, which may be compared in appearance to a very small fox, is of an ashy grey colour, darker on the back, and white on the under parts, as well as on the sides of the face, ears, and the middle of the forehead. Its most distinctive feature is to be found in the alternate rings of black and white on the tail, from which it derives its name (pages 261, 279). There is no fringe of long hair round the face.

The ring-tailed lemur is found in the central parts of Madagascar, ranging on the west coast to Mouroundava and on the east to Andrahoumbe. Like the other members of the group, this lemur lives in small parties, and is most active at early morning and evening, sleeping during the night with its bushy tail curled round its body, and likewise taking a siesta during the heat of the day. Unlike the members of the indri group, it is a noisy creature ; and the whereabouts of a troop are discoverable by the loud cries which they are continually uttering.

THE BLACK LEMUR

In captivity this species thrives well, and is generally numerously represented in the Gardens of the Zoological Society in London, although it does not appear that it has ever bred there.

The Rev. G. A. Shaw says that the habitat of the ring-tailed lemur in the south and southwest of Madagascar is among rocks, over which it can travel easily where it is impossible for the people, although bare-footed, to follow. Its palms are

THE BLACK-HEADED VARIETY OF THE FULVOUS LEMUR

long, smooth, level, and leather-like, and enable the animal to find a firm footing on the slippery wet rocks, very much on the same principle as that which assists a fly to walk up a pane of glass. The thumbs on the hind feet are very much smaller in proportion than in the forest-dwelling lemurs, which depend upon their grasping power for their means of progression and spring from tree to tree, rarely, if ever, touching the ground, except in search of water. Hence the ring-tailed lemurs are an exception to the general habits of the Lemuridæ, in that they are not arboreal. There are very few trees near their district, and these few are very stunted and bushy.

THE MUNGOOSE LEMUR

All the remaining species of true lemurs are distinguished from the preceding by their uniformly coloured tails. Among them, the mungoose lemur, or, as it is called by some, the fulvous lemur, (*L. mungos*) was described by Linnæus. It inhabits the north-western coast of Madagascar; and may be known by its black nose and the iron-grey spot on each side of the forehead. The fur is of a somewhat woolly nature and reddish grey in colour; but the face, chin, the middle line of the forehead, and a streak across the crown of the head are black; while the cheeks and the sides of the forehead are grey. There is considerable individual variation in the width of the black band across the head.

According to Dr. Forsyth Major, the mungoose lemur has nothing to do with the Linnæan species of that name, which had been based on the description and figure of George Edwards in his "Gleanings." The only ascertained localities in which the true *Lemur*

RUFFED LEMURS

mungos occur are the neighbourhood of the Bembatoka Bay (on the north-western coast of Madagascar) and the two islands Anjuan and Mohilla of the Comoro Group. The earliest available name for the usually so-called *Lemur mungos*—a very variable species, spread over a great part of Madagascar—seemed to be *Lemur fuscus*. The two species, like all the members of the genus, are easily distinguishable by the characters of their skulls. In this work the name generally in use has been retained.

The red-fronted race of this species (*L. mungos rufifrons*) may be easily recognised by the two small white stripes running across each side of the rump. The general colour of the fur is grey; the throat and under parts are reddish; the nose and the middle of the forehead are black; while the sides of the nose, the cheeks, and a large spot on either side of the forehead are white. The tail is blackish, with a reddish tinge at the root

The white-fronted lemur (*L. mungos albifrons*), which appears to be restricted to the north-east coast of Madagascar, is mainly distinguished from the typical race by its colour; its most distinctive feature being a broad band of white woolly hairs extending across the forehead, and including the base of the ears, the cheeks, and part of the throat and neck. The prevailing colour of the back and flanks is grizzled brown, tinged with red; the long muzzle and face, together with the hands and feet and the end of the tail, being black. The under parts and inner surfaces of the limbs are whitish grey (page 280). This pretty race was first described by Geoffroy St. Hilaire; and was exhibited in the London Zoological Gardens as far back as 1830.

The so-called black-fronted lemur (*L. mungos nigrifrons*) is another race of the same species, also first made known to science by the naturalist last mentioned. In comparing it with the preceding race, Mr. E. T. Bennett, who had the opportunity of seeing living examples of both, observed that "their size is nearly equal, and there is little, if any, difference in their form; but their colours, invariable as we have hitherto found them, furnish sufficient ground for regarding them as distinct. The present animal has the elongated muzzle of the last, but the black colour embraces in it the forehead and sides of the face, as well as the base of the muzzle; and the hair on the former parts, instead of being long and woolly, is short, smooth, and even. While the black is thus extended backwards over the head, it is bounded on the fore part of the muzzle, which, instead of being uniform in colour, as in the preceding species, becomes grizzled towards its extremity, and at last almost white. The general colour of the upper parts of the body is dark ashy grey, most of the hairs terminating in tawny tips, which are so strongly marked on the back as to give a decided tinge. The tail is light grey at the base, and darker towards the tip; the outside of the limbs is of a light ashy grey; the chin, chest, and throat are pure white; and the under parts, together with the inner side of the hind limbs, pale rufous. The hands, which are blackish, have the same tendency to become grizzled as the fore part of the muzzle."

In captivity this and the other races of the species are described as being perfectly tame and good-natured, without any tendency to the petulant and mischievous habits of the smaller monkeys. In a wild state the habits of all the races are doubtless similar.

THE BLACK LEMUR

With the black lemur (*L. macaco*) we come to the first of two well-defined species, differing considerably from those already noticed; this difference being chiefly shown by the presence of a more or less well-marked ruff fringing the cheeks and chin, and frequently also by a fringe of hairs on the margins of the ears. Moreover, all these lemurs are subject to great variation in colour, which in one case appears to be merely individual, while in another it is distinctive of the two sexes.

The black lemur (pages 277, 281) comes from the north-west coast of Madagascar; and the male, on the evidence of which the species was described, is uniformly black, with a well-developed ruff round the cheeks and neck, and a long fringe to the ears. Very different, however, is the female, which was at first described under the name of the white-whiskered lemur (*L. leucomystax*). In this sex the general colour of the fur is brown, with a patch on the lower part of the back, and the ruff round the face and the fringe on the ears white.

A female of this species in the Gardens of the Zoological Society twice gave birth to a young one, and thus afforded an opportunity of seeing the curious manner in which lemurs carry their offspring. The young one born on March 24, 1884, proved to be a female, and was of the same brown colour as its mother. On April 3 in the following year the second young one was born, which was a male, and at the time of birth it was of the black hue of its father. Each was carried lying nearly across the abdomen of its mother, with its tail passed round her, and thus on to its neck, so as to form a firm attachment; and it is believed that, at least in the wild state, the young are at a later period carried on their mother's back.

THE RUFFED LEMUR

Another and the largest of the true lemurs is the ruffed lemur (*L. varius*), which inhabits the north-east coast of Madagascar (page 282). As its name implies, it is remarkable for the great individual variation in the colour of the fur; such variations being apparently independent of sex. Frequently the colour is a mixture of black and white, disposed in patches on different parts of the body, but occasionally white individuals are met with. Other individuals are of a nearly uniform reddish brown colour; this variety having been regarded as a distinct species, the red lemur (*L. ruber*).

A specimen of the red variety in the Gardens of the Zoological Society had the upper surface of the

THE GREY GENTLE LEMUR
Photograph by W. S. Berridge

body bright rufous brown, while the under parts were deep black. The reddish area included the sides of the face, ears, back, and flanks, and the outer surfaces of the limbs; while the black embraced the forehead and face, the throat, chest, and abdomen, the inner surfaces of the limbs, and the hands and feet, with the exception of a narrow stripe of white across their upper surface. On the back of the neck there was a large white patch. The length of the head and body of this animal was two feet, and that of the tail somewhat more.

Here may be briefly noticed two other members of the typical genus. The first of these is the red-bellied lemur (*L. rubriventer*), so named on account of the rufous colour of the under parts. In the female, the so-called *L. flaviventer*, the under surface is yellowish. The second species is the sooty lemur (*L. nigerrimus*), readily distinguishable from the black lemur by the much shorter ear tufts and ruffs. The skulls of both these species show characteristic features.

THE GENTLE LEMURS

The grey gentle lemur (*Hapalemur griseus*), which, like all the members of the group under consideration, is an inhabitant of Madagascar, differs so decidedly from the true lemurs that it has been made the type of a distinct genus (*Hapalemur*), of which it is the representative species.

It may be readily distinguished from the true lemurs by its rounded head and extremely short muzzle, the ears being likewise very short. A peculiar feature is the presence of a small bare patch on the front surface of the forearm, a little above the palm of the hand, which is covered with small spines. The colour is a dark iron grey, with a tinge of yellow, becoming somewhat paler on the under parts and the inner sides of the limbs. The individual hairs are black, with a reddish band near the tips.

The species differs from the true lemurs in being purely nocturnal in its habits. It is chiefly found in bamboo jungles, and subsists mainly on the young tender shoots of these plants, as well as on their leaves. In such jungles its capture is difficult, and hence living examples are rare in menageries. An allied lemur living in the Zoological Society's Gardens in 1870 was described as *H. simus,* and is now known to be a perfectly distinct species, having, however, the same shiny gland-patch on the wrist as the grey gentle lemur. According to a French traveller, the last is known to the natives of Madagascar as the bokombouli.

THE WEASEL LEMUR

The slender, or weasel, lemur is the last representative of the present group, and belongs to a genus (*Lepidolemur*) containing several species which differ from all other lemur-like animals in having, when adult, either no upper front (incisor) teeth at all, or merely a single pair of minute rudimentary ones. This character will at once suffice to distinguish these animals from the gentle lemurs, which they resemble in being of purely nocturnal habits. Further points of distinction are afforded by the greater length of the muzzle and by the ears being bald and somewhat larger. The tail is long and covered with close-set short hair.

The weasel lemur (*Lepidolemur mustelinus*) is chiefly found in the north-west of Madagascar, and is characterised by having no upper front teeth at all when quite full grown. Its head and body together measure about 10 in. in length, and the tail is 14 in. long. There are two other nearly allied species of these lemurs, namely, *L. ruficaudatus* and *L. edwardsi,* both of which are smaller than the typical species, and distinguished from one another chiefly by their colouring. In the first named the general colour of the fur is pale or reddish grey, becoming brown on the head.

The weasel lemur, during its nocturnal rambles, is marvellously active, and capable of taking tremendous leaps among the trees in which it dwells ; its slender build and long limbs being admirably adapted for such a mode of progression. Like the grey gentle lemur, it subsists solely on leaves ; and it is much sought after as an article of food by the natives of Madagascar, to whom it is known by the name of fitili-ki. It is killed by being knocked on the head with a stick while curled up during the day in its nest of leaves, to which it has been tracked down at the end of its nocturnal excursions.

Certain other species, such as *L. microdon,* are broadly distinguished by their inferior bodily size and the much smaller molar teeth. They are rare and little-known animals, passing most of their time in sleep.

THE MOUSE LEMURS

The tiny creatures known as the mouse lemurs (genus *Chirogaleus*) are the first of a group of two genera which differ from all other members of the lemur tribe in that the bones of the upper part of the ankle are enormously elongated, thus causing the whole foot to be much longer than in the preceding groups.

The mouse lemurs themselves are confined to Madagascar ; and include the smallest of the lemurs, some being even smaller than a rat. They have long tails and rather large ears, which are hairy at their base and cannot be folded upon themselves.

The most remarkable feature connected with the mouse lemurs is that they are in the habit of hibernating, or remaining dormant for a portion of the year. But as their quiescent season is during the hottest and driest time, the term æstivating would be more appropriate. By no means all the mouse lemurs become torpid in this fashion ; and it may be fairly presumed that the species in which this habit occurs are those dwelling in the more arid regions.

To prepare for this protracted period of suspended energy, during which they maintain the heat of their bodies by the consumption of their own substance, the mouse lemurs feed so vigorously that when at length the hot season arrives they are in an extremely fat and sleek condition. Curiously enough, the great accumulation of fat is mainly restricted to the region of the base of the tail ; and when they retire at the close of the rainy season, during which food is extremely abundant, their tails are swollen to a prodigious size. The wasting process which goes on during hibernation leaves them with tails shrunken to very small dimensions. In order to make themselves comfortable during

SMITH'S DWARF MOUSE LEMUR
Photograph by Lewis Medland

their long sleep, they prepare snug little nests of twigs and other substances (which, it may be noted, is the custom of dormice in similar circumstances) ; some of their habitations being described as marvels of neat construction. Their food is mainly of a vegetable nature ; although this diet is largely supplemented by insects, and even small birds are said not to come amiss. Having large round eyes, by which they are enabled readily to distinguish even the smallest objects on the darkest nights, they are in the habit of stalking nocturnal moths and beetles when these insects have settled on the boughs of trees, and then rushing upon and seizing them with a final spring.

FORK-HEADED MOUSE LEMUR

The largest and one of the longest-known of the mouse lemurs is the species taking its name from the black streak running down the middle of the back, and dividing on the top of the head so as to form a distinct fork-like mark between the eyes (*Chirogaleus furcifer*). The colour of the remainder of the body is grey, with a black tip to the tail. This species is found in the forests on both the east and west coasts of Madagascar, though more abundant in the latter region. It is known to the natives as the walouvi; and is not one of those species that hibernate.

LESSER MOUSE LEMUR

The lesser mouse lemur (*C. minor*) differs from the preceding in the absence of the dark stripe down the back. The colour is pale reddish grey, with a broad whitish streak up the middle of the face; the cheeks and under parts being also light-coloured, but the slender tail more brown. The species often described as the myoxine mouse lemur (*C. myoxinus*) appears to be very closely allied.

COQUEREL'S MOUSE LEMUR

Coquerel's mouse lemur (*C. coquereli*) is character-ised by the soft woolly nature of the fur, of which the prevailing colour is greyish brown, tinged with gold (page 162). It makes well-formed nests of twigs, dead leaves, and grass, and having a diameter of some 18 in. In this nest it sleeps during the day, to prowl forth at night in search of food.

DWARF MOUSE LEMUR

The smallest member of the genus is the dwarf mouse lemur (*C. smithi*), not infrequently referred to as the Madagascar rat (page 284), because it was described by Buffon under the name of le rat de Madagascar. The head and body of this diminutive species do not exceed 4 in. in length, while the tail measures 6 in. The prevailing colour is pale grey; the chin and under parts being pale yellow, and the outer surface of the ears light brown, while a white streak runs up the nose and between the eyes. The eyes themselves are surrounded by black rims, giving to the face the appearance of wearing a pair of spectacles.

The dwarf mouse lemur builds beautifully con-structed nests of twigs, lined with hair, in the tops of the lofty trees where it delights to dwell. These nests somewhat resemble those of a rook both in form and size, and are used not only as diurnal resting-places, but as cradles for the young. The species is remark-able for the extreme beauty of its brilliant eyes.

THE BROWN MOUSE LEMUR

The brown mouse lemur (*C. milii*), which is one of the hibernating species, takes its scientific name from M. Milius, a former governor of Réunion, by whom two of these creatures were sent to Paris. They were described by Frédéric Cuvier in 1821 as the *maki nain*, or small lemur. The species is some 9 in. in length, exclusive of the long tail; and it is of a greyish brown colour, with black whiskers, and white throat and under parts, the fur being silky. The specimens sent to Paris throve for some time, and became so tame that they were allowed to leave their cages.

Many naturalists restrict the genus *Chirogaleus* to this and a few other allied species, referring those previously mentioned to another genus, *Microcebus*. Two other kinds—namely, *C. samuti* and *C. thomasi*—are regarded as indicating a third generic group, known as *Opolemur*.

THE AYE-AYE

ONE of the most remarkable of all lemuroids, and at the same time of the whole order of Primates, is the aye-aye of Madagascar, which has teeth so utterly different from all other members of the order that it was long considered to belong to the rodents (rats, rabbits, etc.). Till comparatively recent times the aye-aye was regarded as widely sundered from the typical lemurs of the family *Lemuridæ*. But it has been discovered that it possesses a similarly formed drum of the ear, while it is quite evident that its peculiar teeth are merely an adaptation to a particular kind of life. Accordingly, some naturalists have proposed to include it in the family *Lemuridæ*. But it is, on the whole, more convenient to regard it as the representative of a separate family, the *Chiromyidæ*.

The most peculiar feature about the dentition of the fully adult aye-aye (*Chiromys madagascariensis*) is that the front, or incisor, teeth are reduced to a single pair in each jaw, which are curved, and have their extremities brought to a sharp chisel-like edge, admirably adapted for gnawing and rasping hard substances. The structure of these teeth is, in fact, precisely the same as in the front teeth of rats and beavers; their sharp cutting edges being produced by the circumstance that while the body of the tooth is formed of comparatively soft ivory, the front surface is faced with a layer of hard, flinty enamel. It will be obvious that the result of wear in a tooth of this type will be to produce a chisel-like edge. It will further be apparent that such a tooth, if con-tinually employed in rasping away hard substances, would be very quickly worn down altogether, if it were of the same length as ordinary teeth, and not provided with some kind of renewal. This difficulty is obviated by the front teeth remaining open at their lower ends, and undergoing a continual process of growth; so that as their summits are worn away they are pushed farther up from below. In all these points the teeth are precisely similar to those of rodent mammals. A further resemblance to rodents is shown by the absence of tusks, and also by the cheek teeth being separated by a long gap from the incisors, as well as by being reduced in number, and having their crowns with nearly flat surfaces, instead of being surmounted with the sharp cusps found in those of the true lemurs. Indeed, the total number of teeth in the adult aye-aye is only eighteen, these being expressed by the formula $i.\frac{1}{1}$, $c.\frac{0}{0}$, $p.m.\frac{1}{0}$, $m.\frac{3}{3}$, or exactly the same as in many rodents.

If, then, the teeth of the adult aye-aye are so exactly like those of a rodent, the reader may well ask why it is not placed among the rats and beavers, instead of among the lemurs. To this it may be replied that in the young aye-aye the milk, or baby, teeth are very much more like those of the true lemurs; while the anatomy of the skeleton and the soft parts is essentially that of a lemur, and not that of a rodent. The resemblance of the skull and teeth to those of a rodent is, indeed, an excellent instance of what naturalists term an "adaptive" or "parallel resemblance." When two animals belonging to totally different groups have more or less nearly similar habits, it frequently results that they will closely resemble one another in at least some part of their structure, such particular structure being the one best adapted for a particular mode of life. In all such cases a superficial examination of the animals in question will frequently lead to their being referred to one and the same group; while further minute investigations will reveal the fact that their deep-seated internal structure—which alone reveals their true affinities—is very different. Such was the case with the aye-aye, which was at first referred to the rodents, its affinities to the lemurs not having been discovered till a fuller examination.

The aye-aye agrees with the true lemurs in having the great toe of the foot furnished with a flattened nail, and capable of being opposed to the other toes, this feature being alone sufficient to prove that the creature has nothing to do with the rodents. With the exception of this great toe, however, all the toes and fingers, which are very long and narrow, are furnished with narrow and sharply pointed claws. Although both the hands and the feet are large in proportion to the size of the animal, yet the great peculiarity is concentrated in the hands, in which the fingers are much longer than are the toes of the feet. One finger—namely, that corresponding to the human middle finger—is more remarkable than the others, being of great length and extreme slenderness. It is probable that this ghostly middle finger is employed in extracting from their burrows the larvæ which form a portion of the creature's natural diet.

In size the aye-aye may be compared to a cat; its total length being about 3 ft., of which the larger moiety is formed by the bushy tail. The comparison with a cat may be further extended to the short and rounded head and cat-like face of the animal. The rounded ears are, however, relatively larger than those of a cat, and have the peculiarity of being nearly naked. The fur is long and composed of a mixture of longer, stiffish hairs, with an under-coat of shorter and bushier ones. The prevailing colour is dark brown, tending to black, washed with whitish; the throat is yellowish grey, and the under parts show a rufous tinge. Some of the longer hairs on the back are whitish, producing a somewhat mottled appearance in the fur.

THE AYE-AYE

The aye-aye was discovered by the French traveller Pierre Sonnerat so long ago as 1780; and was described in the first year of the nineteenth century by Baron Cuvier, who regarded it as a kind of squirrel. Nothing more was heard of the creature from Sonnerat's time till 1860, when specimens were sent to England and described by Sir Richard Owen. The name is variously stated to be derived from the aye-aye's cry, "haihay, haihay," or from the natives' exclamation of surprise, "hay! hay!" at the sight of the animal, which they also call the haikay. The Rev. G. A. Shaw gives the following interesting account of its habits:

"Being a nocturnal animal, it is very difficult to get any trustworthy information concerning its habits in the wild state, and native reports are altogether contradictory with respect to these matters. Even with reference to its natural food, no satisfactory explanation can be obtained from the people. Many assert positively that it lives on honey; but one I had in captivity would not eat honey in any form, either strained or in the comb, or mixed with various things I thought he might have a fancy for. Others say it lives on fruits and leaves; others that birds and eggs are its natural food. I fancy, from what I saw of my captive, that both these conjectures are nearer the truth; for after a few days, during which it would eat nothing, and it was thought that the proper food had not been offered—but it was, in reality, pining or sulking—it took several fruits which I was able to procure for it. It liked bananas; but it made sorry efforts at eating them, its teeth being so placed that its mouth was clogged with the food. The small fruits of various native shrubs it also devoured, as also rice boiled in milk and sweetened with sugar; but meat, larvæ, moths, beetles, and eggs it would not touch. But I noticed that when I came near its cage with a light, it almost invariably started and went for a little distance in chase of the shadows of the pieces of bananas attached to the wirework in front of its cage; and I think that if I could have procured some small birds it would have, if not devoured them, at any rate killed them for their blood, as some lemurs are known to do. It drank water occasionally, but in such a way as to make it highly probable that it does not drink from streams or pools in the ordinary way. It did not hold its food in its hands, as the lemurs which I have had in captivity have done, but merely used its hands to steady it on the bottom of the cage. But whenever it had eaten, although it did not always clean its hands, it invariably drew each of its long claws through its mouth, as though, in the natural state, these had taken a chief part in procuring its food.

"In some accounts, given by different writers, the haikay is said to be easily tamed, and to be inoffensive. . . . In each of these qualities, I have found, both from native accounts and from the specimen I have kept, that exactly the reverse is the case. It is very savage, and when attacking strikes with its hands with anything but a slow movement. As might be imagined in a nocturnal animal, its movements in the daytime are slow and uncertain; and it may be said to be inoffensive then. When it bit at the wire netting in the front of its cage, I noticed that each of the pair of incisors in either jaw could separate sufficiently to admit the thick wire even down to the gum, the tips of the teeth then standing a considerable distance apart, leading to the supposition that, by some arrangement of the sockets of the teeth, they could be moved so far without breaking. The haikay brings forth one at a birth, in which the long claw is fully developed."

It has been observed that captive aye-ayes are very partial to the juice of the sugar-cane, which they obtain by ripping up the canes with their front teeth; and since sugar-cane grows wild in Madagascar, we may infer that its juice forms a part of the food of these animals in their wild state. It is therefore probable that the diet of the aye-aye is a mixed one, consisting partly of grubs, partly of the juices of plants and partly of fruit.

GALAGOS AND LORISES

THE lemur-like African animals known as galagos, together with the lorises of South-Western Asia and the pottos of Equatorial Africa, were formerly included in the same family as the lemurs, but since they differ from all the latter in the structure of the drum of the ear, they are now assigned a family by themselves (*Lorisidæ*), of which the galagos form one section, or subfamily, the *Galaginæ*, and the lorises and pottos a second, *Lorisinæ*.

THE GALAGOS

The galagos are the only long-tailed lemurs found throughout the greater part of Africa. The name is said to be that by which one of the species is known to the natives of Senegal. They resemble the mouse lemurs in having the bones of the upper half of the ankle greatly elongated, and thus have the same lengthy foot; but this structure has apparently been developed quite independently in the two groups. Although some are much bigger, there are others quite as small as, the smaller mouse lemurs. There is, however, a readily recognised external character by which a galago can be at once distinguished from a mouse lemur. This consists in the large size of the ears, which are quite bare, and have the unique peculiarity that they can be partially folded upon themselves at such times as their owners please, so as to lie nearly flat upon the sides of the head. This may be for the purpose of protecting these delicate organs when passing through thick foliage, especially if wet.

THE GREAT GALAGO

This distinctive peculiarity of the ears is, of course, sufficient to separate a galago from a mouse lemur, and, indeed, from every other kind of lemur. Naturalists are, however, by no means satisfied with distinguishing animals merely by external characters; and they have discovered a feature in the teeth by which a galago differs markedly from a mouse lemur, although, unfortunately, this point of distinction can only be seen in a dried skull. If, however, we take the skull of a mouse lemur we shall find that while the last three upper teeth, or molars, have broad crowns and are alike, the tooth in advance of these, which is the last premolar, has a smaller and simpler crown, of a triangular shape. In a galago's skull, on the contrary, this last upper premolar, although slightly smaller than the molars, has a similarly shaped crown, broad on the inner side, and nearly quadrangular in form.

The galagos are widely distributed over Africa south of the Sahara, one kind being found so far south as Natal, while there are several on the western side and some on the eastern. Like the mouse lemurs, they are essentially nocturnal, and are, of course, confined to those regions where thick forest prevails. When not enjoying their diurnal repose, they are lively and interesting. They subsist on a mixed diet, including fruits, insects, and small birds and their eggs. Some of the smaller species will readily devour locusts and the peculiar leaf-like mantides, or praying insects. When on the ground, the galagos recall the lemurs of the indri group, in that they generally sit in the upright position, and progress by a series of leaps or hops. They usually have two or three young at a birth, and are stated to have bred in captivity in Africa, although we are not aware that they have done so in Europe. Many of them, however, thrive well in menageries, where some have been represented by a considerable number of individuals. It is stated that the galagos resemble the mouse lemurs in building nests, which are situated in the forked branches of trees ; but it is probable that this is only true of the smaller species. They appear, however, to be peculiar in that several individuals will inhabit the same nest, out of which they all rush when suddenly disturbed. Galagos frequently stand upright in a characteristic attitude, with their arms extended horizontally outwards. The total number of teeth, both in the galagos and the mouse lemurs, is the same as in the true lemurs.

THE GREAT GALAGO

With the exception of a closely allied kind from the West Coast, the great, or thick-tailed, galago (*Galago crassicaudata*) of Mozambique and the Lower Zambezi valley is the largest of all the species. In size it is about equal to a cat of average dimensions, and, indeed, the peculiar manner in which it carries its thick bushy tail high above its back is highly suggestive of a pampered Persian cat. This bushy tail is about one-fourth longer than the head and body. The great galago belongs to a group of several species, in which the ears are unusually long, and the muzzle is considerably elongated, while the feet are comparatively broad and short, and the fingers and toes have broad disc-like expansions at their extremities. The colour of the fur is a uniform dark brown (page 287).

Writing of this species, Sir John Kirk observes that " it is confined to the maritime region, so far as I know never penetrating beyond the band of wood generally known as the mangrove forests. By the Portuguese it is named ' rat of the coco-nut palm,' that being its favourite haunt by day, nestling among the fronds ; but if it be disturbed, performing feats of agility, and darting from one palm to another. It will spring with great rapidity, adhering to any object as if it were a lump of wet clay. It has one failing, otherwise its capture would be no easy task. Should a pot of palm-wine be left on the tree, the creature drinks to excess, comes

GARNETT'S GALAGO
From the "Proceedings" of the Zoological Society

down, and rushes about intoxicated. In captivity they are mild ; during the day remaining either rolled up in a ball, or perched half asleep, with ears stowed away, like a beetle's wing under its hard and ornamented case. I had half a dozen squirrels with one in the same cage ; these were good friends, the former creeping under the galago's soft fur and falling asleep. On introducing a few specimens of [elephant] shrew, the galago seized one and bit off its tail, which, however, it did not eat. The food it took was biscuit, rice, orange, banana, guava, and a little cooked meat. Stupid during the day, it became active at night, or just after darkness set in. The rapidity and length of its leaps, which were absolutely noiseless, must give great facilities to its capturing live prey. I never knew it give a loud call, but it would often make a low, chattering noise. It has been observed at the Luabo mouth of the Zambezi, at Quilimane, and at Mozambique. When I had my live specimen at Zanzibar, the natives did not seem to recognise it ; nevertheless, it may be abundant on the mainland."

On the West Coast of Africa, in Angola, the great galago is represented by the closely allied Monteiro's galago (*G. monteiri*), which is of slightly larger size than the East Coast species : the length of the head and body being 12 in., and that of the tail 16 in. Although these two galagos differ mainly or entirely by their coloration, yet, according to Sir John Kirk, the eastern species is confined to the coast region, and it is probable that there is a wide area separating the habitats of the two, which suggests the advisability of regarding them as distinct species. As a rule, Monteiro's galago is of a uniform pale grey colour, with the sides of the nose somewhat darker, and the throat and tail nearly or quite white. The fur is soft, with the component hairs slate-coloured at their roots and white at the tips.

GARNETT'S GALAGO

Garnett's, or, as it is sometimes rather inappropriately called, the black galago (*G. garnetti*) is a species belonging to the same group as the preceding, from which it differs in its inferior size. It is an inhabitant of Eastern Africa, and is of a dark brown colour, tending to yellowish on the under parts, with black ears, and a white streak on each side of the loins. One of these animals, formerly in the Zoological Society's Gardens, when let loose one night in the apartments of the superintendent, exhibited to perfection the leaping habits and extreme agility characteristic of its tribe. It leaped after the manner of a kangaroo, clearing several feet at a spring, and hopping on to the table and other articles of furniture in the room. Strange to say, it exhibited no signs of fear of the dogs and cats with which it was confronted.

ALLEN'S GALAGO

With the West African species we come to the first representative of a group distinguished from the preceding one by the more rounded head, shorter muzzle, and larger eyes, as well by the longer and more slender form of the foot.

Allen's galago (*G. alleni*) is found at Fernando Po and the Gabun, and is characterised by the tail being thick and bushy, and also by the extreme length and slenderness of its fingers and toes. The prevailing colour of the fur is greyish brown, with the forehead, rump, and the root of the tail grey, a tinge of red is on the limbs, the tail is black, and a streak on the nose and all the under parts are whitish. If we examine the skull, it will be noticed that the last molar tooth on each side of the upper jaw is nearly equal in size to the tooth in advance of it. This will be found an important point of distinction between Allen's galago and all the remaining species, in which the last upper molar is much smaller than the tooth in front of it.

The pale-coloured galago (*G. pallida*) of Western Equatorial Africa seems to be only a colour-phase of the present species, and not a distinct form. It was met with by Du Chaillu, who believed that he had discovered a new species The general colour is grey, and the tail unusually long.

THE SENEGAL GALAGO

The longest known of all the group is the Senegal galago (*G. senegalensis*), originally described in 1796 from specimens brought from Senegambia, which may be regarded as its headquarters. Subsequent discoveries have, however, shown that a galago exists on the east side of Africa to the south of the Sudan, which, although described as a distinct species under the name of the Senar galago (*G. senariensis*), is so closely allied to the Senegal galago that it may probably be regarded as a mere local variety or race. Indeed, it is likely that when we are fully acquainted with the zoology of the vast stretch of country lying to the south of the Sahara Desert, it will be found that this galago extends right across Africa.

In addition to the distinctive character of the upper molar teeth already mentioned, the Senegal species has certain marked external features by which it differs from Allen's galago. Thus, in the tail the

MOHOLI GALAGO
Photograph by W. S. Berridge

hairs near the root are pressed down, only those nearer the end spreading out on all sides, so that the whole tail assumes a somewhat club-like form. Then, again, the fingers and toes are considerably thicker and shorter than in Allen's galago. In colour the typical Senegal galago is grey, with the under parts and a streak on the nose white, and the tail, hands, and feet blackish brown. The Senar race appears to have a rather bluer tinge to the fur, with a darker face, and black rings round the eyes; while the tail is described as being relatively longer. It is of comparatively small size, and appears to be common in the forests of Senegal, and in those on the Blue Nile in Kordofan, and the White Nile in Senar. Its chief food consists of various kinds of insects; but it is stated that it will also eat the gum of several sorts of acacia, which also form part of the diet of the baboons of the Sudan. Its habits are said to be similar to those of the other species.

In South Africa the Senegal galago is represented by a form so nearly allied that many naturalists regard the two as only local races of a single species. This southern form is the Moholi galago (*G. senegalensis moholi*), which is an inland race found as far south as Natal, and also met with in Nyasaland and the adjacent districts. A galago from the neighbourhood of Titi, some distance up the Zambezi, has been identified with the Senar race of the Senegal galago, but it would appear probable that it is the southern form. The prevailing colour is brownish, or yellowish grey, becoming darker on the back, and still more so on the tail; while a broad streak on the nose, the cheeks, and the throat are white, and the inner surfaces of the limbs and the under parts are whitish with a faint tinge of yellow. In a male specimen sent to England by Sir Andrew Smith—the original describer of this animal—the fur surrounding the eyes is of the same colour as that on the other parts of the head. In other examples, however, there are dark rings round the eyes. This variability shows that the presence or absence of such rings cannot be regarded as indicating a specific distinction between the Senegal and Senar galagos.

DEMIDOFF'S GALAGO

The smallest member of this group of the present genus is Demidoff's galago (*G. demidoffi*), from the West Coast of Africa, which differs from both the two species that have just been considered in its much more slender and more cylindrical tail and smaller ears. The length of the head and body is 5 in., and that of the tail, 7½ in. The general colour of the soft fur is brown, darker on the sides of the face; the white streak on the nose being narrow; and the chin, throat, and under parts of a reddish grey colour. The so-called *G. murinus*, from Old Calabar, is probably not specifically separable from this galago.

SLOW LORISES

With the lorises of the warmer parts of Asia (genus *Nycticebus*) we come to the second section of the *Lorisidæ* family; this group also including the pottos of Africa. The members of this group may be recognised either by the total absence of the tail, or by its length not exceeding one-third that of the head and body. The only lemur with which these animals could possibly be confounded would therefore be the indri of Madagascar; but, irrespective of its larger size, the latter is at once distinguished by the web uniting the bases of the toes, and the full development of the index finger of the hand. Moreover, the slow lemurs and the pottos may be further distinguished, not only from the indri, but likewise from all other lemurs, by the index finger of the hand being invariably very small, and even rudimentary, and without any trace of a nail. Then, again, all the lorises and pottos are peculiar in having the thumb of the hand and the great toe of the foot very widely separated from the other digits; this divergence being carried to such an extent in the case of the great toe that it is actually directed backwards instead of forwards.

Apart, therefore, from their distribution, there is no difficulty in distinguishing a loris or a potto from a lemur. All the members of the present group have the same number of teeth as the true lemurs, but they differ from galagos in that the bones of the upper part of the ankle are of ordinary proportions, so that the foot is not abnormally lengthened.

The lorises are purely nocturnal, and are well known for the extreme slowness and deliberation of their movements, the latter characteristic having given their distinctive name to the Asiatic representatives of the group. It was probably their deliberate motions, nocturnal habits, and large glaring eyes, that suggested to Linnæus the name of "lemur" for the group generally.

Lorises are distinguished from pottos by having a well-developed but small index finger on the hand, which has the usual three joints, and is provided with a distinct nail. They have no external tail, and are, as already mentioned, strictly confined to

the tropical and subtropical regions of Asia. There are two distinct types of lorises, which are very closely allied, although most naturalists consider it advisable to divide them into two genera. It has been stated that there is a long-tailed slow loris in the Lushai hills of North-Eastern India, but this requires confirmation.

The slow lorises (page 163) have been regarded as referable to a single variable species, *Nycticebus tardigradus*, but, according to Mr. M. W. Lyon, they are divisible into two distinct groups, in one of which a sagittal crest is developed on the skull of the adult, while in the other no such ridge occurs. The second group occurs only in Borneo and Banka, and appears to be further characterised by having only one pair of upper incisor teeth, in place of the two pairs found in the first.

The distinctive features of the slow loris, as the representative of a genus, are that the eyes are not of very enormous size, and are separated from one another by a considerable space; while the general build of the animal — more especially as regards its limbs—is comparatively stout.

The name loris, by which all these lemuroids are commonly designated, is derived from the Dutch word *loeris*, meaning a clown, and appears to have been applied to these animals by the Dutch colonists of the East Indian Islands. To the natives of India the slow loris is known either by the name sharmindi billi, "bashful cat," or lajjar banar, "bashful monkey." It is an animal about the size of a cat; different individuals or races varying considerably in size, so that while some specimens do not measure more than 13 in. in total length, others may reach as much as 15 in., or even more. Its proportions are thick and clumsy, the head being broad and flat, with a slightly projecting and pointed muzzle. The large eyes are perfectly circular, and their pupils can be completely closed by the gradual contraction of the iris, which opens from above and below, so that when the pupil is half concealed it takes the form of a transverse slit. The ears are short, rounded, and partly buried in the fur, and are thus very different from those of the galagos. The hind limbs are only slightly longer than the front pair. With the exception of the muzzle and the hands and feet, the whole of the body is covered with a thick coat of very close and somewhat long woolly fur.

There is a considerable amount of variation in the colour of different local races of the ordinary slow loris, although in all cases there is a dark stripe running down the middle of the back, sometimes

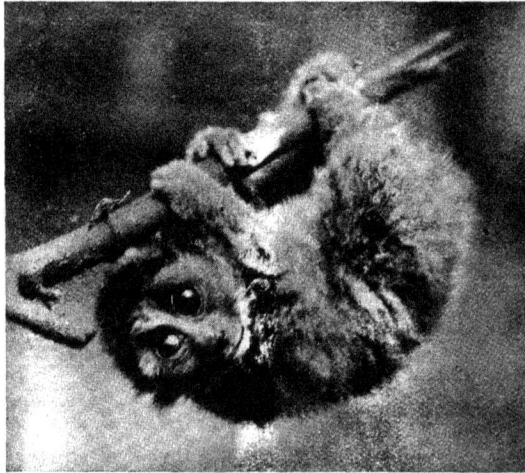

THE SLOW LORIS

Photograph by W. S. Berridge

extending on to the head. In the more common and larger variety, the colour of the fur is ashy grey above, tending to become silvery along the sides of the back, the under parts being lighter, and the rump often having a tinge of red. The stripe on the back is chestnut-coloured, and stops short at the hind part of the crown of the head. The eyes, however, are surrounded by dark rims, between which is the white streak extending upwards from the nose. The ears, together with a small surrounding area, are brown.

In another and generally smaller variety the hue of the upper parts has a distinct tinge of red mingling with the grey, while the stripe on the back is wider, and often of a full brown colour ; but, instead of stopping short at the back of the crown of the head, this band widens out into a large brown patch on the crown, which embraces the ears. The eyes, however, although surrounded by brown rings, are not connected with the patch on the head by a dark-coloured area. There is yet a third variety of this creature, found in Tenasserim, in which the general colour is pale rufescent, while the dark stripe on the back, instead of expanding on the crown of the head, merely splits into a fork, of which each prong joins the dark ring round the eyes.

Slow lorises are found over a large area in the countries lying to the east of the Bay of Bengal. They occur on the north-eastern frontier of India in the provinces of Sylhet and Assam, whence they extend southwards into Burma, Tenasserim, and the Malay Peninsula ; while they are also found in Siam and Cochin China, and the islands of Sumatra, Java, and Borneo.

The food of lorises consists of leaves and young shoots of trees, as well as fruits, various kinds of insects, birds, and their eggs. Lorises have been observed to stand nearly erect upon their feet, and from this advantageous position pounce on insects. They are generally silent, although sometimes uttering a low crackling sound ; but when enraged, and especially if about to bite, they give a kind of fierce growl. The slow loris is tolerably common in the Tenasserim provinces and Arakan ; but, being strictly nocturnal in its habits, is seldom seen. It inhabits the densest forests, and never by choice leaves the trees. Its movements are slow, but it climbs readily, and grasps with great tenacity. If placed on the ground, it can proceed, if frightened, in a wavering kind of trot, the limbs placed at right angles. It sleeps rolled up in a ball, its head and hands buried between its thighs, and wakes up at the dusk of evening to begin its nocturnal rambles. The female bears only one young at a time. Many accounts have been published of the habits of the slow loris in confinement. All observers are agreed that. while these creatures are apt to be fierce when first captured, they soon become docile. They are very susceptible to cold, and, when so affected, are apt to be fractious and petulant.

The Javan slow loris (*N. tardigradus javanicus*), said to be confined to the island from which it derives its name, is distinguished by having four brown bands down the head and face from the crown, one band going to each eye, and one to each ear ; the interspaces being pale and the space between the eyes white. As this type of colouring is only one step in advance on that obtaining in the third variety of the slow loris, it is evident that the Javan loris has no right to rank as a separate species.

SLENDER LORIS

The slender loris (*Loris gracilis*) is the sole species of the genus to which it belongs. It is distinguished from the slow loris by its lighter build of body and longer and more slender limbs, as well as by the greater size of the eyes, which are separated merely by a narrow space. The ears are also somewhat larger than in the slow loris.

The slender loris is a much smaller animal than the slow loris, the length of the head and body being about 8 in. In colour it is a dark earthy grey, with a more or less marked ruddy tinge on the back and outer sides of the limbs, and showing a faint silvery wash, the under parts being much paler. Between the eyes there is the usual narrow white stripe, which spreads out on the forehead ; and the cheeks and region round the eyes are darker than the rest of the body. Some young specimens are decidedly reddish.

This animal is confined to the forests of Southern India and Ceylon, and appears only to be found in

THE SLENDER LORIS

SLENDER LORIS CLIMBING
Photograph by W. S. Berridge

those which are situated at a comparatively slight elevation above the sea-level. This Ceylon race is distinguished by its redder colour from the Indian one, in which the general hue is greyish. The habits of the slender loris are very similar to those of its cousin the slow loris, although its movements are not quite so deliberate. It partakes of similar food, and sleeps rolled up like a ball, with its head between its thighs, and its hand grasping the bough on which it is seated.

I once had occasion to purchase a pair of these animals in the bazaar at Madras, and was surprised to find the number of specimens which were exposed for sale. On the voyage up to Calcutta these pretty little creatures lived mainly on a diet of plantains and rice, supplemented with an occasional cockroach ; but as they passed the whole day in slumber, they could scarcely be reckoned as very lively pets.

Sir J. Emerson Tennent, who states that this loris has acquired the name of the " Ceylon sloth " in Ceylon, observes that its singularly large and intense eyes have attracted the attention of the Sinhalese, "who capture the creature for the purpose of extracting them as charms and love-potions, and this they are said to effect by holding the little animal to the fire till the eyeballs burst. Its Tamil name is thavangu, or 'thin-bodied' ; and hence a deformed child or emaciated person has acquired in the Tamil districts the same epithet. The light-coloured variety of the loris in Ceylon has a spot on the forehead, somewhat resembling the *namam*, or mark worn by the worshippers of Vishnu ; and from this peculiarity it is distinguished as the nama thavangu."

THE POTTOS

In Equatorial Africa the place of the lorises of Asia is taken by certain peculiar species which may be collectively known as pottos, although in its proper application the native name potto appears to be restricted to the first of the two groups. The pottos are distinguished by the index finger of the hand being quite rudimentary, consisting only of a stump without distinct joints, and unprovided with a nail. The typical pottos are further distinguished by possessing a short tail, but since this appendage is rudimentary in the allied group of awantibos, it does not afford any characters by which the African group can be distinguished from their Asiatic relatives. The habits of pottos are very similar to those of lorises, but their movements are still more deliberate and more sluggish.

The typical Bosman's potto (*Perodicticus potto*) takes its name from having been discovered by the Dutch navigator Van Bosman, who met with it on the coast of Guinea, and described it in 1705 under its native name of potto.

It is an animal of somewhat robust build, chiefly characterised by having a tail of about one-third the length of the head and body, the whole body being covered with a thick coat of soft and moderately long hair. The small and rounded ears stand up well above the fur of the head, the large eyes are separated from one another by a considerable interval, and the muzzle is rather broad and not very long. The arms and legs are of nearly equal length. With the exception of the nearly naked nose and chin, which are flesh-coloured, the general colour of the animal is a kind of chestnut, with a black or greyish tinge, the throat and under parts being yellowish brown. The peculiar half-red half-grey tint of the fur on the back is produced by the hairs being slate-coloured at their roots, reddish in the middle, and paler at the tips.

In addition to the loss of the index finger of the hand, the potto presents a curious peculiarity connected with the joints of the backbone in the neck. The spines of these vertebræ, which rise from the upper surface, are so elongated that they actually project beyond the general level of the skin of the back of the neck, where they form a series of little

BOSMAN'S POTTO
Photograph by W. S. Berridge

humps. We are at present unacquainted with the object of this peculiar structural arrangement.

Like lorises, pottos are nocturnal in their habits, sleeping during the whole of the day rolled up in a ball, with the head between the fore legs, and folded into the chest, and supporting themselves in captivity by grasping the bars of the cage with both hands and feet.

Bosman's potto is found over a considerable extent of the West Coast of Africa, having been recorded from Guinea, Sierra Leone, and the Gabun. Unfortunately, however, we have but few details as to its habits in a wild state, this being probably largely due to the creature having been seldom seen by Europeans. A larger species (*P. edwardsi*) inhabits the Gabun district, while a third kind (*P. batesi*) is recorded from the Congo. A fourth species (*P. ibecnus*) is a native of Uganda.

THE AWANTIBOS

The awantibo, or Calabar potto (*Arctocebus calabarensis*), was long believed to be a much rarer animal than Bosman's, but of late years numerous skins have been brought to Europe. It inhabits the regions around the Old Calabar River, flowing into the Bight of Biafra, east of the Niger. The awantibo is distinguished from the potto, not only by its smaller size and more slender build, but also by the tail being reduced to a mere rudiment and by a still further reduction of the index finger, which is represented merely by a little tubercle on the edge of the hand. Moreover, the other fingers of the hand, as well as the toes of the foot, with the exception of the great toe, have their first joints connected by folds of skin. The entire hands and feet are relatively smaller than in Bosman's potto. In the awantibo the spines of the vertebræ do not project through the skin of the neck.

The colour of the typical Calabar awantibo is yellowish brown above, but paler on the under parts, becoming whitish in places ; and the whole length of the body is just over 10 inches.

This animal has only been known to Europeans

THE AWANTIBO

since 1859, when a specimen in spirit was sent home by the Rev. A. Robb, a missionary at Old Calabar, who was unable to obtain any particulars regarding its habits, beyond the fact that it was nocturnal and lived in trees. The first living specimen to reach England arrived at the London Zoological Gardens in 1905.

A second species, the golden awantibo (*A. aureus*), from the Congo, was described in 1902. It has been already mentioned that in both pottos and awantibos the hands and feet are divided into two distinct moieties by the separation of the thumb and great toe from the other digits ; this being most marked in the hand by the loss of the index finger. The hands and feet may accordingly be compared to the feet of a parrot, the structure in both cases being a special adaptation for long-continued grasping without change of position.

THE TARSIERS

THE strange tarsier, although sufficiently nearly related to the lemurs to be included in the same great group, yet differs so markedly as to make it necessary to refer it to a distinct family, the *Tarsiidæ*, its own name being *Tarsius spectrum* (page 164).

This animal takes its first name from the elongation of the bones of the upper part of the ankle (*tarsus*), after the manner noticed in the mouse lemurs (page 284) and galagos (page 287), and its second from its spectre-like appearance. It is a native of various islands in the Malay region, being found, in Celebes, Sumatra, Borneo, the Philippines, and some others. It has never, I believe, been exhibited alive in Europe, and since accounts at first hand from those who have seen animals in their native countries are always valuable, I may quote the description of Dr. F. H. H. Guillemard, who received

a living specimen while at Celebes and, in his "Cruise of the Marchesa," wrote as follows : "The most interesting addition to our menagerie was a tiny lemuroid animal (*Tarsius spectrum*), brought to us by a native, by whom it was said to have been caught upon the mainland. These little creatures, which are of arboreal and nocturnal habits, are about the size of a small rat, and are covered with remarkably thick fur, which is very soft. The tail is long, and covered with hair at the root and tip, while the middle portion of it is nearly bare. The eyes are enormous, and, indeed, seem, together with the equally large ears, to constitute the greater part of the face, for the jaw and nose are very small, and the latter is set on, like that of a pug-dog, almost at a right angle. The hind limb at once attracts attention from the great length of the tarsal [ankle] bones, and the hand is equally noticeable for its length, the curious claws with which it is provided, and the extraordinary disc-shaped palps on the palmar surface of the fingers, which probably enable the animal to retain its hold in almost any position. This weird-looking creature we were unable to keep long in captivity, for we could not get it to eat the cockroaches which were almost the only food with which we could supply it. It remained still by day in its darkened cage, but at night, especially if disturbed, it would spring vertically upwards in an odd mechanical manner, not unlike the hopping of a flea. On the third day it found a grave in a pickle-bottle."

If it be added that the general colour of the fur is usually some shade of brownish fawn, with the face and forehead reddish, and a dark ring round the enormous eyes, the above account gives a very good idea of the general appearance of the tarsier, the length of the body being about 6 in. This account does not show us, however, why this animal should be separated from the typical lemurs as the representative of a separate family ; and we must therefore proceed to the consideration of this point.

Now, the elongation of the bones of the upper half of the ankle evidently allies the tarsier to the galagos and mouse lemurs ; and, if the other characters of the animal approximated to them, there would be no reason why it should not be included in the family. It happens, however, that there are very important differences connected with the teeth, and naturalists largely rely on these in assigning the tarsier to a distinct family. In describing the teeth of the lemurs (page 277), it was seen that the middle pair of incisors in the upper jaw are separated from one another by an intervening space, and it may be added that they are of small size. If, however, we examine the skull of a tarsier, we shall find that these central upper incisors are of large size, and placed quite close to one another. Then, again, we shall find that the upper tusk is much smaller than in the typical lemurs. Moreover, if we examine the lower jaw, it will be seen that the tusk is formed by the canine tooth, instead of being the most anterior of the premolars ; the latter tooth being smaller than either of the two premolars, instead of taking the form and function of a tusk, as in the true lemurs. In having only one pair of lower incisors, the tarsier agrees with the indri ; but, in the presence of three premolars on both sides of each jaw, it resembles the true lemurs. Its whole series of teeth are thirty-four in number—four more than in the indri, and two less than in the true lemurs, and may be expressed by the formula $i.\frac{2}{2}, c.\frac{1}{1}, p.m.\frac{3}{3}, m.\frac{3}{3}$.

GROUP OF TARSIERS

It is therefore clear that the tarsier differs very markedly from ordinary lemurs, being in this respect what naturalists term a more generalised form than the true lemurs. It approximates, indeed, in this particular, very closely to certain extinct lemuroids, referred to on page 295 ; the teeth of the fossil genus *Microchœrus* having the same formula as the one denoting those of the tarsier. It may accordingly be concluded that this animal shows in its teeth signs of affinity with the extinct European lemuroids, which have been lost in the true lemurs and their allies. In regard to the elongation of the upper half of the ankle, the tarsier is, however, evidently a specialised, or highly modified, creature ; and it is probable that the same structural peculiarity did not exist in the Eocene lemuroids.

Another peculiarity of the tarsier is that the two bones of the lower portion of the leg—the tibia and fibula—instead of being quite separate from one another, as in all other lemuroids, are united in their lower half. Then, again, in place of only the toe next the great toe being furnished with a sharp, compressed claw, and all the other toes having flat nails, the middle toe is also provided with a similar compressed and pointed claw.

Reference might also be made to certain features connected with the structure of the skull of the tarsier, and likewise to some peculiarities in the anatomy of its soft parts ; but enough has been said to show what a very remarkable creature it is when properly studied, and to indicate why it is referred to a distinct family. It is, indeed, generalised, or little modified, in regard to its teeth, but highly specialised, or much modified, in respect of the bony skeleton of its legs and feet.

Professor A. A. W. Hubrecht, who has devoted much attention to the embryological development of the tarsier, is of opinion that its separation as a distinct family does not sufficiently express the peculiarities of this wonderful creature, which he regards as the most primitive phase of the Primates, sundered very widely from all the lemuroids, with which it is commonly classed. Dr. Hubrecht even goes so far as to suggest that it is illogical to place the tarsier in the same ordinal group as the lorises, and that it ought to represent an order by itself.

Dr. Guillemard directs special attention to the peculiar leaps made by his captive tarsier ; and this habit of leaping is highly characteristic of the species —as it is of the galagos and mouse lemurs, in which the anklebones are modified in the same manner, although to a less degree. The tarsier is described as progressing in the woods by a series of leaps from bough to bough or along a single bough ; and it doubtless makes use of similar leaps to pounce upon its living prey. In holding on to the boughs of trees, it is much aided by the sucker-like discs forming the extremities of its fingers and toes, alluded to by Dr. Guillemard. Its food consists chiefly of insects and small reptiles, and it does not appear that it ever touches fruits. Tarsiers are rare in their native land, and, instead of going in small parties, are found singly or in pairs. They are looked upon with great dread and horror by the inhabitants of the Malay Islands. According to Mr. Cumming, who once had a female and young tarsier alive, the animal is known to the natives of these islands by the name of the malmag. He also states that only one young is produced at a birth ; and that when the natives capture one of a pair they are sure of securing its fellow. When feeding, the tarsier sits up on its hind quarters and holds its foods in its hands, somewhat after the fashion of a squirrel.

The tarsier of the Philippines, mainly owing to slight difference in colouring, has been ranked as a distinct species, under the name of *Tarsius fuscus*.

EXTINCT LEMUROIDS

ALTHOUGH remains of lemuroids had been known for very many years from the Lower Tertiary rocks both of Hampshire and France, it was long before they were definitely recognised as such, having been at first regarded as belonging to small hoofed mammals. One of these groups of lemuroids, represented by several species of different, though relatively small, dimensions, occurring both in England and France, has been described under the name of *Microchœrus* ; the term meaning " small pig," and having been applied from the supposed affinity of the creature to the hoofed mammals. In France very similar remains have been described under the name of *Necrolemur*. These animals were undoubtedly lemuroids, evidently allied to living forms, their skulls being very like those of the galagos, though their upper premolar teeth more nearly resemble those of the mouse lemurs. Like all early fossil lemuroids, however, they are distinguished from living forms by the circumstance that the place and form of the lower tusk are not taken by the first of the lower premolar teeth. This is very important, since it shows that these ancient lemuroids were less specialised than their living relations, and also removes any difficulty as to the descent of monkeys (in which the lower tusk always remains) from lemuroids.

Another and larger European Tertiary lemuroid, known as the adapis, carries the series one step still farther back, since it has four premolar teeth on both sides of each jaw, whereas no living lemur has more than three of these teeth. Here, then, we have decisive evidence of the approximation of the extinct lemuroids to the inferior orders of mammals, among which four premolar teeth frequently occur. These early Tertiary European lemuroids represent a distinct family group, the *Adapidæ*.

Madagascar has yielded a wonderful series of remains of extinct lemurs, some of which belong to genera still existing, while others indicate totally distinct and, in certain instances, gigantic types. The exploration was carried out under the auspices of the Malagasy Academy, and directed by Dr. Herbert F. Standing. In addition to those of lemuroids, bones of hippopotamuses and crocodiles were met with, as well as a carved hippopotamus tooth and a piece of Chinese pottery. The lemuroid

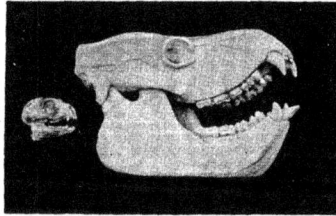

SKULLS OF THE GIANT LEMUR AND A MODERN LEMUR (TO THE LEFT)

remains consisted of between fifty and sixty skulls and a large quantity of limb bones, but no complete skeletons. Besides remains of gigantic species allied to the existing sifakas and indris, but generically distinct, these Malagasy relics have furnished skulls and bones of a large ape-like lemuroid known as *Archæolemur edwardsi*. This and an allied species retain the primitive primate dentition, although there is a reduction in the upper molars. The lower jaw is massive, and shows many monkey-like characters, although in the adult dentition the canine has disappeared, and the anterior premolar, as in existing lemurs, has undertaken the function of that tooth. The upper incisors are large and functionally important, but the lower ones incline forwards much in the same fashion as those of modern lemurs. The dental formula is $i.\frac{2}{2}, c.\frac{1}{0}, p.m.\frac{3}{3}, m.\frac{3}{3}$. This genus retains many features of the higher Primates, which have become lost in the modern sifakas and lemurs.

On account of their gigantic size, special interest attaches to the three species of the genus *Megaladapis*, of which *M. grandidieri* has a skull measuring more than a foot long, so that it must have been a creature as large as a big leopard. In all three species the skull is long, low, and narrow, with a marked constriction behind the sockets of the eyes. The brain shows few convolutions in part, while behind the cerebellum was not covered by the brain itself ; both features indicating degeneration. The dental formula is the same as in modern lemurs, and the genus evidently belongs to the subfamily *Lemuridæ*. The large upper molars are, however, peculiar in having tritubercular crowns, although there is some approximation to this structure in two of the existing genera of lemurs.

Dr. Standing, who does not believe in any close connection between *Microchœrus* and *Adapis*, on the one hand, and modern lemurs, on the other, is of opinion that *Archæolemur* shows signs of affinity with the New World monkeys, and suggests that all the modern lemuroids, with the possible exception of the tarsier, are survivors from an original primate stock which inhabited a great land mass connecting at an earlier period of the earth's history South America, Madagascar, and a part at least of South Africa and India.   R. LYDEKKER